WITHDRAWN FROM
THE LIBRARY

UNIVERSITY OF
WINCHESTER

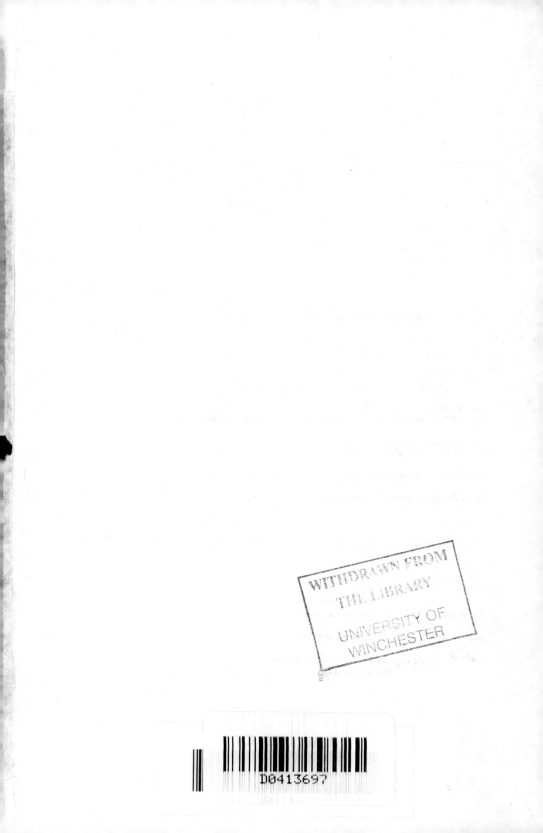

D0413697

OTHER BOOKS PUBLISHED BY BROADWAY PRESS

Backstage Forms, by Paul Carter

Backstage Handbook
An Illustrated Almanac Of Technical Information, by Paul Carter

Reel Exposure
How To Publicize And Promote Today's Motion Pictures, by Steven Jay Rubin

The Skeptical Scenographer
Essays on Theatrical Design and Human Nature, by Beeb Salzer

Stock Scenery Construction Handbook, by Bill Raoul

Stock Scenery Construction Handbook Companion Video Series

Also available, selected technical theatre and design titles from other publishers.

For more information, or to place an order,

call toll free, 800-869-6372 (9-5 Mon-Fri ET)

or visit our Web site: www.broadwaypress.com

PHOTOMETRICS HANDBOOK

PHOTOMETRICS

HANDBOOK

SECOND EDITION

BROADWAY PRESS

Louisville, KY 40241

September 1997

BY

ROBERT C. MUMM

© Copyright 1997 Robert C. Mumm. All rights reserved.

No portion of this book may be reproduced or used in any form, or by any means, without the prior written consent of the publisher.

This book contains information obtained from manufacturer's catalogs and specification sheets and reprinted here with permission. No effort has been made by the author or publisher to independently measure or verify the data given by the manufacturers.

The author and Broadway Press acknowledge that the model names and numbers mentioned in this book are in most cases registered trademarks of the respective companies.

Second Edition
ISBN: 0-911747-37-0

Second printing May, 1999, with corrections on pages 111, 306, 363 and 384.
Manufactured in the United States of America.

Broadway Press
3001 Springcrest Dr.
Louisville, KY 40241-2755
502-426-1211 (voice) 502-423-7467 (fax)
info@broadwaypress.com
www.broadwaypress.com

742.
025   :028/156/
MUM

TABLE OF CONTENTS

INTRODUCTION

FRESNELS

PLANO-CONVEX SPOTLIGHTS

ELLIPSOIDAL REFLECTOR SPOTLIGHTS

MOVING LIGHTS

BEAM PROJECTORS

SCOOPS

FOLLOW SPOTS

STRIPLIGHTS & CYC LIGHTS

PAR-TYPE UNITS AND REFLECTOR LAMPS

APPENDIX

INDEX

Photometrics is not a word you'll find in any standard dictionary. I checked. Webster's Unabridged Dictionary, second edition, has a listing for *photometry* —"the measurement of the intensity of light," and a listing for *photometric* —an adjective meaning "of or by the measurements of a photometer." In this book, and in common usage among lighting designers, photometrics is a noun meaning "the data used to describe the performance of lighting instruments—specifically their intensity and beam spread."

When I began this odyssey (eight years ago), it wasn't intended to be for all of you—it was to be for me. I planned to create a standard page layout for each instrument whose photometric information I had on file. It would be well-organized and easy to use, and probably would have lived in (yet another) three-ring binder.

Why I collected such information is almost unnecessary to explain, because most of us do it. All lighting designers, at one time or another, are faced with designing in an unfamiliar theatre, with equipment they have never before used. I have found myself in this situation too many times, and too many times I have listened to the house electrician or house designer say, "...well, it's not quite a 6x12, but almost... No, I'm afraid I don't know its field angle... Spec sheet? You mean like a catalog page? No, we don't have anything like that."

Hopefully, *The Photometrics Handbook* is making conversations like this obsolete! Now, I just look up the information I need to do my light plot, and I feel pretty confident that when I get to the theatre, the lights will do what I want and expect them to do.

This book is intended to help all of us, as lighting designers, become more familiar with the photometrics of many different instruments. This, in turn, will help us create subtle, elegant designs which use lighting instruments to their full advantage. On the other hand, this book is also intended to assist those designers who have more practical problems to solve. Those who just need to know for instance the field angle of an Altman 1KL6-30 because there are 20 of them at this new theatre and the light plot is due tomorrow.

I sincerely hope readers will find this book easy to use. I have tried to give performance data in a way that is pretty clear and straightforward, and I've tried to provide only the useful information about each instrument and have ignored the rest. Given that we're dealing with a variety of different manufacturers, some of which are no longer around, and given the many different types of instruments, presenting standardized information in a consistent format is a continuing challenge. I'm sure I'll hear from readers who think I've missed the boat—I've never known lighting designers to be shy about what they like and dislike. *So don't hesitate to let me know what you think!*

I would like to briefly thank all those who assisted me, in both small and large ways, in the production of this book: Colin Brown, Hugh Conacher, Tim Fricker, Kevin Griffin, Dan Vilter, Steven J. Haworth, William Jakab, Martha Mountain, J. Michael A Pengelly, Ken Romaine, Martin Sachs, Allan Stichburg, Dale Whittington, Susan Hallman, John Fisher, California Stage and Lighting (hi, Barbara), Four Star Stage Lighting, Tom Ruzika, Cosmo Catalano, Michael Rulf, Brent Stainer, Geoff Yates, (of course) all the luminaire manufacturers for their cooperation, and David Rodger of Broadway Press, my editor/publisher. And a quick note to those who are on this list: if you are not reading a free copy, it is probably because your address has changed since you sent me something...so if you'll drop us a line, a book will wing your way.

I also want to extend thanks to those folks, known and unknown, who use and enjoy the book. It never fails to make me feel good when I hear from people about it, both good and bad. And of course, a continual thanks to those folks, relatives and non-relatives, who form my little extended family, and who have supported me through the proverbial thick and thin...you know who you are, and my thanks are immeasurable. I could even wax poetic and quote Kipling's definition of a true friend, but this is a theatre book, after all...

The rest of this Introduction is a guide to using the book, an explanation of some photometric terminology, and some user-friendly math—so read on, Macduff!

Bob C. Mumm
June, 1997

Comments, questions, submissions of catalog pages and the like will best reach me at this address:

Bob C. Mum
% Broadway Press
3001 Springcrest Drive
Louisville, KY 40241

Correspondence can also reach me at my e-mail address: beam1138@ix.netcom.com

HOW TO USE THIS BOOK

A Few Words of Caution

This book gathers together in one place photometric information about most of the theatrical lighting instruments you'll find hanging in theatres in the United States. The information is taken from manufacturer's catalogs and spec sheets. Although we did cross-check certain pieces of information *mathematically* (which is admittedly less precise than a light meter, under these circumstances) *no attempt has been made to independently measure or verify the data given by the manufacturers*. We couldn't. It isn't really practical for me at the moment, although it is an interesting idea. Therefore, a word of caution is in order: catalogs and spec sheets are primarily sales tools. Manufacturers quite naturally want to present performance information about their products which emphasize their best features. They also want to convince customers that their instruments will perform as well as or better than similar instruments made by other companies. Therefore, when you are looking up information in this book it is important to keep in mind that whatever data is printed here is only as good as what is printed in the manufacturer's catalog—and typos do happen—both to them and probably to us.

Another caveat (which several manufacturers asked me to pass along) is that the performance characteristics given in catalogs and spec sheets are based on new instruments in a "lab" environment. Many things will affect how closely your particular instrument matches the photometrics specified in the catalog—its age, how well it has been maintained, the correct alignment of the lamp, the age of the lamp (both of which have a tremendous effect on the light output of the instrument), as well as the care taken by the manufacturer in measuring the performance characteristics in the first place and the care taken to make sure every instrument manufactured performs the same, which is a clear impossibility. For those who desire more information on this topic, I suggest you get hold of a copy of Karl G. Ruling's article on the subject, "Instrument Testing: How Many Should You Test To Get The Real Specs?," in the April, 1994 (Vol. XVIII, No. 3) edition of *Lighting Dimensions* magazine.

I don't want to have disillusioned readers, so I need to make two things clear:

1) you're *not* going to find in this book *every* lighting instrument ever made or used in the United States. (Although we'll keep trying.)

This book does have *most* of the information you will need about *most* of the theatrical lighting instruments hanging in theatres in the U.S. I'm still counting on you, the readers, to submit spec sheets giving performance information on older theatrical instruments. Please send me your stuff care of Broadway Press. The publisher has also offered to reward readers who submit information not found in this first two editions, and some of you reading this now are reading just such a 'freebie' from the first edition. If you are willing to share your catalogs or spec sheets for the benefit of your fellow designers, you can count on a complimentary copy of the next edition.

The second thing I need to make clear is:

2) This collection of photometric data *does not* include *every* piece of information you will ever want (or need) to know about a given instrument.

The previous edition did not include two pieces of information which we now include—the weight of the instrument, and its physical dimensions, where that information is available.

While we are on the subject of changes, it is important to note that we have changed the way we 'sort' the instruments in the ERS category. First they are sorted into three big groups—Short Throw, Medium Throw, and Long Throw—based on an instrument's intensity and related factors, such as unit size. Within these large groups, the pages are sorted further by field angle (30°-39°, 20°-29°, 10°-19°, etc.), then by manufacturer, then by model number. Hopefully, this will make it easier to find newer units with several different lenses, of different focal lengths, or even units with multiple lens diameters. Of course, if you know a particular luminaire's model number, you can still look it up in the index.

We are also including, for the first time, pages on both moving head, and moving mirror type luminaires, colloquially known as "wiggle lights." These presented great challenges because of their complexity and because the features that distinguish one model from another often have little to do with standard photometric specs like beam size and intensity. A year or so back, I asked the readers of Brad Davis' Stagecraft Mailing List to tell me what *they* were looking for on a moving lights page. They were not hesitant in replying, and many of their suggestions have been taken; some were not. For example, we just didn't have the space for a

DMX/channel table. (To join the Stagecraft Mailing List, send an e-mail message to "stagecraft-request@inquo.net," with "subscribe" in the subject line.)

You will also find significantly more units in the fresnel category—we are now including a number of units that are primarily designed for film/TV use, but that have found some place in theatre lighting as well. (Read their photometrics, and you'll see why.)

A last word of caution: photometrics, as it applies to theatrical lighting instruments, is full of inconsistencies. Manufacturers differ in the care they take measuring and reporting the performance of their instruments. Even terminology is not used consistently. For instance, some manufacturers are very careful when they say "beam angle" to mean the center portion of the total beam of light where the intensity does not drop off below 50% of maximum intensity. But other manufacturers have used "beam angle" to describe the angle of the total beam of light. Another confusing issue which has cropped up in the last few years is the issue of cosine distribution. I'll explain a little more about this terminology later in the introduction, but cosine distribution, or flat field focus as it is sometimes called, has certainly added a new level of complexity and confusion to photometric data.

As this edition goes to press, steps are being taken to standardize the way manufacturers measure and publish photometric data, sort of a "nutrition label" for luminaires. As someone who has analyzed and compared hundreds of theatrical lighting instrument, I find this effort to standardize performance data very exciting, and I applaud the efforts of those folks roundly.

A Guided Tour of a Typical Page

A quick look at the Table of Contents will show you that this book is arranged in groups of instruments—fresnels, ellipsoidal reflector spotlights (ERS), follow spots, moving lights, etc. Each page is devoted to information about a single instrument. There are a few pages which list two or more instruments, but only if those instruments have the same photometrics. For instance; two instruments with different model numbers have been combined on one page when the only difference is that one comes with an iris and the other doesn't.

Within each group of instruments the pages are organized first by the type of instrument (this information appears in the black box at the top of the page), then by the manufacturer, then by the particular model number. So, as an example, in the fresnel group, all the 6" instruments are together with those made by Altman Stage Lighting coming first, and the various different Altman model numbers arranged alphabetically.

The following sample page has lettered call-outs (**A** to **O**) that correspond to explanatory paragraphs below.

A The *type* of instrument is found in this box. In chapters with many instruments (there are over 200 pages in the ellipsoidal chapter) a smaller grouping is also included in this box. For instance, **FRESNEL—6" Lens** or **ERS—Short Throw—40° to 49°**. An asterisk "*" here (or anywhere else on the page) indicates a footnote which will be found at the bottom of the page. (A note: you will not find a section in this book on a relatively popular type of instrument: the PAR can. There is however a section on Reflector Lamps. The reason is obvious to anyone familiar with PAR cans: the photometrics of the instrument are entirely dependent on the reflector and lens found in the lamp. PAR cans aren't the only theatrical lighting instruments which use reflector lamps; they can be found in striplights, follow spots and even several ellipsoidal spots.)

It also brought up a dilemma: what to do with the newer ETC 'Source Four PAR' fixture, which isn't really a PAR can. You will find it with the reflector lamps (p.579), because that is what it is intended to replace.

There are a few other types of instruments which you won't find in this book: special effects projectors, Linnebach projectors, laser generators, searchlights, color wash units (programmable), etc. While these instruments are widely used, they didn't warrant including in this book because either their photometrics were too complicated, irrelevant, or universally unavailable.

B Much of the information in this book is arranged alphabetically by *manufacturer*. In the case of two companies, this presents a problem. Both Strand Lighting and Colortran have had a few name changes over the years: Strand Lighting has also been Century Lighting, Century Strand, and Strand Century. Colortran has been Berkey-Colortran, Colortran, and Lee Colortran. The name listed on a page in this book is the name used on the catalog sheet from which I got the photometric information. There will undoubtedly be instances where an instrument you are looking for is listed in this book under a name

TORTILLA PROJECTOR - 3"*

B **FRANK'S STAGE LIGHTS, INC.**
C Model No. BS

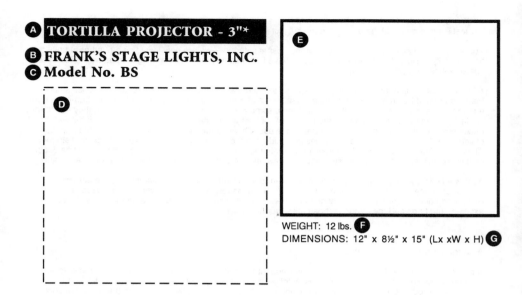

WEIGHT: 12 lbs. **F**
DIMENSIONS: 12" x 8½" x 15" (Lx xW x H) **G**

PHOTOMETRICS CHART

(Performance data for this unit are measured using A-21 - 100 W. lamp.) **H**

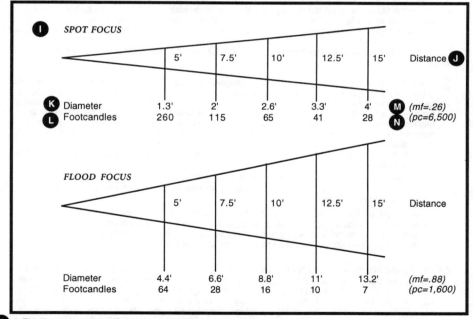

I *SPOT FOCUS*

	5'	7.5'	10'	12.5'	15'	Distance **J**
K Diameter	1.3'	2'	2.6'	3.3'	4'	**M** *(mf=.26)*
L Footcandles	260	115	65	41	28	**N** *(pc=6,500)*

FLOOD FOCUS

	5'	7.5'	10'	12.5'	15'	Distance
Diameter	4.4'	6.6'	8.8'	11'	13.2'	*(mf=.88)*
Footcandles	64	28	16	10	7	*(pc=1,600)*

O * This is a most unusual instrument - I've never seen another like it. **(II)**

P xii 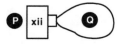 **Q**

different from the one you expect. In some cases when a company changed its name, it changed model numbers, but in most cases, the new name was given to existing instruments without changing model numbers. So, for instance, you might be looking for a Strand Century 6x9 ellipsoidal, but in this book you won't find this instrument by looking alphabetically under "S" because the catalog sheet we used for this instrument came from Century Lighting. I'm sorry for the awkwardness of this situation, but in some cases you are going to have to do a little extra hunting to find the right information.

C *Model number* information is also arranged alphabetically if a manufacturer has more than one of a particular type of instrument. The sort order might be a little confusing to some so let me explain: model numbers are treated as words, not as numbers, because so many model numbers include letters. Numerals are lower on the sort order so they will come first. A couple examples will explain: model number 307 comes before model number 70 because if you think of the model numbers as words, a word beginning with 3 comes before one beginning with 7; model number 6E6x9 comes before Q6E6x9 because numbers are lower in the sort order than letters.

Several pieces of information about the instrument are listed below the manufacturer's name and the model number. The information changes somewhat for each different type of luminaire, but in general, this is where we list the specified beam spread, and information about the luminaire's internal features—lamps, gobos, iris, shutters, color, etc. In the case of moving lights, this list also includes pan/tilt specs and information about data connectors.

D Two terms are common to all theatrical luminaires—beam angle and field angle. All the instruments in this book take light radiating uniformly from a light source and focus it in one direction by means of reflectors and (usually) lenses. This focusing of the light produces a cone shaped beam of light which can be described in terms of its angle. A narrow cone of light is described by a small angle and a wide cone or beam of light is described by a large angle. *Beam angle* is a term used to describe the center portion or "hot spot" of a beam of light where the light intensity does not drop below 50% of the maximum intensity. *Field angle* describes the "whole beam of light." More precisely, a field angle is that portion of the cone of light where the light intensity does not drop below 10% of the maximum intensity. One manufacturer uses the term "cut off angle" which presumably describes the cone of light extending to the point at which the intensity drops to 0% of maximum. The more standard measurement is the field angle.

The following are explanations of the pieces of information found on the top part of each page, broken down by the instrument types:

D **Fresnel** section

SPOT BEAM SPREAD is the angle of the light beam when the unit is adjusted for spot focus. Some manufacturers specify the beam angle, some the field angle, some both. Often times, TV units (in particular; but there are other examples, too) don't include field angle information. It makes sense: The beam angle, or hot spot, is often all of the beam that is 'used' in TV or film lighting.

FLOOD BEAM SPREAD is the angle of the light beam when the unit is adjusted for flood focus.

LAMP BASE TYPE is a piece of information included for most instruments in the book. It is useful when ordering lamps to be sure what you're ordering (or what you have in stock) will fit in that particular instrument. Generally I have used the common name of the lamp base— Medium Prefocus, Mogul Bipost, etc. In a few cases, there is not a common name used in this country. In those cases you will find strange names like "GY9.5" which is the ANSI/IEC designation of this particular lamp base. To help shed some light on the subject, I have included sketches of most lamp bases on page 609 in the appendix.

STANDARD LAMP is the lamp used when the photometric measurements were taken. Whenever possible I have used the ANSI code to describe the lamp and have also given its wattage. This is a little redundant because the ANSI code describes a lamp of a specific wattage, but for those who aren't glib with their ANSI codes, I thought that being reminded about the wattage would be helpful. In some cases, the lamp used to measure the instrument's performance is no longer manufactured and it is so noted.

OTHER LAMPS describes some of the lamps which can be used other than the standard lamp. I selected one lamp from each of a few different wattages, even if the manufacturer had given more. This selection was partly because of space limitations, and partly because if given a choice between two lamps of the same wattage, the one which will last longer is generally more useful, even though it may be a little less bright. It is important to note that these "other lamps" are only those specified by the manufacturer, and are not simply made up by little ol' me. Following the ANSI code and the wattage is another little piece of information, abbreviated: *(cf)*. Cf stands for conversion factor, (or correction factor) and is a handy little number if you want to compare the light intensity of the standard lamp with one of the other lamps. To use the correction factor, simply multiply the given footcandle information for the standard lamp by the correction factor and the results will be the footcandles of the other lamp at the same throw distance. More information on the correction factor is given later in this Introduction. Some of you will be interested in conversion factors from the older FEL-type Medium 2-Pin lamps to the new HX lamps. There are two reasons you won't see this addressed. The first is that we list only *the manufacturer's recommended alternatives.* The second is that while the HX and the FEL share the same base, they don't share *all* the same characteristics, and a luminaire designed for one type of lamp may not perform well with the other.

ACCESSORIES AVAILABLE repeats the accessories listed by the manufacturer on the catalog page. I don't vouch for the completeness of this information because in many cases, my source material was rather sketchy on this topic, but it is often helpful to know something about the available accessories for a particular instrument.

Ⓓ Plano-Convex Spotlight section

(see above: the same terms are used in this section as in the fresnel section)

Ⓓ ERS section

BEAM ANGLE (see definition of beam angle above) *Cosine distribution,* cosine focus, or flat field focus are terms relevant to the beam angle. In recent years, manufacturers have taken note of the fact that it is can be useful to focus the light source (the lamp) within its reflector so that the light output is more even across the center portion of the beam. In other words, a lamp can be adjusted to reduce the intense concentration of light at the center of the beam by spreading it out over a larger area. Some manufacturers call this lamp focus cosine distribution, others call it flat field focus, but generally they all agree that the term means that particular focus where 50% of the maximum intensity is present at 2/3 of the diameter of the total beam of light. In cosine focus, an instrument has a somewhat wider beam angle and the light intensity is less but more even. Unless otherwise noted, in this book, photometrics are assumed to be measured with the lamp adjusted for maximum intensity or peak focus. Where the catalog sheet gives cosine focus information, I have included it as well.

FIELD ANGLE (see definition of field angle in paragraph above fresnel section)

LAMP BASE TYPE (see description in fresnel section above)

LAMP MOUNT describes the relationship of the lamp filament to the reflector. "Axial" describes a lamp mount which positions the axis of the coiled filament along the long axis of the reflector. This positioning minimizes the area of the filament, thus reducing the possibility of an image of the filament being projected by the instrument, and also, with today's more compact filaments, provides a better environment optically—more of a 'point-source.' In the days when filaments and lamp envelopes were much larger and weaker, it was more important to maximize the unobstructed reflector area at center.

STANDARD LAMP (see description in fresnel section above)

OTHER LAMPS (see description in fresnel section above)

INTEGRAL PATTERN SLOT is only relevant to the ERS group. Slots to accept pattern holders for projected patterns (or gobos, or cookies) are standard on most contemporary ellipsoidal spotlights. It was not always so, and this piece of information lets you know if the instrument comes with a built-in pattern slot. Of course, you can always pop the unit open, and trap a gobo against the gate, but...

ACCESSORIES AVAILABLE (see description in fresnel section above)

Ⓓ ERS Zoom section

BEAM ANGLE, NARROW is the narrowest beam angle for this instrument

BEAM ANGLE, WIDE is the widest beam angle for this instrument

FIELD ANGLE, NARROW is the narrowest field angle for this instrument

FIELD ANGLE, WIDE is the widest field angle for this instrument

STANDARD LAMP (see description in fresnel section above)

OTHER LAMPS (see description in fresnel section above)

INTEGRAL PATTERN SLOT (see description in ERS section above)

ACCESSORIES AVAILABLE (see description in fresnel section above)

D **Moving Light** section

SPOT BEAM SPREAD and FLOOD BEAM SPREAD refer to the angles of light of a unit at spot (or narrow) focus, or at flood (wide) focus. Very similar to the beam angle information in the ERS Zoom section. Not all moving lights are zoom-type units, though, so read this information carefully.

BEAM IRIS RANGE refers to the minimum and maximum size of the beam of light using the iris. This information is often not available.

PAN/TILT RANGE gives the maximum range of both pan and tilt in degrees. Please note that while some fixtures are listed as 360°, or even more, this does not imply that it is capable of continuous rotation. Most fixtures will not endlessly rotate, but must, at times, rotate 'the long way around' to get back to 0°.

ACCURACY refers to the "resolution" of the fixture. Particularly when using DMX to address a fixture, the unit does not actually move continuously, but rather in a series of steps. Often, this number is a simple derivative of the Pan/Tilt range, in other words, if the Pan range is 360°, and there are 256 DMX steps used to address Pan, then this accuracy figure will be 1.4°. On some fixtures, this number changes depending on the protocol used to address the fixture. Using the manufacturers own proprietary protocol, as many as 1000 steps (or more) can be used to address a function, (which in the case above would result in a 'degrees per step' of .36°) while using a single DMX channel, only 256 steps are available, resulting in a second figure of 1.4°. Additionally, some fixtures are capable of both 8 and 16 bit resolution. This is a fancy way of saying "lots and lots of steps; more than your console can put out." To further complicate matters, some consoles, while putting out DMX, do so in 100 steps, where each "console step" encompasses several "DMX steps." AMX to DMX conversion has the same problem. So it pays to know how your console addresses this resolution/accuracy issue.

GOBOS refers to the number of gobos normally placed in a fixture, and how they are placed (i.e. on one wheel, on two wheels, etc.). Information on prisms or other effects will also be found here.

COLOR refers to the type of color media used in the fixture. Many color mixing fixtures use three different moving dichroics, most often Cyan, Magenta, and Amber, to mix beam color with. Some fixtures use single dichroic filters, one color per filter, and rotate them in and out of the beam to change colors. Some fixtures use color wheels, where single dichroics are mounted akin to gobos. Some fixtures even allow you to interchange the two. (See why the moving lights pages were tough?)

DIFFUSION refers to the style of diffuser used in the fixture, and is as specific and descriptive as possible.

LAMP is the type of lamp used in this fixture. Unlike most other theatrical luminaires, moving lights generally accept only one type of lamp.

DATA CONNECTORS refers to the type of connector used to connect the fixture to the DMX source or bus—generally some form of XLR connector. It may be 3-pin, 4-pin or 5-pin. I also list the "pin-outs," which tell you what sort of signal the fixture expects to see on the various pins.

D **Beam Projector** section

SPOT BEAM SPREAD (see description in fresnel section above)

FLOOD BEAM SPREAD (see description in fresnel section above)

LAMP BASE TYPE (see description in fresnel section above)

REFLECTOR DIAMETER is the nominal size of the unit's reflector

STANDARD LAMP (see description in fresnel section above)

OTHER LAMPS (see description in fresnel section above)

ACCESSORIES AVAILABLE (see description in fresnel section above)

D **Scoop** section

BEAM ANGLE (see definition of beam angle in paragraph above fresnel section)

FIELD ANGLE (see definition of field angle in paragraph above fresnel section)

FOCUSABLE is a feature found on some scoops and refers to the unit's ability to create a wider or narrower beam of light by moving the lamp forward or backward within the reflector.

LAMP BASE TYPE (see description in fresnel section above)

STANDARD LAMP (see description in fresnel section above)

OTHER LAMPS (see description in fresnel section above)

ACCESSORIES AVAILABLE (see description in fresnel section above)

D **Follow Spot** section

MIN. SPOT W/ IRIS is the size of the beam (sometimes field angle, sometimes the beam angle) with the iris in its minimum spot setting.

SPOT BEAM SPREAD is the narrowest beam spread with the iris fully open. Not all follow spots have adjustable lenses, in which case this term is just "BEAM SPREAD."

FLOOD BEAM SPREAD is the widest beam spread with the iris fully open. Some follow spots use a spread lens to achieve an even wider beam spread for illuminating large stage areas.

COLOR FRAMES refers to the number of frames or gel holders on the boomerang (if it is included as standard equipment).

FADER refers to whether or not there is some type of device for dimming the beam. Faders fall into three categories—a "*dowser*" is a mechanical device often consisting of two or more pivoting, scalloped-edge shutters positioned near the focal plane, a "*blackout frame*" is an opaque frame in the boomerang which can be swung across the beam of light, and a "*dimmer*" is an electrical device to limit the flow of electricity to the lamp.

SHUTTERS refers (generally) to a pair of vertically-moving shutters, which can be used to quickly "chop off" the beam, or to "strip" it, creating a horizontal rectangle of light (commonly used on small choruses, or for curtain call purposes). It can, however, refer to some other way of quickly shutting off the beam, such as a 4-way shutter. In follow spot terminology, horizontal shutters are commonly known as a guillotine, cutters, or choppers.

FAN COOLED refers to whether or not the unit has a fan to provide forced air cooling of the lamp housing.

STANDARD LAMP (see description in fresnel section above)

NOTES is the place where I listed special features, especially operational controls and access to re-lamping.

D **Striplight / Cyc Light** section

BEAM ANGLE (see definition of beam angle in paragraph above fresnel section)

FIELD ANGLE (see definition of field angle in paragraph above fresnel section)

STANDARD CONFIGS. describes the length, number of circuits and total number of lamps for each standard unit. Nearly all manufacturers will make striplights in whatever custom configurations a customer wants, however most also have a few standard units which they sell.

LAMP CENTERS refers simply to the distance from one lamp center to the next.

LAMP BASE TYPE (see description in fresnel section above)

OTHER LAMPS (see description in fresnel section above)

NOTES lists or explains features unique to that unit.

E For all but four or five instruments a *photo* is reproduced in this box for visual reference. In those few cases where there is a note, "No Photo Available," either I couldn't find a photo of the instrument or the only photo I could find was a fourth generation photocopy which just didn't reproduce well enough. If you have one of these fixtures, snap a Polaroid and send it in for a free copy of the next edition.

F *Weight* is the weight given by the manufacturer, and may or may not include c-clamp, connectors, etc.

G *Dimensions* is also pretty self-explanatory; it is listed by length, then width, then height. Length is the overall length of the unit, width is also an overall dimension. Height is measured from the bottom of the unit (unit in hanging configuration) to the top of the yoke, or c-clamp, depending on what is listed by manufacturer. In the case of moving fixtures, where possible I have listed the area required for full movement, a sphere, if you will, surrounding the fixture, which must be clear for movement to occur.

H This information, just above the photometrics chart, describes the lamp used to measure the performance data. In nearly all cases, it is the same information found under *Standard Lamp*.

(I) Many instruments are capable of throwing a beam of light which varies from spot focus to flood focus, or from narrow to wide focus (as is the case with so called zoom ellipsoidal spots). When more than one set of photometric data is given, each chart is identified with a very brief description (i.e. Flood Focus, or Spot Focus). In the moving lights section, this area may also contain multiple lens trains.

Photometric data is really very straightforward. It's simply the size of the beam and intensity of the light at various throw distances. Each manufacturer has its own way of presenting photometrics, but they all are describing the same simple information. The exception (there's always one) is striplights and cyc lights. The photometrics of striplights is complicated by the fact that strips are actually several individual instruments ganged together. Another complicating factor is that the shape of the reflector on most cyc lights is designed to throw light unevenly, so as to compensate for the close proximity to the cyc or backdrop. You will notice in the striplight section, some of the photometric charts are radically different from the rest of the book. This is because some manufacturers try to give performance information in unique ways. I applaud the efforts of manufacturers to show the important characteristics of their cyc lights—it is more important to show that the instrument will wash a cyc with even light, than it is to show the traditional photometric information (beam width and intensity at various distances). Where it was necessary, I have reproduced the performance charts as I found them in the catalogs even though they are radically different from the rest of the book. It would be nice, however, if the manufacturers would get together and decide on a standard format for representing performance data for cyc lights and striplights.

(J) *Distance* describes the throw distance from the instrument to the subject.

(K) *Diameter* is the diameter of the beam at the distance shown.

(L) *Footcandles* describes the intensity of the light at the distance shown, measured in footcandles.

(M) *(mf)* stands for "multiplication factor," a number which, when multiplied by the throw distance, gives you the diameter of the beam at *any* distance—not just at the given distances. A quick word of caution: the multiplication factors given in this book are derived from the distance and diameter data. As such, if the distance and diameter data describes the field angle, then using the multiplication factor will also yield field diameters. However if the distance/diameter data describes the beam angle, which in a few cases it does, then the multiplication factor will give you the diameter of the *beam angle* at any throw distance. There is more information later on this multiplication factor.

(N) *(pc)* stands for "peak candela." Also known as "candlepower," beam center candlepower (BCCP), and center beam candlepower (CBCP), peak candela is the maximum intensity of a lighting instrument and is a function of the lumens of the lamp and the optics of the instrument. In some cases, manufacturers have provided this peak candela figure, but in most cases, the *(pc)* number you see in this book has been calculated from the given footcandle data. More on this topic as well, later in this Introduction.

(O) Asterisk notes, or footnotes, amplify or explain some information found elsewhere on the page.

(P) Page numbers are accompanied by the USITT Lighting Graphic Standard symbol for the type of instrument found on the page. The page number icon is a generic symbol *only*, and shouldn't be taken as the one to use when drafting. Hopefully, this little symbol will make it easier to flip through the book looking for a particular type of instrument. (The complete, and current as of 1997, USITT Lighting Graphic Standard can be found on pages 610-612 in this edition.)

(Q) This is a tortilla: corn, about ten inches in diameter—perfect for home-style tacos. When properly prepared, two tortillas overlap each other slightly, with the interior filled with tasty, succulent cuts of roasted meats. Never mind the ground beef stuff.

PHOTOMETRIC MATH

I must confess, mathematics and I have always lived in different neighborhoods. So, when I started this book, I was blissfully ignorant of many of the various formulas and calculations which apply to photometrics. Fortunately, David Rodger, my editor/publisher, has nothing but audacity when it comes to learning new information. He discovered two articles on the topic of calculating photometric information which have been enormously helpful to me in two ways. Firstly, I now had some tools to check the accuracy of the manufacturer's data. (I was surprised at the number of typographical errors David and I uncovered. Most, if not all, of these errors are noted.) Second, these formulas make it possible to extrapolate information not actually given in the manufacturer's catalog. For instance, many instruments in this book have the following footnote, "* Beam and field angles not specified in manufacturer's catalog. This field angle is calculated from given distance and diameter information." What this means is that using the information provided by the manufacturer about the diameter of the beam at various throw distances, I was able to calculate the *angle* described by the given data.

I strongly recommend you take a look at these two articles: "Math Applications in Theatrical Lighting Design," by Dennis Madigan, in the Winter, 1989 issue of *Theatre Design & Technology* (Vol. XXV, no. 4). The second is also in *TD&T* and is entitled "Computerized Beam Section Calculation," by Douglas J. Rathbun; Winter 1991 issue (Vol. XXVII, no. 1).

The Madigan article deals with the basic math involved; even I could easily follow it, and as I said, math I and don't exactly live next door to each other. It provides a variety of useful formulas for discovering a real throw distance based on height and horizontal distance (i.e. from a section and ground plan). He tells you how to use multiplying factors, (he refers you to Richard Pilbrow's stage lighting text to find them; hopefully now you will look in this book instead) as well as discovering what field angle you need based on individual circumstances. Madigan also covers illumination, and gives a very handy little formula for discovering beam diameter of an instrument which is not striking its subject head-on, but is instead hitting it at an angle—which is, of course, what it almost always does. Fortunately, all you need here is a good calculator, capable of trig functions like sine, cosine and tangent.

Rathbun's article describes setting up a computerized "calculator" based on Madigan's formulas. Using this calculator, a lighting designer can easily calculate the various pieces of photometric data required to do a light plot. Rathbun describes setting up this calculator using the built-in calculating capabilities of a spreadsheet program like Lotus 1-2-3 or Excel.

What follows on the next several pages is an explanation of how to calculate and use various pieces of photometric information. (If you are interested, I suggest you take a look at Lighting and Electronics Inc.'s Website (www.le-us.com). Glen Cunningham has a number of very nice calculators available for your use, in a variety of file formats.)

Using Correction Factors *(cf)*

Throughout the book you will see *(cf=____)* following the description of each "other lamp." This correction factor number is simply the results of dividing the initial lumen output of the "other lamp" by the initial lumen output of the "standard lamp." This ratio of one lamp to the other gives us a little tool for easily finding the intensity of just about any lamp we might want to use, even though the manufacturer gives intensity data for only one lamp (the "standard lamp"). It is important to note, that once again, this will yield a number that is *mathematically* correct; but does *not necessarily* represent reality. Take it with a grain of salt, and perhaps a little fresh cilantro salsa.

To use the correction factor (cf), simply multiply the footcandle information of the standard lamp (as given in the Photometrics Chart) by the correction factor of the lamp of your choice. This will give you the correct footcandles for your lamp at whatever distance you choose.

Example:

The standard lamp is an FEL lamp (1 KW., Medium 2-Pin base) and you have footcandle information about the instrument using this FEL lamp. As given in the GE *Stage/Studio Lamp Catalog (SS-123)* the initial lumens, or maximum output, of an FEL lamp is 27,500. Let's say you want to use a different lamp, say an EHG (750 W., Medium 2-Pin base). The initial lumens listed for this lamp is 15,400. Using these two initial lumens numbers, we can calculate the correction factor for the EHG lamp.

15,400 ÷ 27,500 = .56

To find the intensity of your lighting instrument using the EHG lamp instead of the FEL lamp, simply multiply the given footcandles for the FEL lamp by .56. According to the manufacturer's catalog (or the same information which you can now find in this book), let's say the instrument gives off 100 footcandles at 20 feet throw distance. Using the EHG lamp, the instrument would produce only 56 footcandles at 20 feet:

.56 x 100 fc = 56 fc

A word of caution: initial lumens data is only approximate and most lamp catalogs say so. For this reason, a correction factor ratio between two lamps is not necessarily always going to be the same. In this book, you will find *(cf)* data for the same two lamps which appears inconsistent. For instance EHG is listed as having a correction factor (relative to the FEL lamp) of anywhere from .55 to .59. Why? Because where catalogs gave correction factor information, and many did, I chose to give that information, rather than calculate my own. It is a general rule in this book that wherever possible, the information you see on the page came from a catalog or spec sheet. So, where correction factors vary slightly, it is safe to assume that the initial lumens data used to calculate the correction factor was slightly different.

Using Multiplying Factors (mf)

At the end of each row of diameter figures in the photometric charts, you will find (mf=____). This number can be used just like the lamp correction factor to find beam diameters at any throw distance. The photometric charts in this book give diameter information at a few even throw distances—10 feet, 20 feet, 30 feet, etc. But what if you really need to know the beam diameter at 27 feet? You could guess at a figure between the given diameters for 20 feet and 30 feet, or you could multiply 27 feet by the multiplying factor to get the exact diameter.

Like the correction factor, the multiplying factor is simply the result of dividing the diameter by the throw distance.

Example:

You look up the photometrics chart for your particular instrument, either in this book or in a catalog, and you find that your instrument has a field diameter of 20 feet at a throw distance of 100 feet. What is the multiplying factor?

mf = 20' diameter ÷ 100' distance

mf = .20

Using this multiplying factor, you can easily find the field diameter at any distance, say 27 feet:

diameter = throw distance x multiplying factor

diameter = 27' x .20

diameter = 5.4'

Using Peak Candelas (pc)

Some catalogs include a rather large number, tens of thousands to millions, which is variously labeled as candlepower, beam center candlepower (BCCP) or peak candela. What is this number? If you took a measurement with a photometer dead center right next to the front lens of your instrument, the reading, in lumens, would be the peak candela. Once you have the peak candela or maximum intensity of an instrument, to get the illumination at any given distance it's just a matter of getting out your calculator and dusting off the Inverse Square Law. (You remember, light decreases in proportion to the distance squared.)

Footcandles = peak candela ÷ (distance x distance)

Example:

Let's say the peak candela of your instrument is 340,000.

Footcandles = 340,000 ÷ (27' x 27')

Footcandles = 340,000 ÷ (729')

Footcandles = 466.39

What if you don't know the peak candela of your instrument? By working backwards, you can take whatever footcandle information is given and derive a satisfactory peak candela figure. Here we go:

Rearrange the above equation to solve for peak candela and you come up with the new equation

Peak Candela = footcandles x throw distance squared

Let's say the catalog tells you that at 40 feet this instrument has a maximum intensity of 212 footcandles. Using the new equation we get:

Peak Candela = 212 x (40 x 40)

Peak Candela = 212 x 1,600

Peak Candela = 339,200

Is 339,200 close enough to use as a peak candela figure? Let's plug 339,200 into the original equation, instead of 340,000.

Footcandles = 339,200 ÷ (27' x 27')

Footcandles = 339,200 ÷ (729)

Footcandles = 465.29

The difference is only 1.1 footcandles which is an insignificant amount.

Another word of caution: when you calculate your own peak candela from given footcandle information, remember that footcandle information is (or should be) measured data. As such it is not going to have mathematical consistency, no matter how carefully the measurements are taken. In fact, focusing the instrument for different distances will result in a somewhat different peak candela for each distance. Also—because one or two footcandles really doesn't make much difference, manufacturers always round off footcandles in the photometric charts. So my advice is to figure peak candela using more than one set of footcandle/distance numbers, then average the results.

Calculating Diameters from Angles.

I'll give you one more handy way to calculate photometric information, then you have to take a look at Dennis Madigan's article and Douglas J. Rathbun's article for yourself. What if you don't have a photometrics chart showing diameters at various distances? What if all you know about an instrument is that it has a field angle of 30°? How can you figure out what the field diameter is, at say, a 20 foot throw? The calculations are a little more involved—but thanks again to Dennis Madigan, we have a nifty formula.

Diameter = 2 x (distance x tangent of half of the field angle)

To find out the tangent of an angle you can look it up in a trigonometric table, or use a calculator which has trig functions built into it. Either way, tan 15° is .2679. Remember you want *half* the field angle, which in this case is 15°, not the whole field angle. So, back to the formula:

Diameter = 2 x 20' x .2679

Diameter = 10.7160

For those of you who feel a bit squeamish about trig tables, there is a shortcut. In the appendix, you will find a table which gives lots of angles and their corresponding multiplying factors. So if you know the field angle, or the beam angle for that matter, just look up the multiplying factor for that angle and use it to calculate the field or beam diameter at any throw distance. Easy.

There is another table in the appendix which I think you will find helpful. Remember in Lighting 101 when you learned that you couldn't just measure the throw distance on the light plot? Throw distance is actually the hypotenuse of the right triangle formed by the vertical height (which you can measure on the section)and the horizontal distance (which you measure on the plot). Well, with the help of my then-wife Maggie, who was much more skillful with spreadsheets than I, we came up with a table which gives throw distances from the unit to the *head height* of a subject at various positions. Once you measure the horizontal and vertical distances, you can simply look up the throw distance. (Remember to measure the horizontal on the plot, and not the section—the section doesn't take into account the distance off center a particular unit may be.)

Whew. If you have gotten this far, you know almost as much about photometrics as I do, so congratulations. The rest of the book is a heck of a lot easier read.

ALTMAN STAGE LIGHTING
Model No. 100

SPOT BEAM SPREAD: 10° beam, 23.5° field

FLOOD BEAM SPREAD: 34° beam, 55° field

LAMP BASE TYPE: D.C. Bayonet Candelabra

STANDARD LAMP: ESP - 150 W.

OTHER LAMPS: ESR - 100 W. *(cf=.73)*

ACCESSORIES AVAILABLE:
Color Frame (3⅞" sq.), Pipe Clamp, 2-way Barn Door, Snoot, Motorized Color Wheel, Picture Frame Spot Lens Replacement

WEIGHT: (not specified in manufacturer's catalog)

SIZE: 4¼" x 5¼" x 7½" (L x W x H)

PHOTOMETRICS CHART
(Performance data for this unit are measured using ESP - 150 W. lamp.)

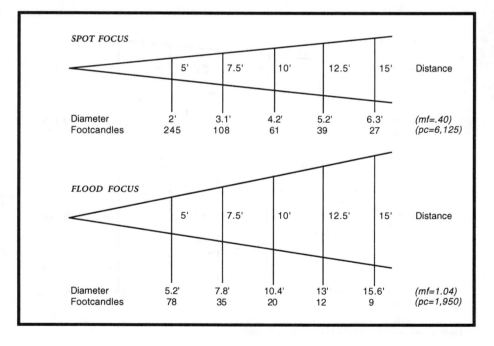

SPOT FOCUS

	5'	7.5'	10'	12.5'	15'	Distance
Diameter	2'	3.1'	4.2'	5.2'	6.3'	*(mf=.40)*
Footcandles	245	108	61	39	27	*(pc=6,125)*

FLOOD FOCUS

	5'	7.5'	10'	12.5'	15'	Distance
Diameter	5.2'	7.8'	10.4'	13'	15.6'	*(mf=1.04)*
Footcandles	78	35	20	12	9	*(pc=1,950)*

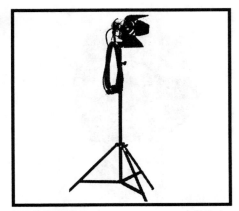

FRESNEL — 3" Lens

ARRIFLEX CORP.
Model No. 531300

SPOT BEAM SPREAD: 16° beam*

FLOOD BEAM SPREAD: 60° beam*

LAMP BASE TYPE: 2-Pin Prefocus

STANDARD LAMP: FKW - 300 W.

OTHER LAMPS: CP81 - 300 W. (220 v.)

ACCESSORIES AVAILABLE:
Color Frame (5" sq.), 4-way Barn Door, Top Hat

WEIGHT: 6.5 lbs.

SIZE: (not specified in manufacturer's catalog)

PHOTOMETRICS CHART

(Performance data for this unit are measured using FKW - 300 W. lamp)

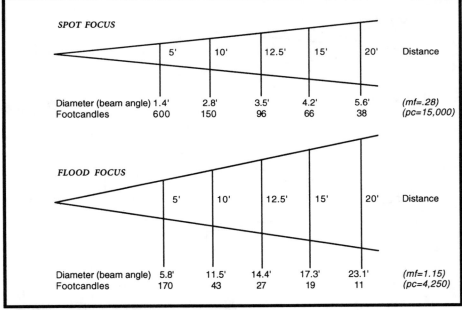

SPOT FOCUS

	5'	10'	12.5'	15'	20'	Distance
Diameter (beam angle)	1.4'	2.8'	3.5'	4.2'	5.6'	(mf=.28)
Footcandles	600	150	96	66	38	(pc=15,000)

FLOOD FOCUS

	5'	10'	12.5'	15'	20'	Distance
Diameter (beam angle)	5.8'	11.5'	14.4'	17.3'	23.1'	(mf=1.15)
Footcandles	170	43	27	19	11	(pc=4,250)

* No field angle data given in manufacturer's catalog.

BERKEY COLORTRAN
Model No. 214-002*

SPOT BEAM SPREAD: 10° beam, 20° field

FLOOD BEAM SPREAD: 36° beam, 48° field

LAMP BASE TYPE: D.C. Bayonet Candelabra

STANDARD LAMP: 176-188 - 150 W.**

OTHER LAMPS: 176-189 - 200 W.**

ACCESSORIES AVAILABLE:
Color Frame, 4-way Barn Door

WEIGHT: 2.3 lbs.

SIZE: 5" x 5½" x 7¼" (L x W x H)

PHOTOMETRICS CHART
(Performance data for this unit are measured using 176-188 - 150 W. lamp.)

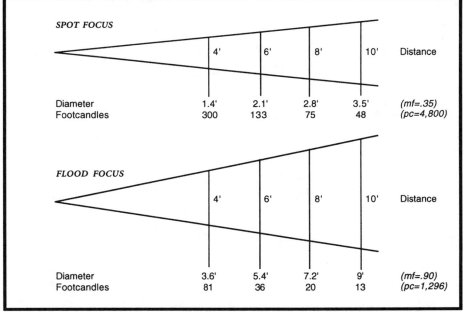

SPOT FOCUS

	4'	6'	8'	10'	Distance
Diameter	1.4'	2.1'	2.8'	3.5'	*(mf=.35)*
Footcandles	300	133	75	48	*(pc=4,800)*

FLOOD FOCUS

	4'	6'	8'	10'	Distance
Diameter	3.6'	5.4'	7.2'	9'	*(mf=.90)*
Footcandles	81	36	20	13	*(pc=1,296)*

* This model can be identified with model numbers -002, -005, -006, or -007 depending the wire leads and connectors supplied.

** These numbers (176-188 and 176-189) are the Colortran catalog numbers for these lamps. No ANSI code is given in the Colortran data sheet. The GE Stage & Studio Lamp catalog offers as a substitute for the 176-188, a 150 W. lamp with ANSI code ETC. No substitute is offered for 176-189. Correction factor *(cf)* information is not listed because manufacturer's catalog does not give initial lumens data for these lamps.

FRESNEL — 3" Lens

CENTURY LIGHTING
Model No. 1211

SPOT BEAM SPREAD: 9° beam, 16° field

FLOOD BEAM SPREAD: 35° beam, 50° field

LAMP BASE TYPE: D.C. Bayonet Candelabra

STANDARD LAMP: 150 G16-1/2 DC - 150 W.*

OTHER LAMPS: ETC - 150 W. *(cf=1.12)***
ESR - 100 W. *(cf=.72)***

ACCESSORIES AVAILABLE:
(None specified in manufacturer's catalog.)

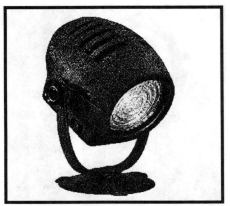

WEIGHT: (not specified in manufacturer's catalog)

SIZE: 5½" x 5½" x 7⅛" (L x W x H)

PHOTOMETRICS CHART
(Performance data for this unit are measured using 150 G16-1/2 DC - 150 W. lamp.)

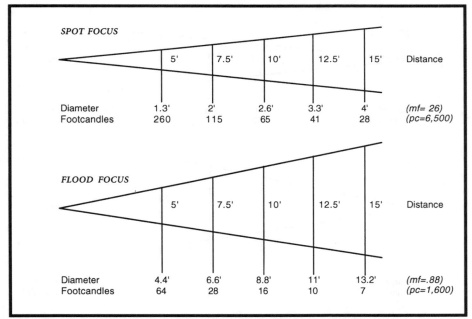

SPOT FOCUS

	5'	7.5'	10'	12.5'	15'	Distance
Diameter	1.3'	2'	2.6'	3.3'	4'	*(mf= 26)*
Footcandles	260	115	65	41	28	*(pc=6,500)*

FLOOD FOCUS

	5'	7.5'	10'	12.5'	15'	Distance
Diameter	4.4'	6.6'	8.8'	11'	13.2'	*(mf=.88)*
Footcandles	64	28	16	10	7	*(pc=1,600)*

* This is an obsolete lamp but it is listed as the standard lamp since the manufacturer uses this lamp to describe the photometric performance.

** These other lamps are currently available substitutes for the lamps suggested in the manufacturer's catalog (150 G16-1/2 DC and 100 G16-1/2 DC). Correction factor *(cf)* information is based on initial lumens data for the 150 G16-1/2 DC lamp as given in the Century catalog.

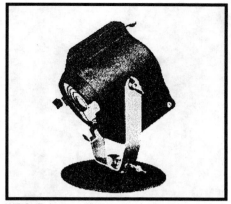

CENTURY LIGHTING
Model No. 523

SPOT BEAM SPREAD: 9° beam, 16° field

FLOOD BEAM SPREAD: 31° beam, 50° field

LAMP BASE TYPE: D.C. Bayonet Candelabra

STANDARD LAMP: 150 G16-1/2 DC - 150 W.*

OTHER LAMPS: ETC - 150 W. *(cf=1.12)***
ESR - 100 W. *(cf=.72)***

ACCESSORIES AVAILABLE:
Color Frame, 2 or 4-way Barn Door, Snoot

WEIGHT: (not specified in manufacturer's catalog)

SIZE: 6½" x 6" x 8½" (L x W x H)

PHOTOMETRICS CHART
(Performance data for this unit are measured using 150 G16-1/2 DC - 150 W. lamp.)

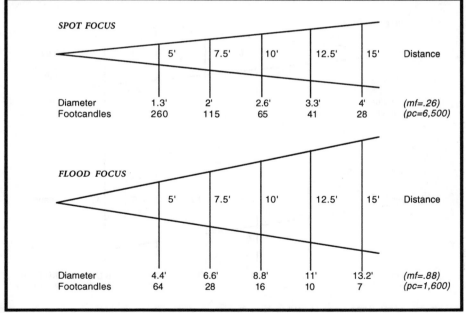

SPOT FOCUS

	5'	7.5'	10'	12.5'	15'	Distance
Diameter	1.3'	2'	2.6'	3.3'	4'	*(mf=.26)*
Footcandles	260	115	65	41	28	*(pc=6,500)*

FLOOD FOCUS

	5'	7.5'	10'	12.5'	15'	Distance
Diameter	4.4'	6.6'	8.8'	11'	13.2'	*(mf=.88)*
Footcandles	64	28	16	10	7	*(pc=1,600)*

* This is an obsolete lamp but it is listed as the standard lamp since the manufacturer uses this lamp to describe the photometric performance.

** These other lamps are currently available substitutes for the lamps suggested in the manufacturer's catalog (150 G16-1/2 DC and 100 G16-1/2 DC). Correction factor *(cf)* information is based on initial lumens data for the 150 G16-1/2 DC lamp as given in the Century catalog.

5

FRESNEL — 3" Lens

CENTURY STRAND
Model No. 3141, "Tini-Spot"

SPOT BEAM SPREAD: 8° beam, 16° field

FLOOD BEAM SPREAD: 38° beam, 58° field

LAMP BASE TYPE: D.C. Bayonet Candelabra

STANDARD LAMP: ESP - 150 W.

OTHER LAMPS: ESS - 250 W. *(cf=1.79)*

ACCESSORIES AVAILABLE:
 Color Frame, 2 or 4-way Barn Door, Snoot

WEIGHT: (not specified in manufacturer's catalog)

SIZE: 6½" x 6" x 8½" (L x W x H)

PHOTOMETRICS CHART
(Performance data for this unit are measured using ESP - 150 W. lamp.)

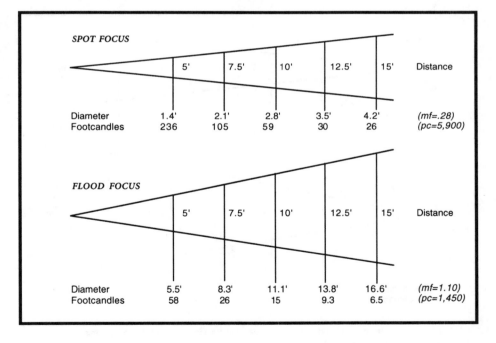

SPOT FOCUS

	5'	7.5'	10'	12.5'	15'	Distance
Diameter	1.4'	2.1'	2.8'	3.5'	4.2'	*(mf=.28)*
Footcandles	236	105	59	30	26	*(pc=5,900)*

FLOOD FOCUS

	5'	7.5'	10'	12.5'	15'	Distance
Diameter	5.5'	8.3'	11.1'	13.8'	16.6'	*(mf=1.10)*
Footcandles	58	26	15	9.3	6.5	*(pc=1,450)*

KLIEGL BROS.
Model No. 3603

SPOT BEAM SPREAD: 22° field*

FLOOD BEAM SPREAD: 64° field*

LAMP BASE TYPE: D.C. Bayonet Candelabra

STANDARD LAMP: ESP -150 W.

OTHER LAMPS: ESR - 100 W.   *(cf=.68)*

ACCESSORIES AVAILABLE:
Color Frame, 8-way Barn Door, Snoot, Stand
Adaptor, 4" Table Base

WEIGHT: 3 lbs.

SIZE: 4⅞" x 5¾" x 7½" (L x W x H)

PHOTOMETRICS CHART
(Performance data for this unit are measured using ESP -150 W. lamp.)

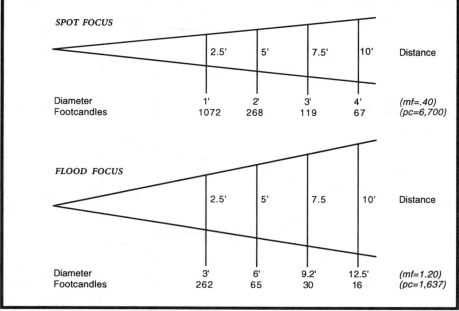

SPOT FOCUS	2.5'	5'	7.5'	10'	Distance
Diameter	1'	2'	3'	4'	*(mf=.40)*
Footcandles	1072	268	119	67	*(pc=6,700)*
FLOOD FOCUS	2.5'	5'	7.5	10'	Distance
Diameter	3'	6'	9.2'	12.5'	*(mf=1.20)*
Footcandles	262	65	30	16	*(pc=1,637)*

* No beam angle data given in manufacturer's catalog.

LIGHTING & ELECTRONICS
Model No. 60-03*

SPOT BEAM SPREAD: 10° beam, 21° field

FLOOD BEAM SPREAD: 30° beam, 50° field

LAMP BASE TYPE: D.C. Bayonet Candelabra

STANDARD LAMP: ESP - 150 W.

OTHER LAMPS: ESR - 100 W. *(cf=.68)*
FEV - 200 W. *(cf=1.96)*

ACCESSORIES AVAILABLE:
Color Frame

WEIGHT: (not specified in manufacturer's catalog)

SIZE: (not specified in manufacturer's catalog)

PHOTOMETRICS CHART

(Performance data for this unit are measured using ESP - 150 W. lamp.)

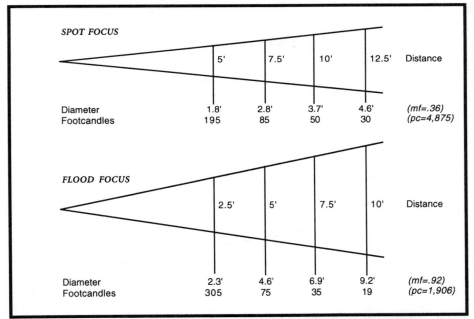

SPOT FOCUS

	5'	7.5'	10'	12.5'	Distance
Diameter	1.8'	2.8'	3.7'	4.6'	*(mf=.36)*
Footcandles	195	85	50	30	*(pc=4,875)*

FLOOD FOCUS

	2.5'	5'	7.5'	10'	Distance
Diameter	2.3'	4.6'	6.9'	9.2'	*(mf=.92)*
Footcandles	305	75	35	19	*(pc=1,906)*

* The instrument pictured on this page has the same model number (60-03) as the instrument pictured on the following page. The model on this page is an earlier version and is distinguished by its screw feed focus.

LIGHTING & ELECTRONICS
Model No. 60-03*

SPOT BEAM SPREAD: 10° beam, 23.5° field

FLOOD BEAM SPREAD: 34° beam, 55° field

LAMP BASE TYPE: D.C. Bayonet Candelabra

STANDARD LAMP: ESR - 100 W.

OTHER LAMPS: (none specified in mfg. catalog)

ACCESSORIES AVAILABLE:
Color Frame, Barn Doors, Mounting Flange, Snoot

WEIGHT: 2 lbs. 8 oz.

SIZE: 4½" x 4" x 5" (L x W x H)

PHOTOMETRICS CHART

(Performance data for this unit are measured using ESR - 100 W. lamp.)

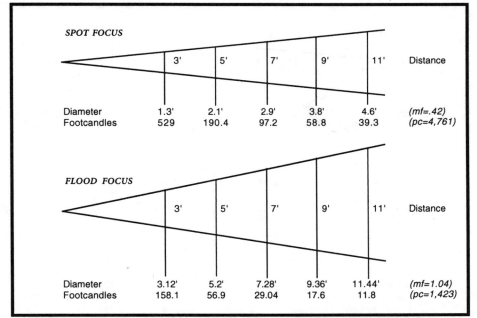

SPOT FOCUS

	3'	5'	7'	9'	11'	Distance
Diameter	1.3'	2.1'	2.9'	3.8'	4.6'	*(mf=.42)*
Footcandles	529	190.4	97.2	58.8	39.3	*(pc=4,761)*

FLOOD FOCUS

	3'	5'	7'	9'	11'	Distance
Diameter	3.12'	5.2'	7.28'	9.36'	11.44'	*(mf=1.04)*
Footcandles	158.1	56.9	29.04	17.6	11.8	*(pc=1,423)*

* The instrument pictured on this page has the same model number (60-03) as the instrument pictured on the previous page. The model on this page is a later version and is distinguished by its slide focus.

LIGHTING & ELECTRONICS
Model No. 60-03*

SPOT BEAM SPREAD: 9° beam, 18° field

FLOOD BEAM SPREAD: 44° beam, 64° field

LAMP BASE TYPE: D. C. Bayonet Candelabra

STANDARD LAMP: ESP - 150 W.

OTHER LAMPS: ESR - 100 W. *(cf=.71)*
               FEV - 200 W. *(cf=1.96)*

ACCESSORIES AVAILABLE:
Color Frame, Barn Door, Top Hat

WEIGHT: 3 lbs.

SIZE: 4¾" x 4⅛" x 5" (L x W x H)

PHOTOMETRICS CHART
(Performance data for this unit are measured using ESP - 150 W. lamp.)

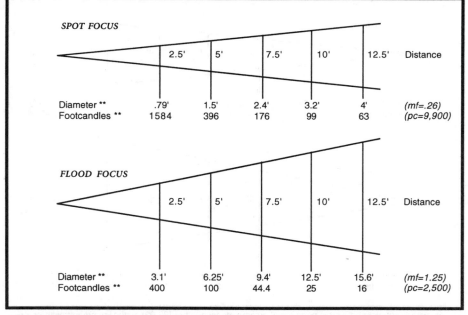

SPOT FOCUS

	2.5'	5'	7.5'	10'	12.5'	Distance
Diameter **	.79'	1.5'	2.4'	3.2'	4'	*(mf=.26)*
Footcandles **	1584	396	176	99	63	*(pc=9,900)*

FLOOD FOCUS

	2.5'	5'	7.5'	10'	12.5'	Distance
Diameter **	3.1'	6.25'	9.4'	12.5'	15.6'	*(mf=1.25)*
Footcandles **	400	100	44.4	25	16	*(pc=2,500)*

* Photometrics on this page are for the standard configuration with a bayonet base socket. See next page for information about this same instrument equipped with a 2-Pin Prefocus base socket.

** The diameter and footcandle information in this chart was calculated using multiplying factor (*mf*) and peak candela (*pc*) data given in catalog. See Introduction for more information about calculating photometric data.

LIGHTING & ELECTRONICS
Model No. 60-03*

SPOT BEAM SPREAD: 15° beam, 30° field

FLOOD BEAM SPREAD: 48° beam, 66° field

LAMP BASE TYPE: 2-Pin Prefocus

STANDARD LAMP: EKB - 420 W.

OTHER LAMPS: JDC120V - 300 W. *(cf=.67)*

ACCESSORIES AVAILABLE:
Color Frame, Barn Door, Top Hat

WEIGHT: 3 lbs.

SIZE: 4¾" x 4⅛" x 5" (L x W x H)

PHOTOMETRICS CHART
(Performance data for this unit are measured using EKB - 420 W. lamp.)

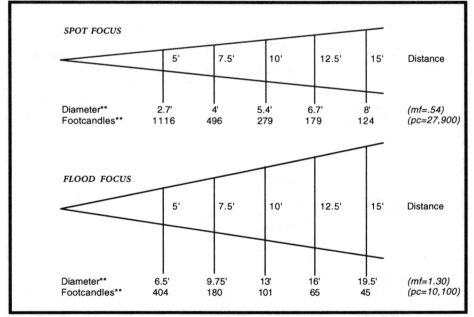

SPOT FOCUS

	5'	7.5'	10'	12.5'	15'	Distance
Diameter**	2.7'	4'	5.4'	6.7'	8'	*(mf=.54)*
Footcandles**	1116	496	279	179	124	*(pc=27,900)*

FLOOD FOCUS

	5'	7.5'	10'	12.5'	15'	Distance
Diameter**	6.5'	9.75'	13'	16'	19.5'	*(mf=1.30)*
Footcandles**	404	180	101	65	45	*(pc=10,100)*

* Photometrics on this page are for an optional configuration with a 2-Pin Prefocus base socket. See previous page for information about this same instrument equipped with the standard bayonet base socket.
** The diameter and footcandle information in this chart was calculated using multiplying factor (*mf*) and peak candela (*pc*) data given in catalog. See Introduction for more information about calculating photometric data.

FRESNEL — 3" Lens

STRAND CENTURY
Model No. 3142

SPOT BEAM SPREAD: 8.5° beam, 16° field

FLOOD BEAM SPREAD: 38° beam, 58° field

LAMP BASE TYPE: D.C. Bayonet Candelabra

STANDARD LAMP: ESS - 250 W.

OTHER LAMPS: ESR - 100 W. *(cf=.38)*
ESP - 150 W. *(cf=.56)*
FEV - 200 W. *(cf=1.10)*

ACCESSORIES AVAILABLE:
Color Frame, 4-way Barn Door, Snoot

WEIGHT: 4 lbs.

SIZE: 5⅞" x 8⅞" (L x H)

PHOTOMETRICS CHART
(Performance data for this unit are measured using ESS - 250 W. lamp.)

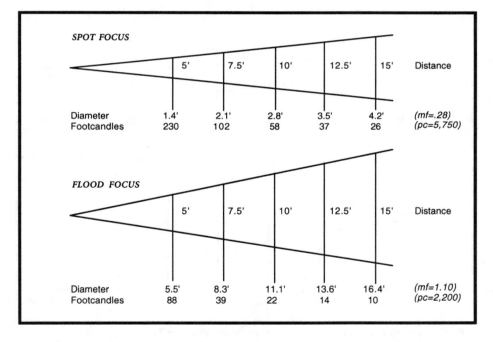

SPOT FOCUS

	5'	7.5'	10'	12.5'	15'	Distance
Diameter	1.4'	2.1'	2.8'	3.5'	4.2'	*(mf=.28)*
Footcandles	230	102	58	37	26	*(pc=5,750)*

FLOOD FOCUS

	5'	7.5'	10'	12.5'	15'	Distance
Diameter	5.5'	8.3'	11.1'	13.6'	16.4'	*(mf=1.10)*
Footcandles	88	39	22	14	10	*(pc=2,200)*

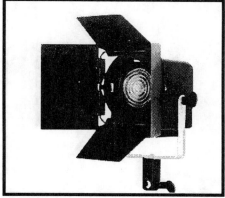

STRAND LIGHTING
Model No. 3101, "Mizar"

SPOT BEAM SPREAD: 14° beam*

FLOOD BEAM SPREAD: 43° beam*

LAMP BASE TYPE: D.C. Bayonet Candelabra

STANDARD LAMP: FEV - 200 W.

OTHER LAMPS: ESR - 100 W. *(cf=.33)*
ESP - 150 W. *(cf=.51)*

ACCESSORIES AVAILABLE:
Color Frame, 4-way Barn Door, Table Stand, Various
Scrims for TV and Photography Use.

WEIGHT: 4 lbs.

SIZE: 5⅛" x 6¼" x 10" (L x W x H)

PHOTOMETRICS CHART
(Performance data for this unit are measured using FEV - 200 W. lamp.)

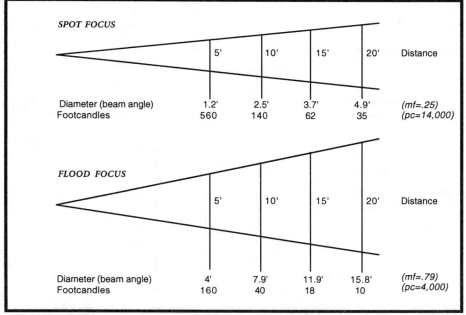

SPOT FOCUS					
	5'	10'	15'	20'	Distance
Diameter (beam angle)	1.2'	2.5'	3.7'	4.9'	*(mf=.25)*
Footcandles	560	140	62	35	*(pc=14,000)*

FLOOD FOCUS					
	5'	10'	15'	20'	Distance
Diameter (beam angle)	4'	7.9'	11.9'	15.8'	*(mf=.79)*
Footcandles	160	40	18	10	*(pc=4,000)*

* No field angle data given in manufacturer's catalog.

STRAND LIGHTING
Model No. 3102, "Mizar"

SPOT BEAM SPREAD: 14° beam*

FLOOD BEAM SPREAD: 47.5° beam*

LAMP BASE TYPE: GY9.5 2 pin

STANDARD LAMP: CP82 (FRB) - 500W.

OTHER LAMPS: CP81 (FKW) - 300 W. *(cf=.53)*

ACCESSORIES AVAILABLE:
Color Frame, 4-way Barn Door, Table Stand, Various Scrims for TV and Photography Use.

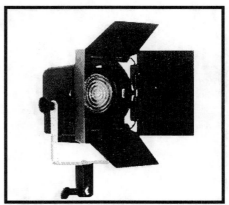

WEIGHT: 4 lbs.

SIZE: 5⅛" x 6¼" x 10" (L x W x H)

PHOTOMETRICS CHART
(Performance data for this unit are measured using CP82 (FRB) - 500 W. lamp.)

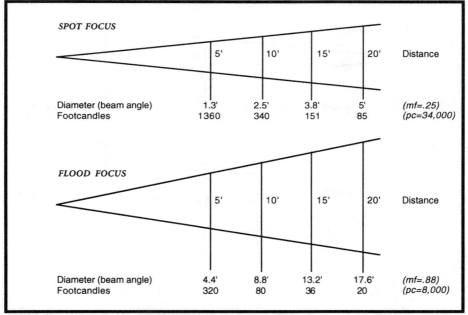

SPOT FOCUS

	5'	10'	15'	20'	Distance
Diameter (beam angle)	1.3'	2.5'	3.8'	5'	*(mf=.25)*
Footcandles	1360	340	151	85	*(pc=34,000)*

FLOOD FOCUS

	5'	10'	15'	20'	Distance
Diameter (beam angle)	4.4'	8.8'	13.2'	17.6'	*(mf=.88)*
Footcandles	320	80	36	20	*(pc=8,000)*

* No field angle data given in manufacturer's catalog.

TIMES SQUARE LIGHTING
Model No. C3

SPOT BEAM SPREAD: 18° field*

FLOOD BEAM SPREAD: 60° field*

LAMP BASE TYPE: D.C. Bayonet Candelabra

STANDARD LAMP: ETC - 150 W.

OTHER LAMPS:

ACCESSORIES AVAILABLE:
Color Frame, Barn Door, Snoot, Pipe Clamp, Table
Base, Adaptors for Tracklight Use.

WEIGHT: (not specified in manufacturer's catalog)

SIZE: 5¼" x 5⅜" x 6⅝" (L x W x H)

PHOTOMETRICS CHART
(Performance data for this unit are measured using ETC - 150 W. lamp.)

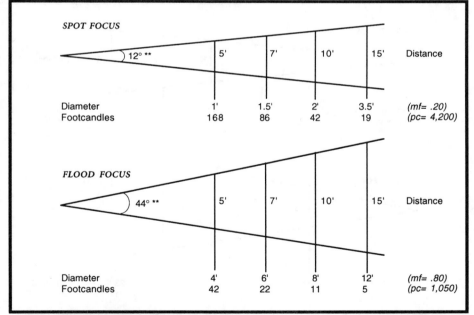

SPOT FOCUS

	12° **	5'	7'	10'	15'	Distance
Diameter		1'	1.5'	2'	3.5'	(mf= .20)
Footcandles		168	86	42	19	(pc= 4,200)

FLOOD FOCUS

	44° **	5'	7'	10'	15'	Distance
Diameter		4'	6'	8'	12'	(mf= .80)
Footcandles		42	22	11	5	(pc= 1,050)

* No beam angle data given in manufacturer's catalog.
** Using the given distances and diameters, the angle at spot focus calculates to be 12°, while at flood focus
the angle calculates to be approximately 44°. Ordinarily these angles should be the same as the given "field
angles" at spot and flood focus. See Introduction for more information about calculating photometric data.

ARRIFLEX CORP.
Model No. 531600

SPOT BEAM SPREAD: 16° beam*

FLOOD BEAM SPREAD: 60° beam*

LAMP BASE TYPE: 2-Pin Prefocus

STANDARD LAMP: FRK - 650 W.

OTHER LAMPS: FRG - 500 W. *(cf=.74)*
                 FKW - 300 W. *(cf=.45)*

ACCESSORIES AVAILABLE:
Color Frame (6⅝" sq.), 4-way Barn Door, Top Hat

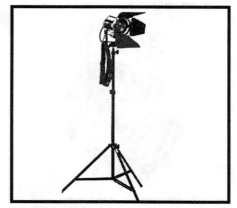

WEIGHT: 7.1 lbs.

SIZE: (not specified in manufacturer's catalog)

PHOTOMETRICS CHART

(Performance data for this unit are measured using FRK - 650 W. lamp)

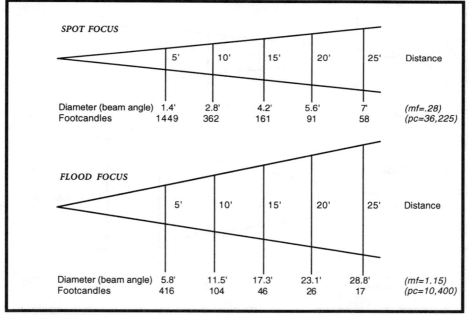

SPOT FOCUS

	5'	10'	15'	20'	25'	Distance
Diameter (beam angle)	1.4'	2.8'	4.2'	5.6'	7'	*(mf=.28)*
Footcandles	1449	362	161	91	58	*(pc=36,225)*

FLOOD FOCUS

	5'	10'	15'	20'	25'	Distance
Diameter (beam angle)	5.8'	11.5'	17.3'	23.1'	28.8'	*(mf=1.15)*
Footcandles	416	104	46	26	17	*(pc=10,400)*

* No field angle data given in manufacturer's catalog.

CCT LIGHTING
Model No. Z0642

SPOT BEAM SPREAD: 8.5° beam, 18° field

FLOOD BEAM SPREAD: 58° beam, 67° field

LAMP BASE TYPE: GY9.5

STANDARD LAMP: FMR - 600 W.

OTHER LAMPS: (none specified in mfg. catalog)

ACCESSORIES AVAILABLE:
Color Frame (5" sq.), 4-way Barn Door

WEIGHT: 5 lbs.

SIZE: 9½" x 7" x 10½" (L x W x H)

PHOTOMETRICS CHART
(Performance data for this unit are measured using FMR - 600 W. lamp.)

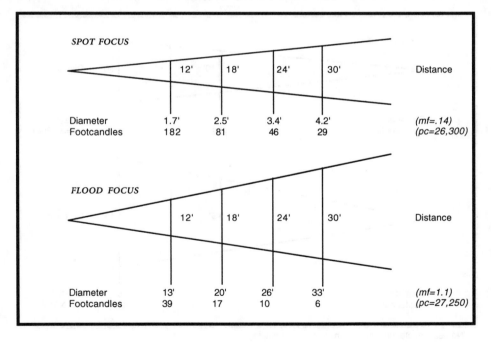

SPOT FOCUS

	12'	18'	24'	30'	Distance
Diameter	1.7'	2.5'	3.4'	4.2'	*(mf=.14)*
Footcandles	182	81	46	29	*(pc=26,300)*

FLOOD FOCUS

	12'	18'	24'	30'	Distance
Diameter	13'	20'	26'	33'	*(mf=1.1)*
Footcandles	39	17	10	6	*(pc=27,250)*

FRESNEL — 4.5" Lens

KLIEGL BROS.
Model No. 3604

SPOT BEAM SPREAD: 18° field*

FLOOD BEAM SPREAD: 54° field*

LAMP BASE TYPE: Medium 2-Pin

STANDARD LAMP: BSP - 500 W.

OTHER LAMPS: BWM - 750 W. *(cf=1.83)*

ACCESSORIES AVAILABLE:
Color Frame, 8-way Barn Door, Snoot, Stand
Adaptor, 6" Table Stand

WEIGHT: 9 lbs.
SIZE: 7⅞" x 7¾" x 16¼" (L x W x H)

PHOTOMETRICS CHART
(Performance data for this unit are measured using BSP - 500 W. lamp.)

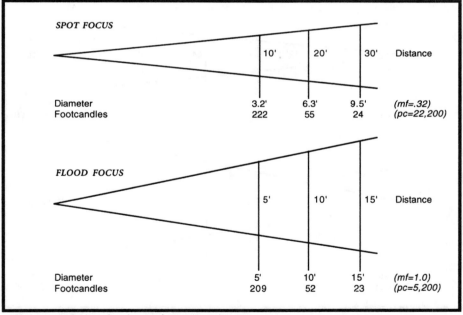

SPOT FOCUS

	10'	20'	30'	Distance
Diameter	3.2'	6.3'	9.5'	*(mf=.32)*
Footcandles	222	55	24	*(pc=22,200)*

FLOOD FOCUS

	5'	10'	15'	Distance
Diameter	5'	10'	15'	*(mf=1.0)*
Footcandles	209	52	23	*(pc=5,200)*

* No beam angle data given in manufacturer's catalog.

LIGHTING & ELECTRONICS
Model No. 60-04

SPOT BEAM SPREAD: 8° beam, 30° field

FLOOD BEAM SPREAD: 26° beam, 50° field

LAMP BASE TYPE: Med 2-Pin

STANDARD LAMP: FLK/LL - 575 W.

OTHER LAMPS: HX400 - 400 W. *(cf=.78)*
EHD - 500 W. *(cf=.83)*
FLK - 575 W. *(cf=1.3)*
HX602 - 575 W. *(cf=1.5)*

WEIGHT: 4 lbs.

SIZE: 7¾" x 6⅝" x 6⅝" (L x W x H)

ACCESSORIES AVAILABLE:
Color Frame, Barn Door

PHOTOMETRICS CHART

(Performance data for this unit are measured using FLK/LL - 575 W. lamp.)

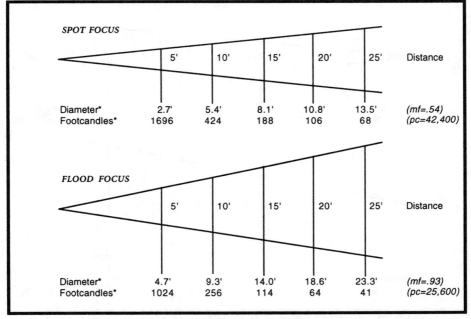

SPOT FOCUS

	5'	10'	15'	20'	25'	Distance
Diameter*	2.7'	5.4'	8.1'	10.8'	13.5'	*(mf=.54)*
Footcandles*	1696	424	188	106	68	*(pc=42,400)*

FLOOD FOCUS

	5'	10'	15'	20'	25'	Distance
Diameter*	4.7'	9.3'	14.0'	18.6'	23.3'	*(mf=.93)*
Footcandles*	1024	256	114	64	41	*(pc=25,600)*

* The diameter and footcandle information in this chart was calculated using multiplying factor (*mf*) and peak candela (*pc*) data given in catalog. See Introduction for more information about calculating photometric data.

FRESNEL — 4.5" Lens

STRAND CENTURY
Model No. 3242

SPOT BEAM SPREAD: 8° beam, 14° field

FLOOD BEAM SPREAD: 36° beam, 45° field

LAMP BASE TYPE: Medium Prefocus

STANDARD LAMP: BTL - 500 W.

OTHER LAMPS: BTM - 500 W. *(cf=1.2)*
750T17 - 500 W. *(cf=.95)*

ACCESSORIES AVAILABLE:
Color Frame

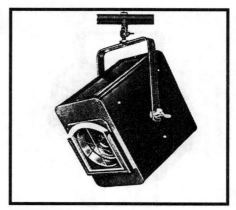

WEIGHT: 8¼ lbs.

SIZE: 11½" x 9" x 12½" (L x W x H)

PHOTOMETRICS CHART
(Performance data for this unit are measured using BTL - 500 W. lamp.)

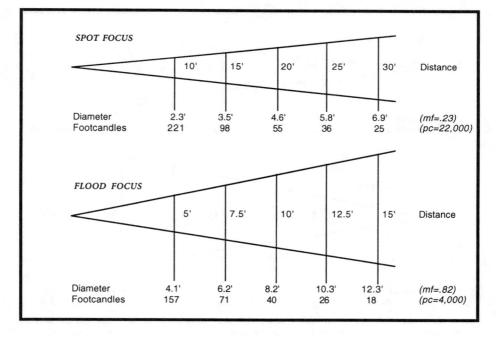

SPOT FOCUS

	10'	15'	20'	25'	30'	Distance
Diameter	2.3'	3.5'	4.6'	5.8'	6.9'	*(mf=.23)*
Footcandles	221	98	55	36	25	*(pc=22,000)*

FLOOD FOCUS

	5'	7.5'	10'	12.5'	15'	Distance
Diameter	4.1'	6.2'	8.2'	10.3'	12.3'	*(mf=.82)*
Footcandles	157	71	40	26	18	*(pc=4,000)*

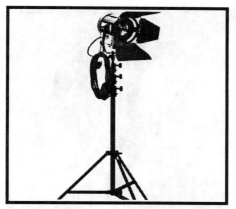

WEIGHT: 11.2 lbs.

SIZE: (not specified in manufacturer's catalog)

ARRIFLEX CORP.
Model No. 531100

SPOT BEAM SPREAD: 12° beam*

FLOOD BEAM SPREAD: 60° beam*

LAMP BASE TYPE: Medium Bipost

STANDARD LAMP: EGT - 1 KW.

OTHER LAMPS: EGR - 750 W. *(cf=.74)*

ACCESSORIES AVAILABLE:
Color Frame (7¾" sq.), 4-way and 8-way Barn Doors, Top Hat

PHOTOMETRICS CHART
(Performance data for this unit are measured using EGT - 1 KW. lamp)

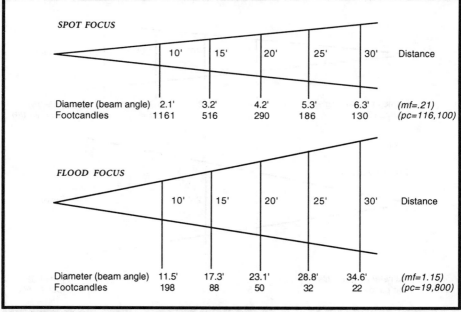

SPOT FOCUS

	10'	15'	20'	25'	30'	Distance
Diameter (beam angle)	2.1'	3.2'	4.2'	5.3'	6.3'	*(mf=.21)*
Footcandles	1161	516	290	186	130	*(pc=116,100)*

FLOOD FOCUS

	10'	15'	20'	25'	30'	Distance
Diameter (beam angle)	11.5'	17.3'	23.1'	28.8'	34.6'	*(mf=1.15)*
Footcandles	198	88	50	32	22	*(pc=19,800)*

* No field angle data given in manufacturer's catalog.

STRAND LIGHTING
Model No. 3201, "Bambino" *

SPOT BEAM SPREAD: 12° beam**

FLOOD BEAM SPREAD: 63° beam**

LAMP BASE TYPE: Medium Bipost

STANDARD LAMP: EGT - 1 KW.

OTHER LAMPS: EGR - 750W. *(cf=.74)*
EGN - 500W. *(cf=.46)*

ACCESSORIES AVAILABLE:
Color Frame, 8-way Barn Door, Various Scrims for TV
and Photography Use.

WEIGHT: 11.7 lbs.

SIZE: 9¾" x 9" x 8¾" (L x W x H)

PHOTOMETRICS CHART
(Performance data for this unit are measured using EGT - 1 KW. lamp.)

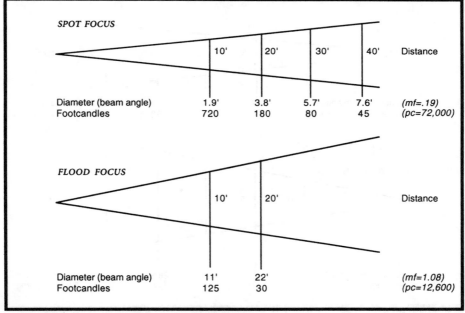

SPOT FOCUS

	10'	20'	30'	40'	Distance
Diameter (beam angle)	1.9'	3.8'	5.7'	7.6'	*(mf=.19)*
Footcandles	720	180	80	45	*(pc=72,000)*

FLOOD FOCUS

	10'	20'	Distance
Diameter (beam angle)	11'	22'	*(mf=1.08)*
Footcandles	125	30	*(pc=12,600)*

* This model comes with a 5/8" socket for stand mounting.
** No field angle data given in manufacturer's catalog.

ALTMAN STAGE LIGHTING
Model No. 1 KAF-MEBP

SPOT BEAM SPREAD: 6.4° beam, 12.9° field

FLOOD BEAM SPREAD: 70.5° beam, 76.5° field

LAMP BASE TYPE: Medium Prefocus

STANDARD LAMP: EGT - 1KW.

OTHER LAMPS: EGN - 500 W. *(cf=.46)*
EGR - 750 W. *(cf=.75)*

ACCESSORIES AVAILABLE:
Color Frame, Pin Spot Adaptor, 4-way Barn Door,
Snoot, Motorized/Non Motorized Color Wheels

WEIGHT: (not specified in manufacturer's catalog)

SIZE: 12½" x 10" x 16⅝" (L x W x H)

PHOTOMETRICS CHART
(Performance data for this unit are measured using EGT - 1 KW. lamp.)

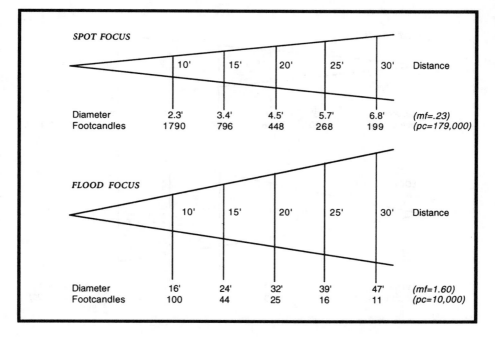

SPOT FOCUS

	10'	15'	20'	25'	30'	Distance
Diameter	2.3'	3.4'	4.5'	5.7'	6.8'	*(mf=.23)*
Footcandles	1790	796	448	268	199	*(pc=179,000)*

FLOOD FOCUS

	10'	15'	20'	25'	30'	Distance
Diameter	16'	24'	32'	39'	47'	*(mf=1.60)*
Footcandles	100	44	25	16	11	*(pc=10,000)*

FRESNEL — 6" Lens

ALTMAN STAGE LIGHTING
Model No. 1 KAF-MEPF

SPOT BEAM SPREAD: 6.4° beam, 12.2° field

FLOOD BEAM SPREAD: 69.2° beam, 73.8° field

LAMP BASE TYPE: Medium Prefocus

STANDARD LAMP: BTR - 1KW.

OTHER LAMPS: EEX - 300 W.
            BTL - *(cf=.26)*
            BTN - 500 W.
            (cf=.40)

ACCESSORIES AVAILABLE:
Color Frame, Pin Spot Adaptor, 4-way Barn Door,
Snoot, Motorized/Non Motorized Color Wheels

WEIGHT: (not specified in manufacturer's catalog)

SIZE: 12½" x 10" x 16⅝" (L x W x H)

PHOTOMETRICS CHART

(Performance data for this unit are measured using BTR - 1 KW. lamp.)

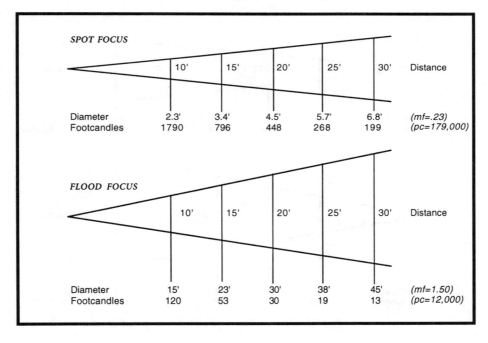

SPOT FOCUS

	10'	15'	20'	25'	30'	Distance
Diameter	2.3'	3.4'	4.5'	5.7'	6.8'	*(mf=.23)*
Footcandles	1790	796	448	268	199	*(pc=179,000)*

FLOOD FOCUS

	10'	15'	20'	25'	30'	Distance
Diameter	15'	23'	30'	38'	45'	*(mf=1.50)*
Footcandles	120	53	30	19	13	*(pc=12,000)*

ALTMAN STAGE LIGHTING
Model Nos. 65Q (slide focus)
165Q (screw feed focus)

SPOT BEAM SPREAD: 8.8° beam, 16° field

FLOOD BEAM SPREAD: 59° beam, 70° field

LAMP BASE TYPE: Medium Prefocus

STANDARD LAMP: BTN - 750 W.

OTHER LAMPS: BTL/BTM - 500 W. *(cf=.65)*
BTP - 750 W.
BTR - *(cf=1.18)*

WEIGHT: (not specified in manufacturer's catalog)

SIZE: 8¾" x 8¼" x 13¾" (L x W x H) (65Q)
10¼" x 9½" x 14¼" (L x W x H) (165Q)

ACCESSORIES AVAILABLE:
Color Frame (7½" sq.), Pin Spot Adaptor, 4-way Barn Door, Snoot, Motorized/Non Motorized Color Wheels

PHOTOMETRICS CHART
(Performance data for these units are measured using BTN - 750 W. lamp.)

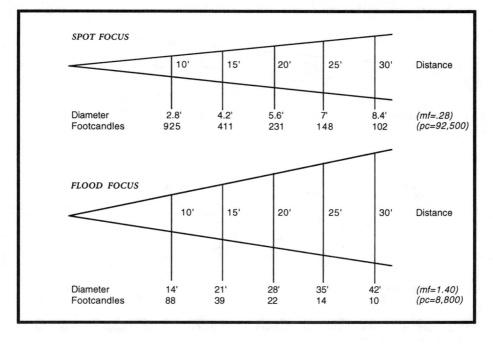

SPOT FOCUS

	10'	15'	20'	25'	30'	Distance
Diameter	2.8'	4.2'	5.6'	7'	8.4'	*(mf=.28)*
Footcandles	925	411	231	148	102	*(pc=92,500)*

FLOOD FOCUS

	10'	15'	20'	25'	30'	Distance
Diameter	14'	21'	28'	35'	42'	*(mf=1.40)*
Footcandles	88	39	22	14	10	*(pc=8,800)*

BERKEY COLORTRAN
Model No. 100-142*

SPOT BEAM SPREAD: 10.5° beam, 19° field

FLOOD BEAM SPREAD: 44° beam, 54° field

LAMP BASE TYPE: Medium Prefocus

STANDARD LAMP: BTR - 1 KW.

OTHER LAMPS: BTC - 500 W. *(cf=.5)*
BTN - 750 W. *(cf=.6)*

ACCESSORIES AVAILABLE:
Color Frame, 4-way Barn Door, Top Hat

WEIGHT: 10 lbs.

SIZE: 13¾" x 10⅜" x 16" (L x W x H)

PHOTOMETRICS CHART

(Performance data for this unit are measured using BTR - 1 KW. lamp.)

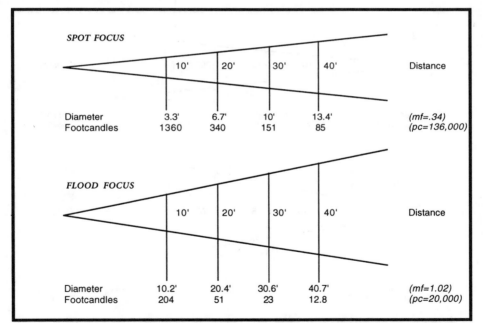

SPOT FOCUS

	10'	20'	30'	40'	Distance
Diameter	3.3'	6.7'	10'	13.4'	*(mf=.34)*
Footcandles	1360	340	151	85	*(pc=136,000)*

FLOOD FOCUS

	10'	20'	30'	40'	Distance
Diameter	10.2'	20.4'	30.6'	40.7'	*(mf=1.02)*
Footcandles	204	51	23	12.8	*(pc=20,000)*

* This model can be identified with model numbers -141, -142, -145, -146 or -147 depending on the wire leads and connectors supplied.

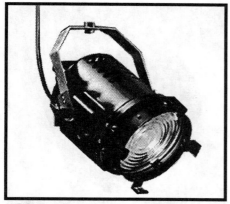

BERKEY COLORTRAN
Model No. 100-412*

SPOT BEAM SPREAD: 10.5° beam, 19° field

FLOOD BEAM SPREAD: 4° beam, 54° field

LAMP BASE TYPE: Medium 2-Pin

STANDARD LAMP: BWN - 1 KW.

OTHER LAMPS: BWM - 750 W.   *(cf=.74)*

ACCESSORIES AVAILABLE:
Color Frame, 4-way Barn Door

WEIGHT: 10 lbs.

SIZE: 11½" x 11" x 18" (L x W x H)

PHOTOMETRICS CHART
(Performance data for this unit are measured using BWN - 750 W. lamp.)

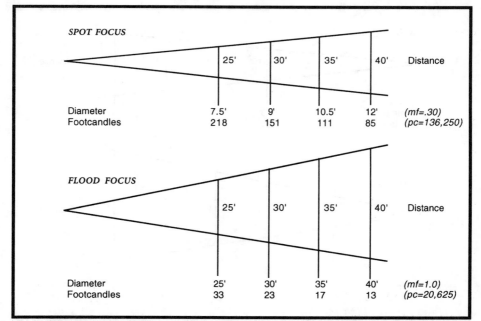

SPOT FOCUS

	25'	30'	35'	40'	Distance
Diameter	7.5'	9'	10.5'	12'	*(mf=.30)*
Footcandles	218	151	111	85	*(pc=136,250)*

FLOOD FOCUS

	25'	30'	35'	40'	Distance
Diameter	25'	30'	35'	40'	*(mf=1.0)*
Footcandles	33	23	17	13	*(pc=20,625)*

* This instrument can be identified with model numbers -412, -415, -416, or -417, depending on the wire leads and connectors supplied.

BERKEY COLORTRAN
Model No. 214-012*

SPOT BEAM SPREAD: 7° beam, 17° field

FLOOD BEAM SPREAD: 40° beam, 53° field

LAMP BASE TYPE: Medium Prefocus

STANDARD LAMP: BTL - 500 W.

OTHER LAMPS: BTN - 750 W. *(cf=1.55)*
                BTR - 1 KW. *(cf=2.50)*

ACCESSORIES AVAILABLE:
Color Frame, 4-way Barn Door

WEIGHT: 8.3 lbs.

SIZE: 9¼" x 11¼" x 19" (L x W x H)

PHOTOMETRICS CHART
(Performance data for this unit are measured using BTL - 500 W. lamp.)

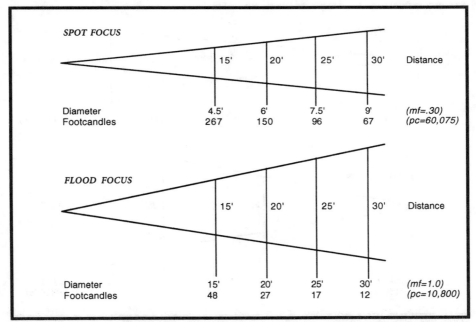

SPOT FOCUS

	15'	20'	25'	30'	Distance
Diameter	4.5'	6'	7.5'	9'	*(mf=.30)*
Footcandles	267	150	96	67	*(pc=60,075)*

FLOOD FOCUS

	15'	20'	25'	30'	Distance
Diameter	15'	20'	25'	30'	*(mf=1.0)*
Footcandles	48	27	17	12	*(pc=10,800)*

* This instrument can be identified with model numbers -012, -015, -016, or -017, depending on the wire leads and connectors supplied.

CCT LIGHTING
Model No. Z0050

SPOT BEAM SPREAD: 6° beam, 10° field

FLOOD BEAM SPREAD: 57° beam, 67° field

LAMP BASE TYPE: Medium Prefocus

STANDARD LAMP: BTR - 1 KW.

OTHER LAMPS: BTL - 500 W. *(cf=.5)*
BTN - 750 W. *(cf=.6)*

ACCESSORIES AVAILABLE:
Color Frame, 4-way Barn Door

WEIGHT: 15 lbs.

SIZE: 12" x 8½" x 17½" (L x W x H)

PHOTOMETRICS CHART

(Performance data for this unit are measured using BTR - 1 KW. lamp.)

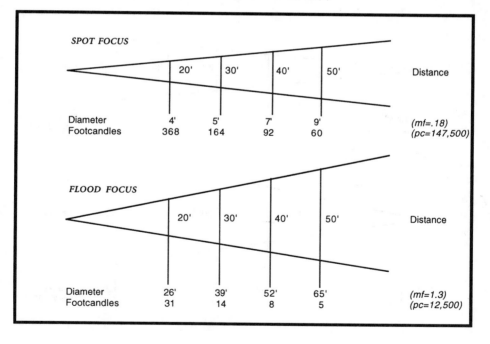

SPOT FOCUS

	20'	30'	40'	50'	Distance
Diameter	4'	5'	7'	9'	*(mf=.18)*
Footcandles	368	164	92	60	*(pc=147,500)*

FLOOD FOCUS

	20'	30'	40'	50'	Distance
Diameter	26'	39'	52'	65'	*(mf=1.3)*
Footcandles	31	14	8	5	*(pc=12,500)*

FRESNEL — 6" Lens

CENTURY LIGHTING
Model No. 500

SPOT BEAM SPREAD: 9° beam, 16° field

FLOOD BEAM SPREAD: 41° beam, 52° field

LAMP BASE TYPE: Medium Prefocus

STANDARD LAMP: BFE - 750 W.

OTHER LAMPS: 250T20/47 - 250 W. *(cf=.27)*
DNW -   500 W. *(cf=.59)*

ACCESSORIES AVAILABLE:
Color Frame (7½" sq.), 2 or 4-way Barn Door,
Snoot

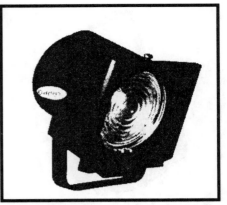

WEIGHT: (not specified in manufacturer's catalog)

SIZE: 8⅞" x 8⅞" x 18" (L x W x H)

PHOTOMETRICS CHART

(Performance data for this unit are measured using BFE - 750 W. lamp.)

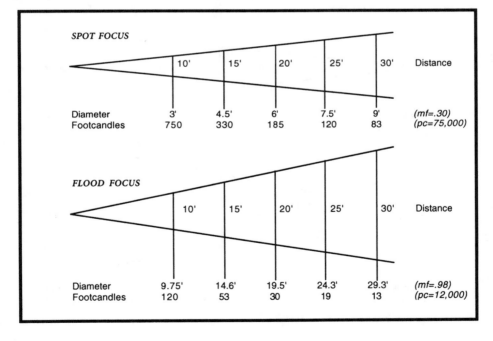

SPOT FOCUS

	10'	15'	20'	25'	30'	Distance
Diameter	3'	4.5'	6'	7.5'	9'	*(mf=.30)*
Footcandles	750	330	185	120	83	*(pc=75,000)*

FLOOD FOCUS

	10'	15'	20'	25'	30'	Distance
Diameter	9.75'	14.6'	19.5'	24.3'	29.3'	*(mf=.98)*
Footcandles	120	53	30	19	13	*(pc=12,000)*

FRESNEL — 6" Lens

CENTURY LIGHTING
Model No. 550

SPOT BEAM SPREAD: 9° beam, 16° field

FLOOD BEAM SPREAD: 41° beam, 52° field

LAMP BASE TYPE: Medium Prefocus

STANDARD LAMP: BFE - 750 W.

OTHER LAMPS: 250T20/47 - 250 W. *(cf=.27)*
DNW - 500 W. *(cf=.59)*

ACCESSORIES AVAILABLE:
Color Frame, 2 or 4-way Barn Door, Snoot

WEIGHT: (not specified in manufacturer's catalog)

SIZE: 8⅞" x 9½" x 18½" (L x W x H)

PHOTOMETRICS CHART

(Performance data for this unit are measured using BFE - 750 W. lamp.)

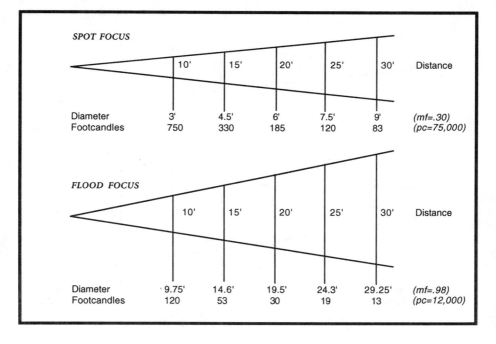

SPOT FOCUS

	10'	15'	20'	25'	30'	Distance
Diameter	3'	4.5'	6'	7.5'	9'	*(mf=.30)*
Footcandles	750	330	185	120	83	*(pc=75,000)*

FLOOD FOCUS

	10'	15'	20'	25'	30'	Distance
Diameter	9.75'	14.6'	19.5'	24.3'	29.25'	*(mf=.98)*
Footcandles	120	53	30	19	13	*(pc=12,000)*

CENTURY LIGHTING
Model No. 553

SPOT BEAM SPREAD: 8° beam, 16° field

FLOOD BEAM SPREAD: 52° beam, 60° field

LAMP BASE TYPE: Recessed Single-Contact

STANDARD LAMP: FAD - 650 W.

OTHER LAMPS: EHP - 300 W. *(cf=.30)*
EHR - 400 W. *(cf=.45)*

ACCESSORIES AVAILABLE:
Color Frame, 2 or 4-way Barn Door, Snoot

WEIGHT: (not specified in manufacturer's catalog)

SIZE: 8⅞" x 9½" x 18½" (L x W x H)

PHOTOMETRICS CHART
(Performance data for this unit are measured using FAD - 650 W. lamp.)

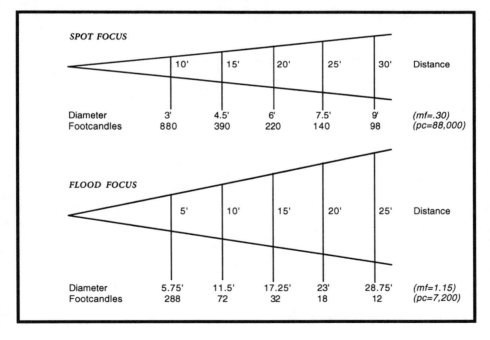

SPOT FOCUS

	10'	15'	20'	25'	30'	Distance
Diameter	3'	4.5'	6'	7.5'	9'	*(mf=.30)*
Footcandles	880	390	220	140	98	*(pc=88,000)*

FLOOD FOCUS

	5'	10'	15'	20'	25'	Distance
Diameter	5.75'	11.5'	17.25'	23'	28.75'	*(mf=1.15)*
Footcandles	288	72	32	18	12	*(pc=7,200)*

CENTURY STRAND
Model No. 3312

SPOT BEAM SPREAD: 8° beam, 16° field

FLOOD BEAM SPREAD: 54° beam, 60° field

LAMP BASE TYPE: Medium Prefocus

STANDARD LAMP: BTN - 750 W.

OTHER LAMPS: BTL - 500 W. *(cf=.65)*
BTP - 750 W. *(cf=1.18)*

ACCESSORIES AVAILABLE:
Color Frame, 2 or 4-way Barn Door, Snoot

WEIGHT: (not specified in manufacturer's catalog)

SIZE: 10" x 10" x 14" (L x W x H)

PHOTOMETRICS CHART
(Performance data for this unit are measured using BTN - 750 W. lamp.)

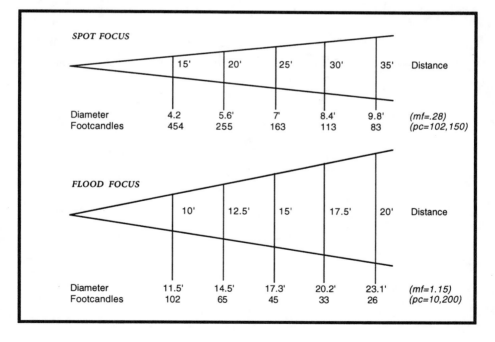

SPOT FOCUS

	15'	20'	25'	30'	35'	Distance
Diameter	4.2	5.6'	7'	8.4'	9.8'	*(mf=.28)*
Footcandles	454	255	163	113	83	*(pc=102,150)*

FLOOD FOCUS

	10'	12.5'	15'	17.5'	20'	Distance
Diameter	11.5'	14.5'	17.3'	20.2'	23.1'	*(mf=1.15)*
Footcandles	102	65	45	33	26	*(pc=10,200)*

CENTURY STRAND
Model No. 33420

SPOT BEAM SPREAD: 8° beam, 17° field

FLOOD BEAM SPREAD: 32° beam, 45.5° field

LAMP BASE TYPE: Medium Prefocus

STANDARD LAMP: BTN - 750 W.

OTHER LAMPS: BTL - 500 W. *(cf=.65)*
　　　　　　　　BTP - 750 W. *(cf=1.18)*

ACCESSORIES AVAILABLE:
Color Frame, 2 or 4-way Barn Door, Snoot

WEIGHT: (not specified in manufacturer's catalog)

SIZE: 8⅞" x 9½" x 12½" (L x W x H)

PHOTOMETRICS CHART
(Performance data for this unit are measured using BTN - 750 W. lamp.)

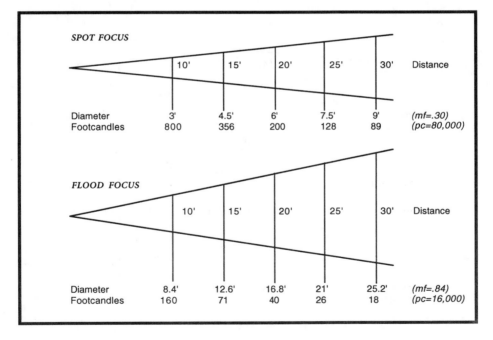

SPOT FOCUS

	10'	15'	20'	25'	30'	Distance
Diameter	3'	4.5'	6'	7.5'	9'	*(mf=.30)*
Footcandles	800	356	200	128	89	*(pc=80,000)*

FLOOD FOCUS

	10'	15'	20'	25'	30'	Distance
Diameter	8.4'	12.6'	16.8'	21'	25.2'	*(mf=.84)*
Footcandles	160	71	40	26	18	*(pc=16,000)*

COLORTRAN
Model No. 100-511*

SPOT BEAM SPREAD: 6.9° beam, 13.9° field

FLOOD BEAM SPREAD: 60.1° beam, 68.8° field

LAMP BASE TYPE: Medium Bipost

STANDARD LAMP: EGT - 1 KW.

OTHER LAMPS: EGR - 750 W. *(cf=.74)*
EGN - 500 W. *(cf=1.46)*

ACCESSORIES AVAILABLE:
Color Frame, 8-way Barn Door, Top Hat

WEIGHT: 11 lbs.

SIZE: 13" x 10½" x 22¾" (L x W x H)

PHOTOMETRICS CHART

(Performance data for this unit are measured using EGT - 1 KW. lamp.)

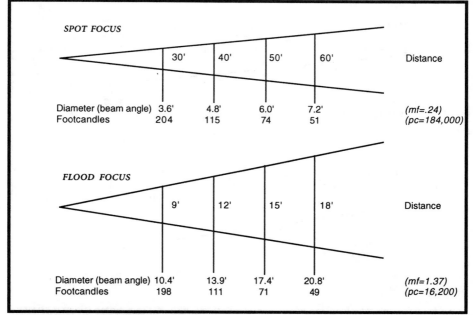

SPOT FOCUS

	30'	40'	50'	60'	Distance
Diameter (beam angle)	3.6'	4.8'	6.0'	7.2'	*(mf=.24)*
Footcandles	204	115	74	51	*(pc=184,000)*

FLOOD FOCUS

	9'	12'	15'	18'	Distance
Diameter (beam angle)	10.4'	13.9'	17.4'	20.8'	*(mf=1.37)*
Footcandles	198	111	71	49	*(pc=16,200)*

* This model can be identified with model numbers -512, -514, -515, -516 or -517 depending on the wire leads and connectors supplied.

COLORTRAN
Model No. 213-202*

SPOT BEAM SPREAD: 6.9° beam, 13.9° field

FLOOD BEAM SPREAD: 61.6° beam, 72.5° field

LAMP BASE TYPE: Medium Prefocus

STANDARD LAMP: BTN - 750 W.

OTHER LAMPS: BTL - 500 W. *(cf=.65)*
BTP - 750 W. *(cf=1.18)*

ACCESSORIES AVAILABLE:
Color Frame, 4-way Barn Door, Snoot

WEIGHT: 9.75 lbs.

SIZE: 10⅜" x 10⅜" x 19¼" (L x W x H)

PHOTOMETRICS CHART
(Performance data for this unit are measured using BTN - 750 W. lamp.)

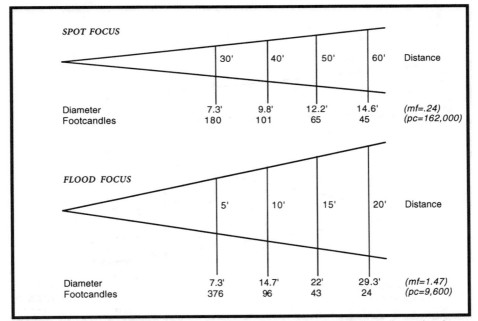

SPOT FOCUS

	30'	40'	50'	60'	Distance
Diameter	7.3'	9.8'	12.2'	14.6'	*(mf=.24)*
Footcandles	180	101	65	45	*(pc=162,000)*

FLOOD FOCUS

	5'	10'	15'	20'	Distance
Diameter	7.3'	14.7'	22'	29.3'	*(mf=1.47)*
Footcandles	376	96	43	24	*(pc=9,600)*

* This instrument can be identified with model numbers -202, -205, -206, or -207, depending on the wire leads and connectors supplied.

ELECTRO CONTROLS
Model Nos. 3466 & 7466A*

SPOT BEAM SPREAD: 8.5° beam, 22° field

FLOOD BEAM SPREAD: 51° beam, 63° field

LAMP BASE TYPE: Medium Prefocus

STANDARD LAMP: BTR - 1 KW.

OTHER LAMPS: BTL - 500 W. *(cf=.40)*
            BTN - 750 W. *(cf=.62)*

ACCESSORIES AVAILABLE:
Color Frame (7½" sq.), 2 or 4-way Barn Door, Snoot

WEIGHT: (not specified in manufacturer's catalog)

SIZE: 8¼" x 11⅜" (L x W)

PHOTOMETRICS CHART

(Performance data for this unit are measured using BTR - 1 KW. lamp.)

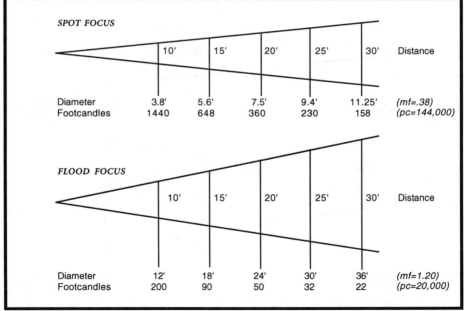

SPOT FOCUS

	10'	15'	20'	25'	30'	Distance
Diameter	3.8'	5.6'	7.5'	9.4'	11.25'	*(mf=.38)*
Footcandles	1440	648	360	230	158	*(pc=144,000)*

FLOOD FOCUS

	10'	15'	20'	25'	30'	Distance
Diameter	12'	18'	24'	30'	36'	*(mf=1.20)*
Footcandles	200	90	50	32	22	*(pc=20,000)*

*These two model numbers were used at different times for the same instrument. (The instrument pictured on this page has a slide focus mechanism, which is the only thing which makes it different from the instrument on the following page, model 3467 / 7467A, which has a screw feed mechanism.)

FRESNEL — 6" Lens

ELECTRO CONTROLS
Model Nos. 3467 & 7467A*

SPOT BEAM SPREAD: 8.5° beam, 22° field

FLOOD BEAM SPREAD: 51° beam, 63° field

LAMP BASE TYPE: Medium Prefocus

STANDARD LAMP: BTR - 1 KW.

OTHER LAMPS: BTL - 500 W. *(cf=.40)*
　　　　　　　　BTN - 750 W. *(cf=.62)*

ACCESSORIES AVAILABLE:
Color Frame (7½" sq.), 2 or 4-way Barn Door, Snoot

WEIGHT: (not specified in manufacturer's catalog)

SIZE: 8¾" x 11⅜" (L x W)

PHOTOMETRICS CHART
(Performance data for this unit are measured using BTR - 1 KW. lamp.)

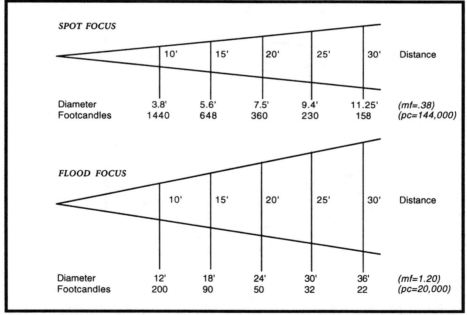

SPOT FOCUS

	10'	15'	20'	25'	30'	Distance
Diameter	3.8'	5.6'	7.5'	9.4'	11.25'	*(mf=.38)*
Footcandles	1440	648	360	230	158	*(pc=144,000)*

FLOOD FOCUS

	10'	15'	20'	25'	30'	Distance
Diameter	12'	18'	24'	30'	36'	*(mf=1.20)*
Footcandles	200	90	50	32	22	*(pc=20,000)*

*These two model numbers were used at different times for the same instrument. (The instrument pictured on this page has a screw feed mechanism, which is the only thing which makes it different from the instrument on the predeeding page, model 3466 / 7466A, which has a slide focus mechanism.)

KLIEGL BROS.
Model No. 3606

SPOT BEAM SPREAD: 18° field*

FLOOD BEAM SPREAD: 62° field*

LAMP BASE TYPE: Medium 2-Pin

STANDARD LAMP: EHD - 500 W.

OTHER LAMPS: EHC - 500 W. *(cf=1.23)*
EHG - 750 W. *(cf=1.45)*

ACCESSORIES AVAILABLE:
Color Frame, 8-way Barn Door, Snoot

WEIGHT: 13 lbs.

SIZE: 13½" x 10" x 20½" (L x W x H)

PHOTOMETRICS CHART
(Performance data for this unit are measured using EHD - 500 W. lamp.)

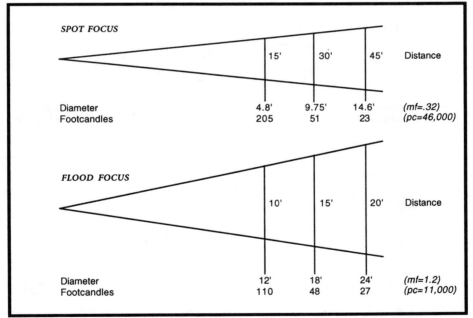

SPOT FOCUS

	15'	30'	45'	Distance
Diameter	4.8'	9.75'	14.6'	*(mf=.32)*
Footcandles	205	51	23	*(pc=46,000)*

FLOOD FOCUS

	10'	15'	20'	Distance
Diameter	12'	18'	24'	*(mf=1.2)*
Footcandles	110	48	27	*(pc=11,000)*

* No beam angle data given in manufacturer's catalog.

FRESNEL — 6" Lens

LEE COLORTRAN
Model No. 213-512*

SPOT BEAM SPREAD: 6.1° beam, 12.5° field

FLOOD BEAM SPREAD: 64.5° beam, 74.6° field

LAMP BASE TYPE: Medium Prefocus

STANDARD LAMP: BTR - 1 KW.

OTHER LAMPS: BTL - 500 W.   *(cf=.40)*
              BTN - 750 W.   *(cf=.62)*

ACCESSORIES AVAILABLE:
Color Frame, 8-way Barn Door, Dichroic Filter,
Scrim Set for TV use, Snoot

WEIGHT: 11 lbs.

SIZE: 13" x 10½" x 22¾" (L x W x H)

PHOTOMETRICS CHART

(Performance data for this unit are measured using BTR - 1 KW. lamp.)

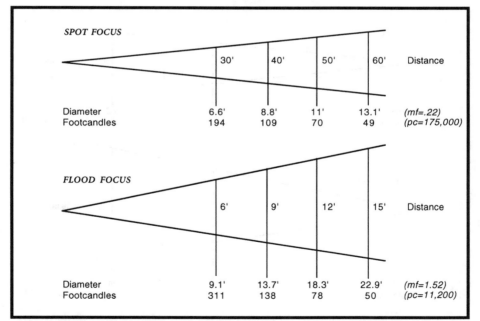

SPOT FOCUS					
	30'	40'	50'	60'	Distance
Diameter	6.6'	8.8'	11'	13.1'	*(mf=.22)*
Footcandles	194	109	70	49	*(pc=175,000)*

FLOOD FOCUS					
	6'	9'	12'	15'	Distance
Diameter	9.1'	13.7'	18.3'	22.9'	*(mf=1.52)*
Footcandles	311	138	78	50	*(pc=11,200)*

* This instrument can be identified with model numbers -512, -515, -516, or -517, depending on the wire leads and connectors supplied.

LIGHTING & ELECTRONICS
Model Nos. 60-06* (slide focus)
60-07* (screw feed)

SPOT BEAM SPREAD: 11° beam, 17° field

FLOOD BEAM SPREAD: 38° beam, 53° field

LAMP BASE TYPE: Medium Prefocus

STANDARD LAMP: BFE - 750 W.

OTHER LAMPS: DNW -   500 W. *(cf=.59)*
1M/T20P/SP - 1 KW. *(cf=1.38)*

WEIGHT: (not specified in manufacturer's catalog)

SIZE: 9¾" (L)

ACCESSORIES AVAILABLE:
Color Frame

PHOTOMETRICS CHART
(Performance data for this unit are measured using BFE - 750 W. lamp.)

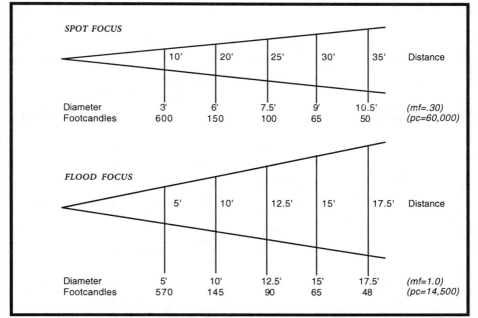

SPOT FOCUS

	10'	20'	25'	30'	35'	Distance
Diameter	3'	6'	7.5'	9'	10.5'	*(mf=.30)*
Footcandles	600	150	100	65	50	*(pc=60,000)*

FLOOD FOCUS

	5'	10'	12.5'	15'	17.5'	Distance
Diameter	5'	10'	12.5'	15'	17.5'	*(mf=1.0)*
Footcandles	570	145	90	65	48	*(pc=14,500)*

* These instruments have the same model numbers (60-06 and 60-07) as instruments manufactured later having slightly different photometrics. See following page.

LIGHTING & ELECTRONICS
Model Nos. 60-06* (slide focus)
60-07* (screw feed)

SPOT BEAM SPREAD: 7° beam, 16° field

FLOOD BEAM SPREAD: 50° beam, 64° field

LAMP BASE TYPE: Medium Prefocus

STANDARD LAMP: BTN - 750 W.

OTHER LAMPS: BTL - 500 W.   *(cf=.65)*
BTP - 750 W.   *(cf=1.18)*

ACCESSORIES AVAILABLE:
Color Frame, Barn Doors, Snoot

WEIGHT: 7 lbs.

SIZE: 9¾" x 7¾" (L x W)

PHOTOMETRICS CHART
(Performance data for this unit are measured using BTN - 750 W. lamp.)

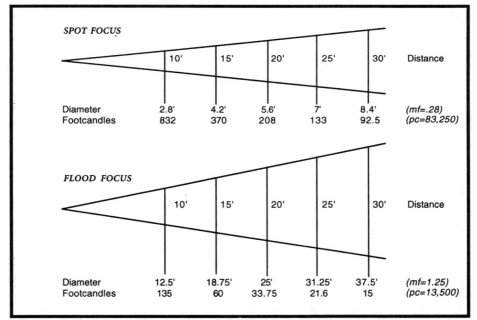

SPOT FOCUS

	10'	15'	20'	25'	30'	Distance
Diameter	2.8'	4.2'	5.6'	7'	8.4'	*(mf=.28)*
Footcandles	832	370	208	133	92.5	*(pc=83,250)*

FLOOD FOCUS

	10'	15'	20'	25'	30'	Distance
Diameter	12.5'	18.75'	25'	31.25'	37.5'	*(mf=1.25)*
Footcandles	135	60	33.75	21.6	15	*(pc=13,500)*

* These instruments have the same model numbers (60-06 and 60-07) as instruments manufactured earlier, having slightly different photometrics. See previous page.

LIGHTING & ELECTRONICS
Model No. 60-10

SPOT BEAM SPREAD: 8° beam, 10° field

FLOOD BEAM SPREAD: 57° beam, 71° field

LAMP BASE TYPE: Medium Prefocus

STANDARD LAMP: BTR - 1 KW.

OTHER LAMPS: BTN - 750 W. *(cf=.62)*
BTL - 500 W. *(cf=.39)*

ACCESSORIES AVAILABLE:
Color Frame, Barn Doors

WEIGHT: 12 lbs.

SIZE: 12" x 7¾" x 9" (L x W x H)

PHOTOMETRICS CHART

(Performance data for this unit are calculated from manufacturer's beam angle and peak candela data using BTR - 1 KW. lamp.)

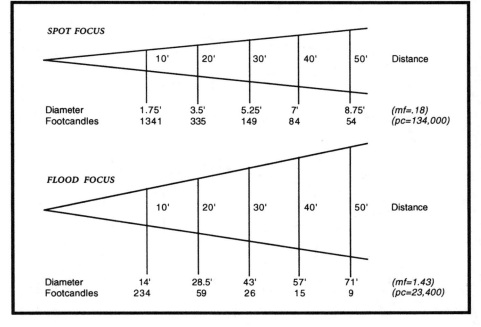

SPOT FOCUS

	10'	20'	30'	40'	50'	Distance
Diameter	1.75'	3.5'	5.25'	7'	8.75'	*(mf=.18)*
Footcandles	1341	335	149	84	54	*(pc=134,000)*

FLOOD FOCUS

	10'	20'	30'	40'	50'	Distance
Diameter	14'	28.5'	43'	57'	71'	*(mf=1.43)*
Footcandles	234	59	26	15	9	*(pc=23,400)*

STRAND CENTURY
Model No. 3380

SPOT BEAM SPREAD: 6° beam, 13° field

FLOOD BEAM SPREAD: 55° beam, 65° field

LAMP BASE TYPE: Medium Prefocus

STANDARD LAMP: BTR - 1 KW.

OTHER LAMPS: BTL - 500 W. *(cf=.40)*
BTN - 750 W. *(cf=.62)*

ACCESSORIES AVAILABLE:
Color Frame, 8-way Barn Door, Snoot

WEIGHT: 10.5 lbs.

SIZE: 12⅛" x 10¼" x 12¼" (L x W x H)

PHOTOMETRICS CHART

(Performance data for this unit are measured using BTR - 1 KW. lamp.)

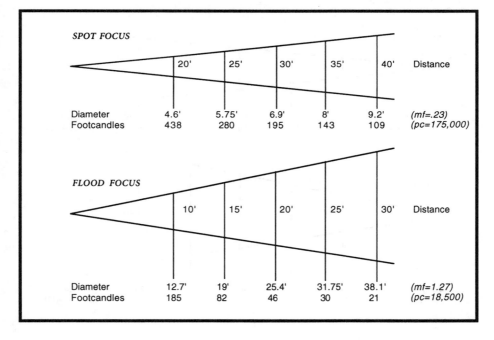

SPOT FOCUS

	20'	25'	30'	35'	40'	Distance
Diameter	4.6'	5.75'	6.9'	8'	9.2'	*(mf=.23)*
Footcandles	438	280	195	143	109	*(pc=175,000)*

FLOOD FOCUS

	10'	15'	20'	25'	30'	Distance
Diameter	12.7'	19'	25.4'	31.75'	38.1'	*(mf=1.27)*
Footcandles	185	82	46	30	21	*(pc=18,500)*

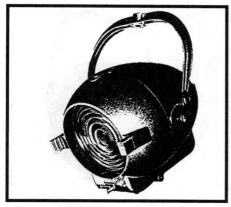

STRAND ELECTRIC
Model No. Patt 123

SPOT BEAM SPREAD: 10° beam, 16° field

FLOOD BEAM SPREAD: 33° beam, 43° field

LAMP BASE TYPE: Medium Prefocus

STANDARD LAMP: 500 T20/48 *

OTHER LAMPS: 250 T20/47 *(cf=.42)*

ACCESSORIES AVAILABLE:
Color Frame (6½" sq.), 4-way Barn Door

WEIGHT: 5.75 lbs.

SIZE: 10½" x 10½" x 12½" (L x W x H)

PHOTOMETRICS CHART

(Performance data for this unit are measured using 500 T20/48 lamp.)*

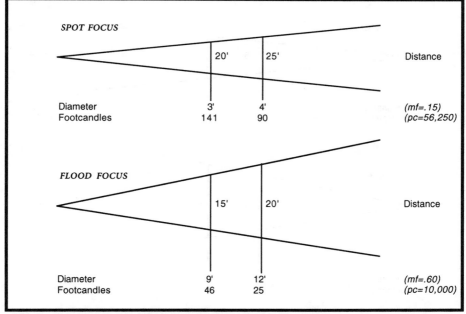

SPOT FOCUS	20'	25'	Distance
Diameter	3'	4'	*(mf=.15)*
Footcandles	141	90	*(pc=56,250)*

FLOOD FOCUS	15'	20'	Distance
Diameter	9'	12'	*(mf=.60)*
Footcandles	46	25	*(pc=10,000)*

*This lamp is comparable to a BTL; both are rated at 11,000 initial lumens.

FRESNEL — 6" Lens

STRAND ELECTRIC
Model No. Patt 123W

SPOT BEAM SPREAD: 16° beam, 26° field

FLOOD BEAM SPREAD: 42° beam, 51° field

LAMP BASE TYPE: Medium Prefocus

STANDARD LAMP: 500 T20/48 *

OTHER LAMPS: 250 T20/47 *(cf=.42)*

ACCESSORIES AVAILABLE:
Color Frame (6½" sq.), 4-way Barn Door

WEIGHT: 5 ½⅕ lbs.

SIZE: 10½" x 10½" x 12½" (L x W x H)

PHOTOMETRICS CHART
(Performance data for this unit are measured using 500 T20/48 lamp.)*

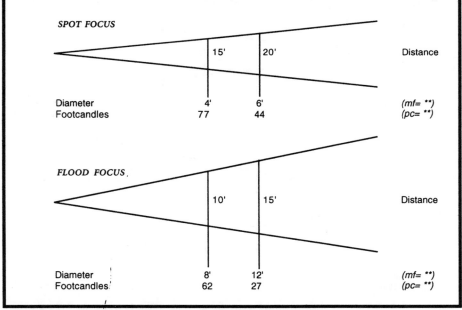

*This lamp is comparable to a BTL; both are rated at 11,000 initial lumens.

** Diameter and footcandle data as given in manufacturer's catalog is not consistent enough to determine a multiplying factor *(mf)* or a peak candela *(pc)* figure for all throw distances. See Introduction for more information about calculating photometric data.

WEIGHT: 11 lbs.

SIZE: 11½" x 11¾" x 20" (L x W x H)

STRAND LIGHTING
Model No. 3301, "Polaris"

SPOT BEAM SPREAD: 6.5° beam, 14° field

FLOOD BEAM SPREAD: 57.5° beam, 66.5° field

LAMP BASE TYPE: Medium Bipost

STANDARD LAMP: EGT - 1 KW.

OTHER LAMPS: EGN - 500 W. *(cf=.46)*
EGR - 750 W. *(cf=.75)*

ACCESSORIES AVAILABLE:
Color Frame, 8-way Barn Door, Various Scrims and Wireguard for TV use, Operating Pole (for focusing), Snoot

PHOTOMETRICS CHART

(Performance data for this unit are measured using EGT - 1KW. lamp.)

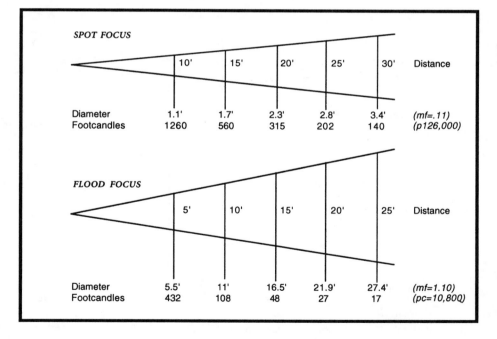

SPOT FOCUS

	10'	15'	20'	25'	30'	Distance
Diameter	1.1'	1.7'	2.3'	2.8'	3.4'	*(mf=.11)*
Footcandles	1260	560	315	202	140	*(p126,000)*

FLOOD FOCUS

	5'	10'	15'	20'	25'	Distance
Diameter	5.5'	11'	16.5'	21.9'	27.4'	*(mf=1.10)*
Footcandles	432	108	48	27	17	*(pc=10,80Q)*

47

STRAND LIGHTING
Model No. 3302, "Bambino" *

SPOT BEAM SPREAD: 10° beam**

FLOOD BEAM SPREAD: 59° beam**

LAMP BASE TYPE: Mogul Bipost

STANDARD LAMP: CYX - 2 KW.

OTHER LAMPS: CXZ - 1.5 KW. *(cf=.75)*
CYV - 1 KW. *(cf=.48)*

ACCESSORIES AVAILABLE:
Color Frame, 8-way Barn Door, Various Scrims for TV
and Photography Use.

WEIGHT: 17.9 lbs.

SIZE: 11" x 11¾" x 15" (L x W x H)

PHOTOMETRICS CHART
(Performance data for this unit are measured using CYX - 2 KW. lamp.)

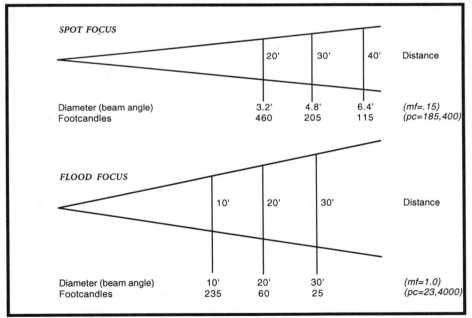

SPOT FOCUS				
	20'	30'	40'	Distance
Diameter (beam angle)	3.2'	4.8'	6.4'	*(mf=.15)*
Footcandles	460	205	115	*(pc=185,400)*

FLOOD FOCUS				
	10'	20'	30'	Distance
Diameter (beam angle)	10'	20'	30'	*(mf=1.0)*
Footcandles	235	60	25	*(pc=23,4000)*

* This model comes with a 5/8" socket for stand mounting.
** No field angle data given in manufacturer's catalog.

STRAND LIGHTING
Model No. 3382, "Quartet F"

SPOT BEAM SPREAD: 9° beam, 16° field

FLOOD BEAM SPREAD: 42° beam, 53° field

LAMP BASE TYPE: GY9.5

STANDARD LAMP: FRE - 650 W.

OTHER LAMPS: (none specified in mfg. catalog)

ACCESSORIES AVAILABLE:
Color Frame, Rotatable Barn Door

WEIGHT: 6.8 lbs.

SIZE: 11½" x 9¾" x 11½" (L x W x H)

PHOTOMETRICS CHART

(Performance data for this unit are measured using FRE - 650 W. lamp.)

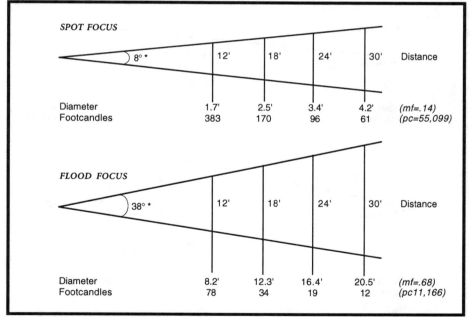

SPOT FOCUS

	8° *	12'	18'	24'	30'	Distance
Diameter		1.7'	2.5'	3.4'	4.2'	*(mf=.14)*
Footcandles		383	170	96	61	*(pc=55,099)*

FLOOD FOCUS

	38° *	12'	18'	24'	30'	Distance
Diameter		8.2'	12.3'	16.4'	20.5'	*(mf=.68)*
Footcandles		78	34	19	12	*(pc11,166)*

* Using the given distances and diameters, the angle at spot focus calculates to be 8°, while at flood focus the angle calculates to be 38°. Ordinarily these angles should be the same as the given "field angles" at spot and flood focus. The diameter/distance data in the manufacturer's data sheet are described as "narrowest focus" and "widest focus" which may explain this discrepancy. See Introduction for more information about calculating photometric data.

TIMES SQUARE LIGHTING
Model No. 6PC

SPOT BEAM SPREAD: 17° field*

FLOOD BEAM SPREAD: 73.7° field*

LAMP BASE TYPE: Medium Prefocus

STANDARD LAMP: BTN - 750 W.

OTHER LAMPS: BTL - 500 W. *(cf=.65)*
             BTP - 750 W. *(cf=1.18)*

ACCESSORIES AVAILABLE:
Color Frame, Barn Door, Snoot, Pipe Clamp, Table
Base, Adaptors for tracklight use.

WEIGHT: 6 lbs.

SIZE: 10½" x 9½" (L x H)

PHOTOMETRICS CHART
(Performance data for this unit are measured using BTN - 750 W. lamp.)

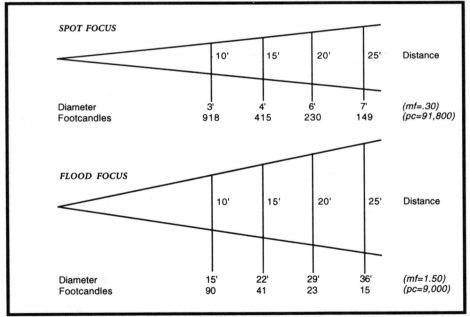

SPOT FOCUS

	10'	15'	20'	25'	Distance
Diameter	3'	4'	6'	7'	*(mf=.30)*
Footcandles	918	415	230	149	*(pc=91,800)*

FLOOD FOCUS

	10'	15'	20'	25'	Distance
Diameter	15'	22'	29'	36'	*(mf=1.50)*
Footcandles	90	41	23	15	*(pc=9,000)*

* No beam or field angle data given in manufacturer's catalog. The angles provided here are calculated from
manufacturer's diameter and distance data. See Introduction for more information about calculating
photometric data.

TIMES SQUARE LIGHTING
Model No. Q6P

SPOT BEAM SPREAD: 33° field*

FLOOD BEAM SPREAD: 71.5° field*

LAMP BASE TYPE: Mini-can Screw

STANDARD LAMP: EVR - 500 W.

OTHER LAMPS: EHT - 250 W. *(cf=.48)*
EHV - 325 W. *(cf=.75)*
400QCL/MC - 400 W. *(cf=.79)*
EYT - 750 W. *(cf=1.78)*

ACCESSORIES AVAILABLE:
Color Frame, Barn Door, Snoot, Pipe Clamp, Table Base, Adaptors for tracklight use.

WEIGHT: 6 lbs.

SIZE: 10½" x 9½" (L x H)

PHOTOMETRICS CHART

(Performance data for this unit are measured using EVR - 500 W. lamp.)

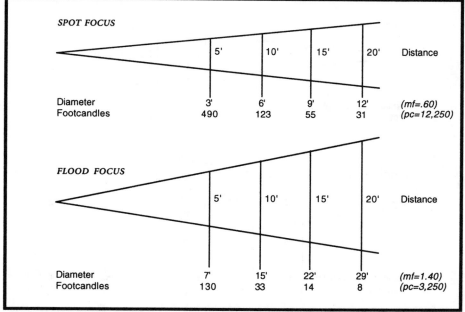

SPOT FOCUS

	5'	10'	15'	20'	Distance
Diameter	3'	6'	9'	12'	*(mf=.60)*
Footcandles	490	123	55	31	*(pc=12,250)*

FLOOD FOCUS

	5'	10'	15'	20'	Distance
Diameter	7'	15'	22'	29'	*(mf=1.40)*
Footcandles	130	33	14	8	*(pc=3,250)*

* No beam or field angle data given in manufacturer's catalog. The angles provided here are calculated from manufacturer's diameter and distance data. See Introduction for more information about calculating photometric data.

FRESNEL — 7" Lens

ARRIFLEX CORP.
Model No. 531200*

SPOT BEAM SPREAD: 15° beam**

FLOOD BEAM SPREAD: 60° beam**

LAMP BASE TYPE: Mogul Bipost

STANDARD LAMP: CYX - 2 KW.

OTHER LAMPS: CXZ - 1.5 KW. *(cf=.75)*
CYV - 1 KW. *(cf=.48)*

ACCESSORIES AVAILABLE:
Color Frame (9" sq.), 4- and 8-way Barn Door, Top
Hat

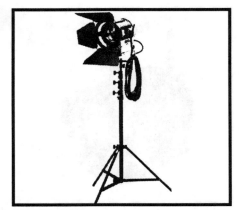

WEIGHT: 17.5 lbs.

SIZE: (not specified in manufacturer's catalog)

PHOTOMETRICS CHART
(Performance data for this unit are measured using CYX - 2 KW. lamp)

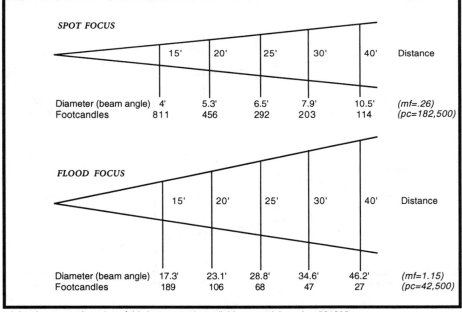

SPOT FOCUS

	15'	20'	25'	30'	40'	Distance
Diameter (beam angle)	4'	5.3'	6.5'	7.9'	10.5'	*(mf=.26)*
Footcandles	811	456	292	203	114	*(pc=182,500)*

FLOOD FOCUS

	15'	20'	25'	30'	40'	Distance
Diameter (beam angle)	17.3'	23.1'	28.8'	34.6'	46.2'	*(mf=1.15)*
Footcandles	189	106	68	47	27	*(pc=42,500)*

* A pole operated version of this instrument is available as model number 531205.

** No field angle data given in manufacturer's catalog.

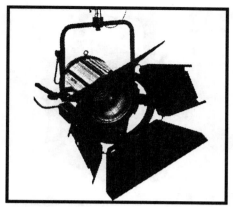

ARRIFLEX CORP.
Model No. 532100*

SPOT BEAM SPREAD: 11° beam**
FLOOD BEAM SPREAD: 62° beam**
LAMP BASE TYPE: Medium Bipost
STANDARD LAMP: EGT - 1 KW.
OTHER LAMPS: EGR - 750 W. *(cf=.74)*

ACCESSORIES AVAILABLE:
Color Frame (9" sq.), 4- and 8-way Barn Door, Top Hat

WEIGHT: 15.2 lbs.
SIZE: (not specified in manufacturer's catalog)

PHOTOMETRICS CHART
(Performance data for this unit are measured using EGT - 1 KW. lamp)

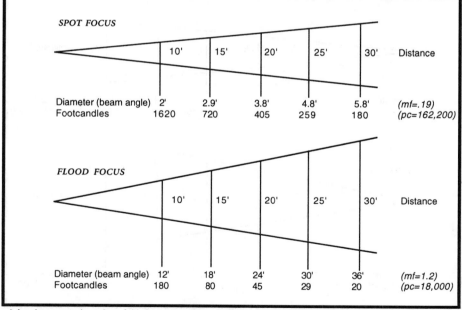

SPOT FOCUS

	10'	15'	20'	25'	30'	Distance
Diameter (beam angle)	2'	2.9'	3.8'	4.8'	5.8'	*(mf=.19)*
Footcandles	1620	720	405	259	180	*(pc=162,200)*

FLOOD FOCUS

	10'	15'	20'	25'	30'	Distance
Diameter (beam angle)	12'	18'	24'	30'	36'	*(mf=1.2)*
Footcandles	180	80	45	29	20	*(pc=18,000)*

* A pole operated version of this instrument is available as model number 532105.
** No field angle data given in manufacturer's catalog.

ALTMAN STAGE LIGHTING
Model Nos.75 (slide focus)
175 (screw feed focus)

SPOT BEAM SPREAD: 9.5° beam, 18° field

FLOOD BEAM SPREAD: 40° beam, 49° field

LAMP BASE TYPE: Mogul Prefocus

STANDARD LAMP: BVT - 1 KW.

OTHER LAMPS: BVV - 1 KW. *(cf=1.20)*

ACCESSORIES AVAILABLE:
Color Frame, 4-way Barn Door, Snoot,
Motorized/Non-Motorized Color Wheels

WEIGHT: (not specified in manufacturer's catalog)

SIZE: 11" x 11" x 18" (L x W x H) (75)
      11" x 13½" x 18" (L x W x H) (175)

PHOTOMETRICS CHART
(Performance data for this unit are measured using BVT - 1 KW. lamp.)

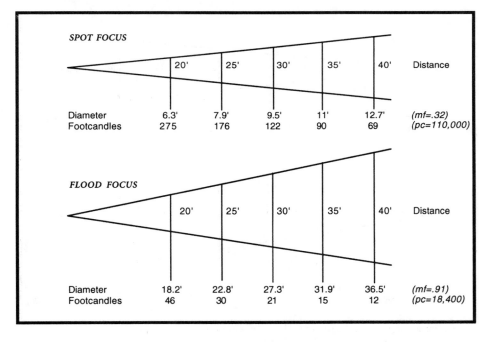

SPOT FOCUS

	20'	25'	30'	35'	40'	Distance
Diameter	6.3'	7.9'	9.5'	11'	12.7'	*(mf=.32)*
Footcandles	275	176	122	90	69	*(pc=110,000)*

FLOOD FOCUS

	20'	25'	30'	35'	40'	Distance
Diameter	18.2'	22.8'	27.3'	31.9'	36.5'	*(mf=.91)*
Footcandles	46	30	21	15	12	*(pc=18,400)*

BERKEY COLORTRAN
Model No. 100-362*

SPOT BEAM SPREAD: 6.5° beam, 12.5° field

FLOOD BEAM SPREAD: 36° beam, 44° field

LAMP BASE TYPE: Mogul Bipost

STANDARD LAMP: CXZ - 1.5 KW.

OTHER LAMPS: CYV - 1 KW. *(cf=.74)*
CYX - 2 KW. *(cf=1.53)*

ACCESSORIES AVAILABLE:
Color Frame, 4-way Barn Door

NOTE: The focus mechanism on this unit is an unusual ring which rotates around the nose of the instrument.

WEIGHT: 19.3 lbs.

SIZE: 11" x 12½" x 26" (L x W x H)

PHOTOMETRICS CHART

(Performance data for this unit are measured using CXZ - 1.5 KW. lamp.)

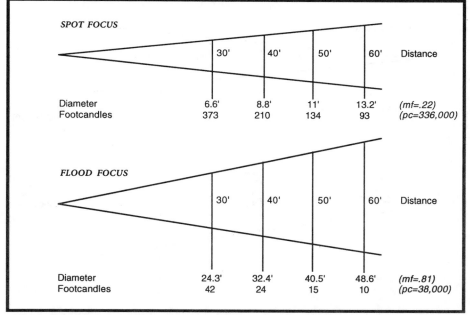

SPOT FOCUS

	30'	40'	50'	60'	Distance
Diameter	6.6'	8.8'	11'	13.2'	*(mf=.22)*
Footcandles	373	210	134	93	*(pc=336,000)*

FLOOD FOCUS

	30'	40'	50'	60'	Distance
Diameter	24.3'	32.4'	40.5'	48.6'	*(mf=.81)*
Footcandles	42	24	15	10	*(pc=38,000)*

* This instrument can be identified with model numbers -362, -365, -366, or -367, depending on the wire leads and connectors supplied.

CCT LIGHTING
Model No. Z0080

SPOT BEAM SPREAD: 7° beam, 12.5° field

FLOOD BEAM SPREAD: 50° beam, 58° field

LAMP BASE TYPE: GY16

STANDARD LAMP: CP43 - 2 KW.

OTHER LAMPS: (none specified in mfg. catalog)

ACCESSORIES AVAILABLE:
Color Frame, 4-way Barn Door

WEIGHT: 28 lbs.

SIZE: 16" x 10½" x 21" (L x W x H)

PHOTOMETRICS CHART
(Performance data for this unit are measured using CP43 - 2 KW. lamp.)

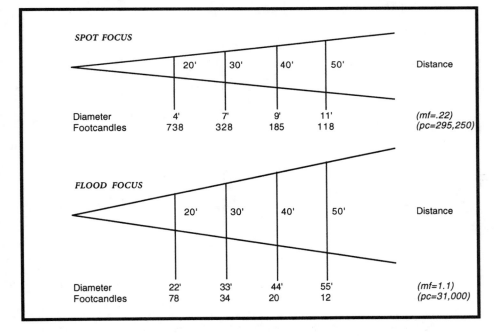

SPOT FOCUS						Distance
	20'	30'	40'	50'		
Diameter	4'	7'	9'	11'		*(mf=.22)*
Footcandles	738	328	185	118		*(pc=295,250)*

FLOOD FOCUS						Distance
	20'	30'	40'	50'		
Diameter	22'	33'	44'	55'		*(mf=1.1)*
Footcandles	78	34	20	12		*(pc=31,000)*

CENTURY LIGHTING
Model No. 3486

SPOT BEAM SPREAD: 12° beam, 20° field

FLOOD BEAM SPREAD: 47° beam, 72° field

LAMP BASE TYPE: Medium 2-Pin

STANDARD LAMP: FEL - 1 KW.

OTHER LAMPS: EHH - 1 KW. *(cf=.81)*

ACCESSORIES AVAILABLE:
Color Frame, 2 or 4-way Barn Door, Snoot

WEIGHT: (not specified in manufacturer's catalog)

SIZE: 20¾" x 15¾" x 19" (L x W x H)

PHOTOMETRICS CHART
(Performance data for this unit are measured using FEL - 1 KW. lamp.)

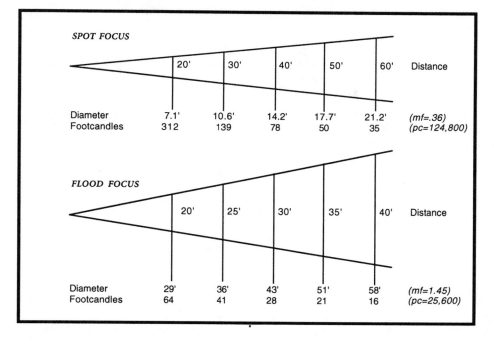

SPOT FOCUS

	20'	30'	40'	50'	60'	Distance
Diameter	7.1'	10.6'	14.2'	17.7'	21.2'	*(mf=.36)*
Footcandles	312	139	78	50	35	*(pc=124,800)*

FLOOD FOCUS

	20'	25'	30'	35'	40'	Distance
Diameter	29'	36'	43'	51'	58'	*(mf=1.45)*
Footcandles	64	41	28	21	16	*(pc=25,600)*

CENTURY LIGHTING
Model No. 570

SPOT BEAM SPREAD: 7° beam, 12° field

FLOOD BEAM SPREAD: 28° beam, 40° field

LAMP BASE TYPE: Mogul Prefocus

STANDARD LAMP: 1500 G40/21 - 1.5 KW.

OTHER LAMPS: 1M/G40/20 - 1 KW. *(cf=.67)*
2M/G48/5 - 2 KW. *(cf=1.55)**

ACCESSORIES AVAILABLE:
Color Frame, 2 or 4-way Barn Door, Snoot

WEIGHT: (not specified in manufacturer's catalog)

SIZE: 14⅛" x 12¾" x 19" (L x W x H)

PHOTOMETRICS CHART
(Performance data for this unit are measured using 1500 G40/21 - 1.5 KW. lamp.)

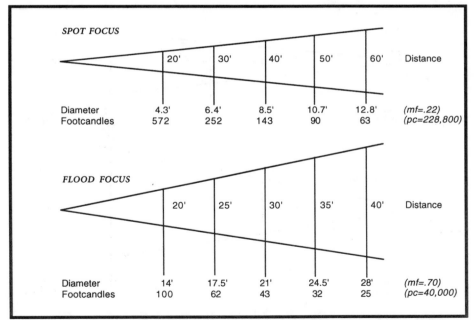

SPOT FOCUS

	20'	30'	40'	50'	60'	Distance
Diameter	4.3'	6.4'	8.5'	10.7'	12.8'	*(mf=.22)*
Footcandles	572	252	143	90	63	*(pc=228,800)*

FLOOD FOCUS

	20'	25'	30'	35'	40'	Distance
Diameter	14'	17.5'	21'	24.5'	28'	*(mf=.70)*
Footcandles	100	62	43	32	25	*(pc=40,000)*

* Not only the illumination changes when this lamp is used. The manufacturer's data sheet gives the field and beam angles for the G48 bulb as follows: spot beam spread: 9° beam, 16° field; flood beam spread: 21° beam, 33° field;

CENTURY STRAND
Model No. 3413

SPOT BEAM SPREAD: 9˚ beam, 16˚ field

FLOOD BEAM SPREAD: 21˚ beam, 33˚ field

LAMP BASE TYPE: Mogul Prefocus

STANDARD LAMP: BVW - 2 KW.

OTHER LAMPS: BVT - 1 KW. *(cf=.39)*

ACCESSORIES AVAILABLE:
Color Frame, 2 or 4-way Barn Door, Snoot

WEIGHT: (not specified in manufacturer's catalog)

SIZE: 14⅛" x 12¾" x 19" (L x W x H)

PHOTOMETRICS CHART
(Performance data for this unit are measured using BVW - 2 KW. lamp.)

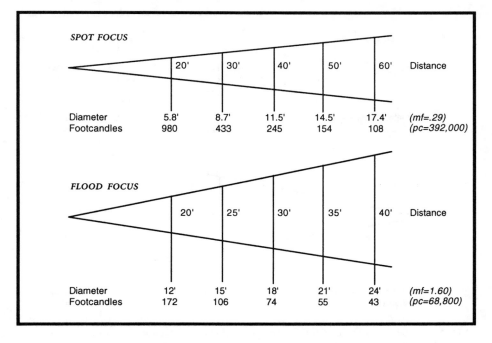

SPOT FOCUS

	20'	30'	40'	50'	60'	Distance
Diameter	5.8'	8.7'	11.5'	14.5'	17.4'	*(mf=.29)*
Footcandles	980	433	245	154	108	*(pc=392,000)*

FLOOD FOCUS

	20'	25'	30'	35'	40'	Distance
Diameter	12'	15'	18'	21'	24'	*(mf=1.60)*
Footcandles	172	106	74	55	43	*(pc=68,800)*

CENTURY STRAND
Model No. Patt 223

SPOT BEAM SPREAD: 7.5° beam, 15° field

FLOOD BEAM SPREAD: 70° beam, 82° field

LAMP BASE TYPE: Medium Prefocus

STANDARD LAMP: BFE - 750 W.

OTHER LAMPS: 1M/T20/SP - 1 KW. *(cf=1.6)**

ACCESSORIES AVAILABLE:
Color Frame, 4-way Barn Door

WEIGHT: 15 lbs.

SIZE: 12½" x 13½" x 15½" (L x W x H)

PHOTOMETRICS CHART
(Performance data for this unit are measured using BFE - 750 W. lamp.)

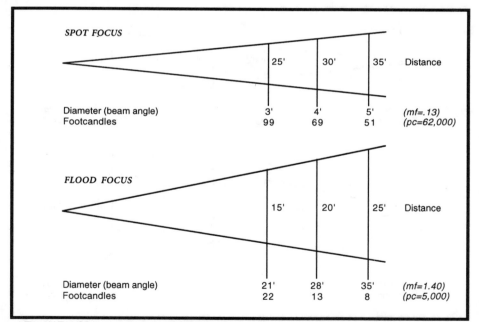

SPOT FOCUS

	25'	30'	35'	Distance
Diameter (beam angle)	3'	4'	5'	*(mf=.13)*
Footcandles	99	69	51	*(pc=62,000)*

FLOOD FOCUS

	15'	20'	25'	Distance
Diameter (beam angle)	21'	28'	35'	*(mf=1.40)*
Footcandles	22	13	8	*(pc=5,000)*

* The 1M/T20/SP is an obsolete lamp. Correction factor *(cf)* information is based on illumination data for both lamps in the manufacturer's catalog.

COLORTRAN
Model No. 100-526*

SPOT BEAM SPREAD: 8.1° beam, 15.1° field

FLOOD BEAM SPREAD: 58.5° beam, 67° field

LAMP BASE TYPE: Mogul Bipost

STANDARD LAMP: CYX - 2 KW.

OTHER LAMPS: CXZ - 1.5 KW. *(cf=.75)*
              CYV - 1 KW. *(cf=.48)*

ACCESSORIES AVAILABLE:
Color Frame, 8-way Barn Door, Top Hat

WEIGHT: 19.5 lbs.

SIZE: 15¹¹⁄₁₆" x 13⅜" x 27¹¹⁄₁₆" (L x W x H)

PHOTOMETRICS CHART
(Performance data for this unit are measured using CYX - 2 KW. lamp.)

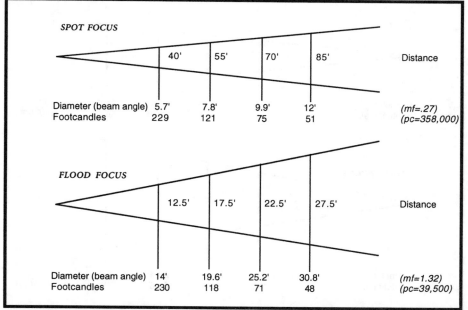

SPOT FOCUS

	40'	55'	70'	85'	Distance
Diameter (beam angle)	5.7'	7.8'	9.9'	12'	*(mf=.27)*
Footcandles	229	121	75	51	*(pc=358,000)*

FLOOD FOCUS

	12.5'	17.5'	22.5'	27.5'	Distance
Diameter (beam angle)	14'	19.6'	25.2'	30.8'	*(mf=1.32)*
Footcandles	230	118	71	48	*(pc=39,500)*

* This model can be identified with model numbers -522, -524, -525 or -527 depending on the wire leads and connectors supplied.

COLORTRAN
Model No. 213-212*

SPOT BEAM SPREAD: 8.6° beam, 15.3° field

FLOOD BEAM SPREAD: 50.8° beam, 59.1° field

LAMP BASE TYPE: Mogul Prefocus

STANDARD LAMP: BVW - 2 KW.

OTHER LAMPS: BVT - 1 KW. *(cf=.46)*
CWZ - 1.5 KW. *(cf=.77)*

ACCESSORIES AVAILABLE:
Color Frame, 4-way Barn Door, Snoot

WEIGHT: 20 lbs.

SIZE: 12¾" x 12½" x 25¼" (L x W x H)

PHOTOMETRICS CHART
(Performance data for this unit are measured using BVW - 2 KW. lamp.)

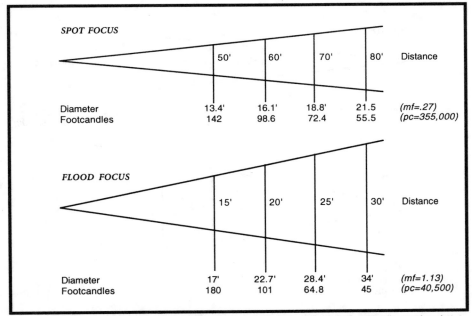

SPOT FOCUS

	50'	60'	70'	80'	Distance
Diameter	13.4'	16.1'	18.8'	21.5	*(mf=.27)*
Footcandles	142	98.6	72.4	55.5	*(pc=355,000)*

FLOOD FOCUS

	15'	20'	25'	30'	Distance
Diameter	17'	22.7'	28.4'	34'	*(mf=1.13)*
Footcandles	180	101	64.8	45	*(pc=40,500)*

* This instrument can be identified with model numbers -212, -215, -216, or -217, depending on the wire leads and connectors supplied.

ELECTRO CONTROLS
Model Nos. 3485 & 7485A*

SPOT BEAM SPREAD: 5.4° beam, 11.9° field

FLOOD BEAM SPREAD: 64.6° beam, 71° field

LAMP BASE TYPE: Medium Prefocus

STANDARD LAMP: EGJ - 1 KW.

OTHER LAMPS: EGE - 500 W. *(cf=.38)*
EGG - 750 W. *(cf=.57)*
EGM - 1 KW. *(cf=.78)*

ACCESSORIES AVAILABLE:
Color Frame (10" sq.), 2 or 4-way Barn Door,
Snoot

WEIGHT: (not specified in manufacturer's catalog)

SIZE: 14⅛" x 13³⁄₁₆" x 18½" (L x W x H)

PHOTOMETRICS CHART
(Performance data for this unit are measured using EGJ - 1 KW. lamp.)

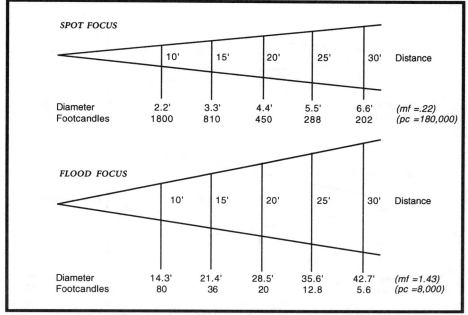

SPOT FOCUS

	10'	15'	20'	25'	30'	Distance
Diameter	2.2'	3.3'	4.4'	5.5'	6.6'	*(mf =.22)*
Footcandles	1800	810	450	288	202	*(pc =180,000)*

FLOOD FOCUS

	10'	15'	20'	25'	30'	Distance
Diameter	14.3'	21.4'	28.5'	35.6'	42.7'	*(mf =1.43)*
Footcandles	80	36	20	12.8	5.6	*(pc =8,000)*

* These two model numbers were used at different times for the same instrument.

ELECTRO CONTROLS
Model Nos. 3485A & 7485B*

SPOT BEAM SPREAD: 6.8° beam, 14° field

FLOOD BEAM SPREAD: 59.7° beam, 70° field

LAMP BASE TYPE: Medium Prefocus

STANDARD LAMP: BTR - 1 KW.

OTHER LAMPS: BTL - 500 W. *(cf=.40)*
BTN - 1 KW. *(cf=.62)*

ACCESSORIES AVAILABLE:
Color Frame (10" sq.), 2 or 4-way Barn Door,
Snoot

WEIGHT: (not specified in manufacturer's catalog)

SIZE: 14⅛" x 13³⁄₁₆" x 18½" (L x W x H)

PHOTOMETRICS CHART
(Performance data for this unit are measured using BTR - 1 KW. lamp.)

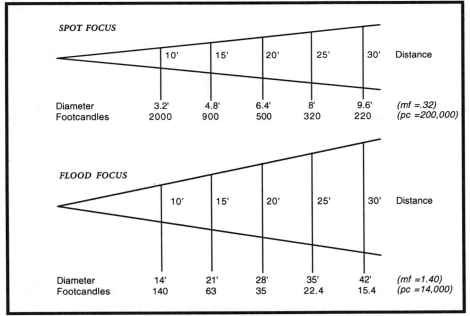

SPOT FOCUS

	10'	15'	20'	25'	30'	Distance
Diameter	3.2'	4.8'	6.4'	8'	9.6'	*(mf =.32)*
Footcandles	2000	900	500	320	220	*(pc =200,000)*

FLOOD FOCUS

	10'	15'	20'	25'	30'	Distance
Diameter	14'	21'	28'	35'	42'	*(mf =1.40)*
Footcandles	140	63	35	22.4	15.4	*(pc =14,000)*

* These two model numbers were used at different times for the same instrument.

KLIEGL BROS.
Model No. 3609

SPOT BEAM SPREAD: 14° field*

FLOOD BEAM SPREAD: 54° field*

LAMP BASE TYPE: Recessed Single-Contact

STANDARD LAMP: DWT - 1 KW.

OTHER LAMPS: FER - 1 KW. *(cf=1.18)*
FEY - 2 KW. *(cf=2.44)*

ACCESSORIES AVAILABLE:
Color Frame, 8-way Barn Door, Snoot, Table Base, Scenery Clamp

WEIGHT: 20 lbs.

SIZE: 15" x 13⅜" x 25⁷⁄₁₆" (L x W x H)

PHOTOMETRICS CHART

(Performance data for this unit are measured using DWT - 1 KW. lamp.)

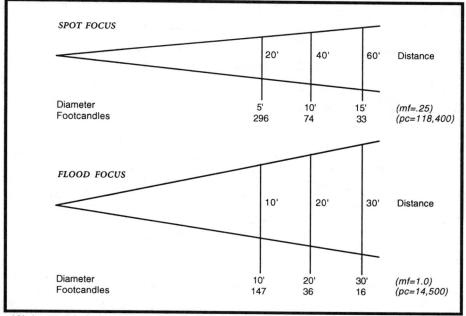

SPOT FOCUS				
	20'	40'	60'	Distance
Diameter	5'	10'	15'	*(mf=.25)*
Footcandles	296	74	33	*(pc=118,400)*

FLOOD FOCUS				
	10'	20'	30'	Distance
Diameter	10'	20'	30'	*(mf=1.0)*
Footcandles	147	36	16	*(pc=14,500)*

* No beam angle data given in manufacturer's catalog.

KLIEGL BROS. / RDS
Model No. 3801

SPOT BEAM SPREAD: 7.6° beam, 13.3° field

FLOOD BEAM SPREAD: 54° beam, 62° field

LAMP BASE TYPE: Medium Bipost

STANDARD LAMP: EGT - 1 KW.

OTHER LAMPS: (none specified in mfg. catalog)

ACCESSORIES AVAILABLE:
Color Frame, Barn Door

WEIGHT: 14 lbs.

SIZE: 14" x 14" x 20" (L x W x H)

PHOTOMETRICS CHART
(Information about lamp used to measure performance data is not provided in manufacturer's catalog.)

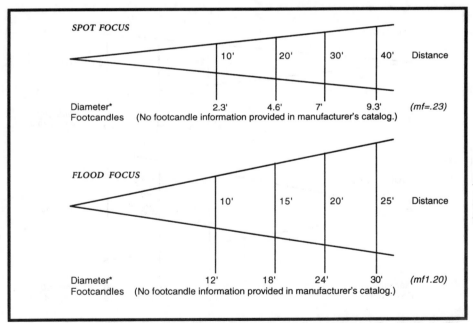

SPOT FOCUS

Distance	10'	20'	30'	40'
Diameter*	2.3'	4.6'	7'	9.3'	(mf=.23)
Footcandles	(No footcandle information provided in manufacturer's catalog.)

FLOOD FOCUS

Distance	10'	15'	20'	25'
Diameter*	12'	18'	24'	30'	(mf1.20)
Footcandles	(No footcandle information provided in manufacturer's catalog.)

* This performance data is calculated from field angles given in manufacturer's catalog. See Introduction for more information about calculating photometric data.

KLIEGL BROS. / RDS
Model No. 3802

SPOT BEAM SPREAD: 9.1° beam, 15.7° field

FLOOD BEAM SPREAD: 54° beam, 64° field

LAMP BASE TYPE: Mogul Bipost

STANDARD LAMP: CYX - 2 KW.

OTHER LAMPS: (none specified in mfg. catalog)

ACCESSORIES AVAILABLE:
Color Frame, Barn Door

WEIGHT: 16.5 lbs.

SIZE: 14" x 14" x 20" (L x W x H)

PHOTOMETRICS CHART

(Information about lamp used to measure performance data is not provided in manufacturer's catalog.)

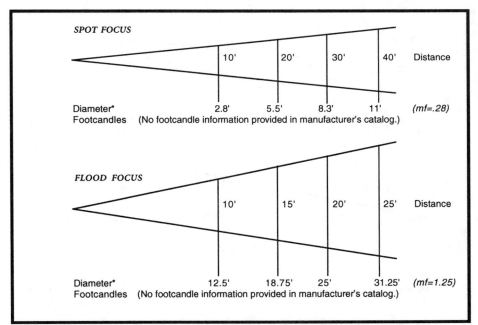

SPOT FOCUS

| | 10' | 20' | 30' | 40' | Distance |

| Diameter* | 2.8' | 5.5' | 8.3' | 11' | (mf=.28) |
| Footcandles | (No footcandle information provided in manufacturer's catalog.) |

FLOOD FOCUS

| | 10' | 15' | 20' | 25' | Distance |

| Diameter* | 12.5' | 18.75' | 25' | 31.25' | (mf=1.25) |
| Footcandles | (No footcandle information provided in manufacturer's catalog.) |

* This performance data is calculated from field angles given in manufacturer's catalog. See Introduction for more information about calculating photometric data.

KLIEGL BROS.
Model No. 44NO8TVG

SPOT BEAM SPREAD: 7.2° x 14.25° *

FLOOD BEAM SPREAD: 22.7° x 33.4° *

LAMP BASE TYPE: Mogul Prefocus

STANDARD LAMP: 1M/G40/23 - 1 KW.**

OTHER LAMPS: 1500/G40/21 - 1.5 KW.
2M/G48/5 - 2 KW.

ACCESSORIES AVAILABLE:
Color Frame, Barn Door

WEIGHT: 15 lbs.

SIZE: 13" x 12½" x 21" (L x W x H)

PHOTOMETRICS CHART

(Information about lamp used to measure performance data is not provided in manufacturer's catalog.)

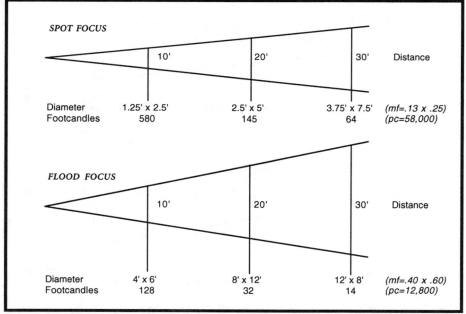

SPOT FOCUS

	10'	20'	30'	Distance
Diameter	1.25' x 2.5'	2.5' x 5'	3.75' x 7.5'	(mf=.13 x .25)
Footcandles	580	145	64	(pc=58,000)

FLOOD FOCUS

	10'	20'	30'	Distance
Diameter	4' x 6'	8' x 12'	12' x 8'	(mf=.40 x .60)
Footcandles	128	32	14	(pc=12,800)

* Beam and field angles for this "oval beam" fresnel not specified in manufacturer's catalog. These angles are calculated from given distance and beam size data. See Introduction for more information about calculating photometric data.
** This lamp is no longer available.

KLIEGL BROS.
Model No. 44N8TVG

SPOT BEAM SPREAD: 20° *

FLOOD BEAM SPREAD: 30° *

LAMP BASE TYPE: Mogul Prefocus

STANDARD LAMP: 1M/G40/23 - 1 KW.**

OTHER LAMPS: 1500/G40/21 - 1.5 KW.
2M/G48/5 - 2 KW.

ACCESSORIES AVAILABLE:
Color Frame, Barn Door

WEIGHT: 15 lbs.

SIZE: 13" x 12½" x 21" (L x W x H)

PHOTOMETRICS CHART

(Information about lamp used to measure performance data is not provided in manufacturer's catalog.)

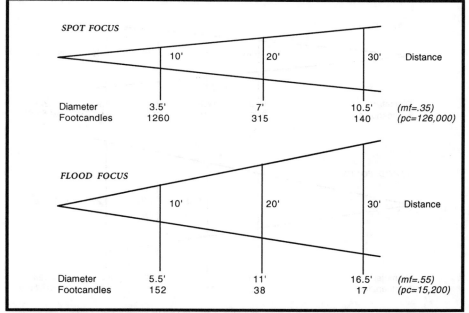

SPOT FOCUS				
	10'	20'	30'	Distance
Diameter	3.5'	7'	10.5'	*(mf=.35)*
Footcandles	1260	315	140	*(pc=126,000)*

FLOOD FOCUS				
	10'	20'	30'	Distance
Diameter	5.5'	11'	16.5'	*(mf=.55)*
Footcandles	152	38	17	*(pc=15,200)*

* Beam and field angles for this "oval beam" fresnel not specified in manufacturer's catalog. These angles are calculated from given distance and beam size data. See Introduction for more information about calculating photometric data.
** This lamp is no longer available.

LEE COLORTRAN
Model No. 213-522*

SPOT BEAM SPREAD: 8.4° beam, 15.3° field

FLOOD BEAM SPREAD: 55.7° beam, 63.5° field

LAMP BASE TYPE: Mogul Bipost

STANDARD LAMP: BVW - 2 KW.

OTHER LAMPS: BVT - 1 KW. *(cf=.39)*
CWZ - 1500 W. *(cf=.65)*

ACCESSORIES AVAILABLE:
Color Frame, 8-way Barn Door, Snoot, Scrim Set
for TV use.

WEIGHT: 19.5 lbs.

SIZE: 15¾" x 13⅜" x 27¾" (L x W x H)

PHOTOMETRICS CHART
(Performance data for this unit are measured using BVW - 2 KW. lamp.)

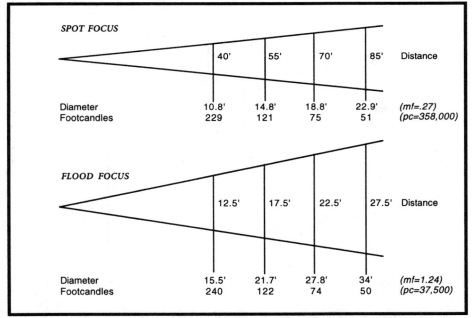

SPOT FOCUS

	40'	55'	70'	85'	Distance
Diameter	10.8'	14.8'	18.8'	22.9'	*(mf=.27)*
Footcandles	229	121	75	51	*(pc=358,000)*

FLOOD FOCUS

	12.5'	17.5'	22.5'	27.5'	Distance
Diameter	15.5'	21.7'	27.8'	34'	*(mf=1.24)*
Footcandles	240	122	74	50	*(pc=37,500)*

* This instrument can be identified with model numbers -522, -525, -526, or -527, depending on the wire leads and connectors supplied.

LIGHTING & ELECTRONICS
Model Nos.60-08 (slide focus)
60-09 (screw feed)

SPOT BEAM SPREAD: 9° beam, 19° field

FLOOD BEAM SPREAD: 52° beam, 64° field

LAMP BASE TYPE: Mogul Prefocus

STANDARD LAMP: BVV - 1 KW.

OTHER LAMPS: BVT - 1 KW. *(cf=.84)*
CWZ - 1.5 KW. *(cf=1.40)*

ACCESSORIES AVAILABLE:
Barn Doors, Snoot

WEIGHT: 25 lbs.

SIZE: 12" x 10¼" x 12½" (L x W x H)

PHOTOMETRICS CHART
(Performance data for this unit are measured using BVV - 1 KW. lamp.)

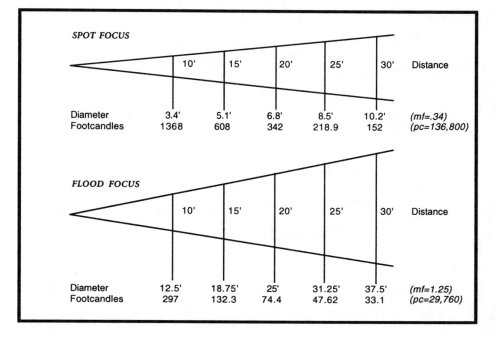

SPOT FOCUS

	10'	15'	20'	25'	30'	Distance
Diameter	3.4'	5.1'	6.8'	8.5'	10.2'	*(mf=.34)*
Footcandles	1368	608	342	218.9	152	*(pc=136,800)*

FLOOD FOCUS

	10'	15'	20'	25'	30'	Distance
Diameter	12.5'	18.75'	25'	31.25'	37.5'	*(mf=1.25)*
Footcandles	297	132.3	74.4	47.62	33.1	*(pc=29,760)*

FRESNEL — 8" Lens

STRAND CENTURY
Model No. 3480

SPOT BEAM SPREAD: 8° beam, 14° field

FLOOD BEAM SPREAD: 62° beam, 68° field

LAMP BASE TYPE: Mogul Prefocus

STANDARD LAMP: BVW - 2 KW.

OTHER LAMPS: BVT - 1 KW. *(cf=.39)*
  CWZ - 1.5 KW. *(cf=.65)*

ACCESSORIES AVAILABLE:
Color Frame, 8-way Barn Door, Snoot

WEIGHT: 17 lbs.

SIZE: 14⅜" x 12¾" x 16¾" (L x W x H)

PHOTOMETRICS CHART

(Performance data for this unit are measured using BVW - 2 KW. lamp.)

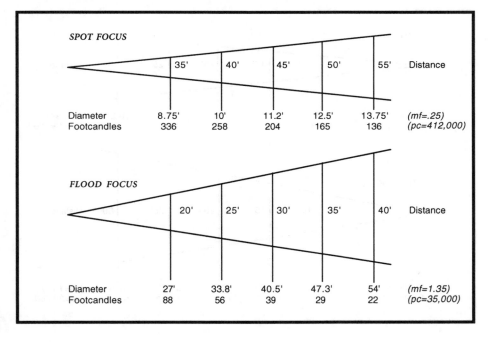

SPOT FOCUS

	35'	40'	45'	50'	55'	Distance
Diameter	8.75'	10'	11.2'	12.5'	13.75'	*(mf=.25)*
Footcandles	336	258	204	165	136	*(pc=412,000)*

FLOOD FOCUS

	20'	25'	30'	35'	40'	Distance
Diameter	27'	33.8'	40.5'	47.3'	54'	*(mf=1.35)*
Footcandles	88	56	39	29	22	*(pc=35,000)*

TIMES SQUARE LIGHTING
Model No. Q8PC

SPOT BEAM SPREAD: 7° field*

FLOOD BEAM SPREAD: 77° field*

LAMP BASE TYPE: Mogul Prefocus

STANDARD LAMP: BVT - 1 KW.

OTHER LAMPS: CWZ - 1.5 KW. *(cf=1.67)*
BVW - 2 KW. *(cf=2.57)*

ACCESSORIES AVAILABLE:
Color Frame, Barn Door, Snoot

WEIGHT: 15 lbs.

SIZE: 13" x 19" (L x H)

PHOTOMETRICS CHART
(Performance data for this unit are measured using BVT - 1 KW. lamp.)

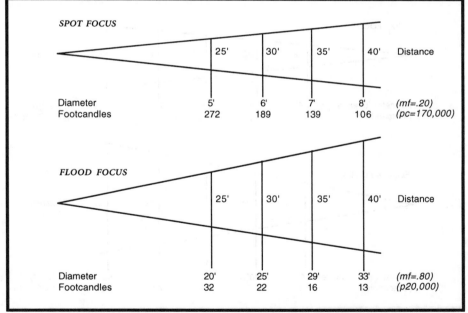

SPOT FOCUS	25'	30'	35'	40'	Distance
Diameter	5'	6'	7'	8'	*(mf=.20)*
Footcandles	272	189	139	106	*(pc=170,000)*

FLOOD FOCUS	25'	30'	35'	40'	Distance
Diameter	20'	25'	29'	33'	*(mf=.80)*
Footcandles	32	22	16	13	*(p20,000)*

* No beam or field angle data given in manufacturer's catalog. The angles provided here are calculated from manufacturer's diameter and distance data. See Introduction for more information about calculating photometric data.

ARRIFLEX CORP.
Model No. 531500*

SPOT BEAM SPREAD: 13° beam**

FLOOD BEAM SPREAD: 62° beam**

LAMP BASE TYPE: Mogul Bipost

STANDARD LAMP: DPY - 5 KW.

OTHER LAMPS: (none specified in mfg. catalog.)

ACCESSORIES AVAILABLE:
Color Frame (13" sq.), Outrigger Color Frame, 4-
and 8-way Barn Door, Top Hat

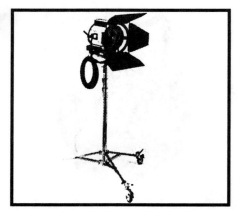

WEIGHT: 34 lbs.

SIZE: (not specified in manufacturer's catalog)

PHOTOMETRICS CHART
(Performance data for this unit are measured using DPY - 5 KW. lamp)

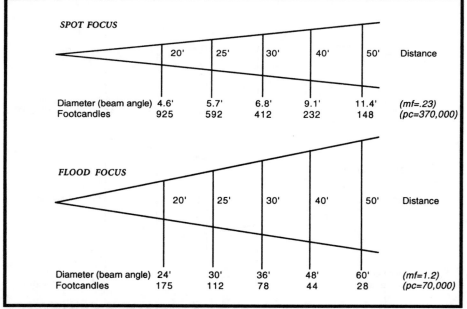

SPOT FOCUS

	20'	25'	30'	40'	50'	Distance
Diameter (beam angle)	4.6'	5.7'	6.8'	9.1'	11.4'	(mf=.23)
Footcandles	925	592	412	232	148	(pc=370,000)

FLOOD FOCUS

	20'	25'	30'	40'	50'	Distance
Diameter (beam angle)	24'	30'	36'	48'	60'	(mf=1.2)
Footcandles	175	112	78	44	28	(pc=70,000)

* A pole operated version of this instrument is available as model number 531505.

** No field angle data given in manufacturer's catalog.

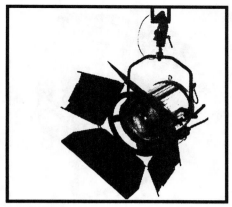

ARRIFLEX CORP.
Model No. 532200*

SPOT BEAM SPREAD: 10° beam**

FLOOD BEAM SPREAD: 62° beam**

LAMP BASE TYPE: Mogul Bipost

STANDARD LAMP: CYX - 2 KW.

OTHER LAMPS: CXZ - 1.5 KW. *(cf=.75)*
CYV - 1 KW. *(cf=.48)*

ACCESSORIES AVAILABLE:
Color Frame (13" sq.), 4- and 8-way Barn Door, Top Hat

WEIGHT: 27 lbs.

SIZE: (not specified in manufacturer's catalog)

PHOTOMETRICS CHART

(Performance data for this unit are measured using CYX - 2 KW. lamp)

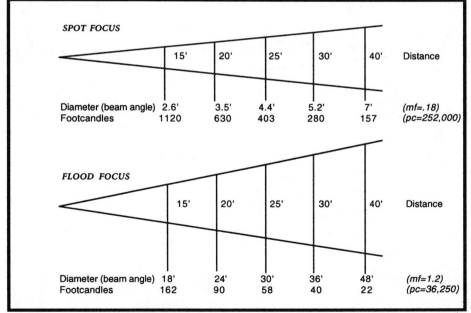

SPOT FOCUS

	15'	20'	25'	30'	40'	Distance
Diameter (beam angle)	2.6'	3.5'	4.4'	5.2'	7'	*(mf=.18)*
Footcandles	1120	630	403	280	157	*(pc=252,000)*

FLOOD FOCUS

	15'	20'	25'	30'	40'	Distance
Diameter (beam angle)	18'	24'	30'	36'	48'	*(mf=1.2)*
Footcandles	162	90	58	40	22	*(pc=36,250)*

* A pole operated version of this instrument is available as model number 532205.
** No field angle data given in manufacturer's catalog.

BERKEY COLORTRAN
Model No. 100-182*

SPOT BEAM SPREAD: 3.2° beam, 18° field

FLOOD BEAM SPREAD: 29° beam, 62° field

LAMP BASE TYPE: Mogul Bipost

STANDARD LAMP: CYX - 2 KW.

OTHER LAMPS: CXZ - 1.5 KW. *(cf=.67)*

ACCESSORIES AVAILABLE:
Color Frame, 4-way Barn Door

WEIGHT: 25 lbs.

SIZE: 16¼" x 14¾" x 17½" (L x W x H)

PHOTOMETRICS CHART
(Performance data for this unit are measured using CYX - 2 KW. lamp.)

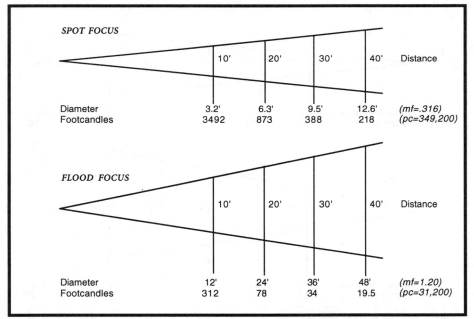

SPOT FOCUS

	10'	20'	30'	40'	Distance
Diameter	3.2'	6.3'	9.5'	12.6'	*(mf=.316)*
Footcandles	3492	873	388	218	*(pc=349,200)*

FLOOD FOCUS

	10'	20'	30'	40'	Distance
Diameter	12'	24'	36'	48'	*(mf=1.20)*
Footcandles	312	78	34	19.5	*(pc=31,200)*

* This instrument can be identified with model numbers -181, -182, -185, -186, or -187, depending on the wire leads and connectors supplied.

BERKEY COLORTRAN
Model No. 100-392

SPOT BEAM SPREAD: 3.2° beam, 16.7° field

FLOOD BEAM SPREAD: 29° beam, 66° field

LAMP BASE TYPE: Mogul Bipost

STANDARD LAMP: CYX - 2KW.

OTHER LAMPS: CXZ - 1.5 KW. *(cf=.67)*

ACCESSORIES AVAILABLE:
Color Frame, 4-way Barn Door

NOTE: The focus mechanism on this unit is an unusual ring which rotates around the nose of the instrument.

WEIGHT: 25 lbs.

SIZE: 13¼" x 15¼" x 34" (L x W x H)

PHOTOMETRICS CHART

(Performance data for this unit are measured using CYX - 2 KW. lamp.)

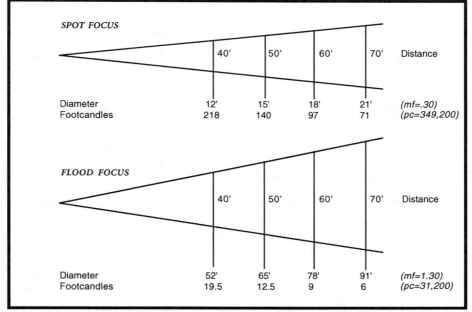

SPOT FOCUS

	40'	50'	60'	70'	Distance
Diameter	12'	15'	18'	21'	*(mf=.30)*
Footcandles	218	140	97	71	*(pc=349,200)*

FLOOD FOCUS

	40'	50'	60'	70'	Distance
Diameter	52'	65'	78'	91'	*(mf=1.30)*
Footcandles	19.5	12.5	9	6	*(pc=31,200)*

* This instrument can be identified with model numbers -392, -395, -396, or -397, depending on the wire leads and connectors supplied.

CENTURY LIGHTING
Model No. 510

SPOT BEAM SPREAD: 16° *

FLOOD BEAM SPREAD: 45° *

LAMP BASE TYPE: Mogul Bipost

STANDARD LAMP: G-48 C13D - 2 KW.

OTHER LAMPS: G-48 C13 - 2 KW.

ACCESSORIES AVAILABLE:
Color Frame

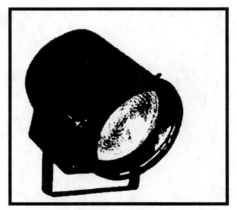

WEIGHT: (not specified in manufacturer's catalog)

SIZE: 14" x 14½" x 26¼" (L x W x H)

PHOTOMETRICS CHART
(Performance data for this unit are measured using G-48 C13D - 2 KW. lamp.)

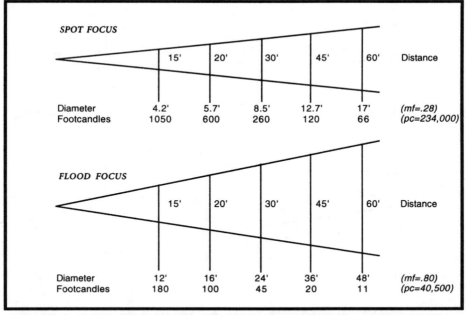

SPOT FOCUS

	15'	20'	30'	45'	60'	Distance
Diameter	4.2'	5.7'	8.5'	12.7'	17'	(mf=.28)
Footcandles	1050	600	260	120	66	(pc=234,000)

FLOOD FOCUS

	15'	20'	30'	45'	60'	Distance
Diameter	12'	16'	24'	36'	48'	(mf=.80)
Footcandles	180	100	45	20	11	(pc=40,500)

* This angle specified in manufacturer's catalog as "where illumination drops to 20 percent of maximum."

CENTURY STRAND
Model No. Patt 243BP*

SPOT BEAM SPREAD: 10° beam, 16° field

FLOOD BEAM SPREAD: 40° beam, 50° field

LAMP BASE TYPE: Mogul Bipost*

STANDARD LAMP: 2M/G48/17 - 2 KW.**

OTHER LAMPS: 1M/G48/11 - 1 KW.**

ACCESSORIES AVAILABLE:
Color Frame (11¾" sq.), 4-way Barn Door

WEIGHT: 31 lbs.

SIZE: 17" x 16" x 19¾" (L x W x H)

PHOTOMETRICS CHART
(Performance data for this unit are measured using 2M/G48/17 - 2 KW. lamp.)

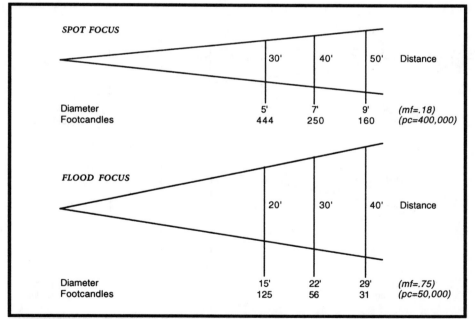

SPOT FOCUS

	30'	40'	50'	Distance
Diameter	5'	7'	9'	(mf=.18)
Footcandles	444	250	160	(pc=400,000)

FLOOD FOCUS

	20'	30'	40'	Distance
Diameter	15'	22'	29'	(mf=.75)
Footcandles	125	56	31	(pc=50,000)

*This unit can be equipped with a Mogul Prefocus lamp base. The model number for this configuration is 243.
**These two lamps are obsolete and no initial lumens data is available; correction factors *(cf)* are thus not available.

COLORTRAN
Model Nos. 100-242* (manual focus)
100-252* (pole focus)

SPOT BEAM SPREAD: 12° beam, 20.4° field

FLOOD BEAM SPREAD: 60° beam, 73° field

LAMP BASE TYPE: Mogul Bipost

STANDARD LAMP: DPY - 5 KW.

OTHER LAMPS: CP29 - 5 KW. (220V) *(cf=.93)*

ACCESSORIES AVAILABLE:
Color Frame, 8-way Barn Door, Combo Stud

WEIGHT: 28 lbs.

SIZE: 16¼" x 16½" x 25" (L x W x H)

PHOTOMETRICS CHART
(Performance data for this unit are measured using DPY - 5 KW. lamp.)

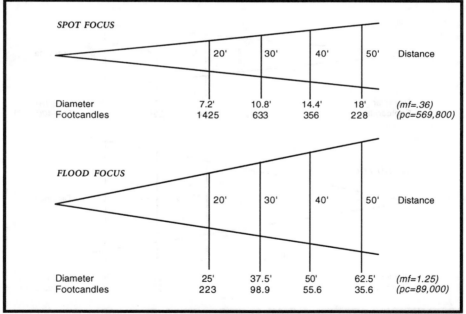

SPOT FOCUS

	20'	30'	40'	50'	Distance
Diameter	7.2'	10.8'	14.4'	18'	*(mf=.36)*
Footcandles	1425	633	356	228	*(pc=569,800)*

FLOOD FOCUS

	20'	30'	40'	50'	Distance
Diameter	25'	37.5'	50'	62.5'	*(mf=1.25)*
Footcandles	223	98.9	55.6	35.6	*(pc=89,000)*

* This instrument can be identified with model numbers -241, -242/-252, -245/-255, -246/-256, or -247/-257, depending on the wire leads and connectors supplied.

KLIEGL BROS.
Model No. 3610

SPOT BEAM SPREAD: 18° field*

FLOOD BEAM SPREAD: 62° field*

LAMP BASE TYPE: Mogul Bipost

STANDARD LAMP: CYX - 2 KW.

OTHER LAMPS: CYV - 1 KW. *(cf=.48)*
CXZ - 1.5 KW. *(cf=.65)*

ACCESSORIES AVAILABLE:
Color Frame, 8-way Barn Door, Table Bases

WEIGHT: 30 lbs.

SIZE: 17" x 16" x 16½" (L x W x H)

PHOTOMETRICS CHART
(Performance data for this unit are measured using CYX - 2 KW. lamp.)

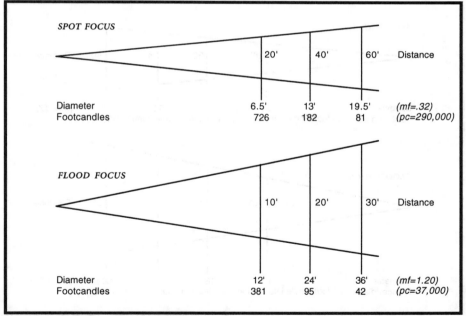

SPOT FOCUS

	20'	40'	60'	Distance
Diameter	6.5'	13'	19.5'	*(mf=.32)*
Footcandles	726	182	81	*(pc=290,000)*

FLOOD FOCUS

	10'	20'	30'	Distance
Diameter	12'	24'	36'	*(mf=1.20)*
Footcandles	381	95	42	*(pc=37,000)*

* No beam angle data given in manufacturer's catalog.

KLIEGL BROS. / RDS
Model No. 3812

SPOT BEAM SPREAD: 7.5° beam, 9.5° field

FLOOD BEAM SPREAD: 50° beam, 62° field

LAMP BASE TYPE: Mogul Prefocus

STANDARD LAMP: CYX - 2 KW.

OTHER LAMPS: (none specified in mfg. catalog)

ACCESSORIES AVAILABLE:
Color Frame, Barn Door

WEIGHT: 25 lbs.
SIZE: 13" x 15" x 18" (L x W x H)

PHOTOMETRICS CHART

(Information about lamp used to measure performance data is not provided in manufacturer's catalog.)

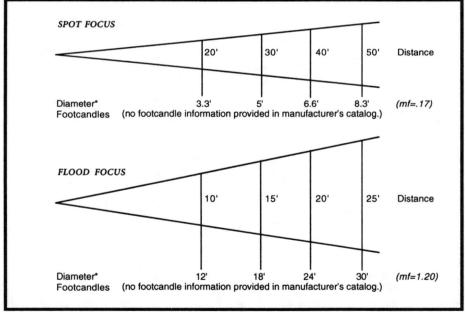

SPOT FOCUS

	20'	30'	40'	50'	Distance
Diameter*	3.3'	5'	6.6'	8.3'	*(mf=.17)*
Footcandles	(no footcandle information provided in manufacturer's catalog.)				

FLOOD FOCUS

	10'	15'	20'	25'	Distance
Diameter*	12'	18'	24'	30'	*(mf=1.20)*
Footcandles	(no footcandle information provided in manufacturer's catalog.)				

* This performance data is calculated from field angles given in manufacturer's catalog. See Introduction for more information about calculating photometric data.

KLIEGL BROS. / RDS
Model No. 3815

SPOT BEAM SPREAD: 9° beam, 15° field

FLOOD BEAM SPREAD: 50° beam, 60° field

LAMP BASE TYPE: Mogul Bipost

STANDARD LAMP: DPY - 5 KW.

OTHER LAMPS: (none specified in mfg. catalog)

ACCESSORIES AVAILABLE:
Color Frame, Barn Door

WEIGHT: 27 lbs.

SIZE: 13" x 18" x 18" (L x W x H)

PHOTOMETRICS CHART

(Information about lamp used to measure performance data is not provided in manufacturer's catalog.)

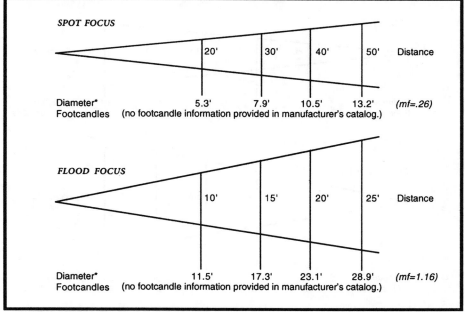

SPOT FOCUS

| | 20' | 30' | 40' | 50' | Distance |

Diameter* 5.3' 7.9' 10.5' 13.2' *(mf=.26)*
Footcandles (no footcandle information provided in manufacturer's catalog.)

FLOOD FOCUS

| | 10' | 15' | 20' | 25' | Distance |

Diameter* 11.5' 17.3' 23.1' 28.9' *(mf=1.16)*
Footcandles (no footcandle information provided in manufacturer's catalog.)

* This performance data is calculated from field angles given in manufacturer's catalog. See Introduction for more information about calculating photometric data.

FRESNEL — 10" Lens

STRAND LIGHTING
Model No. 3501, "Castor"

SPOT BEAM SPREAD: 11.5° beam, 24° field

FLOOD BEAM SPREAD: 53° beam, 64° field

LAMP BASE TYPE: Mogul Bipost

STANDARD LAMP: CYX - 2 KW.

OTHER LAMPS: CYV - 1 KW. *(cf=47)*
CXZ - 1.5 KW. *(cf=.65)*

ACCESSORIES AVAILABLE:
Color Frame, 8-way Barn Door, Various Scrims for TV use.

WEIGHT: 21 lbs.

SIZE: 14" x 16½" x 26⅜" (L x W x H)

PHOTOMETRICS CHART

(Performance data for this unit are measured using CYX - 2 KW. lamp.)

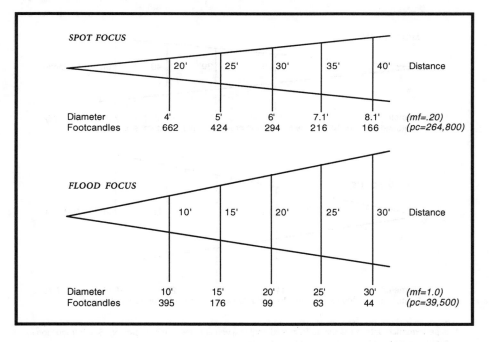

SPOT FOCUS

	20'	25'	30'	35'	40'	Distance
Diameter	4'	5'	6'	7.1'	8.1'	*(mf=.20)*
Footcandles	662	424	294	216	166	*(pc=264,800)*

FLOOD FOCUS

	10'	15'	20'	25'	30'	Distance
Diameter	10'	15'	20'	25'	30'	*(mf=1.0)*
Footcandles	395	176	99	63	44	*(pc=39,500)*

WEIGHT: 24 lbs.

SIZE: 14¼" x 17⅛" x 25⅜" (L x W x H)

STRAND LIGHTING
Model No. 3505, "Bambino"

SPOT BEAM SPREAD: 12° beam, 19° field

FLOOD BEAM SPREAD: 51° beam, 61° field

LAMP BASE TYPE: Mogul Bipost

STANDARD LAMP: DPY - 5 KW.

OTHER LAMPS: CP29 - 5 KW. (220/240 V.) *(cf=.94)*

ACCESSORIES AVAILABLE:
Color Frame, 8-way Barn Door, Various Scrims for
TV/Film use.

PHOTOMETRICS CHART
(Performance data for this unit are measured using DPY - 5 KW. lamp.)

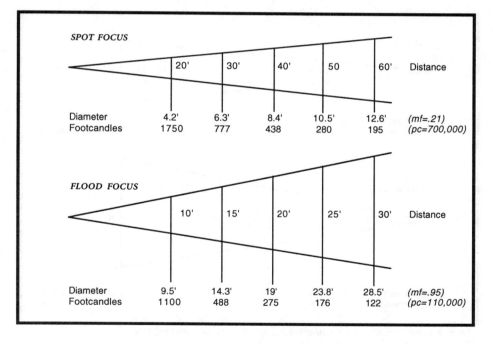

SPOT FOCUS

	20'	30'	40'	50	60'	Distance
Diameter	4.2'	6.3'	8.4'	10.5'	12.6'	*(mf=.21)*
Footcandles	1750	777	438	280	195	*(pc=700,000)*

FLOOD FOCUS

	10'	15'	20'	25'	30'	Distance
Diameter	9.5'	14.3'	19'	23.8'	28.5'	*(mf=.95)*
Footcandles	1100	488	275	176	122	*(pc=110,000)*

FRESNEL — 12" Lens

ARRIFLEX CORP.
Model No. 532500*

SPOT BEAM SPREAD: 13° beam**

FLOOD BEAM SPREAD: 67° beam**

LAMP BASE TYPE: Mogul Bipost

STANDARD LAMP: DPY - 5 KW.

OTHER LAMPS: (none specified in mfg. catalog.)

ACCESSORIES AVAILABLE:
Color Frame (15½" sq.), 4- and 8-way Barn Door,
Top Hat

WEIGHT: 36 lbs.

SIZE: (not specified in manufacturer's catalog)

PHOTOMETRICS CHART
(Performance data for this unit are measured using DPY - 5 KW. lamp)

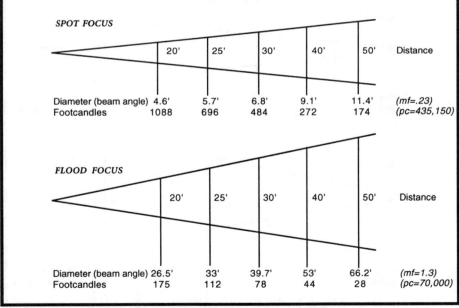

SPOT FOCUS

	20'	25'	30'	40'	50'	Distance
Diameter (beam angle)	4.6'	5.7'	6.8'	9.1'	11.4'	*(mf=.23)*
Footcandles	1088	696	484	272	174	*(pc=435,150)*

FLOOD FOCUS

	20'	25'	30'	40'	50'	Distance
Diameter (beam angle)	26.5'	33'	39.7'	53'	66.2'	*(mf=1.3)*
Footcandles	175	112	78	44	28	*(pc=70,000)*

* A pole operated version of this instrument is available as model number 532505.

** No field angle data given in manufacturer's catalog.

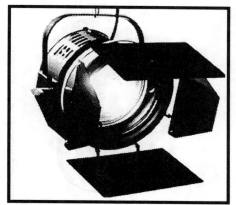

BERKEY COLORTRAN
Model No. 100-402*

SPOT BEAM SPREAD: 9° beam, 17° field

FLOOD BEAM SPREAD: 50° beam, 57.6° field

LAMP BASE TYPE: Mogul Bipost

STANDARD LAMP: DPY - 5 KW.

OTHER LAMPS: 176-160 - 5KW. *(cf=1.17)***

ACCESSORIES AVAILABLE:
Color Frame, 4-way Barn Door

WEIGHT: 36 lbs.

SIZE: 14" x 19" x 30" (L x W x H)

NOTE: The focus mechanism on this unit is an unusual ring which rotates around the nose of the instrument.

PHOTOMETRICS CHART
(Performance data for this unit are measured using DPY - 5 KW. lamp.)

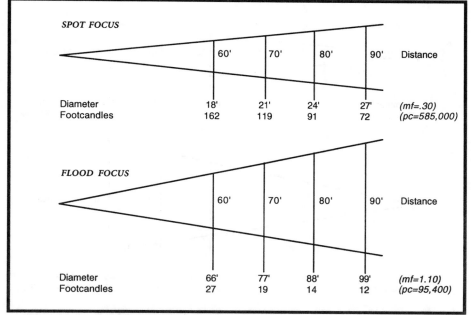

SPOT FOCUS

	60'	70'	80'	90'	Distance
Diameter	18'	21'	24'	27'	(mf=.30)
Footcandles	162	119	91	72	(pc=585,000)

FLOOD FOCUS

	60'	70'	80'	90'	Distance
Diameter	66'	77'	88'	99'	(mf=1.10)
Footcandles	27	19	14	12	(pc=95,400)

* This instrument can be identified with model numbers -402, -405, -406, or -407, depending on the wire leads and connectors supplied.

** This lamp is identiied by its Colortran catalog number. The correction factor *(cf)* is given in mfg. catalog.

KLIEGL BROS.
Model No. 3612

SPOT BEAM SPREAD: 24° field*

FLOOD BEAM SPREAD: 74° field*

LAMP BASE TYPE: Mogul Bipost

STANDARD LAMP: CYX - 2 KW.

OTHER LAMPS: DPY - 5 KW. *(cf=2.42)***

ACCESSORIES AVAILABLE:
Color Frame, 8-way Barn Door, 18" Table Base

WEIGHT: 38 lbs.

SIZE: 18" x 17¼" x 18" (L x W x H)

PHOTOMETRICS CHART
(Performance data for this unit are measured using CYX - 2 KW. lamp.)

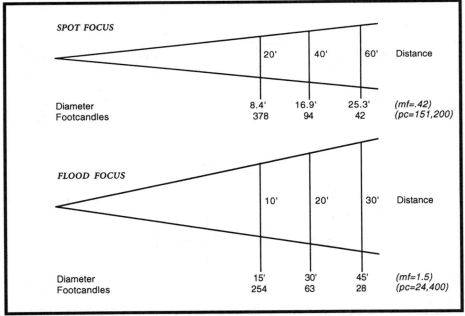

SPOT FOCUS

	20'	40'	60'	Distance
Diameter	8.4'	16.9'	25.3'	*(mf=.42)*
Footcandles	378	94	42	*(pc=151,200)*

FLOOD FOCUS

	10'	20'	30'	Distance
Diameter	15'	30'	45'	*(mf=1.5)*
Footcandles	254	63	28	*(pc=24,400)*

* No beam angle data given in manufacturer's catalog.
**The CYX and the DPY lamps do not have the same LCL (light center length) and are not normally compatible; the LCL of CYX lamp is 5", while the LCL of DPY is 6-1/2".

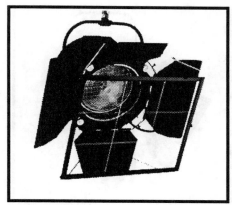

STRAND LIGHTING
Model No. 3601, "Pollux"

SPOT BEAM SPREAD: 13° beam, 27° field

FLOOD BEAM SPREAD: 59° beam, 69° field

LAMP BASE TYPE: Mogul Bipost

STANDARD LAMP: DPY - 5 KW.

OTHER LAMPS: CP29 - 5 KW. (220/240 V.) *(cf=.94)*

ACCESSORIES AVAILABLE:
Color Frame, 8-way Barn Door, Various Scrims for TV use.

WEIGHT: 26 lbs.

SIZE: 14¼" x 20" x 28½" (L x W x H)

PHOTOMETRICS CHART

(Performance data for this unit are measured using DPY - 5 KW. lamp.)

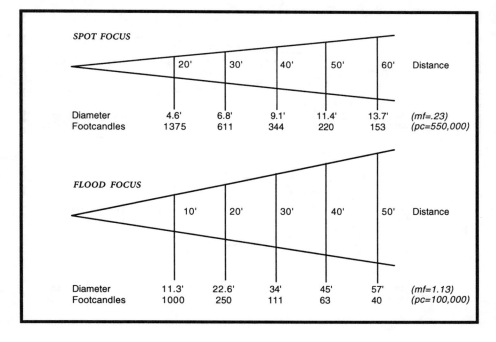

SPOT FOCUS

	20'	30'	40'	50'	60'	Distance
Diameter	4.6'	6.8'	9.1'	11.4'	13.7'	*(mf=.23)*
Footcandles	1375	611	344	220	153	*(pc=550,000)*

FLOOD FOCUS

	10'	20'	30'	40'	50'	Distance
Diameter	11.3'	22.6'	34'	45'	57'	*(mf=1.13)*
Footcandles	1000	250	111	63	40	*(pc=100,000)*

FRESNEL — 14" Lens

STRAND LIGHTING
Model No. 3701, "Vega"

SPOT BEAM SPREAD: 13.5° beam, 24° field

FLOOD BEAM SPREAD: 55.5° beam, 65.5° field

LAMP BASE TYPE: Mogul Bipost

STANDARD LAMP: DTY - 10 KW.

OTHER LAMPS: CP83 - 10 KW. (220/240 V.) *(cf=1)*

ACCESSORIES AVAILABLE:
Color Frame, 8-way Barn Door, Various Scrims for TV/Film use.

WEIGHT: 37 lbs.

SIZE: 17¼" x 19⅞" x 33⅞" (L x W x H)

PHOTOMETRICS CHART
(Performance data for this unit are measured using DTY - 10 KW. lamp.)

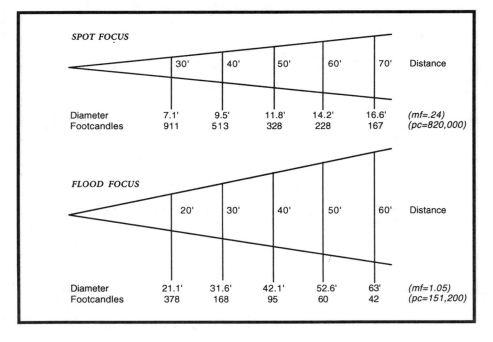

SPOT FOCUS

	30'	40'	50'	60'	70'	Distance
Diameter	7.1'	9.5'	11.8'	14.2'	16.6'	*(mf=.24)*
Footcandles	911	513	328	228	167	*(pc=820,000)*

FLOOD FOCUS

	20'	30'	40'	50'	60'	Distance
Diameter	21.1'	31.6'	42.1'	52.6'	63'	*(mf=1.05)*
Footcandles	378	168	95	60	42	*(pc=151,200)*

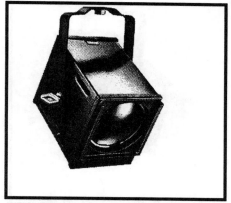

ALTMAN STAGE LIGHTING
Model No. 101, "Box Spot"

SPOT BEAM SPREAD: 19.8° beam, 27° field

FLOOD BEAM SPREAD: 46° beam, 52° field

LAMP BASE TYPE: Medium Screw

STANDARD LAMP: 400G/FL - 400 W.

OTHER LAMPS: 250G/SP - 250 W. *(cf=.66)*
250G/FL - 250 W. *(cf=.54)*
400G/SP - 400 W. *(cf=1.24)*

ACCESSORIES AVAILABLE:
Color Frame, Pin Spot Adaptor, Pin Spot Apertures

WEIGHT: (not specified in manufacturer's catalog)

SIZE: 8⅛" x 6⅞" x 11¼" (L x W x H)

PHOTOMETRICS CHART

(Performance data for this unit are measured using 400G/FL - 400 W. lamp.)

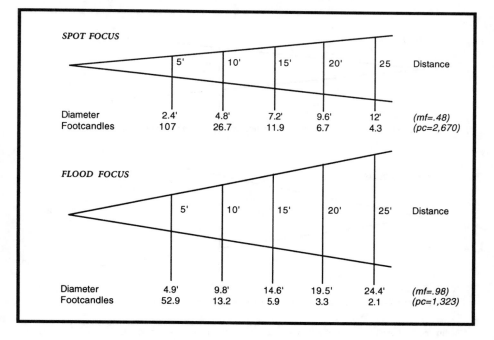

SPOT FOCUS

	5'	10'	15'	20'	25	Distance
Diameter	2.4'	4.8'	7.2'	9.6'	12'	*(mf=.48)*
Footcandles	107	26.7	11.9	6.7	4.3	*(pc=2,670)*

FLOOD FOCUS

	5'	10'	15'	20'	25'	Distance
Diameter	4.9'	9.8'	14.6'	19.5'	24.4'	*(mf=.98)*
Footcandles	52.9	13.2	5.9	3.3	2.1	*(pc=1,323)*

DISPLAY STAGE LIGHTING
Model No. 816

SPOT BEAM SPREAD: 17° *

FLOOD BEAM SPREAD: 62° *

LAMP BASE TYPE: Medium Prefocus

STANDARD LAMP: T-20 - 500 W. **

OTHER LAMPS: G-30 - 250 W.
G-30 - 400 W.

ACCESSORIES AVAILABLE:
Color Frame, Motorized & Non-motorized Color
Wheels, Snoot, Iris, Shutters

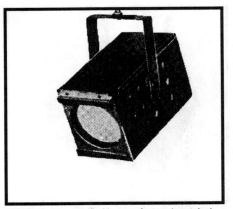

WEIGHT: (not specified in manufacturer's catalog)

SIZE: (not specified in manufacturer's catalog)

PHOTOMETRICS CHART

(Information about lamp used to measure performance data is not provided in manufacturer's catalog.)

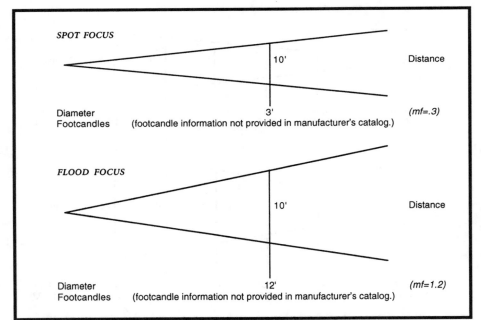

SPOT FOCUS

10'   Distance

Diameter   3'   *(mf=.3)*
Footcandles   (footcandle information not provided in manufacturer's catalog.)

FLOOD FOCUS

10'   Distance

Diameter   12'   *(mf=1.2)*
Footcandles   (footcandle information not provided in manufacturer's catalog.)

* Beam and field angles not specified in manufacturer's catalog. This angle is calculated from given distance and "beam spread" diameter data. See Introduction for more information about calculating photometric data.
** This is an obsolete lamp. Correction factor *(cf)* information is not available for the other lamps (which are also obsolete) because no initial lumens data is available for any of them.

KLIEGL BROS.
Model No. 53

SPOT BEAM SPREAD: 22° *

FLOOD BEAM SPREAD: 53° *

LAMP BASE TYPE: Medium Screw

STANDARD LAMP: 400G/SP - 400 W.

OTHER LAMPS: 250G/SP - 250 W. *(cf=.54)*

ACCESSORIES AVAILABLE:
Color Frame, Motorized & Non-motorized Color Wheels, Alzak Reflector

WEIGHT: 5 lbs.

SIZE: 7¾" x 6¾" x 10" (L x W x H)

PHOTOMETRICS CHART

(Information about lamp used to measure performance data is not provided in manufacturer's catalog.)

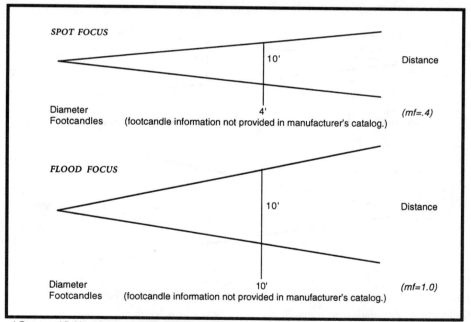

SPOT FOCUS

10'   Distance

Diameter   4'   *(mf=.4)*
Footcandles   (footcandle information not provided in manufacturer's catalog.)

FLOOD FOCUS

10'   Distance

Diameter   10'   *(mf=1.0)*
Footcandles   (footcandle information not provided in manufacturer's catalog.)

* Beam and field angles not specified in manufacturer's catalog. This angle is calculated from given distance and "beam spread" diameter data. See Introduction for more information about calculating photometric data.

KLIEGL BROS.
Model No. 5310

SPOT BEAM SPREAD: 12° *

FLOOD BEAM SPREAD: 62° *

LAMP BASE TYPE: Medium Screw

STANDARD LAMP: 400G/SP - 400 W.

OTHER LAMPS: 250G/SP - 250 W. *(cf=.54)*

ACCESSORIES AVAILABLE:
Color Frame, Motorized & Non-motorized Color
Wheels, Alzak Reflector

WEIGHT: 6 lbs.

SIZE: 10½" x 6¾" x 10" (L x W x H)

PHOTOMETRICS CHART
(Information about lamp used to measure performance data is not provided in manufacturer's catalog.)

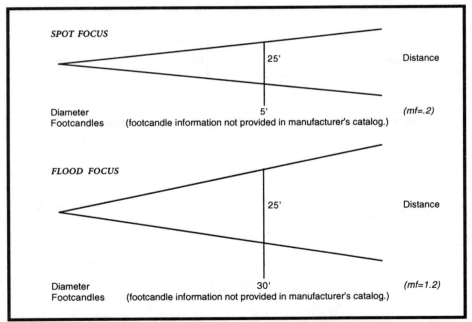

SPOT FOCUS

25'   Distance

Diameter   5'   *(mf=.2)*
Footcandles   (footcandle information not provided in manufacturer's catalog.)

FLOOD FOCUS

25'   Distance

Diameter   30'   *(mf=1.2)*
Footcandles   (footcandle information not provided in manufacturer's catalog.)

* Beam and field angles not specified in manufacturer's catalog. This angle is calculated from given distance and "beam spread" diameter data. See Introduction for more information about calculating photometric data.

KLIEGL BROS.
Model No. 6N19

SPOT BEAM SPREAD: 9° *

FLOOD BEAM SPREAD: 48° *

LAMP BASE TYPE: Mogul Bipost

STANDARD LAMP: 2M/G48/4 - 2 KW.

OTHER LAMPS: (None specified in mfg. catalog.)

ACCESSORIES AVAILABLE:
Color Frame, Motorized & Non-motorized Color Wheels, Iris

WEIGHT: 24 lbs.

SIZE: 20½" x 10" x 15½" (L x W x H)

PHOTOMETRICS CHART
(Information about lamp used to measure performance data is not provided in manufacturer's catalog.)

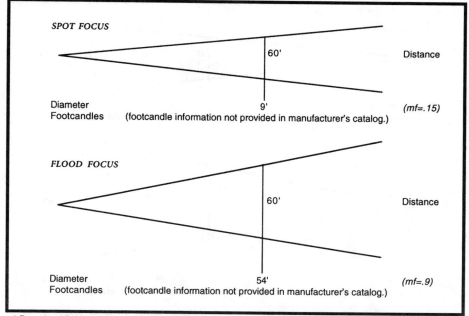

SPOT FOCUS

60'   Distance

Diameter
Footcandles   9'   (mf=.15)
          (footcandle information not provided in manufacturer's catalog.)

FLOOD FOCUS

60'   Distance

Diameter
Footcandles   54'   (mf=.9)
          (footcandle information not provided in manufacturer's catalog.)

* Beam and field angles not specified in manufacturer's catalog. This angle is calculated from given distance and "beam spread" diameter data. See Introduction for more information about calculating photometric data.

Plano-convex Spotlight

KLIEGL BROS.
Model No. 70

SPOT BEAM SPREAD: 12° *

FLOOD BEAM SPREAD: 44° *

LAMP BASE TYPE: Mogul Prefocus

STANDARD LAMP: 1000/G40/PSP - 1 KW.

OTHER LAMPS: 1500/G40/15 - 1.5 KW. **

ACCESSORIES AVAILABLE:
Color Frame, Motorized & Non-motorized Color
Wheels, Iris

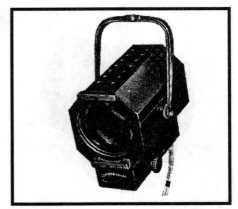

WEIGHT: 14 lbs.

SIZE: 13" x 10" x 16½" (L x W x H)

PHOTOMETRICS CHART

(Information about lamp used to measure performance data is not provided in manufacturer's catalog.)

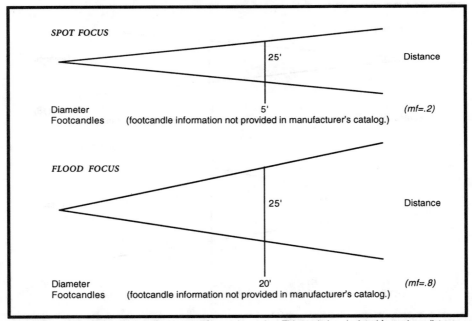

SPOT FOCUS

25'   Distance

Diameter   5'   (mf=.2)
Footcandles   (footcandle information not provided in manufacturer's catalog.)

FLOOD FOCUS

25'   Distance

Diameter   20'   (mf=.8)
Footcandles   (footcandle information not provided in manufacturer's catalog.)

* Beam and field angles not specified in manufacturer's catalog. This angle is calculated from given distance and "beam spread" diameter data. See Introduction for more information about calculating photometric data.
** This is an obsolete lamp. Correction factor *(cf)* information is not available because no initial lumens data is available.

KLIEGL BROS.
Model No. 71

SPOT BEAM SPREAD: 12° *

FLOOD BEAM SPREAD: 48° *

LAMP BASE TYPE: Mogul Prefocus

STANDARD LAMP: 1500/G40/15 - 1.5 KW.

OTHER LAMPS: 2M/G48/4 - 2 KW. **

ACCESSORIES AVAILABLE:
Color Frame, Motorized & Non-motorized Color Wheels, Iris

WEIGHT: 18 lbs.

SIZE: 15" x 11" x 17½" (L x W x H)

PHOTOMETRICS CHART

(Information about lamp used to measure performance data is not provided in manufacturer's catalog.)

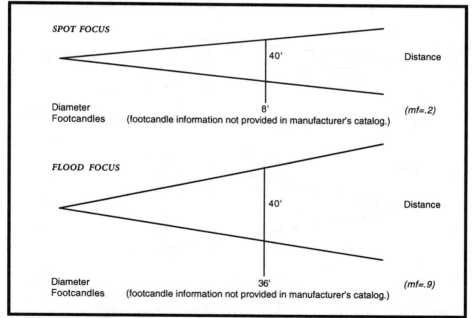

SPOT FOCUS

40'   Distance

Diameter   8'   (mf=.2)
Footcandles   (footcandle information not provided in manufacturer's catalog.)

FLOOD FOCUS

40'   Distance

Diameter   36'   (mf=.9)
Footcandles   (footcandle information not provided in manufacturer's catalog.)

* Beam and field angles not specified in manufacturer's catalog. This angle is calculated from given distance and "beam spread" diameter data. See Introduction for more information about calculating photometric data.
** This is an obsolete lamp. Correction factor *(cf)* information is not available because no initial lumens data is available.

KLIEGL BROS.
Model No. 8N20

SPOT BEAM SPREAD: 8° *

FLOOD BEAM SPREAD: 65° *

LAMP BASE TYPE: Mogul Bipost

STANDARD LAMP: 2M/G48/4 - 2 KW.

OTHER LAMPS: (None specified in mfg. catalog.)

ACCESSORIES AVAILABLE:
Color Frame, Motorized & Non-motorized Color
Wheels, Iris

WEIGHT: 30 lbs.

SIZE: 21½" x 11½" x 16½" (L x W x H)

PHOTOMETRICS CHART

(Information about lamp used to measure performance data is not provided in manufacturer's catalog.)

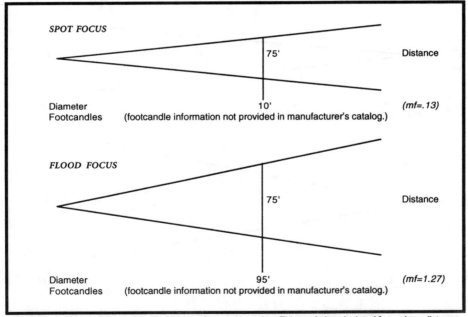

SPOT FOCUS

75'   Distance

Diameter   10'   (mf=.13)
Footcandles   (footcandle information not provided in manufacturer's catalog.)

FLOOD FOCUS

75'   Distance

Diameter   95'   (mf=1.27)
Footcandles   (footcandle information not provided in manufacturer's catalog.)

* Beam and field angles not specified in manufacturer's catalog. This angle is calculated from given distance
and "beam spread" diameter data. See Introduction for more information about calculating photometric data.

KLIEGL BROS.
Model No. 8N24

SPOT BEAM SPREAD: 8° *

FLOOD BEAM SPREAD: 70° *

LAMP BASE TYPE: Mogul Bipost

STANDARD LAMP: 2M/G48/4 - 2 KW.

OTHER LAMPS: (None specified in mfg. catalog.)

ACCESSORIES AVAILABLE:
Color Frame, Motorized & Non-motorized Color Wheels, Iris

WEIGHT: 34 lbs.

SIZE: 26" x 13" x 17" (L x W x H)

PHOTOMETRICS CHART

(Information about lamp used to measure performance data is not provided in manufacturer's catalog.)

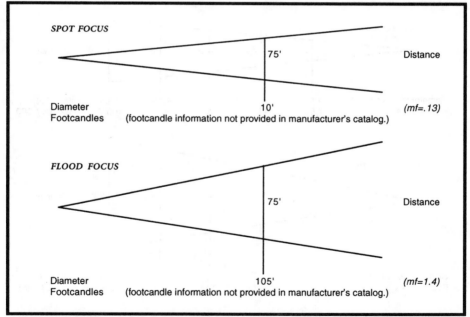

SPOT FOCUS

75'   Distance

Diameter   10'   (mf=.13)
Footcandles   (footcandle information not provided in manufacturer's catalog.)

FLOOD FOCUS

75'   Distance

Diameter   105'   (mf=1.4)
Footcandles   (footcandle information not provided in manufacturer's catalog.)

* Beam and field angles not specified in manufacturer's catalog. This angle is calculated from given distance and "beam spread" diameter data. See Introduction for more information about calculating photometric data.

Plano-convex Spotlight

LUDWIG PANI
Model No. LH-2000

SPOT BEAM SPREAD: 3° beam, 5° field

FLOOD BEAM SPREAD: 43° beam, 48° field

LAMP BASE TYPE: Mogul Bipost

STANDARD LAMP: CP-79 (Thorn) - 2 KW./220 V.

OTHER LAMPS: CP 43 - 2 KW./220 V. *(cf=1.04)*

ACCESSORIES AVAILABLE:
Color Frame (8⅝" sq.), 4-leaf Barn Door,

WEIGHT: 32 lbs.

SIZE: 20¼" x 13⅜" x 19¼" (L x W x H)

PHOTOMETRICS CHART

(Performance data for this unit are measured using Thorn CP-79 - 2 KW. lamp.)

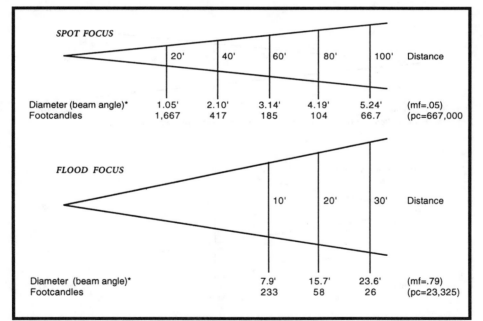

SPOT FOCUS

	20'	40'	60'	80'	100'	Distance
Diameter (beam angle)*	1.05'	2.10'	3.14'	4.19'	5.24'	(mf=.05)
Footcandles	1,667	417	185	104	66.7	(pc=667,000

FLOOD FOCUS

	10'	20'	30'	Distance
Diameter (beam angle)*	7.9'	15.7'	23.6'	(mf=.79)
Footcandles	233	58	26	(pc=23,325)

* Photometric data in manufacturer's catalog is given in metric units, and was converted to English units for this book. Also, these diameter and footcandle figures were calculated from beam angles and peak candela data given in catalog. See Introduction for more information about calculating photometric data.

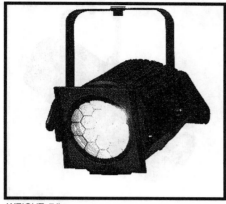

Plano-convex Spotlight

STRAND LIGHTING
Model No. 3383, "Quartet

SPOT BEAM SPREAD: 7.8° beam, 13.3° field

FLOOD BEAM SPREAD: 59° beam, 66° field

LAMP BASE TYPE: GY9.5

STANDARD LAMP: FRE - 650 W.

OTHER LAMPS: (none specified in mfg. catalog)

ACCESSORIES AVAILABLE:
Color Frame, Rotatable Barn Door

WEIGHT: 7 lbs.

SIZE: 11½" x 9¾" x 11½" (L x W x H)

PHOTOMETRICS CHART

(Performance data for this unit are measured using FRE - 650 W. lamp.)

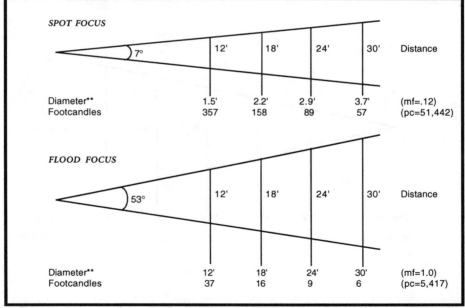

SPOT FOCUS					
7°	12'	18'	24'	30'	Distance
Diameter**	1.5'	2.2'	2.9'	3.7'	(mf=.12)
Footcandles	357	158	89	57	(pc=51,442)

FLOOD FOCUS					
53°	12'	18'	24'	30'	Distance
Diameter**	12'	18'	24'	30'	(mf=1.0)
Footcandles	37	16	9	6	(pc=5,417)

* This instrument is described as having a "prism convex" lens. According to the manufacturer, the lens falls between a fresnel lens and a plano-convex lens.

** The given distance and diameter data yield angles which are close to the beam angles rather than the field angles. See Introduction for more information about calculating photometrics.

ALTMAN STAGE LIGHTING
Model No. 1KL6-50*

BEAM ANGLE: 16° (peak), 31° (flat)

FIELD ANGLE: 50° (peak and flat)

LAMP BASE TYPE: Medium 2-Pin

LAMP MOUNT: Axial

STANDARD LAMP: FEL - 1 KW.

OTHER LAMPS: EHD - 500 W. *(cf=.39)*
EHG - 750 W. *(cf=.56)*

INTEGRAL PATTERN SLOT: Yes

ACCESSORIES AVAILABLE:
Color Frame, Pattern Holder, Snoot, Motorized &
Non-Motorized Color Wheels

WEIGHT: (not specified in manufacturer's catalog)

SIZE: 22⅞" x 17" x 19" (L x W x H)

PHOTOMETRICS CHART
(Photometrics for this unit are measured at peak focus using FEL - 1 KW. lamp.)

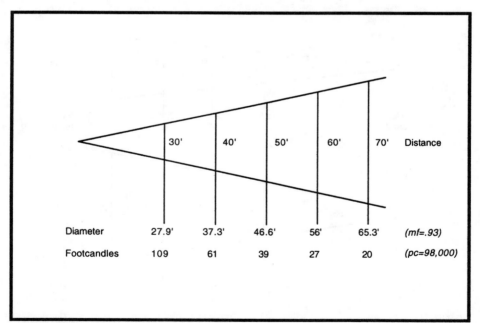

	30'	40'	50'	60'	70'	Distance
Diameter	27.9'	37.3'	46.6'	56'	65.3'	*(mf=.93)*
Footcandles	109	61	39	27	20	*(pc=98,000)*

* With optional iris installed, the model number is 1KL6-I-50.

ALTMAN STAGE LIGHTING
Model No. 360 4.5 x 6.5*

BEAM ANGLE: 25°

FIELD ANGLE: 56°

LAMP BASE TYPE: Medium Prefocus

LAMP MOUNT: Off Axis

STANDARD LAMP: DNT - 750 W.

OTHER LAMPS: DNS - 500 W. *(cf=.65)*
EGE - 500 W. *(cf=.61)*
EGG - 750 W. *(cf=.92)*

INTEGRAL PATTERN SLOT: Yes

ACCESSORIES AVAILABLE:
Color Frame (7½" sq.), Snoot, Motorized &
Non-Motorized Color Wheels

WEIGHT: (not specified in manufacturer's catalog)

SIZE: 17⅝" x 13¼" x 16¾" (L x W x H)

PHOTOMETRICS CHART
(Photometrics for this unit are measured using DNT - 750 W. lamp.)

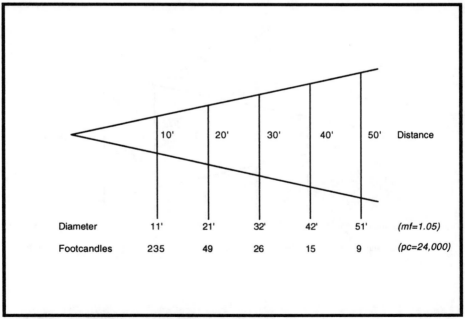

	10'	20'	30'	40'	50'	Distance
Diameter	11'	21'	32'	42'	51'	*(mf=1.05)*
Footcandles	235	49	26	15	9	*(pc=24,000)*

* With optional iris installed, the model number is 360 I 4.5 x 6.5.

ALTMAN STAGE LIGHTING
Model No. 360Q 4.5 x 6.5*

BEAM ANGLE: 22°

FIELD ANGLE: 55°

LAMP BASE TYPE: Medium 2-Pin

LAMP MOUNT: Axial

STANDARD LAMP: EHF - 750 W.

OTHER LAMPS: EHD - 500 W. *(ct=.52)*
EHG - 750 W. *(cf=.75)*

INTEGRAL PATTERN SLOT: Yes

ACCESSORIES AVAILABLE:
Color Frame (7½" sq.), Pattern Holder, Snoot,
Motorized & Non-Motorized Color Wheel

WEIGHT: (not specified in manufacturer's catalog)

SIZE: 18⅛" x 13" x 16¼" (L x W x H)

PHOTOMETRICS CHART
(Photometrics for this unit are measured using EHF - 750 W. lamp.)

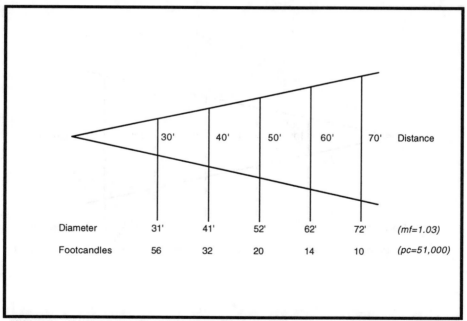

	30'	40'	50'	60'	70'	Distance
Diameter	31'	41'	52'	62'	72'	*(mf=1.03)*
Footcandles	56	32	20	14	10	*(pc=51,000)*

* With optional iris installed, the model number is 360 QI 4.5 x 6.5.

ALTMAN
Model No. S6-50, "Shakespeare"

BEAM ANGLE: 23° (cosine)

FIELD ANGLE: 50°

LAMP BASE TYPE: Medium 2-Pin

LAMP MOUNT: Axial

STANDARD LAMP: HX600 - 575 W.

OTHER LAMPS: EHG - 750 W. *(cf=.93)*
EHD - 500 W. *(cf-.63)*

INTEGRAL PATTERN SLOT: Yes

ACCESSORIES AVAILABLE:
Color Frame (6¼" sq.), Snoot, Pattern Holder

WEIGHT: 18 lbs.

SIZE: 28½" x 12" x 13¾" (L x W x H)

PHOTOMETRICS CHART
(Performance data for this unit are measured at cosine focus using HX600 - 575 W. lamp.)

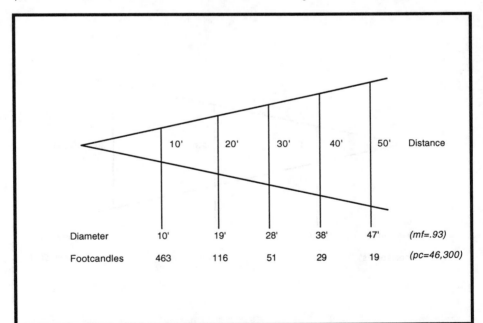

	10'	20'	30'	40'	50'	Distance
Diameter	10'	19'	28'	38'	47'	*(mf=.93)*
Footcandles	463	116	51	29	19	*(pc=46,300)*

CENTURY LIGHTING
Model No. 1580

BEAM ANGLE: (not specified in mfg. catalog)

FIELD ANGLE: 50° *

LAMP BASE TYPE: (not specified in mfg. catalog)

LAMP MOUNT: Off Axis

STANDARD LAMP: 750T12/9 - 750 W.

OTHER LAMPS: (none specified in mfg. catalog)

INTEGRAL PATTERN SLOT: Yes

ACCESSORIES AVAILABLE:
Color Frame (5¾" sq.), Iris

WEIGHT: (not specified in manufacturer's catalog)

SIZE: 14½" x 10" x 20½" (L x W x H)

PHOTOMETRICS CHART

(Information about lamp used to measure performance data is not provided in manufacturer's catalog.)

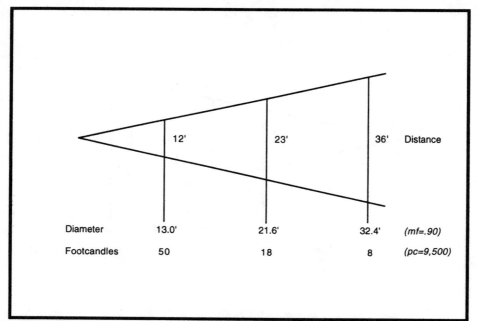

	12'	23'	36'	Distance
Diameter	13.0'	21.6'	32.4'	(mf=.90)
Footcandles	50	18	8	(pc=9,500)

* Beam and field angles not specified in manufacturer's catalog. This field angle is calculated from given distance and field diameter data. See Introduction for more information about calculating photometric data.

CENTURY LIGHTING
Model No. 1581*

BEAM ANGLE: 36°

FIELD ANGLE: 50°

LAMP BASE TYPE: Medium Prefocus

LAMP MOUNT: Off Axis

STANDARD LAMP: DEB - 500 W.

OTHER LAMPS: 250 T12/8 - 250 W.**

INTEGRAL PATTERN SLOT: No

WEIGHT: (not specified in manufacturer's catalog)

SIZE: 14½" x 10" x 20½" (L x W x H)

ACCESSORIES AVAILABLE:
Color Frame (5¾" sq.), Motoried & Non-motorized Color Wheels

PHOTOMETRICS CHART
(Photometrics for this unit are measured using DEB - 500 W. lamp.)

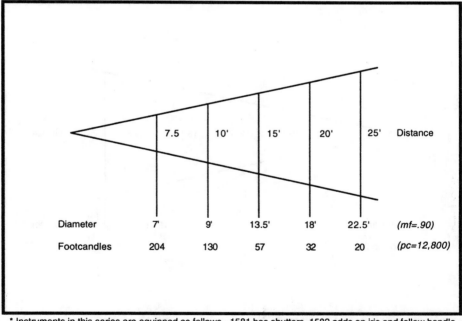

	7.5	10'	15'	20'	25'	Distance
Diameter	7'	9'	13.5'	18'	22.5'	(mf=.90)
Footcandles	204	130	57	32	20	(pc=12,800)

* Instruments in this series are equipped as follows - 1581 has shutters, 1582 adds an iris and follow handle, 1583 adds a pattern holder.

** This is an obsolete lamp. No correction factor (cf) is given because no initial lumens data is available.

CENTURY STRAND
Model No. 2211*

BEAM ANGLE: 38°

FIELD ANGLE: 50°

LAMP BASE TYPE: Medium Prefocus

LAMP MOUNT: Off Axis

STANDARD LAMP: EGE - 500 W.

OTHER LAMPS: EGD - 500 W. *(cf=1.25)*

INTEGRAL PATTERN SLOT: Yes

ACCESSORIES AVAILABLE:
Color Frame, Pattern Holder, Snoot

WEIGHT: (not specified in manufacturer's catalog)

SIZE: (not specified in manufacturer's catalog)

PHOTOMETRICS CHART
(Photometrics for this unit are measured using EGE - 500 W. lamp.)

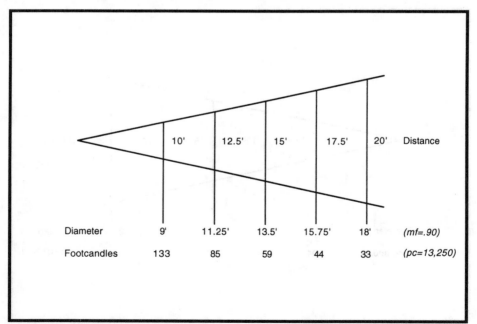

	10'	12.5'	15'	17.5'	20'	Distance
Diameter	9'	11.25'	13.5'	15.75'	18'	*(mf=.90)*
Footcandles	133	85	59	44	33	*(pc=13,250)*

* This unit was also sold as Century 1480.

ELECTRO CONTROLS
Model No. 7363AF 56*

BEAM ANGLE: 27°

FIELD ANGLE: 56°

LAMP BASE TYPE: Medium Prefocus

LAMP MOUNT: Axial

STANDARD LAMP: EGG - 750 W.

OTHER LAMPS: EGF - 750 W. *(cf=1.30)*
EGJ - 1 KW. *(cf=1.75)*

INTEGRAL PATTERN SLOT: Yes

WEIGHT: 17.75 lbs.

SIZE: 19⁹⁄₁₆" x 15½" x 26" (L x W x H)

ACCESSORIES AVAILABLE:
Color Frame (7½" sq.), Pattern Holder, Snoot, Iris
(as an option)

PHOTOMETRICS CHART
(Performance data for this unit are measured using EGG - 750 W. lamp.)

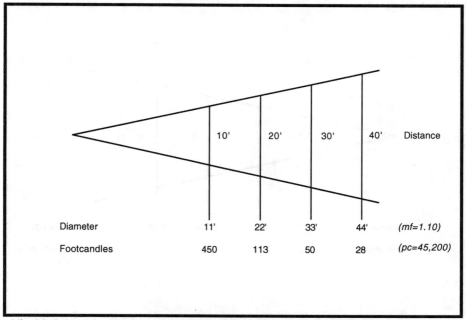

	10'	20'	30'	40'	Distance
Diameter	11'	22'	33'	44'	*(mf=1.10)*
Footcandles	450	113	50	28	*(pc=45,200)*

* If unit includes optional iris the model number is 7363AI 56. Lens systems in the 7363A series are
interchangeable. See Index for page numbers of the other versions of model 7363A.

ELECTRONIC THEATRE CONTROLS (ETC) Model No. 450, "Source Four"

BEAM ANGLE: 33°

FIELD ANGLE: 50°

LAMP BASE TYPE: Medium 2-Pin

LAMP MOUNT: Axial

STANDARD LAMP: HPL 575/115

OTHER LAMPS: HPL 550/77*

INTEGRAL PATTERN SLOT: Yes

ACCESSORIES AVAILABLE:
Color Frame (6¼" sq.), Pattern Holder, Donut, Snoot

WEIGHT: 15.9 lbs.

SIZE: 22½" x 12⅜" x 20⅝" (L x W x H)

PHOTOMETRICS CHART

(Performance data for this unit are measured at cosine focus using HPL 575/115 lamp.)

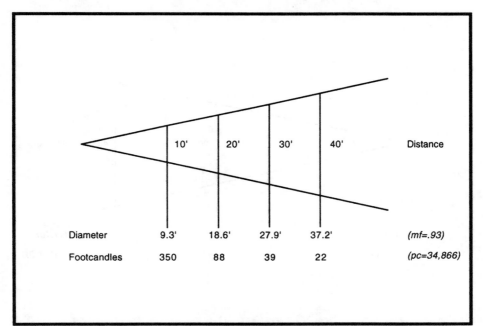

	10'	20'	30'	40'	Distance
Diameter	9.3'	18.6'	27.9'	37.2'	(mf=.93)
Footcandles	350	88	39	22	(pc=34,866)

* The HPL 550/77 lamp is used with an ETC multiplexing system, used to allow multiple units to share a common circuit, with separate dimming of each.

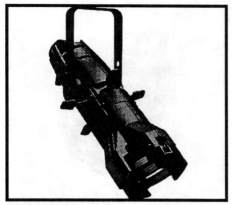

WEIGHT: 12.2 lbs.

SIZE: 19³⁄₁₆" x 9⅝" x 13⅝" (L x W x H)

ELECTRONIC THEATRE CONTROLS (ETC) Model No. 450J, "Source Four, Jr."

BEAM ANGLE: 34°

FIELD ANGLE: 51°

LAMP BASE TYPE: Medium 2-Pin

LAMP MOUNT: Axial

STANDARD LAMP: HPL 575/115

OTHER LAMPS: HP: 375/115 - 375 W.
HPL 375/115X - 375 W.
HPL 575/115X - 575 W.
HPL 550/77 - 550 W.*

INTEGRAL PATTERN SLOT: Yes

ACCESSORIES AVAILABLE:
Color Frame (6¼" sq.), Pattern Holder, Drop-in Iris, Donut, Snoot

PHOTOMETRICS CHART

(Performance data for this unit are measured at cosine focus using HPL 575/115 lamp.)

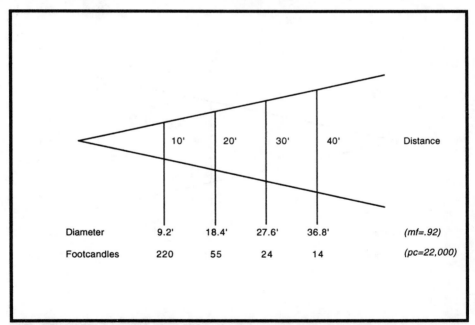

	10'	20'	30'	40'	Distance
Diameter	9.2'	18.4'	27.6'	36.8'	*(mf=.92)*
Footcandles	220	55	24	14	*(pc=22,000)*

* The HPL 550/77 lamp is used with an ETC multiplexing system, used to allow multiple units to share a common circuit, with separate dimming of each.

KLIEGL BROS.
Model No. 1163W

BEAM ANGLE: (not specified in mfg. catalog)

FIELD ANGLE: 53° *

LAMP BASE TYPE: Medium Bipost

LAMP MOUNT: Off Axis

STANDARD LAMP: 500T14/8 - 500 W.

OTHER LAMPS: (none specified in mfg. catalog)

INTEGRAL PATTERN SLOT: No

ACCESSORIES AVAILABLE:
Color Frame

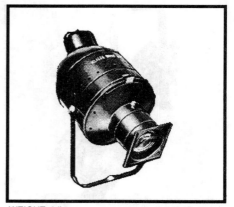

WEIGHT: 8 lbs.

SIZE: (not specified in manufacturer's catalog)

PHOTOMETRICS CHART

(Performance data for this unit are measured using 500T14/8 - 500 W. lamp)

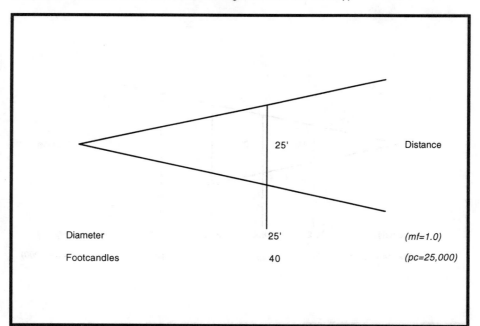

	25'	Distance
Diameter	25'	*(mf=1.0)*
Footcandles	40	*(pc=25,000)*

* Beam and field angles not specified in manufacturer's catalog. This field angle is calculated from given distance and field diameter data. See Introduction for more information about calculating photometric data.

KLIEGL BROS.
Model No. 1164W

BEAM ANGLE: (not specified in mfg. catalog)

FIELD ANGLE: 53° *

LAMP BASE TYPE: Mogul Bipost

LAMP MOUNT: Off Axis

STANDARD LAMP: 1M/T24/12 - 1 KW.

OTHER LAMPS: (none specified in mfg. catalog)

INTEGRAL PATTERN SLOT: No

ACCESSORIES AVAILABLE:
Color Frame

WEIGHT: 18 lbs.

SIZE: (not specified in manufacturer's catalog)

PHOTOMETRICS CHART

(Performance data for this unit are measured using 1M/T24/12 - 1 KW. lamp)

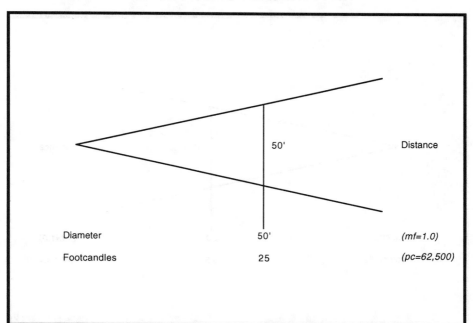

	50'	Distance
Diameter	50'	(mf=1.0)
Footcandles	25	(pc=62,500)

* Beam and field angles not specified in manufacturer's catalog. This field angle is calculated from given distance and field diameter data. See Introduction for more information about calculating photometric data.

KLIEGL BROS.
Model No. 1167W

BEAM ANGLE: (not specified in mfg. catalog)

FIELD ANGLE: 53° *

LAMP BASE TYPE: Mogul Bipost

LAMP MOUNT: Off Axis

STANDARD LAMP: 2M/T30/1 - 1 KW. **

OTHER LAMPS: 1M/T24/5 -   1 KW.
               1550/T24/6 - 1.5 KW.

INTEGRAL PATTERN SLOT: No

ACCESSORIES AVAILABLE:
Color Frame

WEIGHT: 19 lbs.

SIZE: (not specified in manufacturer's catalog)

PHOTOMETRICS CHART
(Performance data for this unit are measured using 2M/T30/1 - 1 KW. lamp)

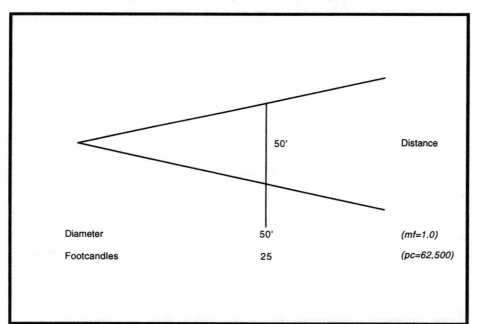

	50'	Distance
Diameter	50'	(mf=1.0)
Footcandles	25	(pc=62,500)

* Beam and field angles not specified in manufacturer's catalog. This field angle is calculated from given distance and field diameter data. See Introduction for more information about calculating photometric data.
** This is an obsolete lamp but it is listed as the standard lamp since the manufacturer uses this lamp to describe the photometric performance. Correction factor *(cf)* information is not available for the other lamps because no initial lumens data is available for the obsolete lamp.

KLIEGL BROS.
Model No. 1555 *

BEAM ANGLE: 21°

FIELD ANGLE: 55° (cosine)

LAMP BASE TYPE: Medium 2-Pin

LAMP MOUNT: Axial

STANDARD LAMP: FEL - 1 KW.

OTHER LAMPS: EHD - 500 W. *(cf=.39)*
EHG - 750 W. *(cf=.56)*

INTEGRAL PATTERN SLOT: Yes

ACCESSORIES AVAILABLE:
Color Frame (7½" sq.), Pattern Holder, Drop-in Iris

WEIGHT: (not specified in manufacturer's catalog)

SIZE: 25½" x 10¼" x 18" (L x W x H)

PHOTOMETRICS CHART

(Performance data for this unit are measured at cosine focus using FEL - 1 KW. lamp.)

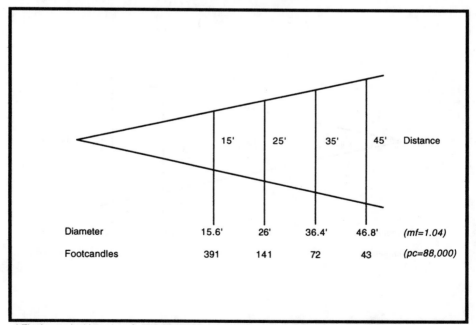

	15'	25'	35'	45'	Distance
Diameter	15.6'	26'	36.4'	46.8'	*(mf=1.04)*
Footcandles	391	141	72	43	*(pc=88,000)*

* The lenses in this series of ellipsoidal spotlights are interchangeable on the same lamp housing. Lens barrels are color-coded to indentify the field angle. This 55° model is color-coded with green.

LEE COLORTRAN
Model No. 650-012*

BEAM ANGLE:17.5° (peak), 33.3° (flat)

FIELD ANGLE: 50° (peak and flat)

LAMP BASE TYPE: Medium 2-Pin

LAMP MOUNT: Axial

STANDARD LAMP: FEL - 1 KW.

OTHER LAMPS: EHD - 500 W. *(cf=.39)*
EHG - 750 W. *(cf=.56)*

INTEGRAL PATTERN SLOT: Yes

ACCESSORIES AVAILABLE:
Color Frame (7½" sq.), Pattern Holder, Iris Kit

WEIGHT: 19.7 lbs.

SIZE: 28" x 15" x 25½" (L x W x H)

PHOTOMETRICS CHART
(Performance data for this unit are measured at peak focus using FEL - 1KW. lamp.)

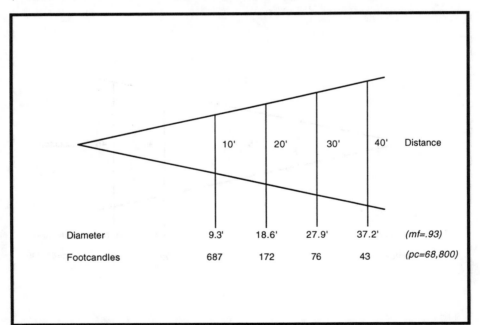

	10'	20'	30'	40'	Distance
Diameter	9.3'	18.6'	27.9'	37.2'	*(mf=.93)*
Footcandles	687	172	76	43	*(pc=68,800)*

* This model can be identified with model numbers -012, -014, -015, or -016 depending on the wire leads and connectors supplied.

LIGHTING & ELECTRONICS
Model No. 61-04

BEAM ANGLE: 38°

FIELD ANGLE: 50°

LAMP BASE TYPE: Medium Prefocus

LAMP MOUNT: Off Axis

STANDARD LAMP: DNS - 500 W.

OTHER LAMPS: DEB - 500 W. *(cf=.82)*
DNT - 750 W. *(cf=1.55)*

INTEGRAL PATTERN SLOT: Yes

ACCESSORIES AVAILABLE:
Color Frame (7½" sq.)

WEIGHT: (not specified in manufacturer's catalog)

SIZE: 11¾" x 7" (L x W)

PHOTOMETRICS CHART
(Photometrics for this unit are measured using DNS - 500 W. lamp.)

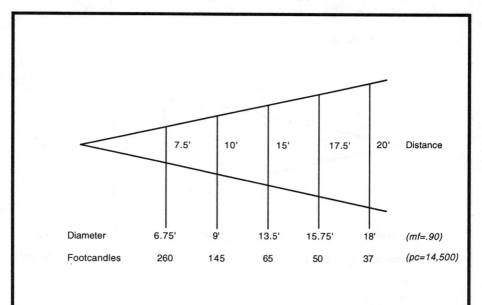

	7.5'	10'	15'	17.5'	20'	Distance
Diameter	6.75'	9'	13.5'	15.75'	18'	*(mf=.90)*
Footcandles	260	145	65	50	37	*(pc=14,500)*

LIGHTING & ELECTRONICS
Model No. AQ1K-50

BEAM ANGLE: 36° at cosine

FIELD ANGLE: 55° at cosine

LAMP BASE TYPE: Medium 2-Pin

LAMP MOUNT: Axial

STANDARD LAMP: FEL - 1KW.

OTHER LAMPS: EHG - 750 W. *(cf=.55)*
EHD - 500 W. *(cf=.36)*

INTEGRAL PATTERN SLOT: Yes

ACCESSORIES AVAILABLE:
Color Frame (7½" sq.), Drop-in Iris

WEIGHT: 21 lbs.

SIZE: 19¹³⁄₁₆" x 7¾" x 7¾" (L x W x H)

PHOTOMETRICS CHART
(Performance data for this unit are measured at the cosine using FEL - 1KW. lamp.)

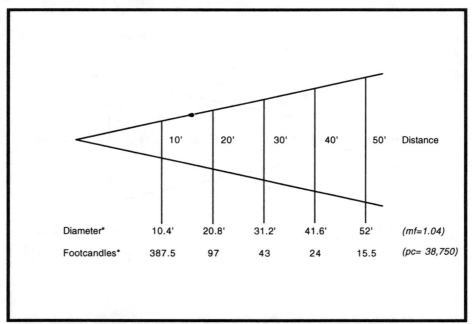

	10'	20'	30'	40'	50'	Distance
Diameter*	10.4'	20.8'	31.2'	41.6'	52'	*(mf=1.04)*
Footcandles*	387.5	97	43	24	15.5	*(pc= 38,750)*

* The diameter and footcandle information in this chart was calculated using multiplying factor (*mf*) and peak candela (*pc*) data given in catalog. See Introduction for more information about calculating photometric data.

LIGHTING & ELECTRONICS
Model No. AQ61-04

BEAM ANGLE: 22°

FIELD ANGLE: 55°

LAMP BASE TYPE: Medium 2-Pin

LAMP MOUNT: Axial

STANDARD LAMP: EHG - 750 W.

OTHER LAMPS: EHD - 500 W. *(cf=.69)*
EHF - 750 W. *(cf=1.32)*
Hx600 - 600 W. *(cf=1.93)*

INTEGRAL PATTERN SLOT: Yes

ACCESSORIES AVAILABLE:
Color Frame (7½" sq.), Pattern Holder, Iris

WEIGHT: 18 lbs.

SIZE: 17¾" x 7⅝" (L x W)

PHOTOMETRICS CHART
(Performance data for this unit measured using EHG - 750 W. lamp.)

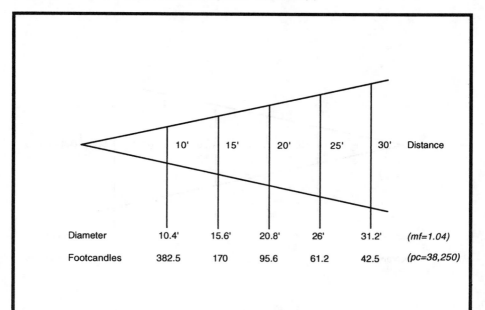

	10'	15'	20'	25'	30'	Distance
Diameter	10.4'	15.6'	20.8'	26'	31.2'	*(mf=1.04)*
Footcandles	382.5	170	95.6	61.2	42.5	*(pc=38,250)*

ALTMAN STAGE LIGHTING
Model No. 3.5 Q 5

BEAM ANGLE: 30°

FIELD ANGLE: 48°

LAMP BASE TYPE: Medium 2-Pin

LAMP MOUNT: Axial

STANDARD LAMP: EHD - 500 W.

OTHER LAMPS: EHC - 500 W. *(cf=1.23)*

INTEGRAL PATTERN SLOT: Yes

ACCESSORIES AVAILABLE:
Color Frame, Pattern Holder, Snoot

WEIGHT: (not specified in manufacturer's catalog)

SIZE: 16½" x 9" x 13½" (L x W x H)

PHOTOMETRICS CHART
(Photometrics for this unit are measured using EHD - 500 W. lamp.)

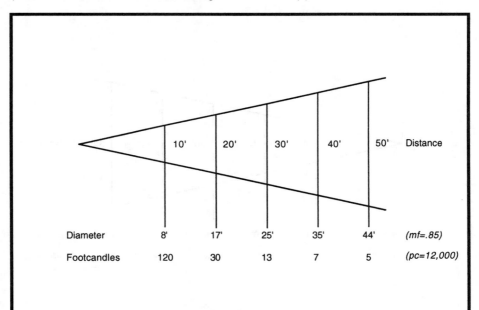

	10'	20'	30'	40'	50'	Distance
Diameter	8'	17'	25'	35'	44'	*(mf=.85)*
Footcandles	120	30	13	7	5	*(pc=12,000)*

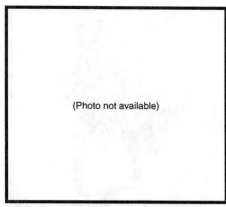

(Photo not available)

WEIGHT: (not specified in manufacturer's catalog)

SIZE: 15¼" x 14½" (L x H)

CAPITOL STAGE LIGHTING
Model No. 6504*

BEAM ANGLE: (not specified in mfg. catalog)

FIELD ANGLE: 46° **

LAMP BASE TYPE: Medium Prefocus

LAMP MOUNT: Off Axis

STANDARD LAMP: DEB - 500 W.

OTHER LAMPS: T12 C13D - 250 W.***

INTEGRAL PATTERN SLOT: Yes (6544 only*)

ACCESSORIES AVAILABLE:
Color Frame (7½" sq.), Pattern Holder, Snoot,
Motorized & Non-Motorized Color Wheels

PHOTOMETRICS CHART

(Photometrics for this unit are measured at peak focus using DEB - 500 W. lamp.)

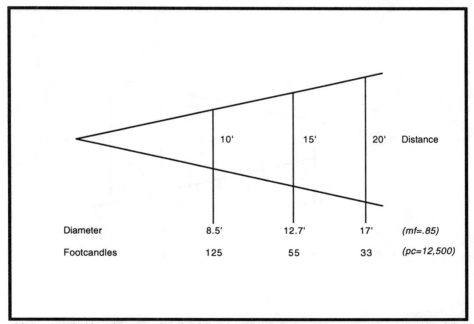

	10'	15'	20'	Distance
Diameter	8.5'	12.7'	17'	*(mf=.85)*
Footcandles	125	55	33	*(pc=12,500)*

* Instruments in this series are equipped as follows - 6504 has shutters, 6524 adds an iris, 6534 adds a follow handle, 6544 has shutters and a template slot.
** Beam and field angles not specified in manufacturer's catalog. This field angle is calculated from given distance and field diameter data. See Introduction for more information about calculating photometric data.
*** This is an obsolete lamp. No correction factor (cf) is given because no initial lumens data is available.

CENTURY LIGHTING
Model No. 1587*

BEAM ANGLE: 32°

FIELD ANGLE: 42°

LAMP BASE TYPE: Medium Prefocus

LAMP MOUNT: Off Axis

STANDARD LAMP: DEB - 500 W.

OTHER LAMPS: 250T12/8 - 250 W.**

INTEGRAL PATTERN SLOT: No

ACCESSORIES AVAILABLE:
Color Frame (5¾" sq.)

WEIGHT: (not specified in manufacturer's catalog)

SIZE: 14½" x 10" x 20½" (L x W x H)

PHOTOMETRICS CHART
(Performance data for this unit are measured using DEB - 500 W. lamp.)

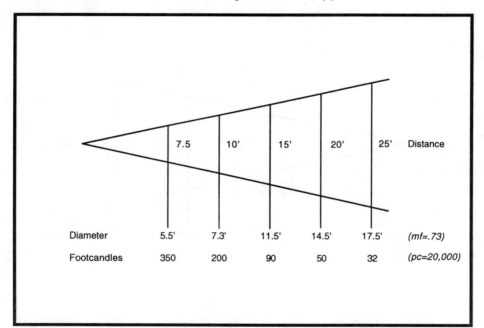

	7.5	10'	15'	20'	25'	Distance
Diameter	5.5'	7.3'	11.5'	14.5'	17.5'	(mf=.73)
Footcandles	350	200	90	50	32	(pc=20,000)

* Instruments in this series are equipped as follows - 1586 has a pattern holder, 1587 has shutters, 1588 has an iris and follow handle.

** This is an obsolete lamp. No correction factor *(cf)* is given because no initial lumens data is available.

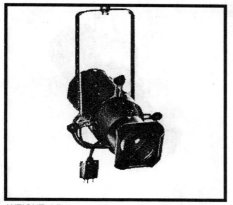

WEIGHT: 9 lbs.

SIZE: 13⅝" x 9⅞" x 16⅜" (L x W x H)

KLIEGL BROS.
Model No. 1340

BEAM ANGLE: (not specified in mfg. catalog)

FIELD ANGLE: 46°

LAMP BASE TYPE: Recessed Single-Contact

LAMP MOUNT: Off Axis

STANDARD LAMP: EHR - 400 W.

OTHER LAMPS: EHP - 300 W. *(cf=.67)*
　　　　　　　　FDA - 400 W. *(cf=1.39)*

INTEGRAL PATTERN SLOT: Yes

ACCESSORIES AVAILABLE:
Color Frame, Pattern Holder, Snoot

PHOTOMETRICS CHART
(Performance data for this unit are measured using EHR - 400 W. lamp.)

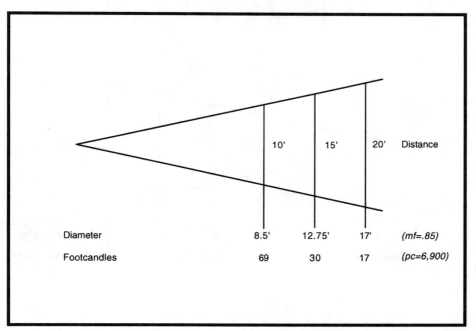

	10'	15'	20'	Distance
Diameter	8.5'	12.75'	17'	*(mf=.85)*
Footcandles	69	30	17	*(pc=6,900)*

LITTLE STAGE LIGHTING
Model No. E306

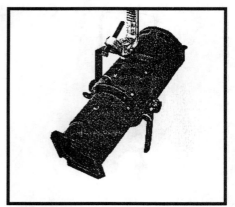

BEAM ANGLE: 20.7°

FIELD ANGLE: 41.7°

LAMP BASE TYPE: Mini-Can Screw

LAMP MOUNT: (not specified in mfg. catalog)

STANDARD LAMP: EVR - 500 W.

OTHER LAMPS: (none specified in mfg. catalog)

INTEGRAL PATTERN SLOT: Yes

ACCESSORIES AVAILABLE:
Color Frame (5" sq.), Pattern Holder

WEIGHT: 9 lbs.

SIZE: 16½" x 10¾" x 9½" (L x W x H)

PHOTOMETRICS CHART
(Performance data for this unit are measured using EVR - 500 W. lamp.)

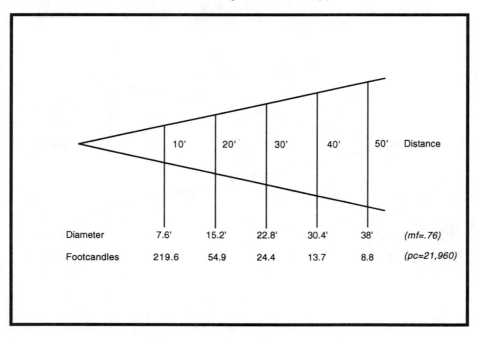

	10'	20'	30'	40'	50'	Distance
Diameter	7.6'	15.2'	22.8'	30.4'	38'	*(mf=.76)*
Footcandles	219.6	54.9	24.4	13.7	8.8	*(pc=21,960)*

WEIGHT: 15.5 lbs.

SIZE: 18" x 12" x 24" (L x W x H)

STRAND CENTURY
Model No. 2204

BEAM ANGLE: 21° (peak), 42° (flat)

FIELD ANGLE: 44° (peak)

LAMP BASE TYPE: Medium 2-Pin

LAMP MOUNT: Axial

STANDARD LAMP: FEL - 1 KW.

OTHER LAMPS: EHD - 500 W. *(cf=.39)*
EHG - 750 W. *(cf=.56)*

INTEGRAL PATTERN SLOT: Yes

ACCESSORIES AVAILABLE:
Color Frame (7½" sq.), Pattern Holder, Snoot, Iris Kit

PHOTOMETRICS CHART

(Performance data for this unit are measured at peak focus using FEL - 1KW. lamp.)

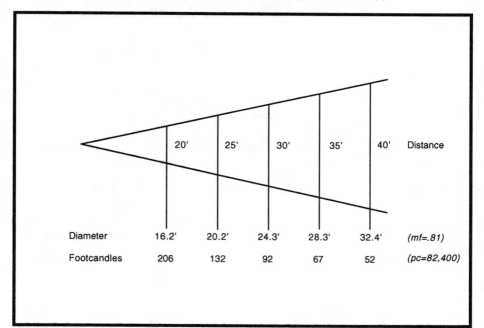

	20'	25'	30'	35'	40'	Distance
Diameter	16.2'	20.2'	24.3'	28.3'	32.4'	*(mf=.81)*
Footcandles	206	132	92	67	52	*(pc=82,400)*

STRAND LIGHTING
Model No. 2250, " Leko"

BEAM ANGLE: 20° peak, 26° cosine

FIELD ANGLE: 45° peak, 53° cosine

LAMP BASE TYPE: Medium 2-Pin

LAMP MOUNT: Axial

STANDARD LAMP: FEL - 1 KW.

OTHER LAMPS: EHD - 500 W. *(cf=.36)*
EHG - 750 W. *(cf=.54)*

INTEGRAL PATTERN SLOT: Yes

ACCESSORIES AVAILABLE:
Color Frame (7½" sq.), Pattern Holder, Drop-in Iris,
High Hat

WEIGHT: 11.9 lbs.

SIZE: 17¼" x 11¾" (L x W)

PHOTOMETRICS CHART
(Performance data for this unit are measured at peak focus using FEL - 1KW. lamp.)

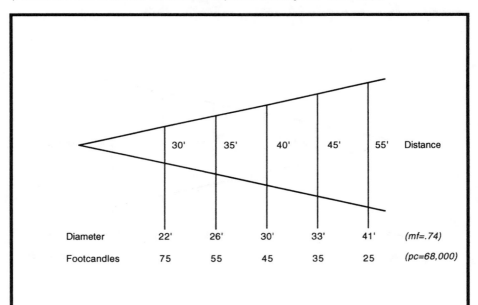

	30'	35'	40'	45'	55'	Distance
Diameter	22'	26'	30'	33'	41'	*(mf=.74)*
Footcandles	75	55	45	35	25	*(pc=68,000)*

TIMES SQUARE LIGHTING
Model No. Q3W

BEAM ANGLE: (not specified in mfg. catalog)

FIELD ANGLE: 40° *

LAMP BASE TYPE: Recessed Single-Contact

LAMP MOUNT: Off Axis

STANDARD LAMP: FAD - 650 W.

OTHER LAMPS: EHP - 300 W. *(cf=.30)*
EHR - 400 W. *(cf=.45)*

INTEGRAL PATTERN SLOT: Yes

ACCESSORIES AVAILABLE:
Color Frame, Pattern Holder, Non-motorized Color Wheel

WEIGHT: (not specified in manufacturer's catalog)

SIZE: (not specified in manufacturer's catalog)

PHOTOMETRICS CHART

(Information about lamp used to measure photometrics is not provided in manufacturer's catalog.)

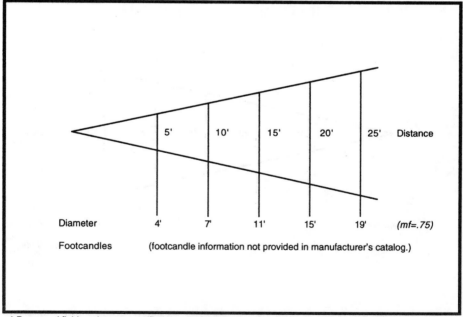

Distance	5'	10'	15'	20'	25'
Diameter	4'	7'	11'	15'	19'

(mf=.75)

Footcandles   (footcandle information not provided in manufacturer's catalog.)

* Beam and field angles not specified in manufacturer's catalog. This field angle is calculated from given distance and field diameter data. See Introduction for more information about calculating photometric data.

ALTMAN STAGE LIGHTING
Model No. 3.5 Q 6

BEAM ANGLE: 21°

FIELD ANGLE: 38°

LAMP BASE TYPE: Medium 2-Pin

LAMP MOUNT: Axial

STANDARD LAMP: EHD - 500 W.

OTHER LAMPS: EHC - 500 W. *(cf=1.23)*

INTEGRAL PATTERN SLOT: Yes

ACCESSORIES AVAILABLE:
Color Frame, Pattern Holder, Snoot

WEIGHT: (not specified in manufacturer's catalog)

SIZE: 14⅝" x 9" x 13½" (L x W x H)

PHOTOMETRICS CHART
(Performance data for this unit are measured using EHD - 500 W. lamp.)

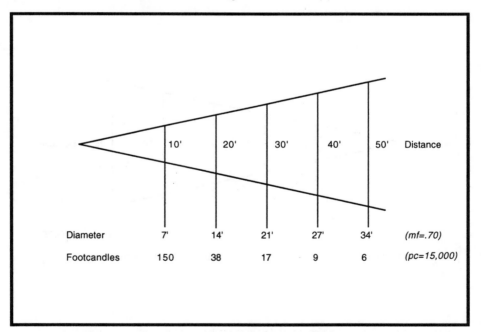

	10'	20'	30'	40'	50'	Distance
Diameter	7'	14'	21'	27'	34'	*(mf=.70)*
Footcandles	150	38	17	9	6	*(pc=15,000)*

WEIGHT: 10.5 lbs.

SIZE: 15" x 9" x 19" (L x W x H)

BERKEY COLORTRAN
Model No. 212-002*

BEAM ANGLE: 18° (peak), 30° (flat)

FIELD ANGLE: 30° (peak), 36° (flat)

LAMP BASE TYPE: Medium 2-Pin

LAMP MOUNT: Axial

STANDARD LAMP: EHD - 500 W.

OTHER LAMPS: EHC - 500 W. *(cf=1.23)*

INTEGRAL PATTERN SLOT: Yes

ACCESSORIES AVAILABLE:
Color Frame, Pattern Holder

PHOTOMETRICS CHART
(Performance data for this unit are measured at peak focus using EHD - 500 W. lamp.)

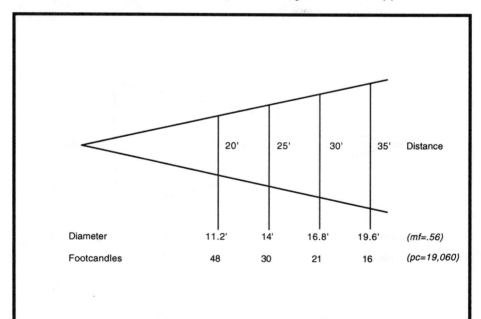

	20'	25'	30'	35'	Distance
Diameter	11.2'	14'	16.8'	19.6'	*(mf=.56)*
Footcandles	48	30	21	16	*(pc=19,060)*

* This model can be identified with model numbers -002, -005, -006, or -007 depending the wire leads and connectors supplied.

CENTURY LIGHTING
Model Nos. 1215 or 2115*

BEAM ANGLE: 24°

FIELD ANGLE: 32°

LAMP BASE TYPE: Recessed Single-Contact

LAMP MOUNT: Off Axis

STANDARD LAMP: EHR - 400 W.

OTHER LAMPS: EHP - 300 W. *(cf=.67)*
FDA - 400 W. *(cf=1.39)*

INTEGRAL PATTERN SLOT: Yes

ACCESSORIES AVAILABLE:
Color Frame, Pattern Holder, Snoot

WEIGHT: (not specified in manufacturer's catalog)

SIZE: 13" x 9" x 17¼" (L x W x H)

PHOTOMETRICS CHART
(Performance data for this unit are measured using EHR - 400 W. lamp.)

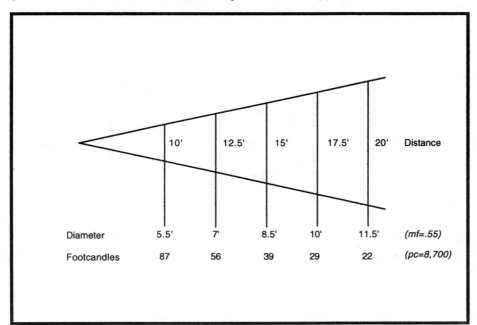

	10'	12.5'	15'	17.5'	20'	Distance
Diameter	5.5'	7'	8.5'	10'	11.5'	*(mf=.55)*
Footcandles	87	56	39	29	22	*(pc=8,700)*

* 2115 is the model number used by Century Strand for this instrument beginning around 1970.

KLIEGL BROS.
Model No. 1341

BEAM ANGLE: (not specified in mfg. catalog)

FIELD ANGLE: 37°

LAMP BASE TYPE: Recessed Single-Contact

LAMP MOUNT: Off Axis

STANDARD LAMP: EHR - 400 W.

OTHER LAMPS: EHP - 300 W. *(cf=.67)*
FDA - 400 W. *(cf=1.39)*

INTEGRAL PATTERN SLOT: Yes

ACCESSORIES AVAILABLE:
Color Frame, Pattern Holder, Snoot

WEIGHT: 9 lbs.

SIZE: 13⅝" x 9⅞" x 16⅜" (L x W x H)

PHOTOMETRICS CHART
(Performance data for this unit are measured using EHR - 400 W. lamp.)

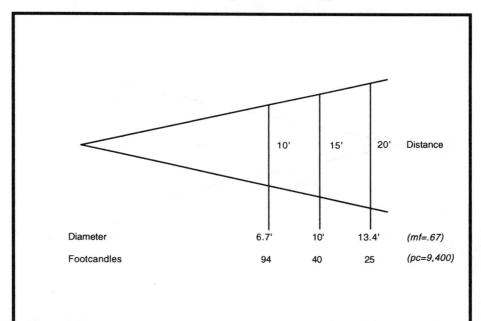

	10'	15'	20'	Distance
Diameter	6.7'	10'	13.4'	*(mf=.67)*
Footcandles	94	40	25	*(pc=9,400)*

KLIEGL BROS.
Model No. 1365W

BEAM ANGLE: (not specified in mfg. catalog)

FIELD ANGLE: 34° *

LAMP BASE TYPE: Medium Bipost

LAMP MOUNT: Off Axis

STANDARD LAMP: 500T14/8 - 500 W.**

OTHER LAMPS: 500T14/7 - 500 W.
750/T14 - 750 W.

INTEGRAL PATTERN SLOT: No

ACCESSORIES AVAILABLE:
Color Frame (6½" sq.)

WEIGHT: 10 lbs.

SIZE: 17" x 13½" x 17" (L x W x H)

PHOTOMETRICS CHART
(Performance data for this unit are measured using 500T14/8 - 500 W. lamp)

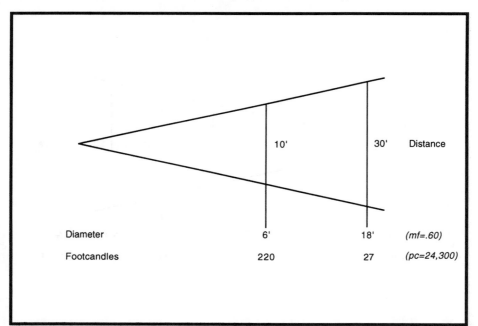

	Distance
	10' 30'
Diameter	6' 18' (mf=.60)
Footcandles	220 27 (pc=24,300)

* Beam and field angles not specified in manufacturer's catalog. This field angle is calculated from given distance and field diameter data. See Introduction for more information about calculating photometric data.
** This is an obsolete lamp but it is listed as the standard lamp since the manufacturer uses this lamp to describe the photometric performance. Correction factor *(cf)* information is not available for the other lamps because no initial lumens data is available for the obsolete lamp.

STRAND CENTURY
Model No. 2225*

BEAM ANGLE: 30°

FIELD ANGLE: 32°

LAMP BASE TYPE: Medium Prefocus

LAMP MOUNT: Off Axis

STANDARD LAMP: BTL - 500 W.

OTHER LAMPS: BTM - 500 W. *(cf=1.2)*

INTEGRAL PATTERN SLOT: Yes

WEIGHT: 7 lbs.

SIZE: 14" x 11¼" x 12¼" (L x W x H)

ACCESSORIES AVAILABLE:
Color Frame, Color Change Wheel, Drop-in Iris

PHOTOMETRICS CHART
(Performance data for this unit are measured using BTL - 500W. lamp.)

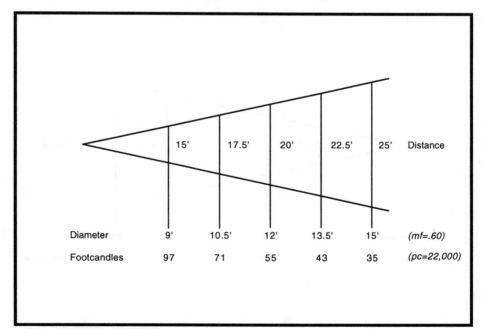

	15'	17.5'	20'	22.5'	25'	Distance
Diameter	9'	10.5'	12'	13.5'	15'	*(mf=.60)*
Footcandles	97	71	55	43	35	*(pc=22,000)*

* This unit also sold as PATT 23F.

STRAND CENTURY
Model No. 2235**

BEAM ANGLE: 36°

FIELD ANGLE: 36°

LAMP BASE TYPE: Medium Prefocus

LAMP MOUNT: Off Axis

STANDARD LAMP: BTL - 500 W.

OTHER LAMPS: BTM - 500 W. *(cf=1.2)*

INTEGRAL PATTERN SLOT: Yes

ACCESSORIES AVAILABLE:
Color Frame, Color Change Wheel, Drop-in Iris

WEIGHT: 7 lbs.

SIZE: 14" x 11¼" x 12¼" (L x W x H)

PHOTOMETRICS CHART
(Performance data for this unit are measured using BTL - 500W. lamp.)

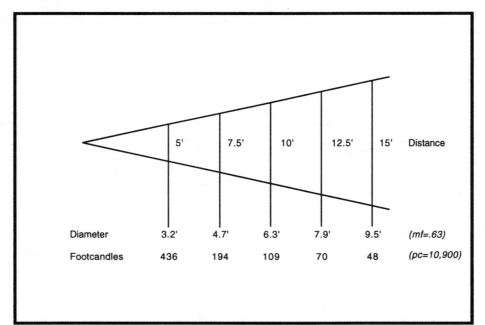

	5'	7.5'	10'	12.5'	15'	Distance
Diameter	3.2'	4.7'	6.3'	7.9'	9.5'	*(mf=.63)*
Footcandles	436	194	109	70	48	*(pc=10,900)*

* Unit uses two 3.5" x 5" lenses.

** This unit also sold as PATT 23W.

**TIMES SQUARE LIGHTING
Model No. Q3M**

BEAM ANGLE: (not specified in mfg. catalog)

FIELD ANGLE: 30°

LAMP BASE TYPE: Recessed Single-Contact

LAMP MOUNT: Off Axis

STANDARD LAMP: FAD - 650 W.

OTHER LAMPS: EHP - 300 W. *(cf=.30)*
EHR - 400 W. *(cf=.45)*

INTEGRAL PATTERN SLOT: Yes

WEIGHT: (not specified in manufacturer's catalog)

SIZE: (not specified in manufacturer's catalog)

ACCESSORIES AVAILABLE:
Color Frame, Pattern Holder, Non-motorized Color Wheel

PHOTOMETRICS CHART
(Information about lamp used to measure photometrics is not provided in manufacturer's catalog.)

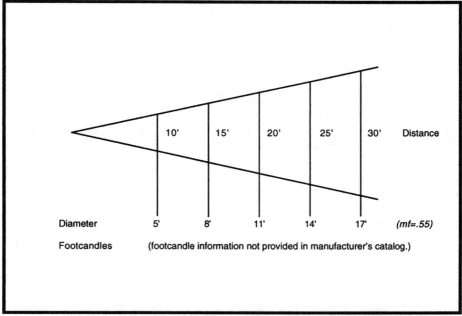

Distance	10'	15'	20'	25'	30'
Diameter	5'	8'	11'	14'	17'	*(mf=.55)*

Footcandles (footcandle information not provided in manufacturer's catalog.)

* Beam and field angles not specified in manufacturer's catalog. This field angle is calculated from given distance and field diameter data. See Introduction for more information about calculating photometric data.

ALTMAN STAGE LIGHTING
Model No. 3.5 Q 08

BEAM ANGLE: 21°

FIELD ANGLE: 28°

LAMP BASE TYPE: Medium 2-Pin

LAMP MOUNT: Axial

STANDARD LAMP: EHD - 500 W.

OTHER LAMPS: EHC - 500 W. *(cf=1.23)*

INTEGRAL PATTERN SLOT: Yes

ACCESSORIES AVAILABLE:
Color Frame, Pattern Holder, Snoot

WEIGHT: (not specified in manufacturer's catalog)

SIZE: 15½" x 9" x 13½" (L x W x H)

PHOTOMETRICS CHART
(Performance data for this unit are measured using EHD - 500 W. lamp.)

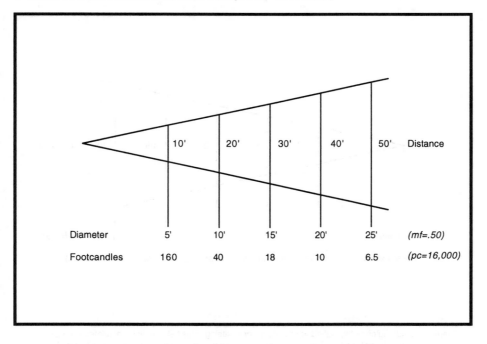

		10'	20'	30'	40'	50' Distance
Diameter		5'	10'	15'	20'	25' *(mf=.50)*
Footcandles		160	40	18	10	6.5 *(pc=16,000)*

ALTMAN STAGE LIGHTING
Model No. 3.5 Q 10

BEAM ANGLE: 21°

FIELD ANGLE: 23°

LAMP BASE TYPE: Medium 2-Pin

LAMP MOUNT: Axial

STANDARD LAMP: EHD - 500 W.

OTHER LAMPS: EHC - 500 W. *(cf=1.23)*

INTEGRAL PATTERN SLOT: Yes

ACCESSORIES AVAILABLE:
Color Frame, Pattern Holder, Snoot

WEIGHT: (not specified in manufacturer's catalog)

SIZE: 16½" x 9" x 13½" (L x W x H)

PHOTOMETRICS CHART
(Performance data for this unit are measured using EHD - 500 W. lamp.)

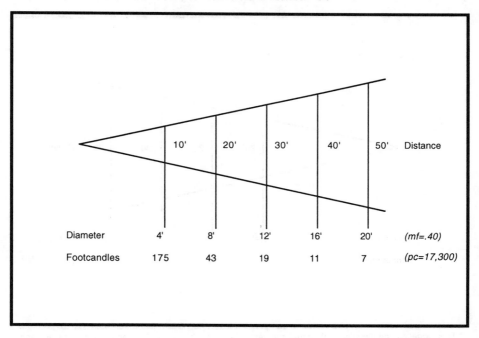

	10'	20'	30'	40'	50'	Distance
Diameter	4'	8'	12'	16'	20'	*(mf=.40)*
Footcandles	175	43	19	11	7	*(pc=17,300)*

BERKEY COLORTRAN
Model No. 212-012*

BEAM ANGLE: 14° (peak), 22° (flat)

FIELD ANGLE: 28° (peak and flat)

LAMP BASE TYPE: Medium 2-Pin

LAMP MOUNT: Axial

STANDARD LAMP: EHD - 500 W.

OTHER LAMPS: EHC - 500 W. *(cf=1.23)*

INTEGRAL PATTERN SLOT: Yes

ACCESSORIES AVAILABLE:
Color Frame, Pattern Holder

WEIGHT: 10.5 lbs.

SIZE: 15" x 9" x 19" (L x W x H)

PHOTOMETRICS CHART

(Performance data for this unit are measured at peak focus using EHD - 500 W. lamp.)

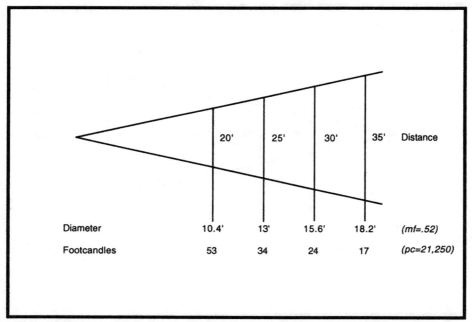

	20'	25'	30'	35'	Distance
Diameter	10.4'	13'	15.6'	18.2'	*(mf=.52)*
Footcandles	53	34	24	17	*(pc=21,250)*

* This model can be identified with model numbers -012, -015, -016, or -017 depending the wire leads and connectors supplied.

BERKEY COLORTRAN
Model No. 212-022*

BEAM ANGLE: 10° (peak), 16° (flat)

FIELD ANGLE: 22° (peak and flat)

LAMP BASE TYPE: Medium 2-Pin

LAMP MOUNT: Axial

STANDARD LAMP: EHD - 500 W.

OTHER LAMPS: EHC - 500 W. *(cf=1.23)*

INTEGRAL PATTERN SLOT: Yes

ACCESSORIES AVAILABLE:
Color Frame, Pattern Holder

WEIGHT: 10.5 lbs.

SIZE: 15" x 9" x 19" (L x W x H)

PHOTOMETRICS CHART
(Performance data for this unit are measured at peak focus using EHD - 500 W. lamp.)

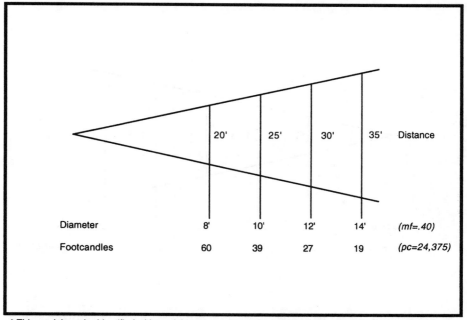

	20'	25'	30'	35'	Distance
Diameter	8'	10'	12'	14'	*(mf=.40)*
Footcandles	60	39	27	19	*(pc=24,375)*

* This model can be identified with model numbers -022, -025, -026, or -027 depending the wire leads and connectors supplied.

CENTURY LIGHTING
Model Nos. 1217 or 2125*

BEAM ANGLE: 18°

FIELD ANGLE: 24°

LAMP BASE TYPE: Recessed Single-Contact

LAMP MOUNT: Off Axis

STANDARD LAMP: EHR - 400 W.

OTHER LAMPS: EHP - 300 W. *(cf=.67)*
FDA - 400 W. *(cf=1.39)*

INTEGRAL PATTERN SLOT: Yes

ACCESSORIES AVAILABLE:
Color Frame, Pattern Holder, Snoot

WEIGHT: (not specified in manufacturer's catalog)

SIZE: 13⅞" x 9" x 17¼" (L x W x H)

PHOTOMETRICS CHART

(Performance data for this unit are measured using EHR - 400 W. lamp.)

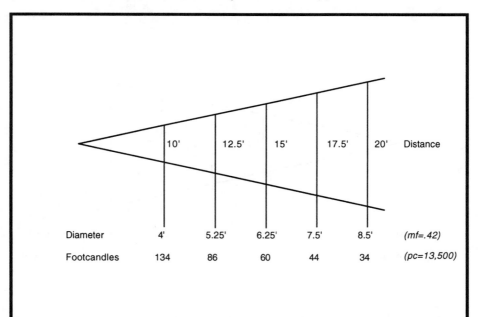

		10'	12.5'	15'	17.5'	20'	Distance
Diameter		4'	5.25'	6.25'	7.5'	8.5'	*(mf=.42)*
Footcandles		134	86	60	44	34	*(pc=13,500)*

* 2125 is the model number used by Century-Strand for this instrument beginning around 1970.

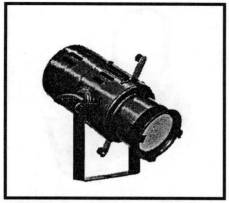

CENTURY LIGHTING
Model Nos. 1219 or 2135*

BEAM ANGLE: 12°

FIELD ANGLE: 20°

LAMP BASE TYPE: Recessed Single-Contact

LAMP MOUNT: Off Axis

STANDARD LAMP: EHR - 400 W.

OTHER LAMPS: EHP - 300 W. *(cf=.67)*
FDA - 400 W. *(cf=1.39)*

INTEGRAL PATTERN SLOT: Yes

ACCESSORIES AVAILABLE:
Color Frame, Pattern Holder, Snoot

WEIGHT: (not specified in manufacturer's catalog)

SIZE: 14⅞" x 9" x 17¼" (L x W x H)

PHOTOMETRICS CHART

(Performance data for this unit are measured using EHR - 400 W. lamp.)

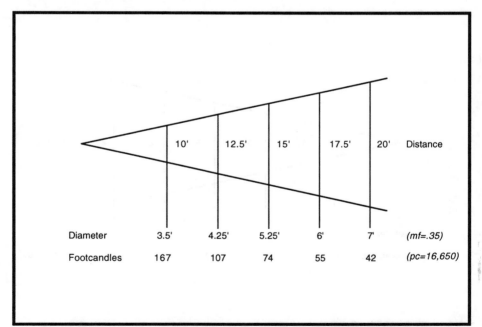

	10'	12.5'	15'	17.5'	20'	Distance
Diameter	3.5'	4.25'	5.25'	6'	7'	*(mf=.35)*
Footcandles	167	107	74	55	42	*(pc=16,650)*

* 2135 is the model number used by Century Strand for this instrument begining around 1970.

141

ELECTRO CONTROLS
Model No. 7345AF 30*

BEAM ANGLE: 16°

FIELD ANGLE: 29.8°

LAMP BASE TYPE: Mini-Can Screw

LAMP MOUNT: Axial

STANDARD LAMP: EVR - 500 W.

OTHER LAMPS: EHT - 250 W. *(cf=.48)*
                EHV -325 W. *(cf=.75)*

INTEGRAL PATTERN SLOT: Yes

ACCESSORIES AVAILABLE:
Color Frame, Pattern Holder, Snoot

WEIGHT: 7⅜ lbs.

SIZE: 16¾" x 6⅜" x 19¹⁵⁄₁₆" (L x W x H)

PHOTOMETRICS CHART
(Performance data for this unit are measured using EVR - 500 W. lamp.)

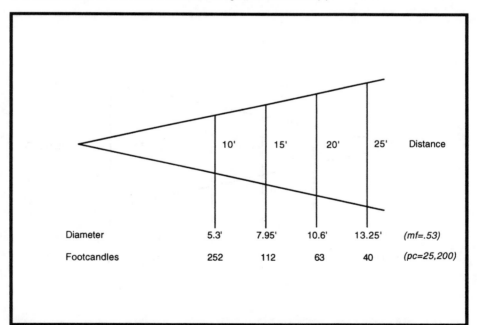

	10'	15'	20'	25'	Distance
Diameter	5.3'	7.95'	10.6'	13.25'	*(mf=.53)*
Footcandles	252	112	63	40	*(pc=25,200)*

* Lens systems in the 7345AF series are interchangeable. See Index for page numbers of the other versions of model 7345AF.

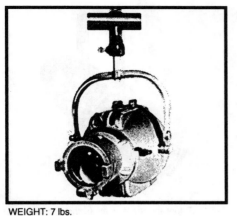

STRAND CENTURY
Model No. 2215*

BEAM ANGLE: 20°

FIELD ANGLE: 20°

LAMP BASE TYPE: Medium Prefocus

LAMP MOUNT: Off Axis

STANDARD LAMP: BTL - 500 W.

OTHER LAMPS: BTM - 500 W. *(cf=1.2)*

INTEGRAL PATTERN SLOT: Yes

WEIGHT: 7 lbs.

SIZE: 14" x 11¼" x 12¼" (L x W x H)

ACCESSORIES AVAILABLE:
Color Frame, Color Change Wheel, Drop-in Iris

PHOTOMETRICS CHART
(Performance data for this unit are measured using BTL - 500W. lamp.)

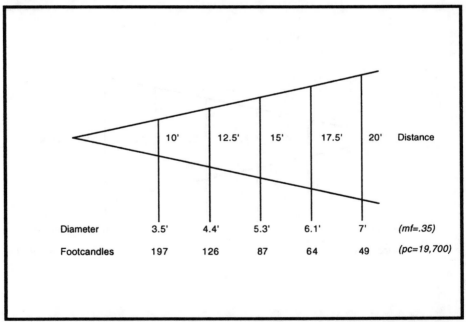

	10'	12.5'	15'	17.5'	20'	Distance
Diameter	3.5'	4.4'	5.3'	6.1'	7'	*(mf=.35)*
Footcandles	197	126	87	64	49	*(pc=19,700)*

* This unit also sold as PATT 23, and in the 1990s, Strand Lighting used model number 2215 for a 15° (6 x 12) ERS. See page 262.

ALTMAN STAGE LIGHTING
Model No. 3.5 Q 12

BEAM ANGLE: 17°

FIELD ANGLE: 18°

LAMP BASE TYPE: Medium 2-Pin

LAMP MOUNT: Axial

STANDARD LAMP: EHD - 500 W.

OTHER LAMPS: EHC - 500 W. *(cf=1.23)*

INTEGRAL PATTERN SLOT: Yes

ACCESSORIES AVAILABLE:
Color Frame, Pattern Holder, Snoot

WEIGHT: (not specified in manufacturer's catalog)

SIZE: 16¾" x 9" x 13½" (L x W x H)

PHOTOMETRICS CHART
(Performance data for this unit are measured using EHD - 500 W. lamp.)

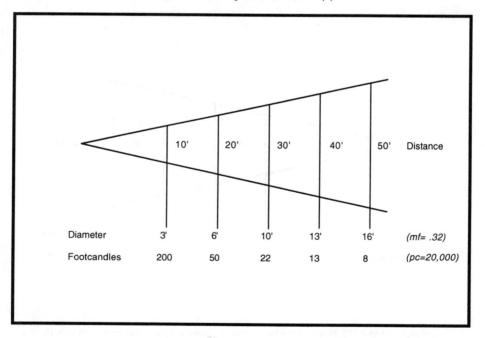

	10'	20'	30'	40'	50'	Distance
Diameter	3'	6'	10'	13'	16'	*(mf= .32)*
Footcandles	200	50	22	13	8	*(pc=20,000)*

KLIEGL BROS.
Model No. 1343

BEAM ANGLE: (not specified in mfg. catalog)

FIELD ANGLE: 19°

LAMP BASE TYPE: Recessed Single-Contact

LAMP MOUNT: Off Axis

STANDARD LAMP: EHR - 400 W.

OTHER LAMPS: EHP - 300 W. *(cf=.67)*
FDA - 400 W. *(cf=1.39)*

INTEGRAL PATTERN SLOT: Yes

ACCESSORIES AVAILABLE:
Color Frame, Pattern Holder, Snoot

WEIGHT: 9 lbs.

SIZE: 17½" x 9⅞" x 16⅜" (L x W x H)

PHOTOMETRICS CHART
(Performance data for this unit are measured using EHR - 400 W. lamp.)

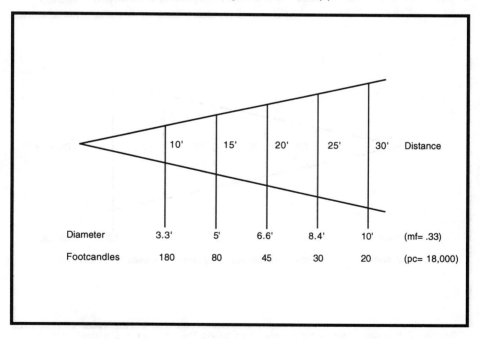

Distance	10'	15'	20'	25'	30'	
Diameter	3.3'	5'	6.6'	8.4'	10'	(mf= .33)
Footcandles	180	80	45	30	20	(pc= 18,000)

TIMES SQUARE LIGHTING
Model No. Q3N

BEAM ANGLE: (not specified in mfg. catalog)

FIELD ANGLE: 13° *

LAMP BASE TYPE: Recessed Single-Contact

LAMP MOUNT: Off Axis

STANDARD LAMP: FAD - 650 W.

OTHER LAMPS: EHP - 300 W. *(cf=.30)*
EHR - 400 W. *(cf=.45)*

INTEGRAL PATTERN SLOT: Yes

ACCESSORIES AVAILABLE:
Color Frame, Pattern Holder, Non-motorized Color
Wheel

WEIGHT: (not specified in manufacturer's catalog)

SIZE: (not specified in manufacturer's catalog)

PHOTOMETRICS CHART

(Information about lamp used to measure photometrics is not provided in manufacturer's catalog.)

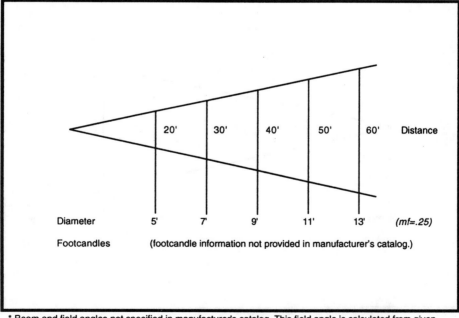

| | 20' | 30' | 40' | 50' | 60' | Distance |

Diameter   5'   7'   9'   11'   13'   *(mf=.25)*

Footcandles   (footcandle information not provided in manufacturer's catalog.)

* Beam and field angles not specified in manufacturer's catalog. This field angle is calculated from given
distance and field diameter data. See Introduction for more information about calculating photometric data.

ALTMAN STAGE LIGHTING
Model No. 1KL6-40

BEAM ANGLE: 14° (peak), 27° (flat)

FIELD ANGLE: 40° (peak and flat)

LAMP BASE TYPE: Medium 2-Pin

LAMP MOUNT: Axial

STANDARD LAMP: FEL - 1 KW.

OTHER LAMPS: EHD - 500 W. *(cf=.39)*
EHG - 750 W. *(cf=.56)*

INTEGRAL PATTERN SLOT: Yes

WEIGHT: (not specified in manufacturer's catalog)

SIZE: 22⅞" x 17" x 19" (L x W x H)

ACCESSORIES AVAILABLE:
Color Frame, Pattern Holder, Snoot, Motorized &
Non-motorized Color Wheels, Iris (as an option)

PHOTOMETRICS CHART
(Performance data for this unit are measured at peak focus using FEL - 1 KW. lamp.)

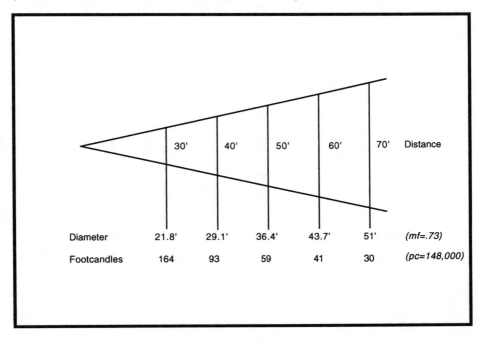

	30'	40'	50'	60'	70'	Distance
Diameter	21.8'	29.1'	36.4'	43.7'	51'	*(mf=.73)*
Footcandles	164	93	59	41	30	*(pc=148,000)*

ALTMAN
Model No. S6-40, "Shakespeare"

BEAM ANGLE: 20° (cosine)

FIELD ANGLE: 38°

LAMP BASE TYPE: Medium 2-Pin

LAMP MOUNT: Axial

STANDARD LAMP: HX600 - 575 W.

OTHER LAMPS: EHG - 750 W. *(cf=.93)*
EHD - 500 W. *(cf-.63)*

INTEGRAL PATTERN SLOT: Yes

ACCESSORIES AVAILABLE:
Color Frame (6¼" sq.), Snoot, Pattern Holder

WEIGHT: 18 lbs.

SIZE: 28½" x 12" x 13¾" (L x W x H)

PHOTOMETRICS CHART
(Performance data for this unit are measured at cosine focus using HX600 - 575 W. lamp.)

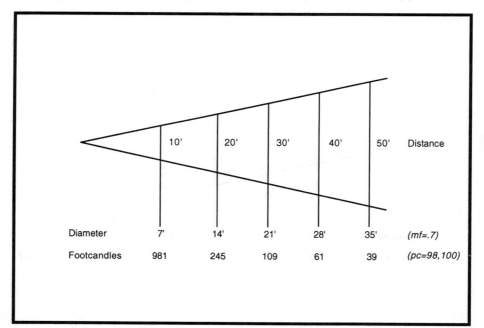

	10'	20'	30'	40'	50'	Distance
Diameter	7'	14'	21'	28'	35'	*(mf=.7)*
Footcandles	981	245	109	61	39	*(pc=98,100)*

BERKEY COLORTRAN
Model No. 212-042*

BEAM ANGLE: 23° (peak and flat)
FIELD ANGLE: 40° (peak and flat)
LAMP BASE TYPE: Medium 2-Pin
LAMP MOUNT: Axial
STANDARD LAMP: EHG - 750 W.
OTHER LAMPS: EHD - 500 W. *(cf=.66)*
INTEGRAL PATTERN SLOT: Yes

WEIGHT: 16 lbs.
SIZE: 20⅝" x 13½" x 23¾" (L x W x H)

ACCESSORIES AVAILABLE:
Color Frame, Pattern Holder

PHOTOMETRICS CHART
(Performance data for this unit are measured at peak focus using EHG - 750 W. lamp.)

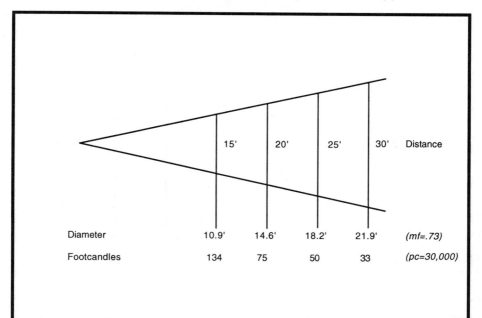

	15'	20'	25'	30'	Distance
Diameter	10.9'	14.6'	18.2'	21.9'	*(mf=.73)*
Footcandles	134	75	50	33	*(pc=30,000)*

* This model can be identified with model numbers -042, -045, -046, or -047 depending on the wire leads and connectors supplied.

BERKEY COLORTRAN
Model No. 213-052*

BEAM ANGLE: 15°

FIELD ANGLE: 40°

LAMP BASE TYPE: Medium 2-Pin

LAMP MOUNT: Axial

STANDARD LAMP: FEL - 1 KW

OTHER LAMPS: EHD - 500 W. *(cf=.38)*
EHG -750 W. *(cf=.56)*

INTEGRAL PATTERN SLOT: Yes

ACCESSORIES AVAILABLE:
Color Frame (7½" sq.), Pattern Holder, Iris Kit

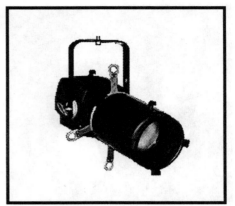

WEIGHT: 20.1 lbs.

SIZE: 27½" x 15" x 24½" (L x W x H)

PHOTOMETRICS CHART
(Performance data for this unit are measured at peak focus using FEL - 1 KW. lamp.)

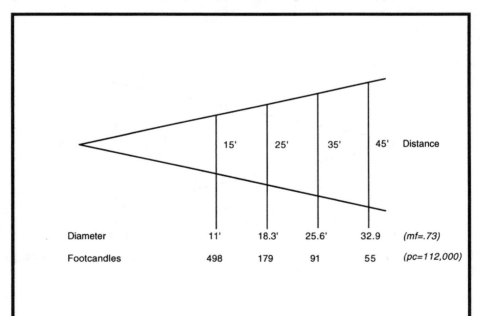

	15'	25'	35'	45'	Distance
Diameter	11'	18.3'	25.6'	32.9	*(mf=.73)*
Footcandles	498	179	91	55	*(pc=112,000)*

* This model can be identified with model numbers -052, -054, -055, or -056 depending on the wire leads and connectors supplied.

CENTURY LIGHTING
Model No. 1590*

BEAM ANGLE: (not specified in mfg. catalog)

FIELD ANGLE: 43° **

LAMP BASE TYPE: (not specified in mfg. catalog)

LAMP MOUNT: Off Axis

STANDARD LAMP: 750T12/8 - 750 W.

OTHER LAMPS: (none specified in mfg. catalog)

INTEGRAL PATTERN SLOT: Yes

WEIGHT: (not specified in manufacturer's catalog)

SIZE: (not specified in manufacturer's catalog)

ACCESSORIES AVAILABLE:
Color Frame (7½" sq.), Motorized and Non-motorized Color Wheels

PHOTOMETRICS CHART

(Information about lamp used to measure performance data is not provided in manufacturer's catalog.)

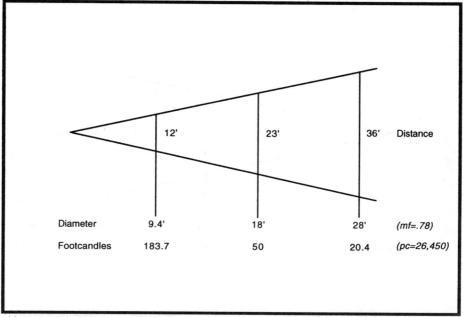

	12'	23'	36'	Distance
Diameter	9.4'	18'	28'	(mf=.78)
Footcandles	183.7	50	20.4	(pc=26,450)

* Instruments in this series are equipped as follow - 1590 has four shutters and template shot, 1591 has four shutters, and 1592 has an iris and handle.
** Beam and field angles not specified in manufacturer's catalog. This field angle is calculated from given distance and field diameter data. See Introduction for more information about calculating photometric data.

CENTURY STRAND
Model Nos. 2321 (w/ shutters)
2322 (w/ iris, handle)*

BEAM ANGLE: 24°

FIELD ANGLE: 40°

LAMP BASE TYPE: Medium Prefocus

LAMP MOUNT: Off Axis

STANDARD LAMP: EGG - 750 W.

OTHER LAMPS: 500T12/8 - 500 W. *(cf=.57)*
EGD - 750 W. *(cf=.83)*

INTEGRAL PATTERN SLOT: Yes (2321 only)

ACCESSORIES AVAILABLE:
Color Frame (7½" sq.), Pattern Holder, Snoot

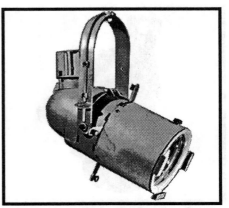

WEIGHT: (not specified in manufacturer's catalog)

SIZE: 19¾" x 11⅜" x 24⅛" (L x W x H)

PHOTOMETRICS CHART
(Performance data for this unit are measured using EGG - 750 W. lamp.)

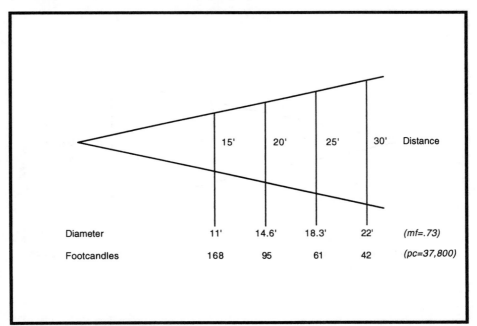

	15'	20'	25'	30'	Distance
Diameter	11'	14.6'	18.3'	22'	*(mf=.73)*
Footcandles	168	95	61	42	*(pc=37,800)*

* This unit was also sold as Century 1490.

WEIGHT: 17.75 lbs.

SIZE: 19⁹⁄₁₆" x 15½" x 26" (L x W x H)

ELECTRO CONTROLS
Model No. 7363AF 45*

BEAM ANGLE: 22°

FIELD ANGLE: 45°

LAMP BASE TYPE: Medium Prefocus

LAMP MOUNT: Axial

STANDARD LAMP: EGG - 750 W.

OTHER LAMPS: EGF - 750 W. *(cf=1.30)*
　　　　　　　　EGJ - 1 KW. *(cf=1.75)*

INTEGRAL PATTERN SLOT: Yes

ACCESSORIES AVAILABLE:
Color Frame (7½" sq.), Pattern Holder, Snoot, Iris (as an option)

PHOTOMETRICS CHART

(Performance data for this unit are measured using EGG - 750 W. lamp.)

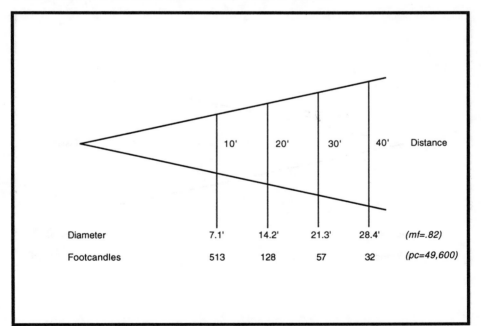

	10'	20'	30'	40'	Distance
Diameter	7.1'	14.2'	21.3'	28.4'	*(mf=.82)*
Footcandles	513	128	57	32	*(pc=49,600)*

* If unit includes optional iris the model number is 7363AI 45. Lens systems in the 7363A series are interchangeable. See Index for page numbers of the other versions of model 7363A.

153

KLIEGL BROS.
Model No. 1165W

BEAM ANGLE: (not specified in mfg. catalog)

FIELD ANGLE: 44° *

LAMP BASE TYPE: Medium Bipost

LAMP MOUNT: Off Axis

STANDARD LAMP: 500T14/8 - 500 W. **

OTHER LAMPS: 500T14/7 - 500 W. *(cf=1.0)*
750/T14 - 750 W. *(cf=1.89)*

INTEGRAL PATTERN SLOT: No

ACCESSORIES AVAILABLE:
Color Frame

WEIGHT: 14 lbs.

SIZE: (not specified in manufacturer's catalog)

PHOTOMETRICS CHART

(Performance data for this unit are measured using 500T14/8 - 500 W. lamp)

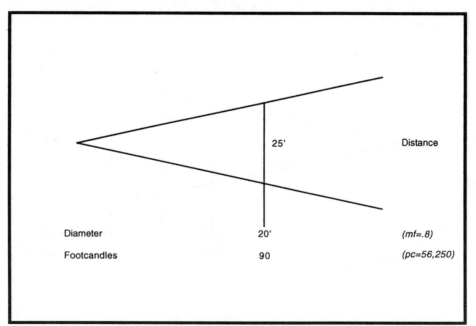

	25'	Distance
Diameter	20'	*(mf=.8)*
Footcandles	90	*(pc=56,250)*

* Beam and field angles not specified in manufacturer's catalog. This field angle is calculated from given distance and field diameter data. See Introduction for more information about calculating photometric data.
** This is an obsolete lamp but it is listed as the standard lamp since the manufacturer uses this lamp to describe the photometric performance. Correction factor *(cf)* information is not available for the other lamps because no initial lumens data is available for the obsolete lamp.

KLIEGL BROS.
Model No. 1554*

BEAM ANGLE: 15°

FIELD ANGLE: 40° (cosine)

LAMP BASE TYPE: Medium 2-Pin

LAMP MOUNT: Axial

STANDARD LAMP: FEL - 1 KW.

OTHER LAMPS: EHD - 500 W. *(cf=.38)*
EHG - 750 W. *(cf=.56)*

INTEGRAL PATTERN SLOT: Yes

ACCESSORIES AVAILABLE:
Color Frame (7½" sq.), Pattern Holder, Drop-in Iris

WEIGHT: 22 lbs.

SIZE: 29" x 16" x 18" (L x W x H)

PHOTOMETRICS CHART
(Performance data for this unit are measured at cosine focus using FEL - 1 KW. lamp.)

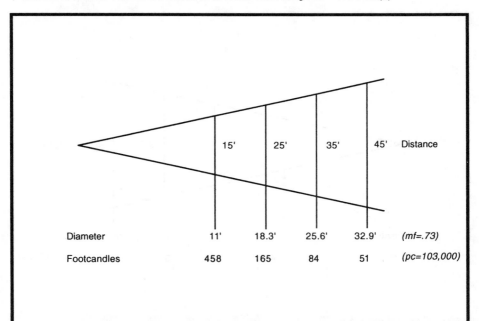

	15'	25'	35'	45'	Distance
Diameter	11'	18.3'	25.6'	32.9'	*(mf=.73)*
Footcandles	458	165	84	51	*(pc=103,000)*

* The lenses in this series of ellipsoidal spotlights are interchangeable on the same lamp housing. Lens barrels are color-coded to indentify the field angle. This 40° model is color-coded with yellow.

LEE COLORTRAN
Model No. 650-022*

BEAM ANGLE: 15° (peak), 26.7° (cosine)

FIELD ANGLE: 40°

LAMP BASE TYPE: Medium 2-Pin

LAMP MOUNT: Axial

STANDARD LAMP: FEL - 1KW

OTHER LAMPS: EHD - 500 W. *(cf=.38)*
EHG - 750 W. *(cf=.56)*

INTEGRAL PATTERN SLOT: Yes

ACCESSORIES AVAILABLE:
Color Frame (7½" sq.), Pattern Holder, Iris Kit

WEIGHT: 20.3 lbs.
SIZE: 28" x 15" x 25½" (L x W x H)

PHOTOMETRICS CHART

(Performance data for this unit are measured at peak focus using FEL - 1 KW. lamp.)

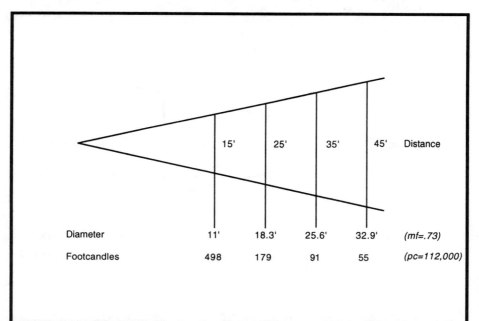

	15'	25'	35'	45'	Distance
Diameter	11'	18.3'	25.6'	32.9'	*(mf=.73)*
Footcandles	498	179	91	55	*(pc=112,000)*

* This model can be identified with model numbers -022, -024, -025, or -026 depending on the wire leads and connectors supplied.

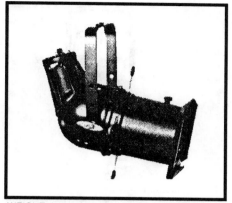

LIGHTING & ELECTRONICS
Model No. 61-06

BEAM ANGLE: 32°

FIELD ANGLE: 40°

LAMP BASE TYPE: Medium Prefocus

LAMP MOUNT: Off Axis

STANDARD LAMP: DNT - 750 W.

OTHER LAMPS: DEB - 500 W. *(cf=.53)*

INTEGRAL PATTERN SLOT: Yes

WEIGHT: (not specified in manufacturer's catalog)

SIZE: 17¼" (L)

ACCESSORIES AVAILABLE:
Color Frame (7½" sq.)

PHOTOMETRICS CHART
(Performance data for this unit are measured using DNT - 750 W. lamp.)

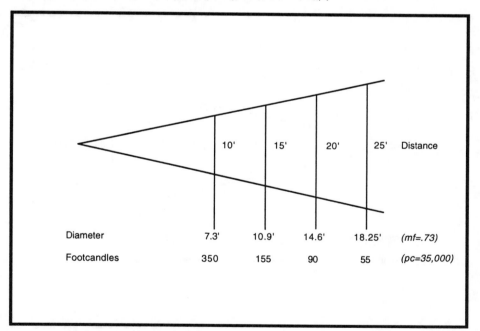

	10'	15'	20'	25'	Distance
Diameter	7.3'	10.9'	14.6'	18.25'	*(mf=.73)*
Footcandles	350	155	90	55	*(pc=35,000)*

157

TIMES SQUARE LIGHTING
Model No. 6E6x9

BEAM ANGLE: (not specified in mfg. catalog)

FIELD ANGLE: 42° *

LAMP BASE TYPE: Medium Prefocus

LAMP MOUNT: Off Axis

STANDARD LAMP: EGG - 750 W.

OTHER LAMPS: EGE - 500 W. *(cf=.66)*
EGJ - 1 KW. *(cf=1.68)*

INTEGRAL PATTERN SLOT: Yes

ACCESSORIES AVAILABLE:
Color Frame (7½" sq.), Color Wheel, Pattern Holder

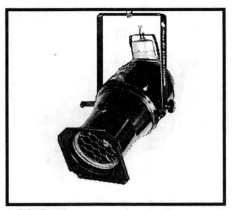

WEIGHT: 12 lbs.
SIZE: 19" x 9½" x 16½" (L x W x H)

PHOTOMETRICS CHART

(Information not provided in manufacturer's catalog about lamp used to measure performance data.)

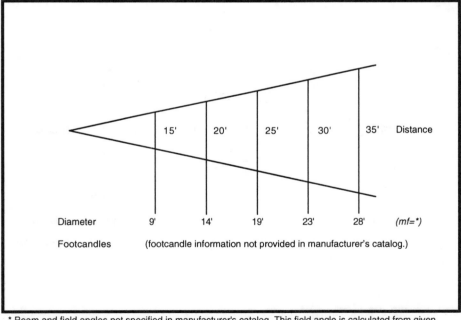

| | 15' | 20' | 25' | 30' | 35' | Distance |

| Diameter | 9' | 14' | 19' | 23' | 28' | *(mf=*)* |

Footcandles   (footcandle information not provided in manufacturer's catalog.)

* Beam and field angles not specified in manufacturer's catalog. This field angle is calculated from given distance and field diameter data. See Introduction for more information about calculating photometric data. Also note, the field diameter data is not consistent. At the given distances, the field angle which is described by the diameter information varies from 34° at 9' to 44° at 28'. Because of this inconsistent data, no diameter multiplying factor *(mf)* is given.

TIMES SQUARE LIGHTING
Model No. Q6E6x9

BEAM ANGLE: (not specified in mfg. catalog.)

FIELD ANGLE: 44° *

LAMP BASE TYPE: Medium 2-Pin

LAMP MOUNT: Axial

STANDARD LAMP: EHG - 750 W.

OTHER LAMPS: EHD - 500 W. *(cf=.68)*
FEL - 1 KW. *(cf=1.79)*

INTEGRAL PATTERN SLOT: Yes

ACCESSORIES AVAILABLE:
Color Frame (7½" sq.), Color Wheel, Pattern Holder

WEIGHT: 12 lbs.

SIZE: 20" x 9½" (L x W)

PHOTOMETRICS CHART

(Information not provided in manufacturer's catalog about lamp used to measure performance data.)

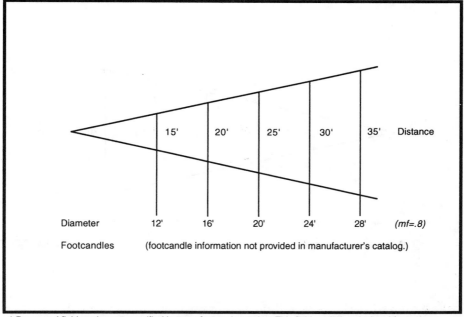

| | 15' | 20' | 25' | 30' | 35' | Distance |

Diameter 12' 16' 20' 24' 28' (mf=.8)

Footcandles (footcandle information not provided in manufacturer's catalog.)

* Beam and field angles not specified in manufacturer's catalog. This field angle is calculated from given distance and field diameter data. See Introduction for more information about calculating photometric data.

ALTMAN STAGE LIGHTING
Model No. 1KL6-30

BEAM ANGLE: 11° (peak), 20° (flat field)

FIELD ANGLE: 30°

LAMP BASE TYPE: Medium 2-Pin

LAMP MOUNT: Axial

STANDARD LAMP: FEL - 1 KW.

OTHER LAMPS: EHD - 500 W. *(cf=.38)*
EHG - 750 W. *(cf=.56)*

INTEGRAL PATTERN SLOT: Yes

ACCESSORIES AVAILABLE:
Color Frame, Pattern Holder, Snoot, Motorized &
Non-motorized Color Wheels, Iris

WEIGHT: (not specified in manufacturer's catalog)

SIZE: 20⅜" x 17" x 19" (L x W x H)

PHOTOMETRICS CHART
(Performance data for this unit are measured using FEL - 1 KW. lamp.)

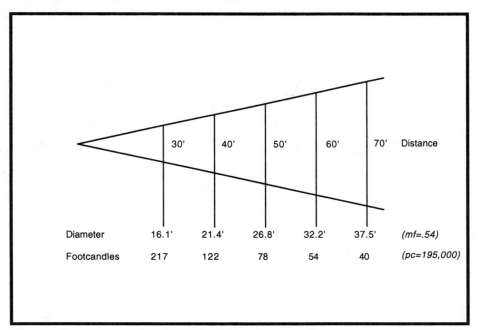

	30'	40'	50'	60'	70'	Distance
Diameter	16.1'	21.4'	26.8'	32.2'	37.5'	*(mf=.54)*
Footcandles	217	122	78	54	40	*(pc=195,000)*

ALTMAN STAGE LIGHTING
Model No. 360 6x9

BEAM ANGLE: 15°

FIELD ANGLE: 35°

LAMP BASE TYPE: Medium Prefocus

LAMP MOUNT: Off Axis

STANDARD LAMP: DNT - 750 W.

OTHER LAMPS: DNS - 500 W. *(cf=.65)*

INTEGRAL PATTERN SLOT: Yes

WEIGHT: (not specified in manufacturer's catalog)

SIZE: 17⅝" x 13¼" x 16¾" (L x W x H)

ACCESSORIES AVAILABLE:
Color Frame (7½" sq.), Pattern Holder, Snoot,
Motorized & Non-motorized Color Wheels, Iris (as
an option)

PHOTOMETRICS CHART
(Performance data for this unit are measured using DNT - 750 W. lamp.)

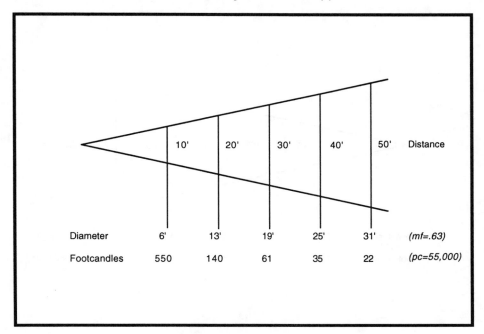

	10'	20'	30'	40'	50'	Distance
Diameter	6'	13'	19'	25'	31'	(mf=.63)
Footcandles	550	140	61	35	22	(pc=55,000)

161

ALTMAN STAGE LIGHTING
Model No. 360Q 6x9

BEAM ANGLE: 16°

FIELD ANGLE: 37°

LAMP BASE TYPE: Medium 2-Pin

LAMP MOUNT: Axial

STANDARD LAMP: EHF - 750 W.

OTHER LAMPS: EHD - 500 W. *(cf=.5)*
EHG - 750 W. *(cf=.75)*

INTEGRAL PATTERN SLOT: Yes

ACCESSORIES AVAILABLE:
Color Frame (7½" sq.), Pattern Holder, Snoot,
Motorized & Non-motorized Color Wheels, Iris (as
an option)

WEIGHT: (not specified in manufacturer's catalog)

SIZE: 18⅛" x 13" x 16¼" (L x W x H)

PHOTOMETRICS CHART
(Performance data for this unit are measured using EHF - 750 W. lamp.)

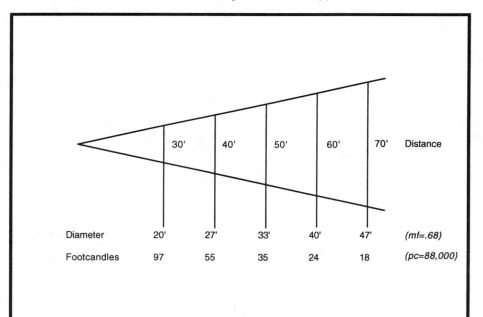

	30'	40'	50'	60'	70'	Distance
Diameter	20'	27'	33'	40'	47'	*(mf=.68)*
Footcandles	97	55	35	24	18	*(pc=88,000)*

ALTMAN
Model No. S6-30, "Shakespeare"

BEAM ANGLE: 13° (cosine)

FIELD ANGLE: 28°

LAMP BASE TYPE: Medium 2-Pin

LAMP MOUNT: Axial

STANDARD LAMP: HX600 - 575 W.

OTHER LAMPS: EHG - 750 W. *(cf=.93)*
EHD - 500 W. *(cf-.63)*

INTEGRAL PATTERN SLOT: Yes

WEIGHT: 18 lbs.

SIZE: 28½" x 12" x 13¾" (L x W x H)

ACCESSORIES AVAILABLE:
Color Frame (6¼" sq.), Snoot, Pattern Holder

PHOTOMETRICS CHART

(Performance data for this unit are measured at cosine focus using HX600 - 575 W. lamp.)

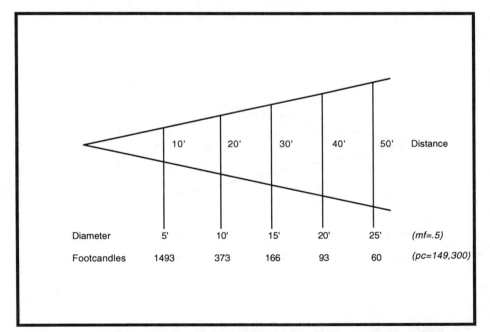

	10'	20'	30'	40'	50'	Distance
Diameter	5'	10'	15'	20'	25'	*(mf=.5)*
Footcandles	1493	373	166	93	60	*(pc=149,300)*

BERKEY COLORTRAN
Model No. 213-062*

BEAM ANGLE: 11°

FIELD ANGLE: 30°

LAMP BASE TYPE: Medium 2-Pin

LAMP MOUNT: Axial

STANDARD LAMP: FEL - 1 KW.

OTHER LAMPS: EHD - 500 W. *(cf=.38)*
EHG - 750 W. *(cf=.56)*

INTEGRAL PATTERN SLOT: Yes

ACCESSORIES AVAILABLE:
Color Frame (7½" sq.), Pattern Holder, Iris Kit

WEIGHT: 20.1 lbs.

SIZE: 27½" x 15" x 24½" (L x W x H)

PHOTOMETRICS CHART
(Performance data for this unit are measured at peak focus using FEL - 1 KW. lamp.)

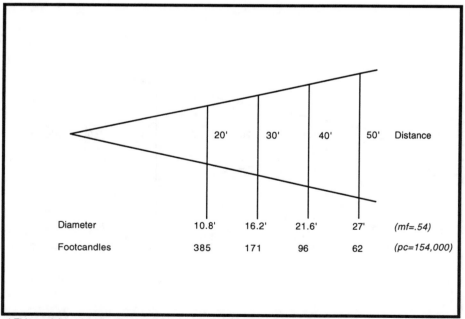

	20'	30'	40'	50'	Distance
Diameter	10.8'	16.2'	21.6'	27'	*(mf=.54)*
Footcandles	385	171	96	62	*(pc=154,000)*

* This model can be identified with model numbers -062, -064, -065, or -066 depending on the wire leads and connectors supplied.

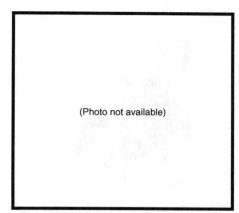

(Photo not available)

WEIGHT: (not specified in manufacturer's catalog)

SIZE: 15¼" x 14½" (L x H)

CAPITOL STAGE LIGHTING
Model No. 650*

BEAM ANGLE: (not specified in mfg. catalog)

FIELD ANGLE: 38° **

LAMP BASE TYPE: Medium Prefocus

LAMP MOUNT: Off Axis

STANDARD LAMP: 750 T12/8 - 750 W.***

OTHER LAMPS: 250T12/8 - 250 W.
500T12/8 - 500 W.

INTEGRAL PATTERN SLOT: No

ACCESSORIES AVAILABLE:
Color Frame (7½" sq.), Pattern Holder, Snoot, Motorized & Non-motorized Color Wheels, Iris (as an option)

PHOTOMETRICS CHART

(Performance data for this unit are measured using 750 T12/8 - 750 W. lamp.)

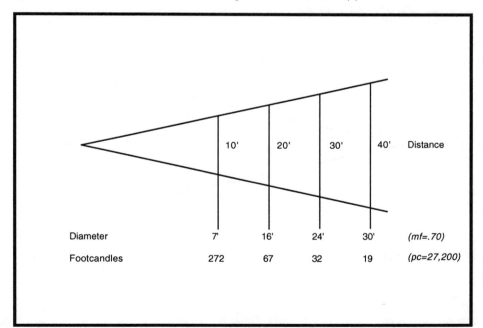

	10'	20'	30'	40'	Distance
Diameter	7'	16'	24'	30'	(mf=.70)
Footcandles	272	67	32	19	(pc=27,200)

* Instruments in this series are equipped as follows - 651 has one 6x9 step lens, 652 has an iris, 653 adds a follow handle, and 654 has a pattern slot.

** Beam and field angles not specified in manufacturer's catalog. This field angle is calculated from given distance and field diameter data. See Introduction for more information about calculating photometric data.

***This is an obsolete lamp but it is listed as the standard lamp since the manufacturer uses this lamp to describe the photometric performance. Correction factor (cf) information is not available for the other lamps because no initial lumens data is available for the obsolete lamp.

CENTURY LIGHTING
Model Nos. 1533 (w/ shutters)
1534 (w/ iris, handle)

BEAM ANGLE: (not specified in manufacturer's catalog)

FIELD ANGLE: 32°

LAMP BASE TYPE: Medium Prefocus

LAMP MOUNT: Off Axis

STANDARD LAMP: DNT - 750 W. *

OTHER LAMPS: DEB - 500 W.

INTEGRAL PATTERN SLOT: No

ACCESSORIES AVAILABLE:
Color Frame (7½" sq.)

WEIGHT: (not specified in manufacturer's catalog)

SIZE: (not specified in manufacturer's catalog)

PHOTOMETRICS CHART
(Performance data for this unit are measured using DNT 750 W. lamp.)

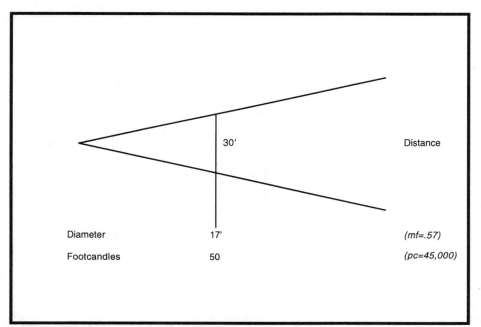

	30'	Distance
Diameter	17'	*(mf=.57)*
Footcandles	50	*(pc=45,000)*

* This is an obsolete lamp but it is listed as the standard lamp since the manufacturer uses this lamp to describe the photometric performance. Correction factor *(cf)* information is not available for the other lamps because no initial lumens data is available for the obsolete lamp.

CENTURY LIGHTING
Model Nos. 1537 (w/ shutters)
1538 (w/ iris, handle)

BEAM ANGLE: (not specified in manufacturer's catalog)

FIELD ANGLE: 36°

LAMP BASE TYPE: Medium Prefocus

LAMP MOUNT: Off Axis

STANDARD LAMP: DNT - 750 W. *

OTHER LAMPS: DEB - 500 W.

INTEGRAL PATTERN SLOT: No

ACCESSORIES AVAILABLE:
Color Frame (7½" sq.)

WEIGHT: (not specified in manufacturer's catalog)

SIZE: (not specified in manufacturer's catalog)

PHOTOMETRICS CHART
(Performance data for this unit are measured using DNT 750 W. lamp.)

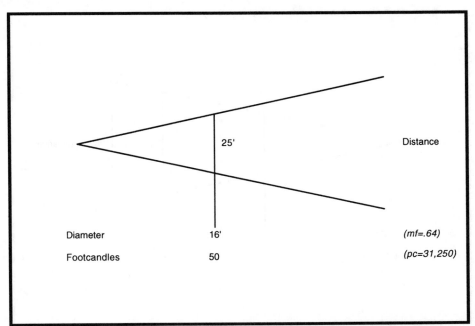

	25'	Distance
Diameter	16'	(mf=.64)
Footcandles	50	(pc=31,250)

* This is an obsolete lamp but it is listed as the standard lamp since the manufacturer uses this lamp to describe the photometric performance. Correction factor *(cf)* information is not available for the other lamps because no initial lumens data is available for the obsolete lamp.

ELECTRO CONTROLS
Model Nos.3201 (w/ shutters)
3202 (w/ iris)

BEAM ANGLE: (not specified in mfg. catalog)

FIELD ANGLE: 36° *

LAMP BASE TYPE: Extended Mog. End Prong

LAMP MOUNT: Off Axis

STANDARD LAMP: 500 PAR 64/NSP - 500 W.**

OTHER LAMPS: Q1000PAR64/NSP-1 KW. *(cf=1.84)*

INTEGRAL PATTERN SLOT: Yes (3201 only)

ACCESSORIES AVAILABLE:
Color Frame (9" sq.), 2-way Barn Door, Iris Kit,
Shutter Kit, Snoot

WEIGHT: 24 lbs.
SIZE: 25" (L)

PHOTOMETRICS CHART

(Performance data for this unit are measured using 500 PAR 64/NSP - 500 W. lamp.)

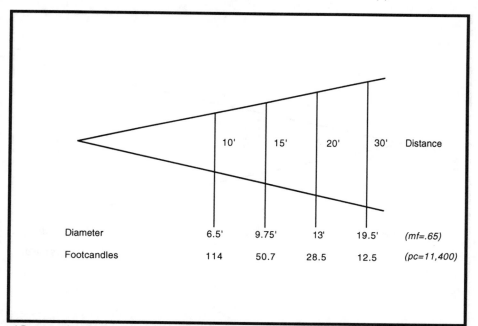

	10'	15'	20'	30'	Distance
Diameter	6.5'	9.75'	13'	19.5'	*(mf=.65)*
Footcandles	114	50.7	28.5	12.5	*(pc=11,400)*

* Beam and field angles not specified in manufacturer's catalog. This field angle is calculated from given distance and field diameter data. See Introduction for more information about calculating photometric data.
** This unusual instrument uses a 500 or 1000 W. PAR 64 lamp for a source, with an 8x18 step lens to gather and concentrate the light, and an 8x4.5 lens to focus it.

ELECTRO CONTROLS
Model Nos. 3203 (w/ shutters)
3204 (w/ iris)

BEAM ANGLE: (not specified in mfg. catalog)

FIELD ANGLE: 37.3° *

LAMP BASE TYPE: Extended Mog. End Prong

LAMP MOUNT: Off Axis

STANDARD LAMP: 500 PAR 64/NSP - 500 W.**

OTHER LAMPS: Q1000PAR64/NSP-1 KW. *(cf=1.82)*

INTEGRAL PATTERN SLOT: Yes (3203 only)

WEIGHT: 24 lbs.

SIZE: 25" x 9" x 12¼" (L x W x H)

ACCESSORIES AVAILABLE:
Color Frame (9" sq.), 2-way Barn Door, Iris Kit, Shutter Kit, Snoot

PHOTOMETRICS CHART

(Performance data for this unit are measured using 500 PAR 64/NSP - 500 W. lamp.)

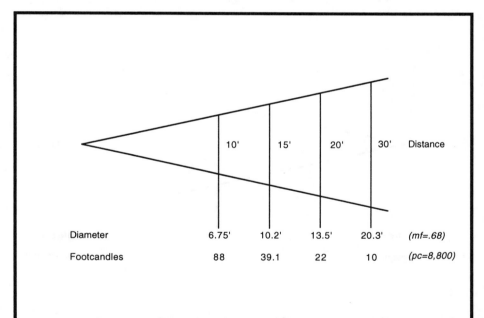

	10'	15'	20'	30'	Distance
Diameter	6.75'	10.2'	13.5'	20.3'	*(mf=.68)*
Footcandles	88	39.1	22	10	*(pc=8,800)*

** Beam and field angles not specified in manufacturer's catalog. This field angle is calculated from given distance and field diameter data. See Introduction for more information about calculating photometric data.
** This unusual instrument uses a 500 or 1000 W. PAR 64 lamp for a source, with an 8x18 step lens to gather and concentrate the light, and a pair of 6x12 lenses to focus it.

169

ELECTRO CONTROLS
Model No. 3345*

BEAM ANGLE: (not specified in mfg. catalog)

FIELD ANGLE: 34°

LAMP BASE TYPE: Mini-Can Screw

LAMP MOUNT: Axial

STANDARD LAMP: EVR - 500 W.

OTHER LAMPS: EHT - 250 W. *(cf=.48)*
EHV - 325 W. *(cf=.75)*

INTEGRAL PATTERN SLOT: Yes

ACCESSORIES AVAILABLE:
Color Frame, Pattern Holder, Snoot

WEIGHT: 7⅜ lbs.

SIZE: 16¾" x 6⅜" x 19¹⁵⁄₁₆" (L x W x H)

PHOTOMETRICS CHART
(Performance data for this unit are measured using EVR - 500 W. lamp.)

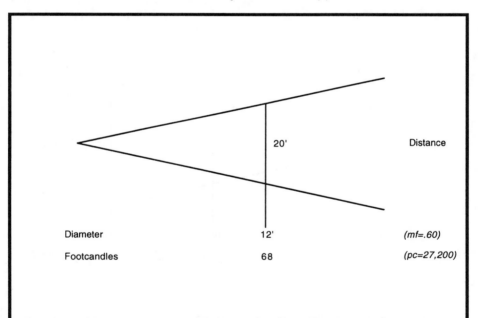

	20'	Distance
Diameter	12'	*(mf=.60)*
Footcandles	68	*(pc=27,200)*

* This model number, 3345, can accept four different lens configurations - two 4.5x7 plano lenses (this page), one 4.5x7 and one 4.5x9 plano lens, one 4.5x3 step lens, and two 4.5x9 plano lenes. See Index for page numbers of the other versions of model 3345.

WEIGHT: (not specified in manufacturer's catalog)

SIZE: (not specified in manufacturer's catalog)

ELECTRO CONTROLS
Model No. 3365*

BEAM ANGLE: (not specified in mfg. catalog)

FIELD ANGLE: 34.7°

LAMP BASE TYPE: Mini-Can Screw

LAMP MOUNT: Axial

STANDARD LAMP: EGJ - 1 KW.

OTHER LAMPS: EGE - 500 W. *(cf=.38)*
EGG - 750 W. *(cf=.57)*

INTEGRAL PATTERN SLOT: Yes

ACCESSORIES AVAILABLE:
Color Frame (7½" sq.), Iris (as an option)

PHOTOMETRICS CHART
(Performance data for this unit are measured using EGJ - 1 KW. lamp.)

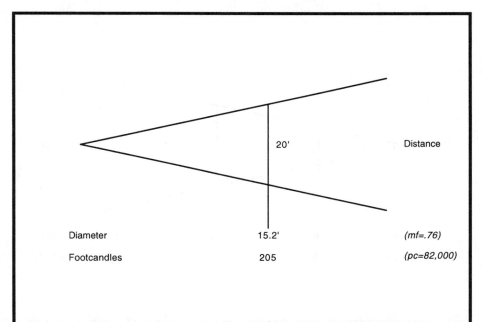

	20'	Distance
Diameter	15.2'	*(mf=.76)*
Footcandles	205	*(pc=82,000)*

* This model number, 3365, can accept 6 different lens configurations: 1 6x6 step lens, 2 6x9 plano lenses (this page), 2 6x12 lenses, 2 6x16 plano lenses, 2 6x20 plano lenses, and 1 6x8 step lens. See Index for the page numbers of other versions of model 3365.

ELECTRO CONTROLS
Model No. 3366*

BEAM ANGLE: 17.5°

FIELD ANGLE: 34.7°

LAMP BASE TYPE: Medium Prefocus

LAMP MOUNT: Axial

STANDARD LAMP: EGJ - 1 KW.

OTHER LAMPS: EGE - 500 W. *(cf=.38)*
EGG - 750 W. *(cf=.57)*

INTEGRAL PATTERN SLOT: No

ACCESSORIES AVAILABLE:
Color Frame (7½" sq.), Iris (as an option)

WEIGHT: (not specified in manufacturer's catalog)
SIZE: 24" x 12" x 17" (L x W x H)

PHOTOMETRICS CHART
(Performance data for this unit are measured using EGJ - 1 KW. lamp.)

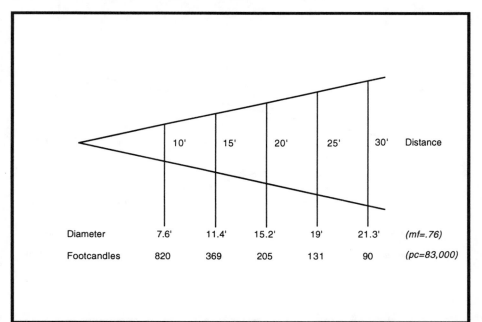

	10'	15'	20'	25'	30'	Distance
Diameter	7.6'	11.4'	15.2'	19'	21.3'	*(mf=.76)*
Footcandles	820	369	205	131	90	*(pc=83,000)*

* This model number, 3366, can accept three different lens configurations - two 6x9 plano lenses (this page), two 6x12 plano lenses, and one 6x6 step lens. See Index for the page numbers of other versions of model 3366.

ELECTRO CONTROLS
Model No. 7345AF 33*

BEAM ANGLE: 16.4°

FIELD ANGLE: 33°

LAMP BASE TYPE: Mini-Can Screw

LAMP MOUNT: Axial

STANDARD LAMP: EVR - 500 W.

OTHER LAMPS: EHT - 250 W. *(cf=.48)*
EHV - 325 W. *(cf=.75)*

INTEGRAL PATTERN SLOT: Yes

ACCESSORIES AVAILABLE:
Color Frame, Pattern Holder, Snoot

WEIGHT: 7⅜ lbs.

SIZE: 16¾" x 6⅜" x 19¹⁵⁄₁₆" (L x W x H)

PHOTOMETRICS CHART
(Performance data for this unit are measured using EVR - 500 W. lamp.)

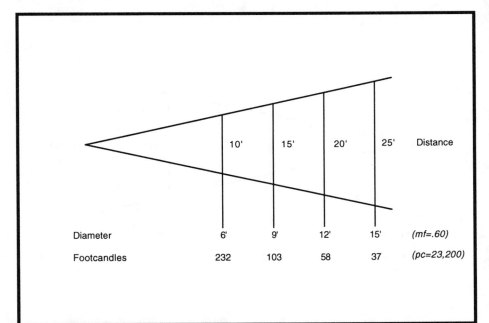

	10'	15'	20'	25'	Distance
Diameter	6'	9'	12'	15'	*(mf=.60)*
Footcandles	232	103	58	37	*(pc=23,200)*

* Lens systems in the 7345AF series are interchangeable. See Index for page numbers of the other versions of model 7345AF.

ELECTRO CONTROLS
Model No. 7363AF 39*

BEAM ANGLE: 21°

FIELD ANGLE: 39°

LAMP BASE TYPE: Medium Prefocus

LAMP MOUNT: Axial

STANDARD LAMP: EGG - 750 W.

OTHER LAMPS: EGF - 750 W. *(cf=1.30)*
EGJ - 1 KW. *(cf=1.75)*

INTEGRAL PATTERN SLOT: Yes

ACCESSORIES AVAILABLE:
Color Frame (7½" sq.), Pattern Holder, Snoot, Iris
(as an option)

WEIGHT: 17.75 lbs.

SIZE: 19⁹⁄₁₆" x 15½" x 26" (L x W x H)

PHOTOMETRICS CHART

(Performance data for this unit are measured using EGG - 750 W. lamp.)

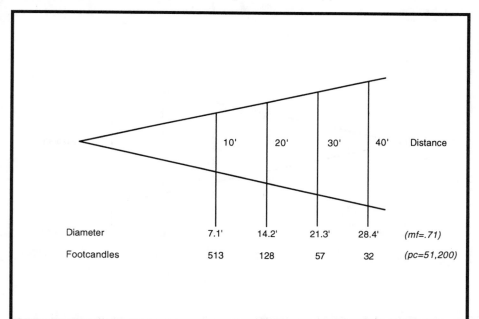

	10'	20'	30'	40'	Distance
Diameter	7.1'	14.2'	21.3'	28.4'	*(mf=.71)*
Footcandles	513	128	57	32	*(pc=51,200)*

* If unit includes optional iris the model number is 7363AI 39. Lens systems in the 7363A series are
interchangeable. See Index for page numbers of the other versions of model 7363A.

ELECTRONIC THEATRE CONTROLS (ETC)
Model No. 436, "Source Four"

BEAM ANGLE: 24°

FIELD ANGLE: 36°

LAMP BASE TYPE: Medium 2-Pin

LAMP MOUNT: Axial

STANDARD LAMP: HPL 575/115 - 575 W.

OTHER LAMPS: HPL 550/77 - 550W.*

INTEGRAL PATTERN SLOT: Yes

WEIGHT: 16.3 lbs.

SIZE: 22½" x 12⅜" x 20⅝" (L x W x H)

ACCESSORIES AVAILABLE:
Color Frame (6¼" sq.), Pattern Holder, Donut, Snoot

PHOTOMETRICS CHART
(Performance data for this unit are measured at cosine focus using HPL 575/115 lamp.)

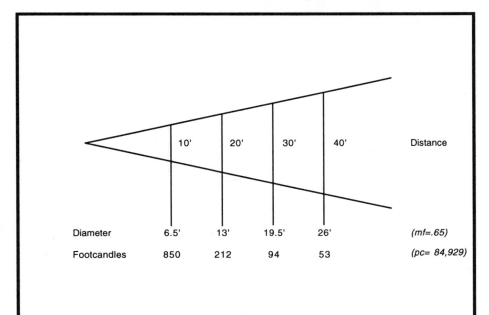

	10'	20'	30'	40'	Distance
Diameter	6.5'	13'	19.5'	26'	*(mf=.65)*
Footcandles	850	212	94	53	*(pc= 84,929)*

* The HPL 550/77 lamp is used with an ETC multiplexing system, used to allow multiple units to share a common circuit, with separate dimming of each.

ELECTRONIC THEATRE CONTROLS (ETC)
Model No. 436J, "Source Four, Jr."

BEAM ANGLE: 24°

FIELD ANGLE: 37°

LAMP BASE TYPE: Medium 2-Pin

LAMP MOUNT: Axial

STANDARD LAMP: HPL 575/115 - 575 W.

OTHER LAMPS: HPL 375/115 - 375 W.
HPL 375/115X - 375 W.
HPL 575/115X - 575 W.
HPL 550/77 - 550W.*

INTEGRAL PATTERN SLOT: Yes

WEIGHT: 12.2 lbs.

SIZE: 19³⁄₁₆" x 9⅝" x 13⅝" (L x W x H)

ACCESSORIES AVAILABLE:
Color Frame (6¼" sq.), Pattern Holder, Drop-in Iris, Donut, Snoot

PHOTOMETRICS CHART

(Performance data for this unit are measured at cosine focus using HPL 575/115 lamp.)

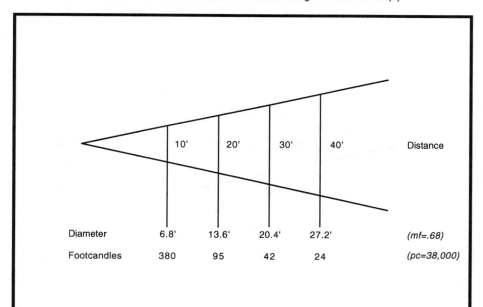

	10'	20'	30'	40'	Distance
Diameter	6.8'	13.6'	20.4'	27.2'	*(mf=.68)*
Footcandles	380	95	42	24	*(pc=38,000)*

* The HPL 550/77 lamp is used with an ETC multiplexing system, used to allow multiple units to share a common circuit, with separate dimming of each.

KLIEGL BROS.
Model No. 1553*

BEAM ANGLE: 10.5°

FIELD ANGLE: 30° (cosine)

LAMP BASE TYPE: Medium 2-Pin

LAMP MOUNT: Axial

STANDARD LAMP: FEL - 1 KW.

OTHER LAMPS: EHD - 500 W. *(cf=.38)*
EHG - 750 W. *(cf=.56)*

INTEGRAL PATTERN SLOT: Yes

ACCESSORIES AVAILABLE:
Color Frame (7½" sq.), Pattern Holder, Drop-in Iris

WEIGHT: 22 lbs.

SIZE: 27" x 16" x 18" (L x W x H)

PHOTOMETRICS CHART
(Performance data for this unit are measured at cosine focus using FEL - 1 KW. lamp.)

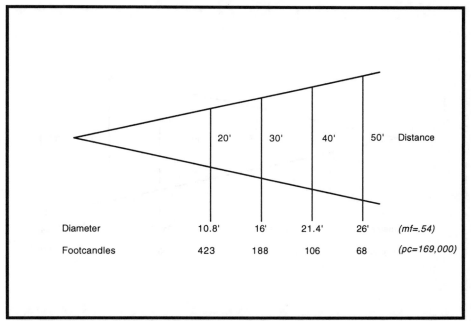

	20'	30'	40'	50'	Distance
Diameter	10.8'	16'	21.4'	26'	*(mf=.54)*
Footcandles	423	188	106	68	*(pc=169,000)*

* The lenses in this series of ellipsoidal spotlights are interchangeable on the same lamp housing. Lens barrels are color-coded to indentify the field angle. This 30° model is color-coded with blue.

LEE COLORTRAN
Model No. 650-032*

BEAM ANGLE: 11° (peak), 20° (cosine)

FIELD ANGLE: 30°

LAMP BASE TYPE: Medium 2-Pin

LAMP MOUNT: Axial

STANDARD LAMP: FEL - 1 KW.

OTHER LAMPS: EHD - 500 W. *(cf=.38)*
EHG - 750 W. *(cf=.56)*

INTEGRAL PATTERN SLOT: Yes

ACCESSORIES AVAILABLE:
Color Frame (7½" sq.), Pattern Holder, Iris Kit

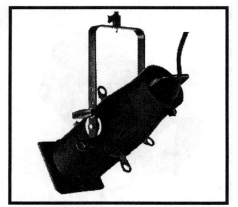

WEIGHT: 20.5 lbs.
SIZE: 28" x 15" x 25½" (L x W x H)

PHOTOMETRICS CHART
(Performance data for this unit are measured at peak focus using FEL - 1 KW. lamp.)

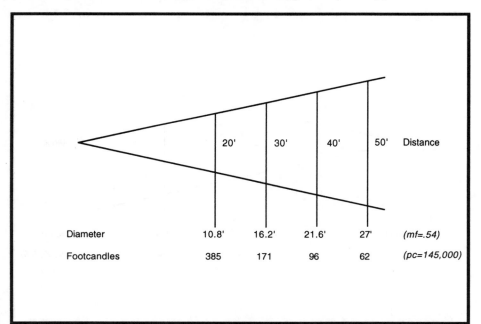

	20'	30'	40'	50'	Distance
Diameter	10.8'	16.2'	21.6'	27'	*(mf=.54)*
Footcandles	385	171	96	62	*(pc=145,000)*

* This model can be identified with model numbers -032, -034, -035, or -036 depending on the wire leads and connectors supplied.

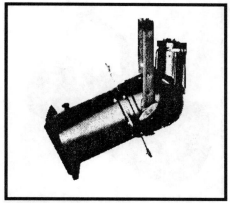

LIGHTING & ELECTRONICS
Model No. 61-12

BEAM ANGLE: 20°

FIELD ANGLE: 30°

LAMP BASE TYPE: Medium Prefocus

LAMP MOUNT: Off Axis

STANDARD LAMP: DNT - 750 W.

OTHER LAMPS: DEB - 500 W. *(cf=.53)*

INTEGRAL PATTERN SLOT: Yes

WEIGHT: (not specified in manufacturer's catalog)

SIZE: 19½" (L)

ACCESSORIES AVAILABLE:
Color Frame (7½" sq.)

PHOTOMETRICS CHART
(Performance data for this unit are measured using DNT - 750 W. lamp.)

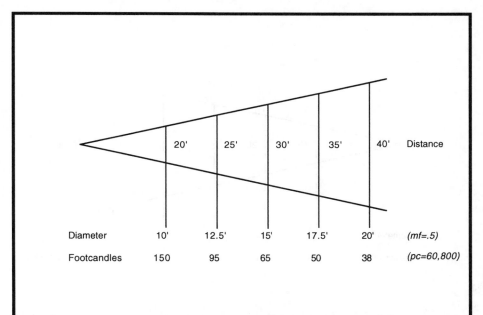

	20'	25'	30'	35'	40'	Distance
Diameter	10'	12.5'	15'	17.5'	20'	*(mf=.5)*
Footcandles	150	95	65	50	38	*(pc=60,800)*

LIGHTING & ELECTRONICS
Model No. AQ1K-40

BEAM ANGLE: 24°

FIELD ANGLE: 37°

LAMP BASE TYPE: Medium 2-Pin

LAMP MOUNT: Axial

STANDARD LAMP: FEL - 1KW.

OTHER LAMPS: EHG - 750 W. *(cf=.55)*
EHD - 500 W. *(cf=.36)*

INTEGRAL PATTERN SLOT: Yes

ACCESSORIES AVAILABLE:
Color Frame (7½" sq.), Drop-in Iris

WEIGHT: 21 lbs.

SIZE: 19¹³⁄₁₆" x 7¾" x 7¾" (L x W x H)

PHOTOMETRICS CHART
(Performance data for this unit are measured at the cosine using FEL - 1KW. lamp.)

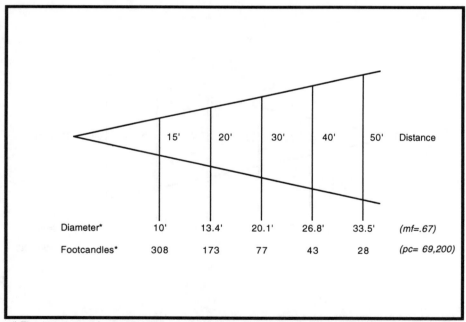

	15'	20'	30'	40'	50'	Distance
Diameter*	10'	13.4'	20.1'	26.8'	33.5'	*(mf=.67)*
Footcandles*	308	173	77	43	28	*(pc= 69,200)*

* The diameter and footcandle information in this chart was calculated using multiplying factor (*mf*) and peak candela (*pc*) data given in catalog. See Introduction for more information about calculating photometric data.

LIGHTING & ELECTRONICS
Model No. AQ61-06

BEAM ANGLE: 16°

FIELD ANGLE: 37°

LAMP BASE TYPE: Medium 2-Pin

LAMP MOUNT: Axial

STANDARD LAMP: EHG - 750 W.

OTHER LAMPS: EHD - 500 W. *(cf=.68)*
EHF - 750 W. *(cf=1.32)*
Hx600 - 600 W. *(cf=1.93)*

INTEGRAL PATTERN SLOT: Yes

ACCESSORIES AVAILABLE:
Color Frame (7½" sq.), Pattern Holder, Iris

WEIGHT: 18 lbs.

SIZE: 17¾" (L)

PHOTOMETRICS CHART
(Performance data for this unit are measured using EHG - 750 W. lamp.)

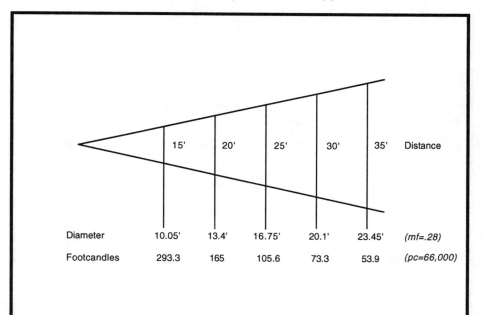

	15'	20'	25'	30'	35'	Distance
Diameter	10.05'	13.4'	16.75'	20.1'	23.45'	*(mf=.28)*
Footcandles	293.3	165	105.6	73.3	53.9	*(pc=66,000)*

LITTLE STAGE LIGHTING
Model No. E109

BEAM ANGLE: 14.7°

FIELD ANGLE: 30.2°

LAMP BASE TYPE: Medium Prefocus

LAMP MOUNT: Off Axis

STANDARD LAMP: 1MT12/2 - 1 KW.

OTHER LAMPS: (none specified in mfg. catalog)

INTEGRAL PATTERN SLOT: Yes

ACCESSORIES AVAILABLE:
Color Frame (6½" sq.), Pattern Holder, Snoot,
Motorized Color Wheel, Iris

WEIGHT: 18 lbs.

SIZE: 19⅜" x 15" x 18⅛" (L x W x H)

PHOTOMETRICS CHART
(Performance data for this unit are measured using 1MT 12/2 - 1 KW. lamp.)

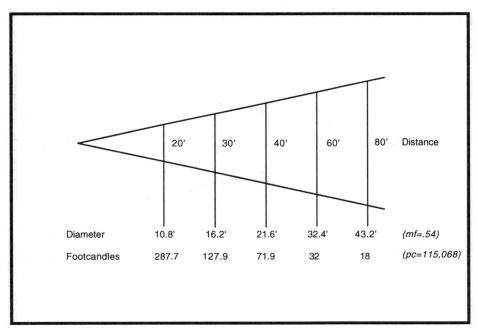

	20'	30'	40'	60'	80'	Distance
Diameter	10.8'	16.2'	21.6'	32.4'	43.2'	*(mf=.54)*
Footcandles	287.7	127.9	71.9	32	18	*(pc=115,068)*

LITTLE STAGE LIGHTING
Model No. E209

BEAM ANGLE: 13.3°

FIELD ANGLE: 30.8°

LAMP BASE TYPE: Medium Prefocus

LAMP MOUNT: Off Axis

STANDARD LAMP: 1MT12/2 - 1 KW.

OTHER LAMPS: (none specified in mfg. catalog)

INTEGRAL PATTERN SLOT: Yes

WEIGHT: 18 lbs.

SIZE: 26¼" x 15" x 18⅛" (L x W x H)

ACCESSORIES AVAILABLE:
Color Frame (6½" sq.), Pattern Holder, Snoot, Motorized Color Wheel, Iris

PHOTOMETRICS CHART
(Performance data for this unit are measured using 1MT 12/2 - 1 KW. lamp.)

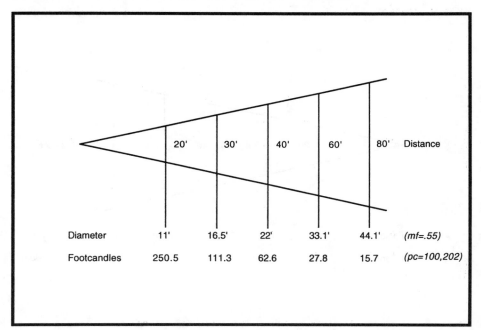

	20'	30'	40'	60'	80'	Distance
Diameter	11'	16.5'	22'	33.1'	44.1'	*(mf=.55)*
Footcandles	250.5	111.3	62.6	27.8	15.7	*(pc=100,202)*

STRAND CENTURY
Model No. 2209

BEAM ANGLE: 17° (peak), 29° (flat)

FIELD ANGLE: 31° (peak)

LAMP BASE TYPE: Medium 2-Pin

LAMP MOUNT: Axial

STANDARD LAMP: FEL - 1 KW.

OTHER LAMPS: EHD - 500 W. *(cf=.38)*
EHG - 750 W. *(cf=.56)*

INTEGRAL PATTERN SLOT: Yes

ACCESSORIES AVAILABLE:
Color Frame (7½" sq.), Pattern Holder, Snoot, Iris Kit

WEIGHT: 17 lbs.

SIZE: 18" x 12" x 24" (L x W x H)

PHOTOMETRICS CHART
(Performance data for this unit are measured at peak focus using FEL - 1KW. lamp.)

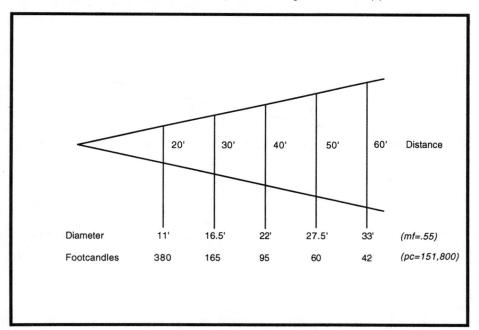

	20'	30'	40'	50'	60'	Distance
Diameter	11'	16.5'	22'	27.5'	33'	*(mf=.55)*
Footcandles	380	165	95	60	42	*(pc=151,800)*

TIMES SQUARE LIGHTING
Model No. 6E6x12

BEAM ANGLE: (not specified in mfg. catalog)

FIELD ANGLE: 30° *

LAMP BASE TYPE: Medium Prefocus

LAMP MOUNT: Off Axis

STANDARD LAMP: EGG - 750 W.

OTHER LAMPS: EGE - 500 W. *(cf=.66)*
　　　　　　　　EGJ - 1 KW. *(cf=1.75)*

INTEGRAL PATTERN SLOT: Yes

ACCESSORIES AVAILABLE:
Color Frame (7½" sq.), Color Wheel, Pattern Holder

WEIGHT: 12 lbs.

SIZE: (not specified in manufacturer's catalog)

PHOTOMETRICS CHART
(Information not provided in manufacturer's catalog about lamp used to measure performance data.)

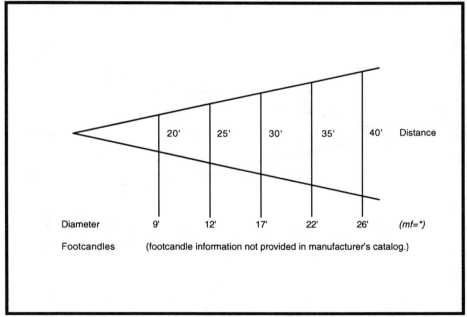

Distance	20'	25'	30'	35'	40'
Diameter	9'	12'	17'	22'	26'	(mf=*)
Footcandles	(footcandle information not provided in manufacturer's catalog.)

* Beam and field angles not specified in manufacturer's catalog. This field angle is calculated from given distance and field diameter data. See Introduction for more information about calculating photometric data. Also note, the field diameter data is not consistent. At the given distances, the field angle which is described by the diameter information varies from 25° at 9' to 36° at 26'. Because of this inconsistent data, no diameter multiplying factor *(mf)* is given.

185

TIMES SQUARE LIGHTING
Model No. Q6x9

BEAM ANGLE: (not specified in mfg. catalog)

FIELD ANGLE: 30° *

LAMP BASE TYPE: Medium 2-Pin

LAMP MOUNT: Axial

STANDARD LAMP: FEL - 1 KW.

OTHER LAMPS: EHD - 500 W. *(cf=.38)*
EHG - 750 W. *(cf=.56)*

INTEGRAL PATTERN SLOT: Yes

ACCESSORIES AVAILABLE:
Color Frame, Pattern Holder, Snoot, Motorized
or Indexing Color Wheels, Iris Kit

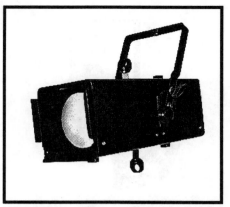

WEIGHT: (not specified in manufacturer's catalog)

SIZE: (not specified in manufacturer's catalog)

PHOTOMETRICS CHART
(Performance data for this unit are measured using FEL - 1 KW. lamp.)

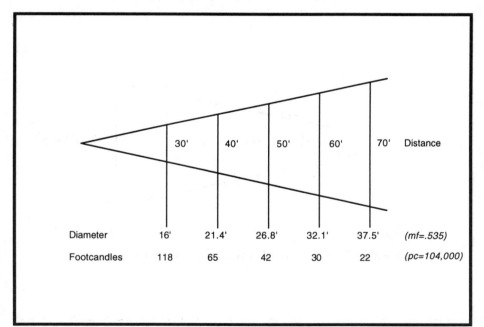

	30'	40'	50'	60'	70'	Distance
Diameter	16'	21.4'	26.8'	32.1'	37.5'	(mf=.535)
Footcandles	118	65	42	30	22	(pc=104,000)

* Beam and field angles not specified in manufacturer's catalog. This field angle is calculated from given distance and field diameter data. See Introduction for more information about calculating photometric data.

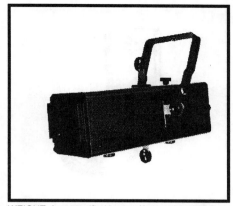

TIMES SQUARE LIGHTING
Model No. Q6x12

BEAM ANGLE: (not specified in mfg. catalog)

FIELD ANGLE: 30° *

LAMP BASE TYPE: Medium 2-Pin

LAMP MOUNT: Axial

STANDARD LAMP: FEL - 1 KW.

OTHER LAMPS: EHD - 500 W. *(cf=.38)*
EHG - 750 W. *(cf=.56)*

INTEGRAL PATTERN SLOT: Yes

WEIGHT: (not specified in manufacturer's catalog)

SIZE: (not specified in manufacturer's catalog)

ACCESSORIES AVAILABLE:
Color Frame, Pattern Holder, Snoot, Motorized or Indexing Color Wheels, Iris Kit

PHOTOMETRICS CHART
(Performance data for this unit are measured using FEL - 1 KW. lamp.)

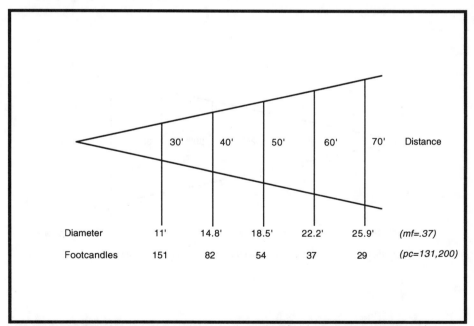

	30'	40'	50'	60'	70'	Distance
Diameter	11'	14.8'	18.5'	22.2'	25.9'	*(mf=.37)*
Footcandles	151	82	54	37	29	*(pc=131,200)*

* Beam and field angles not specified in manufacturer's catalog. This field angle is calculated from given distance and field diameter data. See Introduction for more information about calculating photometric data.

TIMES SQUARE LIGHIING
Model No. Q6E6x12

BEAM ANGLE: (not specified in mfg. catalog)

FIELD ANGLE: 30° *

LAMP BASE TYPE: Medium 2-Pin

LAMP MOUNT: Axial

STANDARD LAMP: EHG - 750 W.

OTHER LAMPS: EHD - 500 W. *(cf=.68)*
              FEL - 1 KW. *(cf=1.79)*

INTEGRAL PATTERN SLOT: Yes

ACCESSORIES AVAILABLE:
Color Frame (7½" sq.), Color Wheel, Pattern Holder

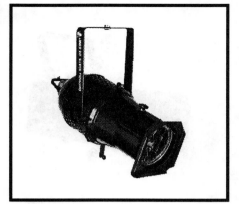

WEIGHT: 12 lbs.

SIZE: (not specified in manufacturer's catalog)

PHOTOMETRICS CHART

(Information not provided in manufacturer's catalog about lamp used to measure performance data.)

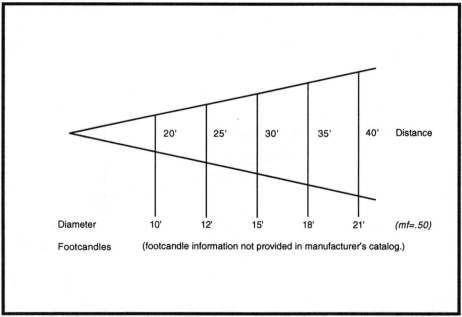

	20'	25'	30'	35'	40'	Distance
Diameter	10'	12'	15'	18'	21'	*(mf=.50)*
Footcandles	(footcandle information not provided in manufacturer's catalog.)					

* Beam and field angles not specified in manufacturer's catalog. This field angle is calculated from given distance and field diameter data. See Introduction for more information about calculating photometric data.

ALTMAN STAGE LIGHTING
Model No. 1KL6-20

BEAM ANGLE: 9.5° (peak), 11.5° (flat field)

FIELD ANGLE: 20°

LAMP BASE TYPE: Medium 2-Pin

LAMP MOUNT: Axial

STANDARD LAMP: FEL - 1 KW.

OTHER LAMPS: EHD - 500 W. *(cf=.39)*
EHG - 750 W. *(cf=.56)*

INTEGRAL PATTERN SLOT: Yes

ACCESSORIES AVAILABLE:
Color Frame, Pattern Holder, Snoot, Motorized &
Non-motorized Color Wheels, Iris

WEIGHT: (not specified in manufacturer's catalog)

SIZE: 22⅞" x 17" x 19" (L x W x H)

PHOTOMETRICS CHART
(Performance data for this unit are measured at peak focus using FEL - 1 KW. lamp.)

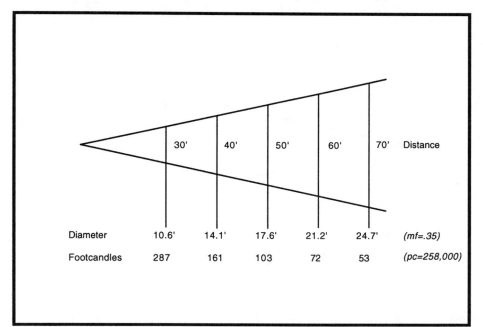

	30'	40'	50'	60'	70'	Distance
Diameter	10.6'	14.1'	17.6'	21.2'	24.7'	*(mf=.35)*
Footcandles	287	161	103	72	53	*(pc=258,000)*

189

ALTMAN STAGE LIGHTING
Model No. 360 6x12

BEAM ANGLE: 15°

FIELD ANGLE: 25°

LAMP BASE TYPE: Medium Prefocus

LAMP MOUNT: Off Axis

STANDARD LAMP: DNT - 750 W.

OTHER LAMPS: DEB - 500 W. *(cf=.53)*

INTEGRAL PATTERN SLOT: Yes

ACCESSORIES AVAILABLE:
Color Frame (7½" sq.), Pattern Holder, Snoot,
Motorized & Non-motorized Color Wheels, Iris

WEIGHT: (not specified in manufacturer's catalog)

SIZE: 19⅝" x 13¼" x 16¾" (L x W x H)

PHOTOMETRICS CHART
(Performance data for this unit are measured using DNT - 750 W. lamp.)

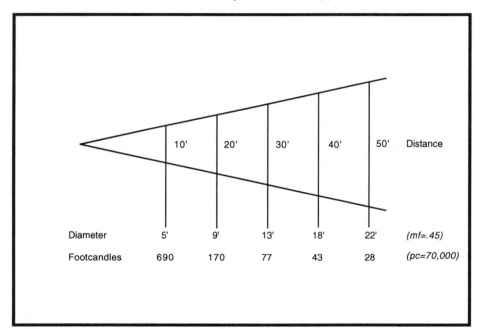

	10'	20'	30'	40'	50'	Distance
Diameter	5'	9'	13'	18'	22'	*(mf=.45)*
Footcandles	690	170	77	43	28	*(pc=70,000)*

ALTMAN STAGE LIGHTING
Model No. 360 6x16

BEAM ANGLE: 14°

FIELD ANGLE: 21°

LAMP BASE TYPE: Medium Prefocus

LAMP MOUNT: Off Axis

STANDARD LAMP: DNT - 750 W.

OTHER LAMPS: DEB - 500 W. *(cf=.53)*
EGE - 500 W. *(cf=.6)*
EGG - 750 W. *(cf=.9)*

INTEGRAL PATTERN SLOT: Yes

ACCESSORIES AVAILABLE:
Color Frame (7½" sq.), Pattern Holder, Snoot,
Motorized & Non-motorized Color Wheels, Iris

WEIGHT: (not specified in manufacturer's catalog)

SIZE: 22½" x 13¼" x 16¾" (L x W x H)

PHOTOMETRICS CHART
(Performance data for this unit are measured using DNT - 750 W. lamp.)

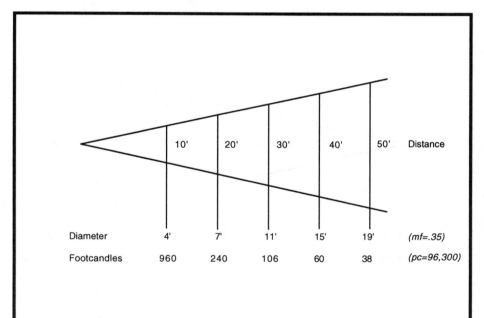

	10'	20'	30'	40'	50'	Distance
Diameter	4'	7'	11'	15'	19'	*(mf=.35)*
Footcandles	960	240	106	60	38	*(pc=96,300)*

ALTMAN STAGE LIGHTING
Model No. 360Q 6x12

BEAM ANGLE: 11°

FIELD ANGLE: 26°

LAMP BASE TYPE: Medium 2-Pin

LAMP MOUNT: Axial

STANDARD LAMP: EHF - 750 W.

OTHER LAMPS: EHD - 500 W. *(cf=.51)*
EHG - 750 W. *(cf=.75)*

INTEGRAL PATTERN SLOT: Yes

ACCESSORIES AVAILABLE:
Color Frame (7½" sq.), Pattern Holder, Snoot,
Motorized & Non-motorized Color Wheels, Iris

WEIGHT: (not specified in manufacturer's catalog)

SIZE: 20⅛" x 13" x 16¼" (L x W x H)

PHOTOMETRICS CHART
(Performance data for this unit are measured using EHF - 750 W. lamp.)

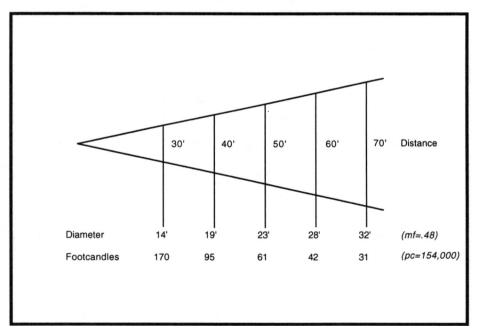

	30'	40'	50'	60'	70'	Distance
Diameter	14'	19'	23'	28'	32'	*(mf=.48)*
Footcandles	170	95	61	42	31	*(pc=154,000)*

ALTMAN STAGE LIGHTING
Model No. 365

BEAM ANGLE: 11°

FIELD ANGLE: 21°

LAMP BASE TYPE: Medium Prefocus

LAMP MOUNT: Off Axis

STANDARD LAMP: DNT - 750 W.

OTHER LAMPS: DNS - 500 W. *(cf=.65)*
EGE - 500 W. *(cf=.61)*
EGG - 750 W. *(cf=.92)*

INTEGRAL PATTERN SLOT: Yes

ACCESSORIES AVAILABLE:
Color Frame (7½" sq.), Pattern Holder, Snoot,
Motorized & Non-motorized Color Wheels

WEIGHT: (not specified in manufacturer's catalog)

SIZE: 19½" x 13¼" x 18½" (L x W x H)

PHOTOMETRICS CHART

(Performance data for this unit are measured using DNT - 750 W. lamp.)

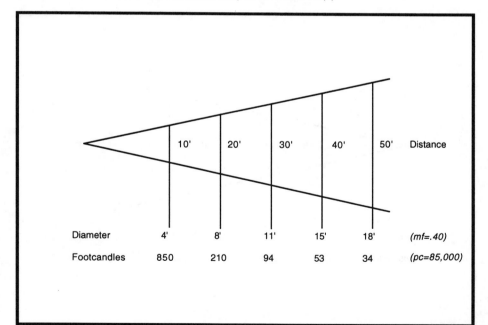

	10'	20'	30'	40'	50'	Distance
Diameter	4'	8'	11'	15'	18'	*(mf=.40)*
Footcandles	850	210	94	53	34	*(pc=85,000)*

ALTMAN STAGE LIGHTING
Model No. 365/6x6

BEAM ANGLE: 12°

FIELD ANGLE: 24°

LAMP BASE TYPE: Medium Prefocus

LAMP MOUNT: Off Axis

STANDARD LAMP: DNT - 750 W.

OTHER LAMPS: DNS - 500 W. *(cf=.65)*
EGC - 500 W. *(cf=.61)*
EGG - 750 W. *(cf=.92)*

INTEGRAL PATTERN SLOT: Yes

ACCESSORIES AVAILABLE:
Color Frame (7½" sq.), Pattern Holder, Snoot,
Motorized & Non-motorized Color Wheels

WEIGHT: (not specified in manufacturer's catalog)

SIZE: 19½" x 13¼" x 18½" (L x W x H)

PHOTOMETRICS CHART

(Performance data for this unit are measured using DNT - 750 W. lamp.)

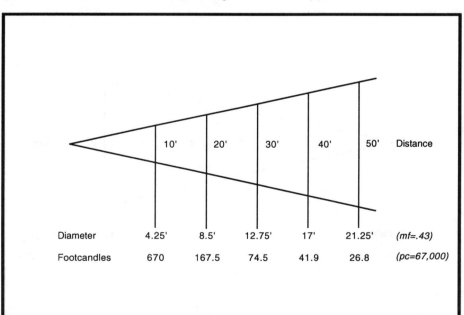

	10'	20'	30'	40'	50'	Distance
Diameter	4.25'	8.5'	12.75'	17'	21.25'	*(mf=.43)*
Footcandles	670	167.5	74.5	41.9	26.8	*(pc=67,000)*

ALTMAN
Model No. S6-20, "Shakespeare"

BEAM ANGLE: 13° (cosine)

FIELD ANGLE: 20°

LAMP BASE TYPE: Medium 2-Pin

LAMP MOUNT: Axial

STANDARD LAMP: HX600 - 575 W.

OTHER LAMPS: EHG - 750 W. *(cf=.93)*
EHD - 500 W. *(cf-.63)*

INTEGRAL PATTERN SLOT: Yes

ACCESSORIES AVAILABLE:
Color Frame (6¼" sq.), Snoot, Pattern Holder

WEIGHT: 18 lbs.

SIZE: 28½" x 12" x 13¾" (L x W x H)

PHOTOMETRICS CHART
(Performance data for this unit are measured at cosine focus using HX600 - 575 W. lamp.)

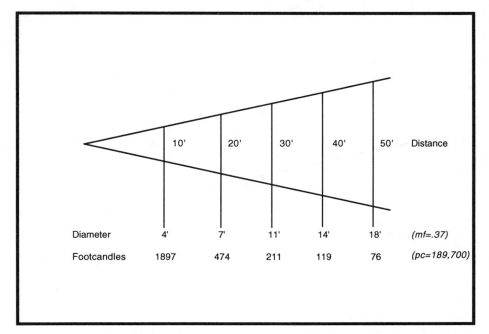

	10'	20'	30'	40'	50'	Distance
Diameter	4'	7'	11'	14'	18'	*(mf=.37)*
Footcandles	1897	474	211	119	76	*(pc=189,700)*

BERKEY COLORTRAN
Model No. 212-052*

BEAM ANGLE: 23° (peak and flat focus)

FIELD ANGLE: 26° (peak and flat focus)

LAMP BASE TYPE: Medium 2-Pin

LAMP MOUNT: Axial

STANDARD LAMP: EHG - 750 W.

OTHER LAMPS: EHD - 500 W. *(cf=.66)*
EHF - 750 W. *(cf=1.3)*

INTEGRAL PATTERN SLOT: Yes

ACCESSORIES AVAILABLE:
Color Frame (7½" sq.), Pattern Holder

WEIGHT: 16 lbs.

SIZE: 20⅝" x 13½" x 23¾" (L x W x H)

PHOTOMETRICS CHART
(Performance data for this unit are measured at peak focus using an EHG - 750 W. lamp.)

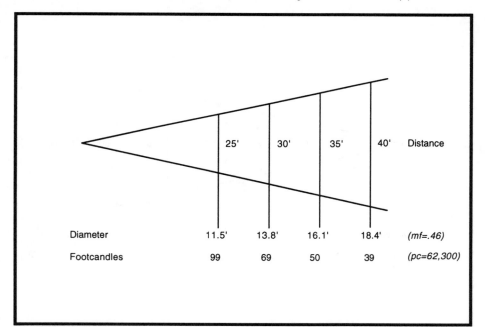

	25'	30'	35'	40'	Distance
Diameter	11.5'	13.8'	16.1'	18.4'	*(mf=.46)*
Footcandles	99	69	50	39	*(pc=62,300)*

* This model can be identified with model numbers -052, -055, -056, or -057 depending on the wire leads and connectors supplied.

BERKEY COLORTRAN
Model No. 213-072*

BEAM ANGLE: 8.3°

FIELD ANGLE: 20°

LAMP BASE TYPE: Medium 2-Pin

LAMP MOUNT: Axial

STANDARD LAMP: FEL - 1 KW.

OTHER LAMPS: EHD - 500 W. *(cf=.39)*
EHG - 750 W. *(cf=.56)*

INTEGRAL PATTERN SLOT: Yes

WEIGHT: 19.5 lbs.

SIZE: 27½" x 15" x 24½" (L x W x H)

ACCESSORIES AVAILABLE:
Color Frame (7½" sq.), Pattern Holder, Iris Kit

PHOTOMETRICS CHART
(Performance data for this unit are measured at peak focus using FEL - 1 KW. lamp.)

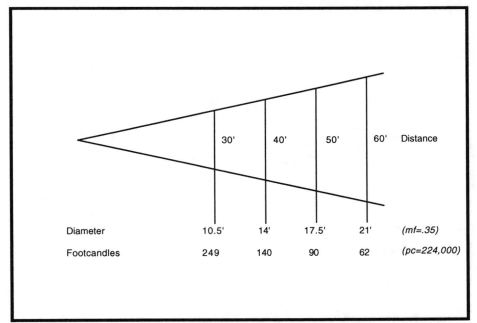

	30'	40'	50'	60'	Distance
Diameter	10.5'	14'	17.5'	21'	*(mf=.35)*
Footcandles	249	140	90	62	*(pc=224,000)*

* This model can be identified with model numbers -072, -075, -076, or -077 depending on the wire leads and connectors supplied.

CAPITOL STAGE LIGHTING
Model No. 851*

BEAM ANGLE: (not specified in mfg. catalog)

FIELD ANGLE: 20° **

LAMP BASE TYPE: Medium Prefocus

LAMP MOUNT: Off Axis

STANDARD LAMP: T12 C13D - 1 KW.

OTHER LAMPS: DEB - 500 W. *(cf=.36)*
DNT - 750 W. *(cf=.68)*

INTEGRAL PATTERN SLOT: Yes (no. 854 only*)

ACCESSORIES AVAILABLE:
Color Frame (10" sq.), Snoot, Motorized &
Non-motorized Color Wheels

(Photo not available)

WEIGHT: (not specified in manufacturer's catalog)

SIZE: 24¾" x 20¼" (L x H)

PHOTOMETRICS CHART

(Performance data for this unit are measured using T12C13D - 1 KW. lamp.)

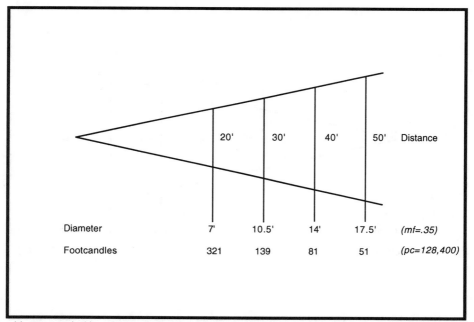

	20'	30'	40'	50'	Distance
Diameter	7'	10.5'	14'	17.5'	*(mf=.35)*
Footcandles	321	139	81	51	*(pc=128,400)*

* Instruments in this series are equipped as follows - 851 has 4-way shutters, 852 has shutters and an iris, 853 adds a follow handle to the iris unit, and 854 has shutters and a template slot.
** Beam and field angles not specified in manufacturer's catalog. This angle is calculated from given distance and diameter data. See Introduction for more information about calculating photometric data.

CENTURY LIGHTING
Model Nos. 1597 (w/ shutters)
1598 (w/ iris, handle)

BEAM ANGLE: (not specified in manufacturer's catalog)

FIELD ANGLE: 24°

LAMP BASE TYPE: Medium Prefocus

LAMP MOUNT: Off Axis

STANDARD LAMP: DNT - 750 W. *

OTHER LAMPS: DEB - 500 W.

INTEGRAL PATTERN SLOT: No

WEIGHT: (not specified in manufacturer's catalog)

SIZE: (not specified in manufacturer's catalog)

ACCESSORIES AVAILABLE:
Color Frame (7½" sq.)

PHOTOMETRICS CHART
(Performance data for this unit are measured using DNT 750 W. lamp.)

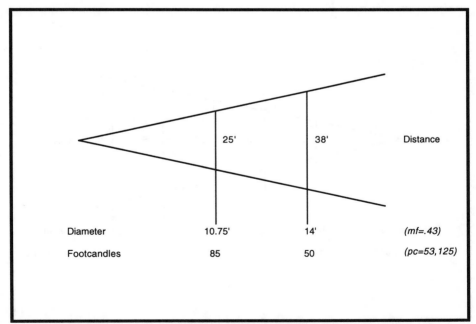

	25'	38'	Distance
Diameter	10.75'	14'	*(mf=.43)*
Footcandles	85	50	*(pc=53,125)*

* This is an obsolete lamp but it is listed as the standard lamp since the manufacturer uses this lamp to describe the photometric performance. Correction factor *(cf)* information is not available for the other lamps because no initial lumens data is available for the obsolete lamp.

CENTURY LIGHTING
Model Nos. 2343 (w/ shutters)
2344 (w/ iris)

BEAM ANGLE: 16°

FIELD ANGLE: 28°

LAMP BASE TYPE: Medium 2-Pin

LAMP MOUNT: Off Axis

STANDARD LAMP: EHG - 750 W.

OTHER LAMPS: EHD - 500 W. *(cf=.69)*
EHF - 750 W. *(cf=1.32)*
FEL - 1 KW. *(cf=1.79)*

INTEGRAL PATTERN SLOT: Yes (2343 only)

ACCESSORIES AVAILABLE:
Color Frame (7½" sq.), Pattern Holder, Snoot

WEIGHT: (not specified in manufacturer's catalog)

SIZE: 16" x 11½" x 24" (L x W x H)

PHOTOMETRICS CHART
(Performance data for this unit are measured using EHG - 750 W. lamp.)

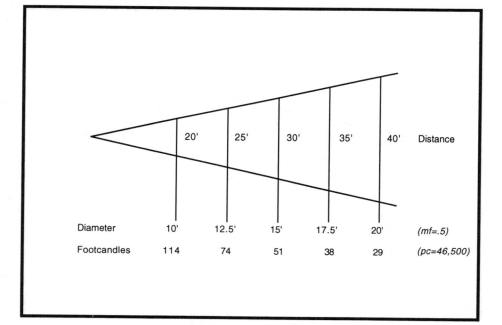

Distance	20'	25'	30'	35'	40'
Diameter	10'	12.5'	15'	17.5'	20'	*(mf=.5)*
Footcandles	114	74	51	38	29	*(pc=46,500)*

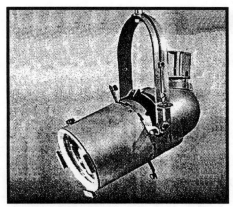

CENTURY STRAND
Model Nos. 2331 (w/ shutters)*
2332 (w/ iris)

BEAM ANGLE: 16°

FIELD ANGLE: 28°

LAMP BASE TYPE: Medium Prefocus

LAMP MOUNT: Off Axis

STANDARD LAMP: EGG - 750 W.

OTHER LAMPS: EGC - 750 W. *(cf=.81)*

INTEGRAL PATTERN SLOT: Yes (2331 only)

WEIGHT: (not specified in manufacturer's catalog)

SIZE: (not specified in manufacturer's catalog)

ACCESSORIES AVAILABLE:
Color Frame (7½" sq.), Pattern Holder, Snoot

PHOTOMETRICS CHART
(Performance data for this unit are measured using EGG - 750 W. lamp.)

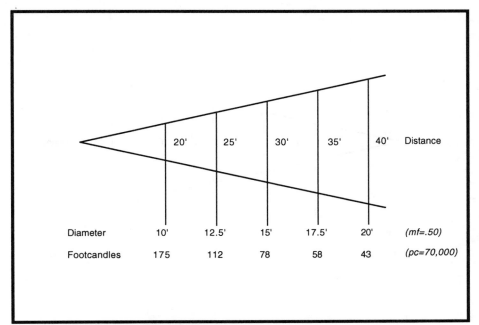

	20'	25'	30'	35'	40'	Distance
Diameter	10'	12.5'	15'	17.5'	20'	(mf=.50)
Footcandles	175	112	78	58	43	(pc=70,000)

• This unit was also sold as Century 1493.

CENTURY STRAND
Model Nos. 2337 (w/ shutters)
2338 (w/ iris)

BEAM ANGLE: 12.5°

FIELD ANGLE: 22.5°

LAMP BASE TYPE: Medium Prefocus

LAMP MOUNT: Axial

STANDARD LAMP: EGJ - 1 KW.

OTHER LAMPS: EGE - 500 W. *(cf=.38)*
EGG - 750 W. *(cf=.57)*

INTEGRAL PATTERN SLOT: Yes (2337 only)

ACCESSORIES AVAILABLE:
Color Frame, Pattern Holder, Snoot

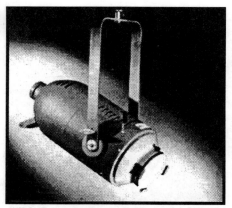

WEIGHT: (not specified in manufacturer's catalog)

SIZE: 34½" x 12" x 27½" (L x W x H)

PHOTOMETRICS CHART
(Performance data for this unit are measured using EGJ - 1 KW. lamp.)

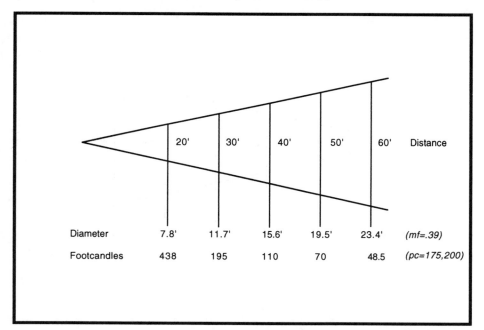

	20'	30'	40'	50'	60'	Distance
Diameter	7.8'	11.7'	15.6'	19.5'	23.4'	*(mf=.39)*
Footcandles	438	195	110	70	48.5	*(pc=175,200)*

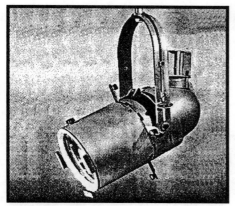

CENTURY STRAND
Model Nos. 2341 (w/ shutters) *
2342 (w/ iris)

BEAM ANGLE: 18°

FIELD ANGLE: 28°

LAMP BASE TYPE: Medium Prefocus

LAMP MOUNT: Off Axis

STANDARD LAMP: EGG - 750 W.

OTHER LAMPS: EGE - 500 W. *(cf=.66)*

INTEGRAL PATTERN SLOT: Yes (2341 only)

ACCESSORIES AVAILABLE:
Color Frame (7½" sq.), Pattern Holder, Snoot

WEIGHT: (not specified in manufacturer's catalog)

SIZE: 16" x 11½" x 24" (L x W x H)

PHOTOMETRICS CHART
(Performance data for this unit are measured using EGG - 750 W. lamp.)

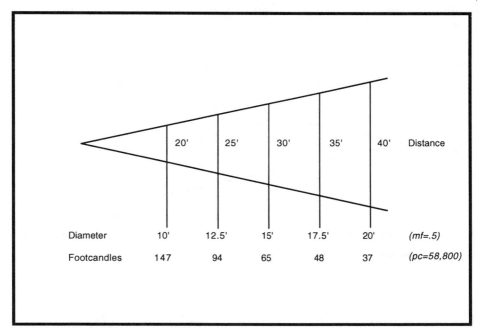

	20'	25'	30'	35'	40'	Distance
Diameter	10'	12.5'	15'	17.5'	20'	*(mf=.5)*
Footcandles	147	94	65	48	37	*(pc=58,800)*

* This unit was also sold as Century 1496.

CENTURY STRAND
Model No. Patt 264W

BEAM ANGLE: 26°

FIELD ANGLE: (not specified in mfg. catalog)

LAMP BASE TYPE: Medium Prefocus

LAMP MOUNT: Off Axis

STANDARD LAMP: Q1000T12/4CL - 1 KW.

OTHER LAMPS: 750T12/CL - 750 W. *(cf=.68)*

INTEGRAL PATTERN SLOT: Yes

ACCESSORIES AVAILABLE:
Color Frame (6½" sq.), 12-leaf Iris

WEIGHT: 16.75 lbs.

SIZE: 22¼" x 12" x 15½" (L x W x H)

PHOTOMETRICS CHART
(Performance data for this unit are measured using Q1000T 12/4CL - 1 KW. lamp.)

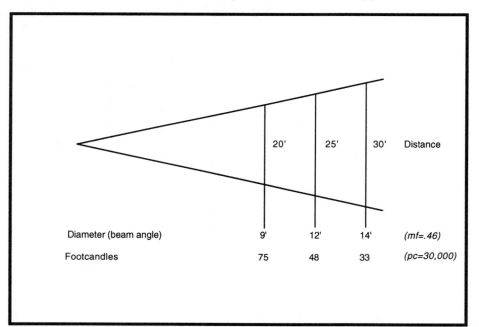

	20'	25'	30'	Distance
Diameter (beam angle)	9'	12'	14'	*(mf=.46)*
Footcandles	75	48	33	*(pc=30,000)*

ELECTRO CONTROLS
Model Nos. 3205 (w/ shutters)
3206 (w/ iris)

BEAM ANGLE: (not specified in mfg. catalog)

FIELD ANGLE: 24.8° *

LAMP BASE TYPE: Extended Mog. End Prong

LAMP MOUNT: Off Axis

STANDARD LAMP: 500 PAR 64/NSP - 500 W.**

OTHER LAMPS: Q1000PAR64/NSP-1 KW. *(cf=1.84)*

INTEGRAL PATTERN SLOT: Yes (3205 only)

ACCESSORIES AVAILABLE:
Color Frame (9" sq.), 2-way Barn Door, Iris Kit,
Shutter Kit, Snoot

WEIGHT: (not specified in manufacturer's catalog)

SIZE: 25" x 12¼" (L x H)

PHOTOMETRICS CHART
(Performance data for this unit are measured using 500 PAR64/NSP - 500 W. lamp.)

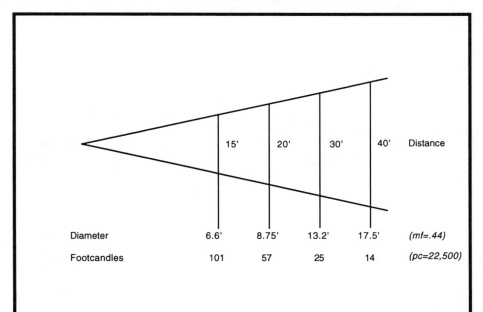

	15'	20'	30'	40'	Distance
Diameter	6.6'	8.75'	13.2'	17.5'	*(mf=.44)*
Footcandles	101	57	25	14	*(pc=22,500)*

*Beam and field angles not specified in manufacturer's catalog. This field angle is calculated from given distance and field diameter data. See Introduction for more information about calculating photometric data.
** This unusual instrument uses a 500 or 1000 W. PAR 64 lamp for a source, with an 8x18 step lens to gather and concentrate the light, and an 8x8 lens to focus it.

ELECTRO CONTROLS
Model Nos. 3207 (w/ shutters)
3208 (w/ iris)

BEAM ANGLE: (not specified in mfg. catalog)

FIELD ANGLE: 26.7° *

LAMP BASE TYPE: Extended Mog. End Prong

LAMP MOUNT: Off Axis

STANDARD LAMP: 500 PAR 64/NSP - 500 W.**

OTHER LAMPS: Q1000PAR64/NSP-1 KW. *(cf=1.82)*

INTEGRAL PATTERN SLOT: Yes (3207 only)

ACCESSORIES AVAILABLE:
Color Frame (9" sq.), 2-way Barn Door, Iris Kit,
Shutter Kit, Snoot

WEIGHT: 24 lbs.
SIZE: 25" (L)

PHOTOMETRICS CHART
(Performance data for this unit are measured using 500 PAR 64/NSP - 500 W. lamp.)

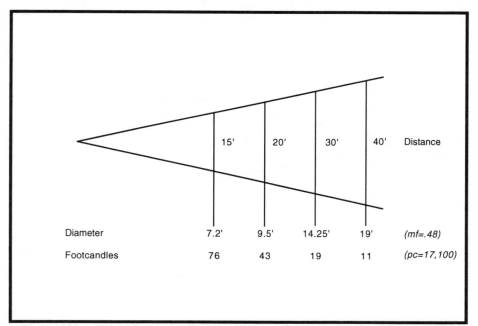

	15'	20'	30'	40'	Distance
Diameter	7.2'	9.5'	14.25'	19'	*(mf=.48)*
Footcandles	76	43	19	11	*(pc=17,100)*

* Beam and field angles not specified in manufacturer's catalog. This field angle is calculated from given distance and field diameter data. See Introduction for more information about calculating photometric data.
** This unusual instrument uses a 500 or 1000 W. PAR 64 lamp for a source, with an 8x18 step lens to gather and concentrate the light, and a pair of 6x16 lenses to focus it.

ELECTRO CONTROLS
Model Nos.3209 (w/ shutters)
3210 (w/ iris)

BEAM ANGLE: (not specified in mfg. catalog)

FIELD ANGLE: 20° *

LAMP BASE TYPE: Extended Mog. End Prong

LAMP MOUNT: Off Axis

STANDARD LAMP: 500 PAR 64/NSP - 500 W.**

OTHER LAMPS: Q1000PAR64/NSP-1 KW. *(cf=1.84)*

INTEGRAL PATTERN SLOT: Yes (3209 only)

ACCESSORIES AVAILABLE:
Color Frame (9" sq.), 2-way Barn Door, Iris Kit,
Shutter Kit, Snoot

WEIGHT: 23 lbs.

SIZE: 28" x 12¼" (L x H)

PHOTOMETRICS CHART
(Performance data for this unit are measured using 500 PAR64/NSP - 500 W. lamp.)

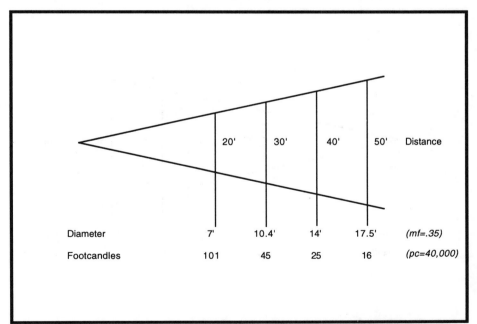

	20'	30'	40'	50'	Distance
Diameter	7'	10.4'	14'	17.5'	*(mf=.35)*
Footcandles	101	45	25	16	*(pc=40,000)*

* Beam and field angles not specified in manufacturer's catalog. This field angle is calculated from given distance and field diameter data. See Introduction for more information about calculating photometric data.
** This unusual instrument uses a 500 or 1000 W. PAR 64 lamp for a source, with an 8x18 step lens to gather and concentrate the light, and an 8x11 lens to focus it.

ELECTRO CONTROLS
Model Nos. 3211 (w/ shutters)
3212 (w/ iris)

BEAM ANGLE: (not specified in mfg. catalog)

FIELD ANGLE: 20.6° *

LAMP BASE TYPE: Extended Mog. End Prong

LAMP MOUNT: Off Axis

STANDARD LAMP: 500 PAR 64/NSP - 500 W.**

OTHER LAMPS: Q1000PAR64/NSP-1 KW. *(cf=1.84)*

INTEGRAL PATTERN SLOT: Yes (3211 only)

ACCESSORIES AVAILABLE:
Color Frame, 2-way Barn Door, Iris Kit, Shutter Kit, Snoot

WEIGHT: (not specified in manufacturer's catalog)

SIZE: (not specified in manufacturer's catalog)

PHOTOMETRICS CHART
(Performance data for this unit are measured using 500 PAR 64/NSP - 500 W. lamp.)

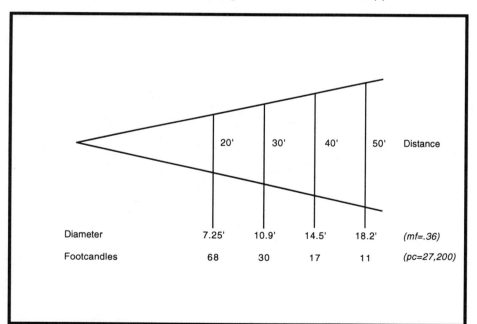

	20'	30'	40'	50'	Distance
Diameter	7.25'	10.9'	14.5'	18.2'	*(mf=.36)*
Footcandles	68	30	17	11	*(pc=27,200)*

* Beam and field angles not specified in manufacturer's catalog. This field angle is calculated from given distance and field diameter data. See Introduction for more information about calculating photometric data.
**This unusual instrument uses a 500 or 1000 W. PAR 64 lamp for a source, with an 8x18 step lens to gather and concentrate the light, and a pair of 6x20 lenses to focus it.

ELECTRO CONTROLS
Model No. 3345*

BEAM ANGLE: (not specified in mfg. catalog)

FIELD ANGLE: 26.5°

LAMP BASE TYPE: Mini-Can Screw

LAMP MOUNT: Axial

STANDARD LAMP: EVR - 500 W.

OTHER LAMPS: EHT - 250 W. *(cf=.48)*
EHV - 325 W. *(cf=.75)*

INTEGRAL PATTERN SLOT: Yes

ACCESSORIES AVAILABLE:
Color Frame, Pattern Holder, Snoot

WEIGHT: 7⅜ lbs.

SIZE: 16¾" x 6⅜" x 19¹⁵⁄₁₆" (L x W x H)

PHOTOMETRICS CHART
(Performance data for this unit are measured using EVR - 500 W. lamp.)

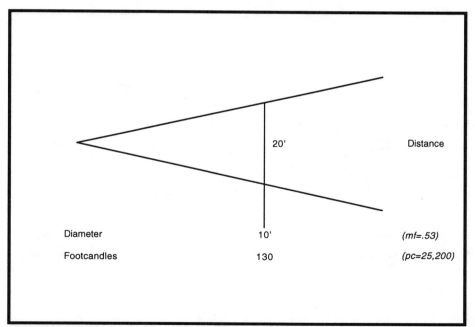

	20'	Distance
Diameter	10'	*(mf=.53)*
Footcandles	130	*(pc=25,200)*

* This model number, 3345, can accept four different lens configurations - two 4.5x7 plano lenses, one 4.5x7 and one 4.5x9 plano lens, one 4.5x3 step lens (this page), and two 4.5x9 plano lenses. See Index for page numbers of the other versions of model 3345.

ELECTRO CONTROLS
Model No. 3345*

BEAM ANGLE: (not specified in mfg. catalog)

FIELD ANGLE: 21°

LAMP BASE TYPE: Mini-Can Screw

LAMP MOUNT: Axial

STANDARD LAMP: EVR - 500 W.

OTHER LAMPS: EHT - 250 W. *(cf=.48)*
EHV - 325 W. *(cf=.75)*

INTEGRAL PATTERN SLOT: Yes

ACCESSORIES AVAILABLE:
Color Frame, Pattern Holder, Snoot

WEIGHT: 7⅜ lbs.

SIZE: 16¾" x 6⅜" x 19¹⁵⁄₁₆" (L x W x H)

PHOTOMETRICS CHART
(Performance data for this unit are measured using EVR - 500 W. lamp.)

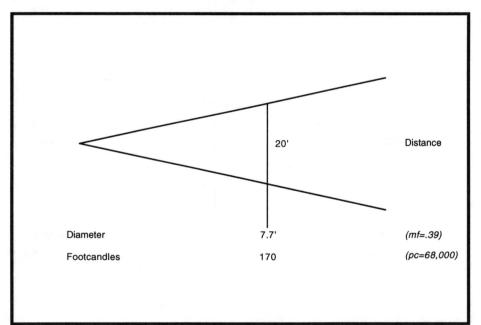

	20'	Distance
Diameter	7.7'	*(mf=.39)*
Footcandles	170	*(pc=68,000)*

* This model number, 3345, can accept four different lens configurations - two 4.5x7 plano lenses, one 4.5x7 and one 4.5x9 plano lens, one 4.5x3 step lens, and two 4.5x9 plano lenses (this page). See Index for page numbers of the other versions of model 3345.

WEIGHT: 7⅜ lbs.

SIZE: 16¾" x 6⅜" x 19¹⁵⁄₁₆" (L x W x H)

ELECTRO CONTROLS
Model No. 3345*

BEAM ANGLE: (not specified in mfg. catalog)

FIELD ANGLE: 27.5°

LAMP BASE TYPE: Mini-Can Screw

LAMP MOUNT: Axial

STANDARD LAMP: EVR - 500 W.

OTHER LAMPS: EHT - 250 W. *(cf=.48)*
EHV - 325 W. *(cf=.75)*

INTEGRAL PATTERN SLOT: Yes

ACCESSORIES AVAILABLE:
Color Frame, Pattern Holder, Snoot

PHOTOMETRICS CHART
(Performance data for this unit are measured using EVR - 500 W. lamp.)

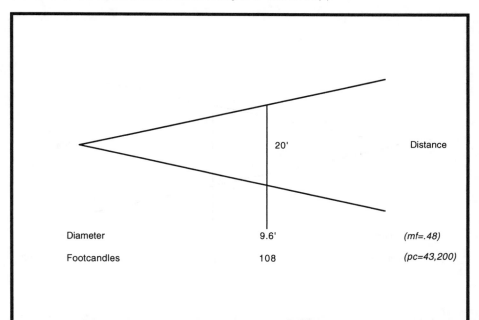

	20'	Distance
Diameter	9.6'	*(mf=.48)*
Footcandles	108	*(pc=43,200)*

* This model number, 3345, can accept four different lens configurations - two 4.5x7 plano lenses, one 4.5x7 and one 4.5x9 plano lens (this page), one 4.5x3 step lens, and two 4.5x9 plano lenses. See Index for page numbers of the other versions of model 3345.

ELECTRO CONTROLS
Model No. 3365*

BEAM ANGLE: (not specified in mfg. catalog)

FIELD ANGLE: 25°

LAMP BASE TYPE: Mini-Can Screw

LAMP MOUNT: Axial

STANDARD LAMP: EGJ - 1 KW.

OTHER LAMPS: EGE - 500 W. *(cf=.38)*
EGG - 750 W. *(cf=.57)*

INTEGRAL PATTERN SLOT: Yes

ACCESSORIES AVAILABLE:
Color Frame (7½" sq.), Iris (as an option)

WEIGHT: (not specified in manufacturer's catalog)

SIZE: (not specified in manufacturer's catalog)

PHOTOMETRICS CHART
(Performance data for this unit are measured using EGJ - 1 KW. lamp.)

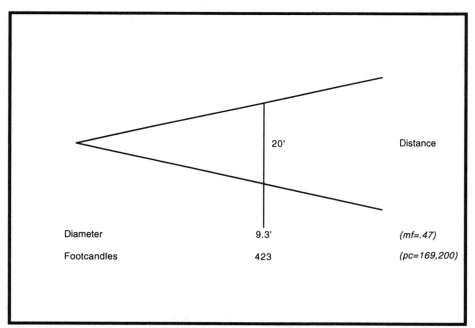

	20'	Distance
Diameter	9.3'	*(mf=.47)*
Footcandles	423	*(pc=169,200)*

* This model number, 3365, can accept six different lens configurations: one 6x6 step lens, two 6x9 plano lenses, two 6x12 lenses (this page), two 6x16 plano lenses, two 6x20 plano lenses, and one 6x8 step lens. See Index for the page numbers of other versions of model 3365.

ELECTRO CONTROLS
Model No. 3365*

BEAM ANGLE: (not specified in mfg. catalog)

FIELD ANGLE: 20°

LAMP BASE TYPE: Mini-Can Screw

LAMP MOUNT: Axial

STANDARD LAMP: EGJ - 1 KW.

OTHER LAMPS: EGE - 500 W. *(cf=.38)*
EGG - 750 W. *(cf=.57)*

INTEGRAL PATTERN SLOT: Yes

ACCESSORIES AVAILABLE:
Color Frame (7½" sq.), Iris (as an option)

WEIGHT: (not specified in manufacturer's catalog)

SIZE: (not specified in manufacturer's catalog)

PHOTOMETRICS CHART
(Performance data for this unit are measured using EGJ - 1 KW. lamp.)

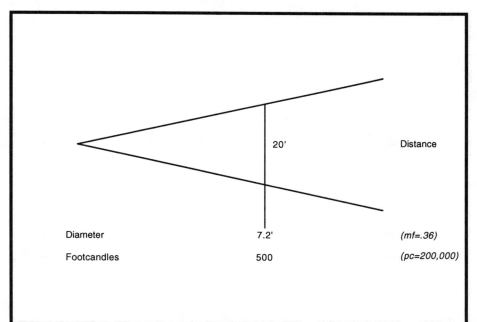

	20'	Distance
Diameter	7.2'	*(mf=.36)*
Footcandles	500	*(pc=200,000)*

* This model number, 3365, can accept six different lens configurations: one 6x6 step lens, two 6x9 plano lenses, two 6x12 lenses, two 6x16 plano lenses (this page), two 6x20 plano lenses, and one 6x8 step lens. See Index for the page numbers of other versions of model 3365.

ELECTRO CONTROLS
Model No. 3366*

BEAM ANGLE: 10°

FIELD ANGLE: 25°

LAMP BASE TYPE: Medium Prefocus

LAMP MOUNT: Axial

STANDARD LAMP: EGJ - 1 KW.

OTHER LAMPS: EGE - 500 W. *(cf=.38)*
EGG - 750 W. *(cf=.57)*

INTEGRAL PATTERN SLOT: No

ACCESSORIES AVAILABLE:
Color Frame (7½" sq.), Iris (as an option)

WEIGHT: 24 lbs.

SIZE: 25" x 9" x 12¼" (L x W x H)

PHOTOMETRICS CHART
(Performance data for this unit are measured using EGJ - 1 KW. lamp.)

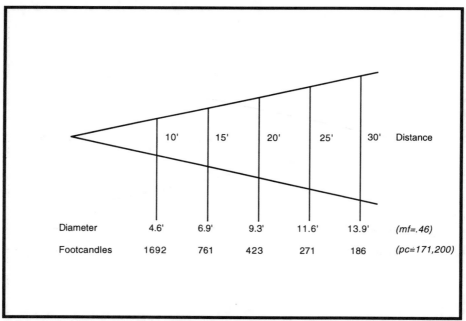

	10'	15'	20'	25'	30'	Distance
Diameter	4.6'	6.9'	9.3'	11.6'	13.9'	*(mf=.46)*
Footcandles	1692	761	423	271	186	*(pc=171,200)*

* This model number, 3366, can accept three different lens configurations - two 6x9 plano lenses, two 6x12 plano lenses (this page), and one 6x6 step lens. See Index for page numbers of the other versions of model 3366.

ELECTRO CONTROLS
Model No. 3386A/4.5

BEAM ANGLE: 18°

FIELD ANGLE: 29°

LAMP BASE TYPE: Medium Prefocus

LAMP MOUNT: Off Axis

STANDARD LAMP: EGJ - 1 KW.

OTHER LAMPS: EGE - 500 W. *(cf=.38)*
EGG - 750 W. *(cf=.57)*

INTEGRAL PATTERN SLOT: Yes

ACCESSORIES AVAILABLE:
Color Frame, Iris

WEIGHT: (not specified in manufacturer's catalog)

SIZE: (not specified in manufacturer's catalog)

PHOTOMETRICS CHART
(Performance data for this unit are measured using EGJ - 1 KW. lamp.)

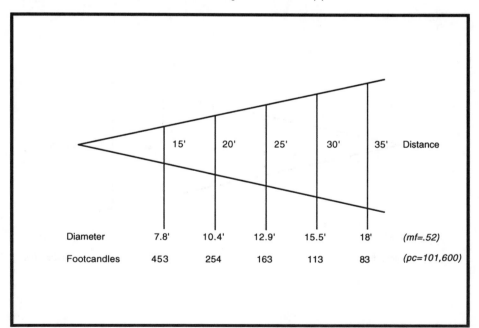

	15'	20'	25'	30'	35'	Distance
Diameter	7.8'	10.4'	12.9'	15.5'	18'	*(mf=.52)*
Footcandles	453	254	163	113	83	*(pc=101,600)*

ELECTRO CONTROLS
Model No. 7345AF 28*

BEAM ANGLE: 12.8°

FIELD ANGLE: 28.2°

LAMP BASE TYPE: Mini-Can Screw

LAMP MOUNT: Axial

STANDARD LAMP: EVR - 500 W.

OTHER LAMPS: EHT - 250 W. *(cf=.48)*
EHV - 325 W. *(cf=.75)*

INTEGRAL PATTERN SLOT: Yes

ACCESSORIES AVAILABLE:
Color Frame, Pattern Holder, Snoot

WEIGHT: 7⅜ lbs.

SIZE: 16¾" x 6⅜" x 19¹⁵⁄₁₆" (L x W x H)

PHOTOMETRICS CHART
(Performance data for this unit are measured using EVR - 500 W. lamp.)

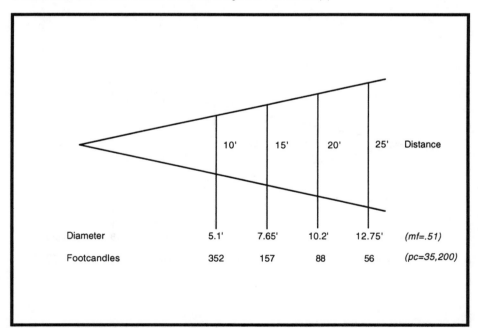

	10'	15'	20'	25'	Distance
Diameter	5.1'	7.65'	10.2'	12.75'	*(mf=.51)*
Footcandles	352	157	88	56	*(pc=35,200)*

* Lens systems in the 7345AF series are interchangeable. See Index for page numbers of the other versions of model 7345AF.

ELECTRO CONTROLS
Model No. 7345AF 29*

BEAM ANGLE: 15.6°

FIELD ANGLE: 29.6°

LAMP BASE TYPE: Mini-Can Screw

LAMP MOUNT: Axial

STANDARD LAMP: EVR - 500 W.

OTHER LAMPS: EHT - 250 W. *(cf=.48)*
EHV - 325 W. *(cf=.75)*

INTEGRAL PATTERN SLOT: Yes

ACCESSORIES AVAILABLE:
Color Frame, Pattern Holder, Snoot

WEIGHT: 7⅜ lbs.

SIZE: 16¾" x 6⅜" x 19¹⁵⁄₁₆" (L x W x H)

PHOTOMETRICS CHART
(Performance data for this unit are measured using EVR - 500 W. lamp.)

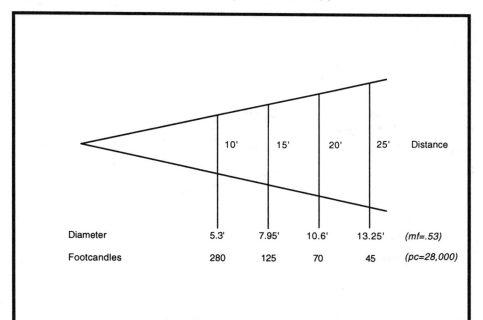

	10'	15'	20'	25'	Distance
Diameter	5.3'	7.95'	10.6'	13.25'	*(mf=.53)*
Footcandles	280	125	70	45	*(pc=28,000)*

* Lens systems in the 7345AF series are interchangeable. See Index for page numbers of the other versions of model 7345AF.

ELECTRONIC THEATRE CONTROLS (ETC)
Model No. 426, "Source Four"

BEAM ANGLE: 17°

FIELD ANGLE: 26°

LAMP BASE TYPE: Medium 2-Pin

LAMP MOUNT: Axial

STANDARD LAMP: HPL 575/115 - 575 W.

OTHER LAMPS: HPL 550/77 - 550W.*

INTEGRAL PATTERN SLOT: Yes

ACCESSORIES AVAILABLE:
Color Frame (6¼" sq.), Pattern Holder, Donut, Snoot

WEIGHT: 15.9 lbs.

SIZE: 22½" x 12⅜" x 20⅓" (L x W x H)

PHOTOMETRICS CHART
(Performance data for this unit are measured at cosine focus using HPL 575/115 lamp.)

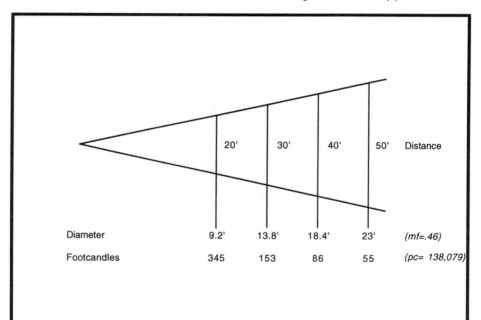

	20'	30'	40'	50'	Distance
Diameter	9.2'	13.8'	18.4'	23'	(mf=.46)
Footcandles	345	153	86	55	(pc= 138,079)

* The HPL 550/77 lamp is used with an ETC multiplexing system, used to allow multiple units to share a common circuit, with separate dimming of each.

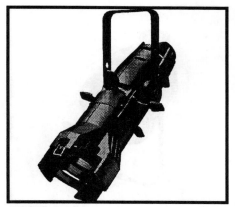

ELECTRONIC THEATRE CONTROLS (ETC)
Model No. 426J, "Source Four, Jr."

BEAM ANGLE: 17°

FIELD ANGLE: 26°

LAMP BASE TYPE: Medium 2-Pin

LAMP MOUNT: Axial

STANDARD LAMP: HPL 575/115 - 575 W.

OTHER LAMPS: HPL 375/115 - 375 W.
HPL 375/115X - 375W.
HPL 575/115X - 575W.
HPL 550/77 - 550W.*

INTEGRAL PATTERN SLOT: Yes

ACCESSORIES AVAILABLE:
Color Frame (6¼" sq.), Pattern Holder, Drop-in Iris, Donut, Snoot

WEIGHT: 12.2 lbs.

SIZE: 19³⁄₁₆" x 9⅝" x 13⅝" (L x W x H)

PHOTOMETRICS CHART

(Performance data for this unit are measured at cosine focus using HPL 575/115 lamp.)

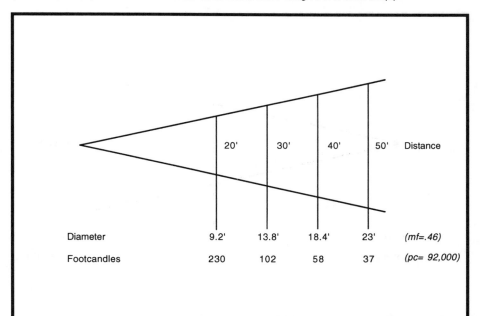

	20'	30'	40'	50'	Distance
Diameter	9.2'	13.8'	18.4'	23'	*(mf=.46)*
Footcandles	230	102	58	37	*(pc= 92,000)*

* The HPL 550/77 lamp is used with an ETC multiplexing system, used to allow multiple units to share a common circuit, with separate dimming of each.

KLIEGL BROS.
Model No. 1163

BEAM ANGLE: (not specified in mfg. catalog)

FIELD ANGLE: 27° *

LAMP BASE TYPE: Medium Bipost

LAMP MOUNT: Off Axis

STANDARD LAMP: 500T14/8 - 500 W.

OTHER LAMPS: (none specified in mfg. catalog)

INTEGRAL PATTERN SLOT: No

ACCESSORIES AVAILABLE:
Color Frame

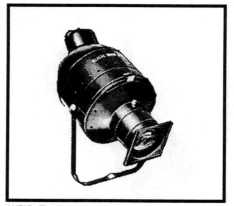

WEIGHT: 8 lbs.

SIZE: (not specified in manufacturer's catalog)

PHOTOMETRICS CHART

(Performance data for this unit are measured using 500T14/8 - 500 W. lamp)

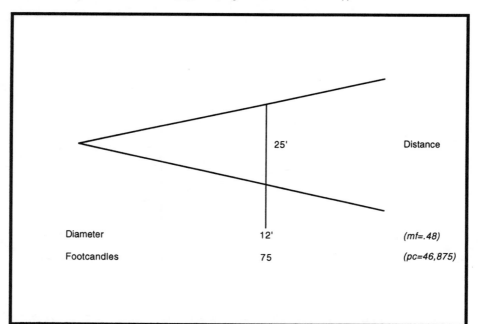

	25'	Distance
Diameter	12'	(mf=.48)
Footcandles	75	(pc=46,875)

* Beam and field angles not specified in manufacturer's catalog. This field angle is calculated from given distance and field diameter data. See Introduction for more information about calculating photometric data.

KLIEGL BROS.
Model No. 1164

BEAM ANGLE: (not specified in mfg. catalog)

FIELD ANGLE: 27° *

LAMP BASE TYPE: Mogul Bipost

LAMP MOUNT: Off Axis

STANDARD LAMP: 1M/T24/12 - 1 KW.

OTHER LAMPS: (none specified in mfg. catalog)

INTEGRAL PATTERN SLOT: No

WEIGHT: 18 lbs.

SIZE: (not specified in manufacturer's catalog)

ACCESSORIES AVAILABLE:
Color Frame

PHOTOMETRICS CHART
(Performance data for this unit are measured using 1M/T24/12 - 1 KW. lamp)

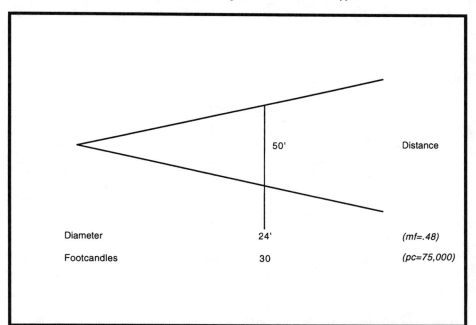

	50'	Distance
Diameter	24'	(mf=.48)
Footcandles	30	(pc=75,000)

* Beam and field angles not specified in manufacturer's catalog. This field angle is calculated from given distance and field diameter data. See Introduction for more information about calculating photometric data.

KLIEGL BROS.
Model No. 1165

BEAM ANGLE: (not specified in mfg. catalog)

FIELD ANGLE: 23° *

LAMP BASE TYPE: Medium Bipost

LAMP MOUNT: Off Axis

STANDARD LAMP: 500T14/8 - 500 W. **

OTHER LAMPS: 500T14/7 - 500 W.
750/T14 - 750 W.

INTEGRAL PATTERN SLOT: No

ACCESSORIES AVAILABLE:
Color Frame

WEIGHT: 14 lbs.

SIZE: (not specified in manufacturer's catalog)

PHOTOMETRICS CHART
(Performance data for this unit are measured using 500T14/8 - 500 W. lamp)

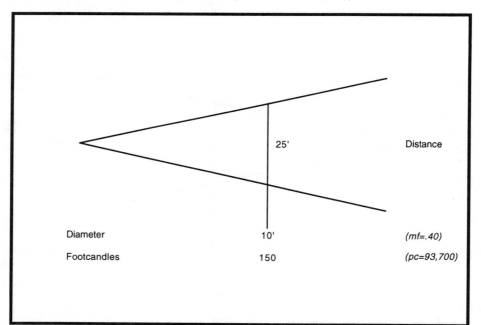

	25'	Distance
Diameter	10'	(mf=.40)
Footcandles	150	(pc=93,700)

* Beam and field angles not specified in manufacturer's catalog. This field angle is calculated from given distance and field diameter data. See Introduction for more information about calculating photometric data.
** This is an obsolete lamp but it is listed as the standard lamp since the manufacturer uses this lamp to describe the photometric performance. Correction factor *(cf)* information is not available for the other lamps because no initial lumens data is available for the obsolete lamp.

KLIEGL BROS.
Model No. 1167

BEAM ANGLE: (not specified in mfg. catalog)

FIELD ANGLE: 27° *

LAMP BASE TYPE: Mogul Bipost

LAMP MOUNT: Off Axis

STANDARD LAMP: 2M/T30/1 - 1 KW. **

OTHER LAMPS: 1M/T24/5 -   1 KW.
1550/T24/6 - 1.5 KW.

INTEGRAL PATTERN SLOT: No

ACCESSORIES AVAILABLE:
Color Frame

WEIGHT: 19 lbs.

SIZE: (not specified in manufacturer's catalog)

PHOTOMETRICS CHART

(Performance data for this unit are measured using 2M/T30/1 - 1 KW. lamp)

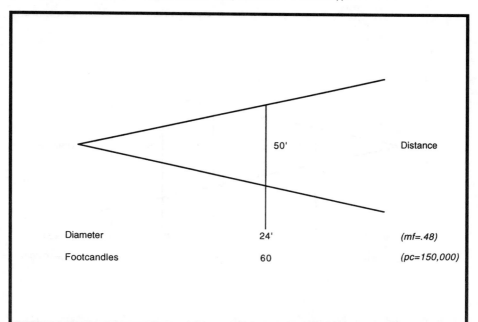

	50'	Distance
Diameter	24'	(mf=.48)
Footcandles	60	(pc=150,000)

* Beam and field angles not specified in manufacturer's catalog. This field angle is calculated from given distance and field diameter data. See Introduction for more information about calculating photometric data.
** This is an obsolete lamp but it is listed as the standard lamp since the manufacturer uses this lamp to describe the photometric performance. Correction factor (cf) information is not available for the other lamps because no initial lumens data is available for the obsolete lamp.

KLIEGL BROS.
Model No. 1355W*

BEAM ANGLE: (not specified in mfg. catalog)

FIELD ANGLE: 27°

LAMP BASE TYPE: Medium 2-Pin

LAMP MOUNT: Off Axis

STANDARD LAMP: EHG - 750 W.

OTHER LAMPS: EHD - 500 W. *(cf=.68)*
                 EHF - 750 W. *(cf=1.32)*

INTEGRAL PATTERN SLOT: Yes

ACCESSORIES AVAILABLE:
Color Frame (6½" sq.), Pattern Holder, Snoot

WEIGHT: 12 lbs.

SIZE: 22" x 13½" x 19½" (L x W x H)

PHOTOMETRICS CHART
(Performance data for this unit are measured using EHG - 750 W. lamp.)

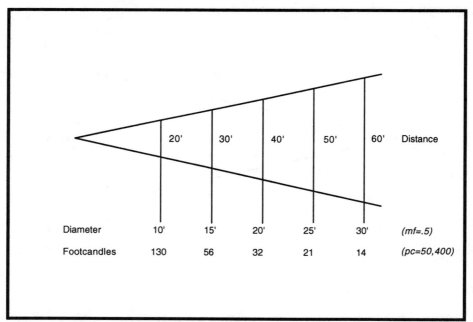

	20'	30'	40'	50'	60'	Distance
Diameter	10'	15'	20'	25'	30'	*(mf=.5)*
Footcandles	130	56	32	21	14	*(pc=50,400)*

* With optional iris installed, the model number is 1355I/W.

WEIGHT: (not specified in manufacturer's catalog)
SIZE: (not specified in manufacturer's catalog)

KLIEGL BROS.
Model No. 1357/6W*

BEAM ANGLE: (not specified in mfg. catalog)
FIELD ANGLE: 27°
LAMP BASE TYPE: Recessed Single-Contact
LAMP MOUNT: Off Axis
STANDARD LAMP: DWT - 1 KW.
OTHER LAMPS: FER - 1 KW. *(cf=1.18)*
INTEGRAL PATTERN SLOT: Yes

ACCESSORIES AVAILABLE:
Color Frame, Pattern Holder, Snoot

PHOTOMETRICS CHART
(Performance data for this unit are measured using DWT - 1 KW. lamp.)

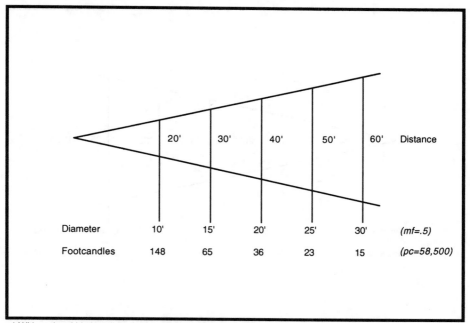

	20'	30'	40'	50'	60'	Distance
Diameter	10'	15'	20'	25'	30'	(mf=.5)
Footcandles	148	65	36	23	15	(pc=58,500)

* With optional iris installed, the model number is 1357I/6W.

KLIEGL BROS.
Model No. 1358/6*

BEAM ANGLE: (not specified in mfg. catalog)

FIELD ANGLE: 22° **

LAMP BASE TYPE: Mogul Bipost

LAMP MOUNT: Off Axis

STANDARD LAMP: BWA - 2 KW.

OTHER LAMPS: CYX - 2 KW. *(cf=1.04)*

INTEGRAL PATTERN SLOT: Yes

ACCESSORIES AVAILABLE:
Color Frame (6½" sq.), Pattern Holder

WEIGHT: 27 lbs.

SIZE: 27" x 18" x 26" (L x W x H)

PHOTOMETRICS CHART
(Performance data for this unit are measured using BWA - 2 KW. lamp.)

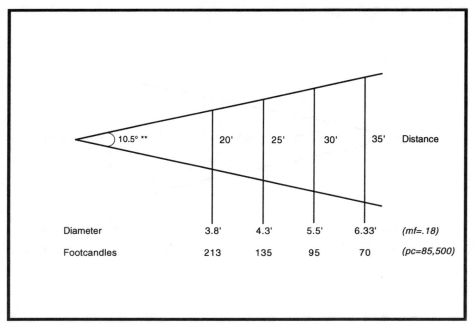

	20'	25'	30'	35'	Distance
10.5° **					
Diameter	3.8'	4.3'	5.5'	6.33'	*(mf=.18)*
Footcandles	213	135	95	70	*(pc=85,500)*

* With optional iris installed, the model number is 1358I/6.
** Using the given distances and diameters, the angle calculates to be 10.5°. Ordinarily this angle should be the same as the given "field angle."

KLIEGL BROS.
Model No. 1366

BEAM ANGLE: (not specified in mfg. catalog)

FIELD ANGLE: 28° *

LAMP BASE TYPE: Mogul Bipost

LAMP MOUNT: Off Axis

STANDARD LAMP: 2M/T30/1 - 2 KW.**

OTHER LAMPS: 1M/T24/5 -   1 KW.
                      1500/T24/6 -1500 W.

INTEGRAL PATTERN SLOT: No

ACCESSORIES AVAILABLE:
Color Frame (10½" sq.)

WEIGHT: 25 lbs.

SIZE: 25¼" x 18" x 24" (L x W x H)

PHOTOMETRICS CHART
(Performance data for this lamp are measured using 2M/T30/1 - 2 KW. lamp)

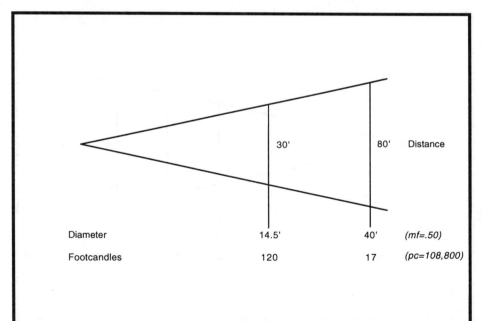

	30'	80'	Distance
Diameter	14.5'	40'	*(mf=.50)*
Footcandles	120	17	*(pc=108,800)*

* Beam and field angles not specified in manufacturer's catalog. This field angle is calculated from given
distance and field diameter data. See Introduction for more information about calculating photometric data.
** This is an obsolete lamp but it is listed as the standard lamp since the manufacturer uses this lamp to
describe the photometric performance. Correction factor *(cf)* information is not available for
 the other lamps because no initial lumens data is available for the obsolete lamp.

KLIEGL BROS.
Model No. 1552*

BEAM ANGLE: 9.5°

FIELD ANGLE: 20° (cosine)

LAMP BASE TYPE: Medium 2-Pin

LAMP MOUNT: Axial

STANDARD LAMP: FEL - 1 KW.

OTHER LAMPS: EHD - 500 W. *(cf=.39)*
                EHG - 750 W. *(cf=.56)*

INTEGRAL PATTERN SLOT: Yes

ACCESSORIES AVAILABLE:
Color Frame (7½" sq.), Pattern Holder, Drop-in Iris

WEIGHT: 22 lbs.

SIZE: 28" x 18" (L x H)

PHOTOMETRICS CHART
(Performance data for this unit are measured at cosine focus using FEL - 1 KW. lamp.)

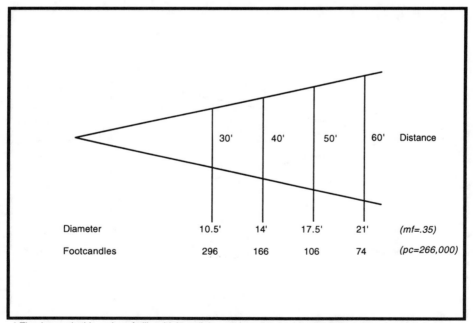

	30'	40'	50'	60'	Distance
Diameter	10.5'	14'	17.5'	21'	*(mf=.35)*
Footcandles	296	166	106	74	*(pc=266,000)*

* The lenses in this series of ellipsoidal spotlights are interchangeable on the same lamp housing. Lens barrels are color-coded to indentify the field angle. This 20° model is color-coded with white.

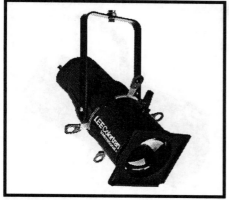

LEE COLORTRAN
Model No. 650-042*

BEAM ANGLE: 8.3° (peak),13.3° (cosine)

FIELD ANGLE: 20°

LAMP BASE TYPE: Medium 2-Pin

LAMP MOUNT: Axial

STANDARD LAMP: FEL - 1 KW.

OTHER LAMPS: EHD - 500 W. *(cf=.39)*
EHG - 750 W. *(cf=.56)*

INTEGRAL PATTERN SLOT: Yes

ACCESSORIES AVAILABLE:
Color Frame (7½" sq.), Pattern Holder, Iris Kit

WEIGHT: 19.7 lbs.

SIZE: 28" x 15" x 25½" (L x W x H)

PHOTOMETRICS CHART
(Performance data for this unit are measured at peak focus using FEL - 1 KW. lamp.)

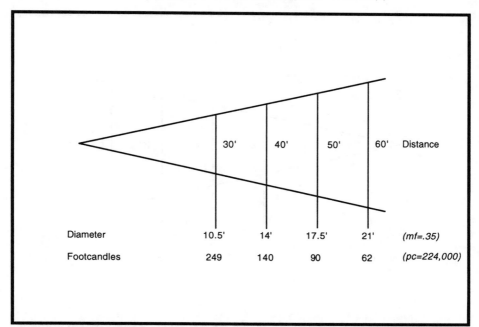

	30'	40'	50'	60'	Distance
Diameter	10.5'	14'	17.5'	21'	*(mf=.35)*
Footcandles	249	140	90	62	*(pc=224,000)*

* This model can be identified with model numbers -042, -044, -045, or -046 depending on the wire leads and connectors supplied.

LIGHTING & ELECTRONICS
Model No. AQ1K-30

BEAM ANGLE: 17°

FIELD ANGLE: 26°

LAMP BASE TYPE: Medium 2-Pin

LAMP MOUNT: Axial

STANDARD LAMP: FEL - 1KW.

OTHER LAMPS: EHG - 750 W. *(cf=.55)*
EHD - 500 W. *(cf=.36)*

INTEGRAL PATTERN SLOT: Yes

ACCESSORIES AVAILABLE:
Color Frame (7½" sq.), Drop-in Iris

WEIGHT: 21 lbs.

SIZE: 21⅝" x 7¾" x 7¾" (L x W x H)

PHOTOMETRICS CHART
(Performance data for this unit are measured at the cosine using FEL - 1KW. lamp.)

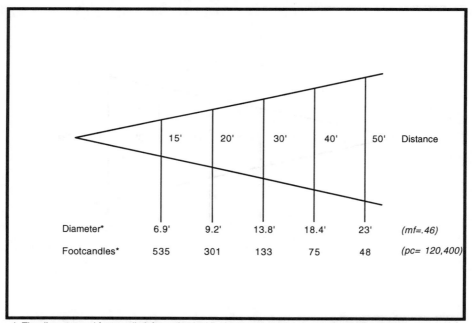

	15'	20'	30'	40'	50'	Distance
Diameter*	6.9'	9.2'	13.8'	18.4'	23'	*(mf=.46)*
Footcandles*	535	301	133	75	48	*(pc= 120,400)*

* The diameter and footcandle information in this chart was calculated using multiplying factor (*mf*) and peak candela (*pc*) data given in catalog. See Introduction for more information about calculating photometric data.

WEIGHT: 18 lbs.
SIZE: 19½" x 7⅝" (L x W)

LIGHTING & ELECTRONICS
Model No. AQ61-12

BEAM ANGLE: 11°

FIELD ANGLE: 26°

LAMP BASE TYPE: Medium 2-Pin

LAMP MOUNT: Axial

STANDARD LAMP: EHG - 750 W.

OTHER LAMPS: EHD - 500 W. *(cf=.68)*
EHF - 750 W. *(cf=1.32)*
Hx600 - 600 W. *(cf=1.93)*

INTEGRAL PATTERN SLOT: Yes

ACCESSORIES AVAILABLE:
Color Frame (7½" sq.), Pattern Holder, Iris

PHOTOMETRICS CHART
(Performance data for this unit are measured using EHG - 750 W. lamp.)

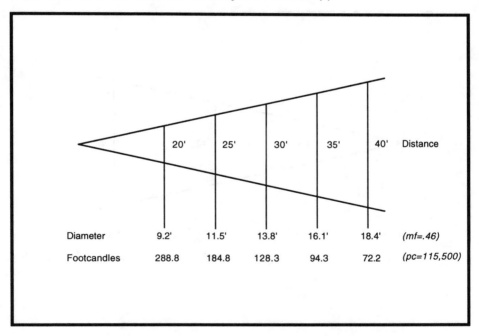

	20'	25'	30'	35'	40'	Distance
Diameter	9.2'	11.5'	13.8'	16.1'	18.4'	*(mf=.46)*
Footcandles	288.8	184.8	128.3	94.3	72.2	*(pc=115,500)*

LITTLE STAGE LIGHTING
Model No. E112

BEAM ANGLE: 12°

FIELD ANGLE: 22°

LAMP BASE TYPE: Medium Prefocus

LAMP MOUNT: Off Axis

STANDARD LAMP: 1MT12/2 - 1 KW.

OTHER LAMPS: (none specified in mfg. catalog)

INTEGRAL PATTERN SLOT: Yes

ACCESSORIES AVAILABLE:
Color Frame (6½" sq.), Pattern Holder, Snoot,
Motorized Color Wheel, Iris

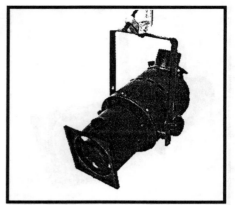

WEIGHT: 18 lbs.

SIZE: 22⅜" x 15" x 18⅛" (L x W x H)

PHOTOMETRICS CHART
(Performance data for this unit are measured using 1MT 12/2 - 1 KW. lamp.)

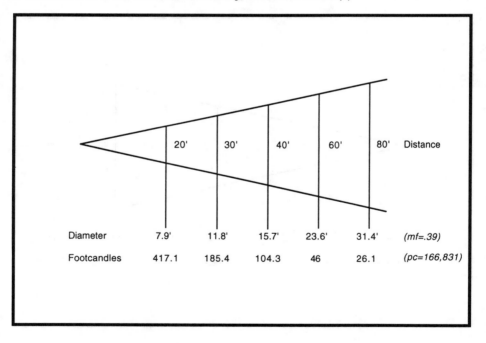

	20'	30'	40'	60'	80'	Distance
Diameter	7.9'	11.8'	15.7'	23.6'	31.4'	*(mf=.39)*
Footcandles	417.1	185.4	104.3	46	26.1	*(pc=166,831)*

LITTLE STAGE LIGHTING
Model No. E212

BEAM ANGLE: 11.2°

FIELD ANGLE: 22.4°

LAMP BASE TYPE: Medium 2-Pin

LAMP MOUNT: Axial

STANDARD LAMP: FEL - 1 KW.

OTHER LAMPS: (none specified in mfg. catalog)

INTEGRAL PATTERN SLOT: Yes

WEIGHT: 18 lbs.

SIZE: 22¼" x 15" x 18⅛" (L x W x H)

ACCESSORIES AVAILABLE:
Color Frame (6½" sq.), Pattern Holder, Snoot,
Motorized Color Wheel, Iris

PHOTOMETRICS CHART
(Performance data for this unit are measured using FEL - 1 KW. lamp.)

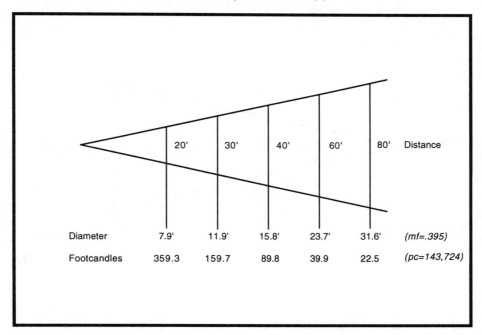

	20'	30'	40'	60'	80'	Distance
Diameter	7.9'	11.9'	15.8'	23.7'	31.6'	*(mf=.395)*
Footcandles	359.3	159.7	89.8	39.9	22.5	*(pc=143,724)*

LITTLE STAGE LIGHTING
Model No. E309

BEAM ANGLE: 13.4°

FIELD ANGLE: 26°

LAMP BASE TYPE: Mini-Can Screw

LAMP MOUNT: (not specified in mfg. catalog)

STANDARD LAMP: EVR - 500 W.

OTHER LAMPS: (none specified in mfg. catalog)

INTEGRAL PATTERN SLOT: Yes

ACCESSORIES AVAILABLE:
Color Frame (5" sq.), Pattern Holder

WEIGHT: (not specified in manufacturer's catalog)

SIZE: 18⅝" x 10¾" x 9½" (L x W x H)

PHOTOMETRICS CHART
(Performance data for this unit are measured using EVR - 500 W. lamp.)

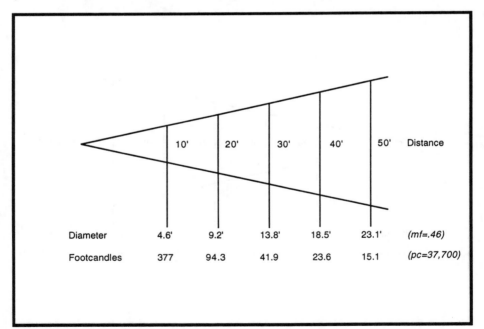

	10'	20'	30'	40'	50'	Distance
Diameter	4.6'	9.2'	13.8'	18.5'	23.1'	*(mf=.46)*
Footcandles	377	94.3	41.9	23.6	15.1	*(pc=37,700)*

WEIGHT: 16 lbs.

SIZE: 18" x 12" x 24" (L x W x H)

STRAND CENTURY
Model Nos. 2212 (w/ shutters)
2213 (plus iris)

BEAM ANGLE: 11° (peak), 21° (cosine)

FIELD ANGLE: 25° (peak)

LAMP BASE TYPE: Medium 2-Pin

LAMP MOUNT: Axial

STANDARD LAMP: FEL - 1 KW.

OTHER LAMPS: EHD - 500 W. *(cf=.38)*
EHG - 750 W. *(cf=.56)*

INTEGRAL PATTERN SLOT: Yes

ACCESSORIES AVAILABLE:
Color Frame (7½" sq.), Pattern Holder, Snoot

PHOTOMETRICS CHART
(Performance data for this unit are measured at peak focus using FEL - 1 KW. lamp.)

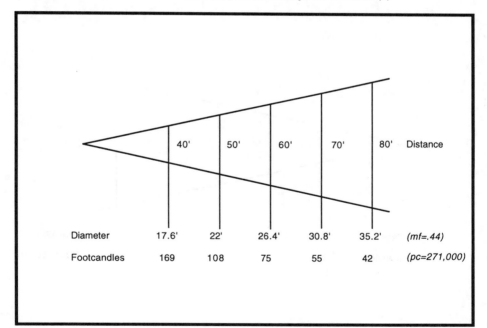

	40'	50'	60'	70'	80'	Distance
Diameter	17.6'	22'	26.4'	30.8'	35.2'	*(mf=.44)*
Footcandles	169	108	75	55	42	*(pc=271,000)*

235

STRAND LIGHTING
Model No. 2230, "Leko"

BEAM ANGLE: 15° peak, 20° cosine

FIELD ANGLE: 27° peak, 30° cosine

LAMP BASE TYPE: Medium 2-Pin

LAMP MOUNT: Axial

STANDARD LAMP: FEL - 1 KW.

OTHER LAMPS: EHD - 500 W. *(cf=.36)*
EHG - 750 W. *(cf=.54)*

INTEGRAL PATTERN SLOT: Yes

ACCESSORIES AVAILABLE:
Color Frame (7½" sq.), Pattern Holder, Drop-in Iris,
High Hat

WEIGHT: 12.3 lbs.

SIZE: 21" x 11¾" (L x W)

PHOTOMETRICS CHART
(Performance data for this unit are measured at peak focus using FEL - 1KW. lamp.)

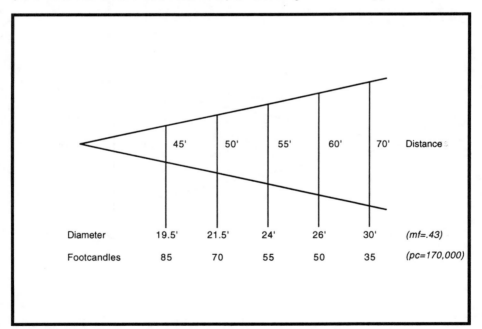

	45'	50'	55'	60'	70'	Distance
Diameter	19.5'	21.5'	24'	26'	30'	*(mf=.43)*
Footcandles	85	70	55	50	35	*(pc=170,000)*

STRAND LIGHTING
Model No. 2240, "Leko"

BEAM ANGLE: 21° peak, 17° cosine

FIELD ANGLE: 29° peak, 38° cosine

LAMP BASE TYPE: Medium 2-Pin

LAMP MOUNT: Axial

STANDARD LAMP: FEL - 1 KW.

OTHER LAMPS: EHD - 500 W. *(cf=.36)*
EHG - 750 W. *(cf=.54)*

INTEGRAL PATTERN SLOT: Yes

WEIGHT: 12.3 lbs.

SIZE: 18¾" x 11¾" (L x W)

ACCESSORIES AVAILABLE:
Color Frame (7½" sq.), Pattern Holder, Drop-in Iris,
High Hat

PHOTOMETRICS CHART
(Performance data for this unit are measured at peak focus using FEL - 1KW. lamp.)

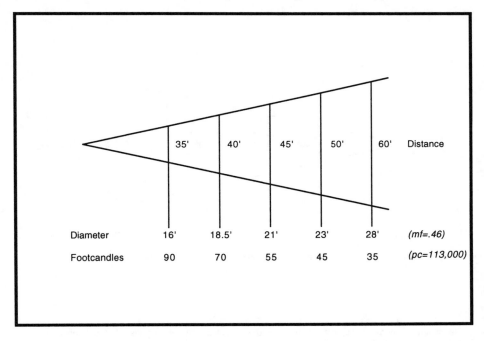

	35'	40'	45'	50'	60'	Distance
Diameter	16'	18.5'	21'	23'	28'	*(mf=.46)*
Footcandles	90	70	55	45	35	*(pc=113,000)*

TIMES SQUARE LIGHTING
Model No. Q6E6x16

BEAM ANGLE: (not specified in mfg. catalog)

FIELD ANGLE: 20° *

LAMP BASE TYPE: Medium 2-Pin

LAMP MOUNT: Axial

STANDARD LAMP: EHG - 750 W.

OTHER LAMPS: EHD - 500 W. *(cf=.69)*
FEL - 1 KW. *(cf=1.79)*

INTEGRAL PATTERN SLOT: Yes

ACCESSORIES AVAILABLE:
Color Frame (7½" sq.), Color Wheel, Pattern Holder

WEIGHT: (not specified in manufacturer's catalog)

SIZE: (not specified in manufacturer's catalog)

PHOTOMETRICS CHART
(Information not provided in manufacturer's catalog about lamp used to measure performance data.)

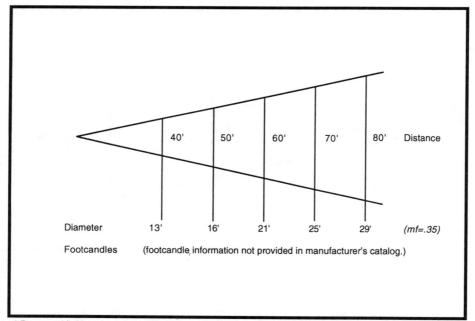

| | 40' | 50' | 60' | 70' | 80' | Distance |

Diameter 13' 16' 21' 25' 29' *(mf=.35)*

Footcandles (footcandle information not provided in manufacturer's catalog.)

* Beam and field angles not specified in manufacturer's catalog. This field angle is calculated from given distance and field diameter data. See Introduction for more information about calculating photometric data.

WEIGHT: (not specified in manufacturer's catalog)
SIZE: 22⅝" x 13" x 16¼" (L x W x H)

ALTMAN STAGE LIGHTING
Model No. 360Q 6x16

BEAM ANGLE: 8.5°

FIELD ANGLE: 19°

LAMP BASE TYPE: Medium 2-Pin

LAMP MOUNT: Axial

STANDARD LAMP: EHF - 750 W.

OTHER LAMPS: EHD - 500 W. *(cf=.5)*
EHG - 750 W. *(cf=.75)*

INTEGRAL PATTERN SLOT: Yes

ACCESSORIES AVAILABLE:
Color Frame (7½" sq.), Pattern Holder, Snoot, Motorized & Non-motorized Color Wheels, Iris

PHOTOMETRICS CHART
(Performance data for this unit are measured using EHF - 750 W. lamp.)

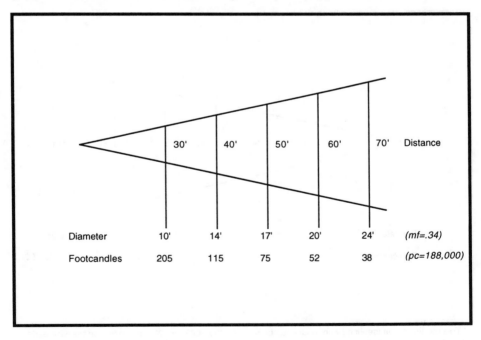

	30'	40'	50'	60'	70'	Distance
Diameter	10'	14'	17'	20'	24'	*(mf=.34)*
Footcandles	205	115	75	52	38	*(pc=188,000)*

BERKEY COLORTRAN
Model No. 213-092*

BEAM ANGLE: 6.8°

FIELD ANGLE: 12°

LAMP BASE TYPE: Medium 2-Pin

LAMP MOUNT: Axial

STANDARD LAMP: FEL - 1 KW.

OTHER LAMPS: EHD - 500 W. *(cf=.39)*
EHG - 750 W. *(cf=.56)*

INTEGRAL PATTERN SLOT: Yes

ACCESSORIES AVAILABLE:
Color Frame (7½" sq.), Pattern Holder, Iris Kit

WEIGHT: 19.1 lbs.

SIZE: 27½" x 15" x 24½" (L x W x H)

PHOTOMETRICS CHART
(Performance data for this unit are measured at peak focus using FEL - 1 KW. lamp.)

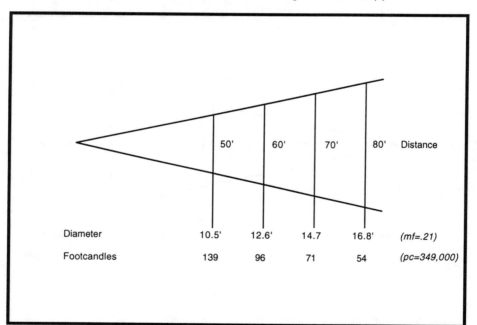

	50'	60'	70'	80'	Distance
Diameter	10.5'	12.6'	14.7'	16.8'	*(mf=.21)*
Footcandles	139	96	71	54	*(pc=349,000)*

* This model can be identified with model numbers -092, -095, -096, or -097 depending on the wire leads and connectors supplied.

CENTURY STRAND
Model No. Patt 264

BEAM ANGLE: 17°

FIELD ANGLE: (not specified in mfg. catalog)

LAMP BASE TYPE: Medium Prefocus

LAMP MOUNT: Off Axis

STANDARD LAMP: DNV - 1 KW.

OTHER LAMPS: DNT - 750 W. *(cf=.62)*

INTEGRAL PATTERN SLOT: Yes

ACCESSORIES AVAILABLE:
Color Frame (6½" sq.), 12-leaf Iris

WEIGHT: 16.5 lbs.

SIZE: 22¼" x 12" x 15½" (L x W x H)

PHOTOMETRICS CHART
(Performance data for this unit are measured using DNV - 1 KW. lamp.)

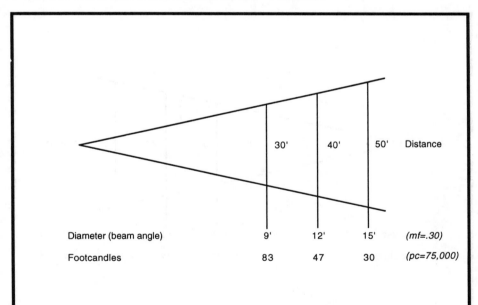

	30'	40'	50'	Distance
Diameter (beam angle)	9'	12'	15'	*(mf=.30)*
Footcandles	83	47	30	*(pc=75,000)*

ELECTRO CONTROLS
Model Nos. 3213 (w/ shutters)
3214 (w/ iris)

BEAM ANGLE: (not specified in mfg. catalog)

FIELD ANGLE: 12° *

LAMP BASE TYPE: Extended Mog. End Prong

LAMP MOUNT: Off Axis

STANDARD LAMP: 500 PAR 64/NSP - 500 W.**

OTHER LAMPS: Q1000PAR64/NSP-1 KW. *(cf=1.84)*

INTEGRAL PATTERN SLOT: Yes (3213 only)

ACCESSORIES AVAILABLE:
Color Frame, 2-way Barn Door, Iris Kit, Shutter Kit, Snoot

WEIGHT: (not specified in manufacturer's catalog)

SIZE: (not specified in manufacturer's catalog)

PHOTOMETRICS CHART

(Performance data for this unit are measured using 500 PAR 64/NSP 500 W. lamp.)

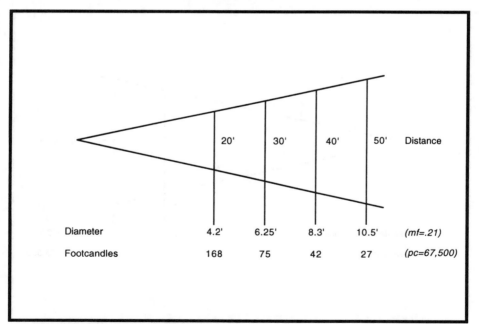

	20'	30'	40'	50'	Distance
Diameter	4.2'	6.25'	8.3'	10.5'	*(mf=.21)*
Footcandles	168	75	42	27	*(pc=67,500)*

* Beam and field angles not specified in manufacturer's catalog. This field angle is calculated from given distance and field diameter data. See Introduction for more information about calculating photometric data.
**This unusual instrument uses a 500 or 1000 W. PAR 64 lamp for a source, with an 8x18 step lens to gather and concentrate the light, and a 6x18 lens to focus it.

ELECTRO CONTROLS
Model No. 3365*

BEAM ANGLE: (not specified in mfg. catalog)

FIELD ANGLE: 18.4°

LAMP BASE TYPE: Mini-Can Screw

LAMP MOUNT: Axial

STANDARD LAMP: EGJ - 1 KW.

OTHER LAMPS: EGE - 500 W. *(cf=.38)*
EGG - 750 W. *(cf=.57)*

INTEGRAL PATTERN SLOT: Yes

ACCESSORIES AVAILABLE:
Color Frame (7½" sq.), Iris (as an option)

WEIGHT: (not specified in manufacturer's catalog)

SIZE: (not specified in manufacturer's catalog)

PHOTOMETRICS CHART
(Performance data for this unit are measured using EGJ - 1 KW. lamp.)

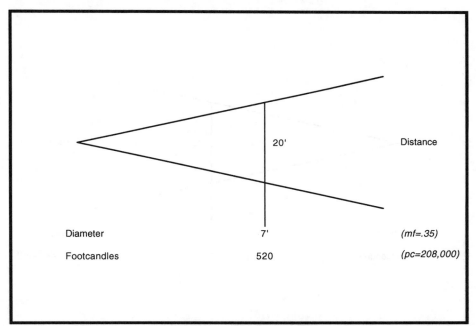

	20'	Distance
Diameter	7'	*(mf=.35)*
Footcandles	520	*(pc=208,000)*

* This model number, 3365, can accept 6 different lens configurations: 1 6x6 step lens, 2 6x9 plano lenses, 2 6x12 lenses, 2 6x16 plano lenses, 2 6x20 plano lenses, and 1 6x8 step lens (this page). See Index for the page numbers of other versions of model 3365.

ELECTRO CONTROLS
Model No. 3365*

BEAM ANGLE: (not specified in mfg. catalog)

FIELD ANGLE: 16°

LAMP BASE TYPE: Mini-Can Screw

LAMP MOUNT: Axial

STANDARD LAMP: EGJ - 1 KW.

OTHER LAMPS: EGE - 500 W. *(cf=.38)*
EGG - 750 W. *(cf=.57)*

INTEGRAL PATTERN SLOT: Yes

ACCESSORIES AVAILABLE:
Color Frame (7½" sq.), Iris (as an option)

WEIGHT: (not specified in manufacturer's catalog)

SIZE: (not specified in manufacturer's catalog)

PHOTOMETRICS CHART
(Performance data for this unit are measured using EGJ - 1 KW. lamp.)

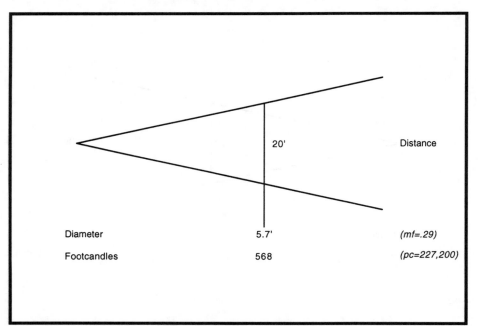

	20'	Distance
Diameter	5.7'	*(mf=.29)*
Footcandles	568	*(pc=227,200)*

* This model number, 3365, can accept 6 different lens configurations: 1 6x6 step lens, 2 6x9 plano lenses, 2 6x12 lenses, 2 6x16 plano lenses, 2 6x20 plano lenses (this page), and 1 6x8 step lens. See Index for the page numbers of other versions of model 3365.

ELECTRONIC THEATRE CONTROLS (ETC) Model No. 419, "Source Four"

BEAM ANGLE: 13°

FIELD ANGLE: 19°

LAMP BASE TYPE: Medium 2-Pin

LAMP MOUNT: Axial

STANDARD LAMP: HPL 575/115 - 575 W.

OTHER LAMPS: HPL 550/77 - 550W.*

INTEGRAL PATTERN SLOT: Yes

WEIGHT: 15.9 lbs.

SIZE: 22½" x 12⅜" x 20⅝" (L x W x H)

ACCESSORIES AVAILABLE:
Color Frame (6¼" sq.), Pattern Holder, Donut, Snoot

PHOTOMETRICS CHART

(Performance data for this unit are measured at cosine focus using HPL 575/110 lamp.)

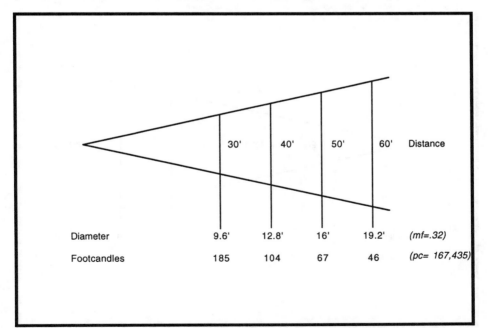

	30'	40'	50'	60'	Distance
Diameter	9.6'	12.8'	16'	19.2'	*(mf=.32)*
Footcandles	185	104	67	46	*(pc= 167,435)*

* The HPL 550/77 lamp is used with an ETC multiplexing system, used to allow multiple units to share a common circuit, with separate dimming of each.

KLIEGL BROS.
Model No. 1355*

BEAM ANGLE: (not specified in mfg. catalog)

FIELD ANGLE: 18°

LAMP BASE TYPE: Medium 2-Pin

LAMP MOUNT: Off Axis

STANDARD LAMP: EHG - 750 W.

OTHER LAMPS: EHD - 500 W. *(cf=.67)*
　　　　　　　 EHF - 750 W. *(cf=1.33)*

INTEGRAL PATTERN SLOT: Yes

ACCESSORIES AVAILABLE:
Color Frame (6½" sq.), Pattern Holder, Snoot, Iris
(as an option)

WEIGHT: 12 lbs.
SIZE: 22" x 9" x 19½"

PHOTOMETRICS CHART
(Performance data for this unit are measured using EHG - 750 W. lamp.)

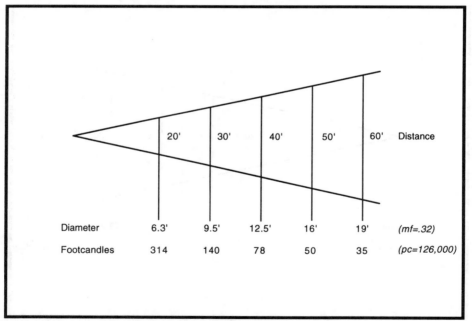

	20'	30'	40'	50'	60'	Distance
Diameter	6.3'	9.5'	12.5'	16'	19'	*(mf=.32)*
Footcandles	314	140	78	50	35	*(pc=126,000)*

* When this instrument is equipped with optional iris the model number is 1355I.

WEIGHT: 21 lbs.
SIZE: 26" x 10" x 27¼" (L x W x H)

KLIEGL BROS.
Model No. 1357/6*

BEAM ANGLE: (not specified in mfg. catalog)

FIELD ANGLE: 18°

LAMP BASE TYPE: Recessed Single-Contact

LAMP MOUNT: Off Axis

STANDARD LAMP: DWT - 1 KW.

OTHER LAMPS: FER - 1 KW. *(cf=1.3)*

INTEGRAL PATTERN SLOT: Yes

ACCESSORIES AVAILABLE:
Color Frame (6½" sq.), Pattern Holder, Snoot, Iris
(as an option)

PHOTOMETRICS CHART
(Performance data for this unit are measured using DWT - 1 KW. lamp.)

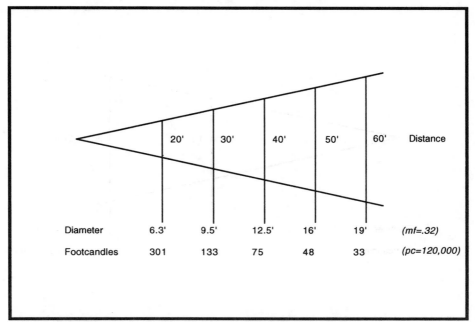

	20'	30'	40'	50'	60'	Distance
Diameter	6.3'	9.5'	12.5'	16'	19'	*(mf=.32)*
Footcandles	301	133	75	48	33	*(pc=120,000)*

* When this instrument is equipped with optional iris the model number is 1357I/6.

247

KLIEGL BROS.
Model No. 1365

BEAM ANGLE: (not specified in mfg. catalog)

FIELD ANGLE: 17° *

LAMP BASE TYPE: Medium Bipost

LAMP MOUNT: Off Axis

STANDARD LAMP: 500T14/8 - 500 W.**

OTHER LAMPS: 500T14/7 - 500 W.
750/T14 - 750 W.

INTEGRAL PATTERN SLOT: No

ACCESSORIES AVAILABLE:
Color Frame (6½" sq.)

WEIGHT: 10 lbs.

SIZE: 19" x 13½" x 17" (L x W x H)

PHOTOMETRICS CHART

(Performance data for this unit are measured using 500T14/8 - 500 W. lamp)

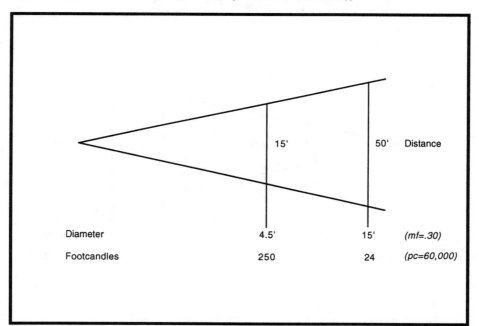

	15'	50'	Distance
Diameter	4.5'	15'	(mf=.30)
Footcandles	250	24	(pc=60,000)

* Beam and field angles not specified in manufacturer's catalog. This field angle is calculated from given distance and field diameter data. See Introduction for more information about calculating photometric data.
** This is an obsolete lamp but it is listed as the standard lamp since the manufacturer uses this lamp to describe the photometric performance. Correction factor *(cf)* information is not available for the other lamp because no initial lumens data is available for the obsolete lamp.

KLIEGL BROS.
Model No. 1368

BEAM ANGLE: (not specified in mfg. catalog)

FIELD ANGLE: 18° *

LAMP BASE TYPE: Mogul Bipost

LAMP MOUNT: Off Axis

STANDARD LAMP: 2M/T30/1 - 2 KW.**

OTHER LAMPS: 1M/T24/5 -   1 KW.
1500/T24/6 -1.5 KW.

INTEGRAL PATTERN SLOT: No

ACCESSORIES AVAILABLE:
Color Frame, Iris (as an option)

WEIGHT: 26 lbs.

SIZE: (not specified in manufacturer's catalog)

PHOTOMETRICS CHART

(Performance data for this unit are measured using 2M/T30/1 - 2 KW. lamp.)

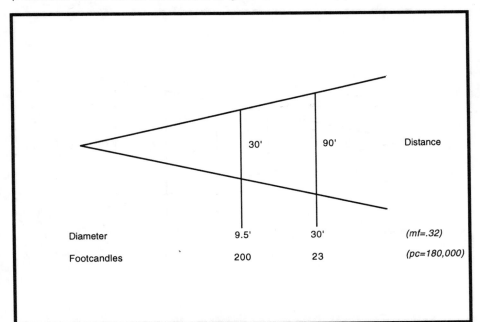

	30'	90'	Distance
Diameter	9.5'	30'	(mf=.32)
Footcandles	200	23	(pc=180,000)

* Beam and field angles not specified in manufacturer's catalog. This field angle is calculated from given distance and field diameter data. See Introduction for more information about calculating photometric data.
** This is an obsolete lamp but it is listed as the standard lamp since the manufacturer uses this lamp to describe the photometric performance. Correction factor *(cf)* information is not available for the other lamps because no initial lumens data is available for the obsolete lamp.

KLIEGL BROS.
Model No. 1551*

BEAM ANGLE: (not specified in mfg. catalog)

FIELD ANGLE: 15° (cosine)

LAMP BASE TYPE: Medium 2-Pin

LAMP MOUNT: Axial

STANDARD LAMP: FEL - 1 KW.

OTHER LAMPS: EHD - 500 W. *(cf=.39)*
               EHG - 750 W. *(cf=.56)*

INTEGRAL PATTERN SLOT: Yes

ACCESSORIES AVAILABLE:
Color Frame (7½" sq.), Pattern Holder, Drop-in Iris

WEIGHT: 22 lbs.
SIZE: 28" x 18" (L x H)

PHOTOMETRICS CHART
(Performance data for this unit are measured at cosine focus using FEL - 1 KW. lamp.)

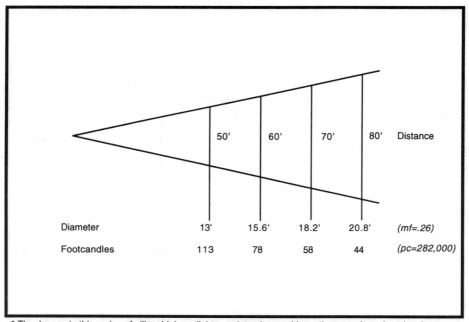

	50'	60'	70'	80'	Distance
Diameter	13'	15.6'	18.2'	20.8'	*(mf=.26)*
Footcandles	113	78	58	44	*(pc=282,000)*

* The lenses in this series of ellipsoidal spotlights are interchangeable on the same lamp housing. Lens barrels are color-coded to indentify the field angle. This 15° model is color-coded with tan.

LEE COLORTRAN
Model No. 650-052*

BEAM ANGLE: 6.8° (peak),10° (cosine)

FIELD ANGLE: 15° (peak & cosine)

LAMP BASE TYPE: Medium 2-Pin

LAMP MOUNT: Axial

STANDARD LAMP: FEL - 1 KW.

OTHER LAMPS: EHD - 500 W. *(cf=.39)*
EHG - 750 W. *(cf=.56)*

INTEGRAL PATTERN SLOT: Yes

ACCESSORIES AVAILABLE:
Color Frame (7½" sq.), Pattern Holder, Iris Kit

WEIGHT: 19.3 lbs.

SIZE: 29½" x 15" x 25½" (L x W x H)

PHOTOMETRICS CHART
(Performance data for this unit are measured at peak focus using FEL - 1 KW. lamp.)

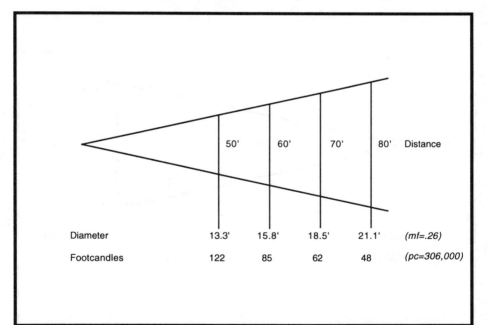

	50'	60'	70'	80'	Distance
Diameter	13.3'	15.8'	18.5'	21.1'	*(mf=.26)*
Footcandles	122	85	62	48	*(pc=306,000)*

* This model can be identified with model numbers -052, -054, -055, or -056 depending on the wire leads and connectors supplied.

LIGHTING & ELECTRONICS
Model No. 61-16

BEAM ANGLE: 12°

FIELD ANGLE: 19°

LAMP BASE TYPE: Medium Prefocus

LAMP MOUNT: Off Axis

STANDARD LAMP: DNT - 750 W.

OTHER LAMPS: DEB - 500 W. *(cf=.53)*
DNY - 1 KW. *(cf=1.47)*

INTEGRAL PATTERN SLOT: Yes

ACCESSORIES AVAILABLE:
Color Frame (7½" sq.)

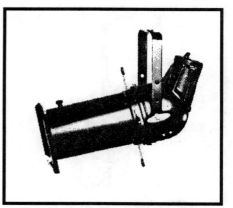

WEIGHT: (not specified in manufacturer's catalog)

SIZE: 21¾" (L)

PHOTOMETRICS CHART

(Performance data for this unit are measured using DNT - 750 W. lamp.)

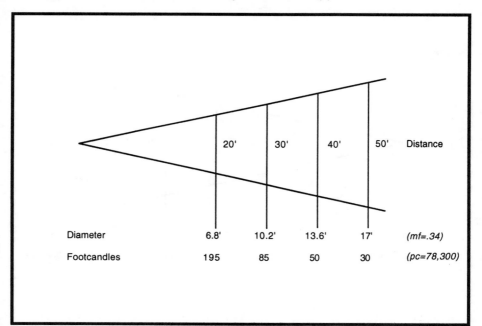

	20'	30'	40'	50'	Distance
Diameter	6.8'	10.2'	13.6'	17'	*(mf=.34)*
Footcandles	195	85	50	30	*(pc=78,300)*

WEIGHT: 21 lbs.

SIZE: 30⅝" x 7¾" x 7¾" (L x W x H)

LIGHTING & ELECTRONICS
Model No. AQ1K-10

BEAM ANGLE: 7°

FIELD ANGLE: 11°

LAMP BASE TYPE: Medium 2-Pin

LAMP MOUNT: Axial

STANDARD LAMP: FEL - 1KW.

OTHER LAMPS: EHG - 750 W. *(cf=.55)*
EHD - 500 W. *(cf=.36)*

INTEGRAL PATTERN SLOT: Yes

ACCESSORIES AVAILABLE:
Color Frame (7½" sq.), Drop-in Iris

PHOTOMETRICS CHART
(Performance data for this unit are measured at the cosine using FEL - 1KW. lamp.)

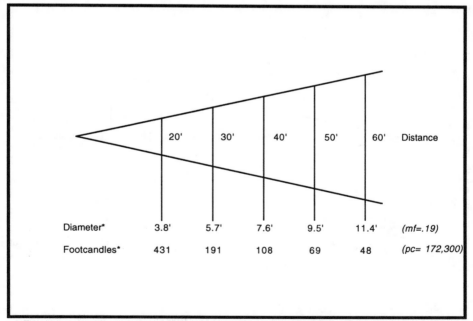

	20'	30'	40'	50'	60'	Distance
Diameter*	3.8'	5.7'	7.6'	9.5'	11.4'	*(mf=.19)*
Footcandles*	431	191	108	69	48	*(pc= 172,300)*

* The diameter and footcandle information in this chart was calculated using multiplying factor (*mf*) and peak candela (*pc*) data given in catalog. See Introduction for more information about calculating photometric data.

LIGHTING & ELECTRONICS
Model No. AQ1K-20

BEAM ANGLE: 12°

FIELD ANGLE: 19°

LAMP BASE TYPE: Medium 2-Pin

LAMP MOUNT: Axial

STANDARD LAMP: FEL - 1KW.

OTHER LAMPS: EHG - 750 W. *(cf=.55)*
EHD - 500 W. *(cf=.36)*

INTEGRAL PATTERN SLOT: Yes

ACCESSORIES AVAILABLE:
Color Frame (7½" sq.), Drop-in Iris

WEIGHT: 21 lbs.

SIZE: 23⅝" x 7¾" x 7¾" (L x W x H)

PHOTOMETRICS CHART
(Performance data for this unit are measured at the cosine using FEL - 1KW. lamp.)

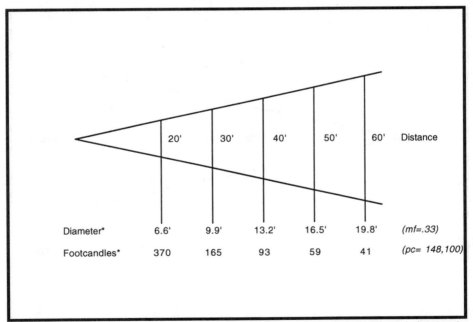

	20'	30'	40'	50'	60'	Distance
Diameter*	6.6'	9.9'	13.2'	16.5'	19.8'	*(mf=.33)*
Footcandles*	370	165	93	59	41	*(pc= 148,100)*

* The diameter and footcandle information in this chart was calculated using multiplying factor (*mf*) and peak candela (*pc*) data given in catalog. See Introduction for more information about calculating photometric data.

WEIGHT: 18 lbs.

SIZE: 21½" (L)

LIGHTING & ELECTRONICS
Model No. AQ61-16

BEAM ANGLE: 8.5°

FIELD ANGLE: 19°

LAMP BASE TYPE: Medium 2-Pin

LAMP MOUNT: Axial

STANDARD LAMP: EHG - 750 W.

OTHER LAMPS: EHD - 500 W. *(cf=.69)*
EHF - 750 W. *(cf=1.32)*
Hx600 - 600 W. *(cf=1.93)*

INTEGRAL PATTERN SLOT: Yes

ACCESSORIES AVAILABLE:
Color Frame (7½" sq.), Pattern Holder, Iris

PHOTOMETRICS CHART

(Performance data for this unit are measured using EHG - 750 W. lamp.)

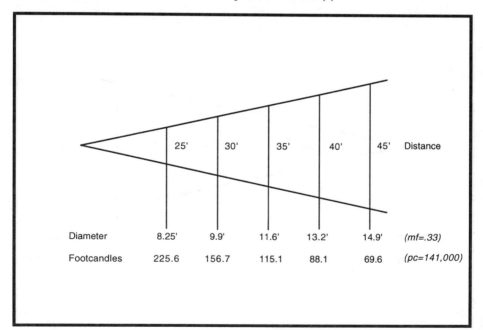

	25'	30'	35'	40'	45'	Distance
Diameter	8.25'	9.9'	11.6'	13.2'	14.9'	*(mf=.33)*
Footcandles	225.6	156.7	115.1	88.1	69.6	*(pc=141,000)*

LITTLE STAGE LIGHTING
Model No. E116

BEAM ANGLE: 10.1°

FIELD ANGLE: 16.3°

LAMP BASE TYPE: Medium Prefocus

LAMP MOUNT: Off Axis

STANDARD LAMP: 1MT12/2 - 1 KW.

OTHER LAMPS: (none specified in mfg. catalog)

INTEGRAL PATTERN SLOT: Yes

ACCESSORIES AVAILABLE:
Color Frame (6½" sq.), Pattern Holder, Snoot,
Motorized Color Wheel, Iris

WEIGHT: (not specified in manufacturer's catalog)

SIZE: 24⅜" x 15" x 18⅛" (L x W x H)

PHOTOMETRICS CHART
(Performance data for this unit are measured using 1MT 12/2 - 1 KW. lamp.)

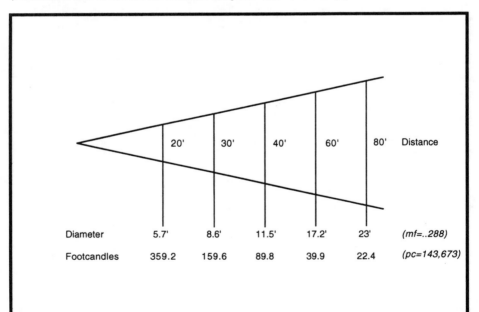

	20'	30'	40'	60'	80'	Distance
Diameter	5.7'	8.6'	11.5'	17.2'	23'	*(mf=..288)*
Footcandles	359.2	159.6	89.8	39.9	22.4	*(pc=143,673)*

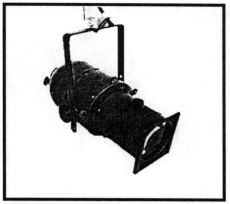

LITTLE STAGE LIGHTING
Model No. E216

BEAM ANGLE: 9.6°

FIELD ANGLE: 16.6°

LAMP BASE TYPE: Medium 2-Pin

LAMP MOUNT: Axial

STANDARD LAMP: FEL - 1 KW.

OTHER LAMPS: (none specified in mfg. catalog)

INTEGRAL PATTERN SLOT: Yes

WEIGHT: (not specified in manufacturer's catalog)

SIZE: 24¼" x 15" x 18⅛" (L x W x H)

ACCESSORIES AVAILABLE:
Color Frame (7½" sq.), Pattern Holder, Snoot,
Motorized Color Wheel, Iris

PHOTOMETRICS CHART

(Performance data for this unit are measured using FEL - 1 KW. lamp.)

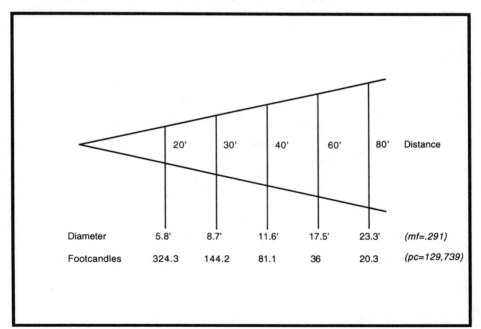

	20'	30'	40'	60'	80'	Distance
Diameter	5.8'	8.7'	11.6'	17.5'	23.3'	*(mf=.291)*
Footcandles	324.3	144.2	81.1	36	20.3	*(pc=129,739)*

LITTLE STAGE LIGHTING
Model No. E312

BEAM ANGLE: 10.6°

FIELD ANGLE: 19°

LAMP BASE TYPE: Mini-Can Screw

LAMP MOUNT: (not specified in mfg. catalog)

STANDARD LAMP: EVR - 500 W.

OTHER LAMPS: (none specified in mfg. catalog)

INTEGRAL PATTERN SLOT: Yes

ACCESSORIES AVAILABLE:
Color Frame (5" sq.), Pattern Holder

WEIGHT: (not specified in manufacturer's catalog)

SIZE: 20¼" x 10¾" x 9½" (L x W x H)

PHOTOMETRICS CHART
(Performance data for this unit are measured using EVR - 500 W. lamp.)

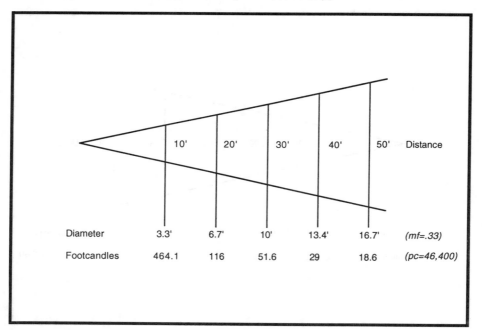

	10'	20'	30'	40'	50'	Distance
Diameter	3.3'	6.7'	10'	13.4'	16.7'	*(mf=.33)*
Footcandles	464.1	116	51.6	29	18.6	*(pc=46,400)*

WEIGHT: 15.5 lbs.

SIZE: 23¾" x 12" x 24" (L x W x H)

STRAND CENTURY
Model Nos. 2111 (w/ shutters)
2112 (plus iris)

BEAM ANGLE: 9° (peak), 10° (cosine)

FIELD ANGLE: 14° (peak)

LAMP BASE TYPE: Medium 2-Pin

LAMP MOUNT: Axial

STANDARD LAMP: FEL - 1 KW.

OTHER LAMPS: EHD - 500 W. *(cf=.38)*
EHG - 750 W. *(cf=.56)*

INTEGRAL PATTERN SLOT: Yes

ACCESSORIES AVAILABLE:
Color Frame (7½" sq.), Pattern Holder, Snoot

PHOTOMETRICS CHART
(Performance data for this unit are measured at peak focus using FEL - 1 KW. lamp.)

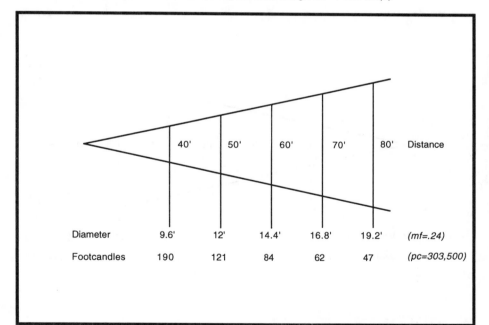

	40'	50'	60'	70'	80'	Distance
Diameter	9.6'	12'	14.4'	16.8'	19.2'	*(mf=.24)*
Footcandles	190	121	84	62	47	*(pc=303,500)*

259

STRAND CENTURY
Model Nos. 2216 (w/ shutters)
2217 (plus iris)

BEAM ANGLE: 7° (peak), 15° (cosine)

FIELD ANGLE: 17°

LAMP BASE TYPE: Medium 2-Pin

LAMP MOUNT: Axial

STANDARD LAMP: FEL - 1 KW.

OTHER LAMPS: EHD - 500 W. *(cf=.36)*
EHG - 750 W. *(cf=.54)*

INTEGRAL PATTERN SLOT: Yes

ACCESSORIES AVAILABLE:
Color Frame (7½" sq.), Pattern Holder, Snoot, Iris
Kit (2216 only)

WEIGHT: 16 lbs.

SIZE: 19¼" x 12" x 24" (L x W x H)

PHOTOMETRICS CHART
(Performance data for this unit are measured at peak focus using FEL - 1 KW. lamp.)

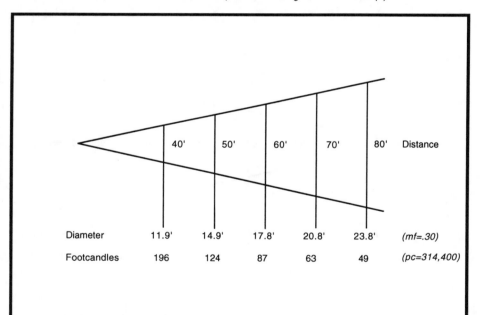

	40'	50'	60'	70'	80'	Distance
Diameter	11.9'	14.9'	17.8'	20.8'	23.8'	*(mf=.30)*
Footcandles	196	124	87	63	49	*(pc=314,400)*

STRAND CENTURY
Model No. 2355*

BEAM ANGLE: 11°

FIELD ANGLE: 11°

LAMP BASE TYPE: Medium Prefocus

LAMP MOUNT: Off Axis

STANDARD LAMP: BTL - 500 W.

OTHER LAMPS: BTM - 500 W. *(cf=1.2)*

INTEGRAL PATTERN SLOT: Yes

WEIGHT: 12 lbs.

SIZE: 19¾" x 11¼" x 12¼" (L x W x H)

ACCESSORIES AVAILABLE:
Color Frame, Removable Iris

PHOTOMETRICS CHART
(Performance data for this unit are measured using BTL - 500W. lamp.)

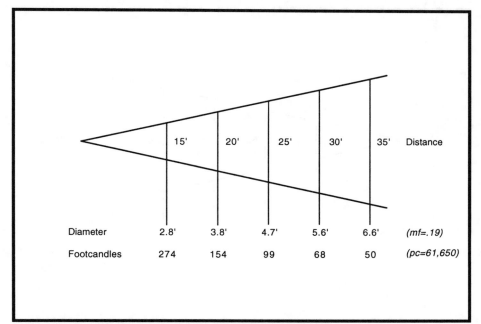

	15'	20'	25'	30'	35'	Distance
Diameter	2.8'	3.8'	4.7'	5.6'	6.6'	*(mf=.19)*
Footcandles	274	154	99	68	50	*(pc=61,650)*

* This unit is equipped with a handle to enable it to be used as a followspot.

Strand Lighting
Model No. 2215,* "Leko"

BEAM ANGLE: 10°

FIELD ANGLE: 14° peak, 15° cosine

LAMP BASE TYPE: Medium 2-Pin

LAMP MOUNT: Axial

STANDARD LAMP: FEL - 1 KW.

OTHER LAMPS: EHD - 500 W. *(cf=.36)*
EHG - 750 W. *(cf=.54)*

INTEGRAL PATTERN SLOT: Yes

ACCESSORIES AVAILABLE:
Color Frame (7½" sq.), Pattern Holder, Drop-in Iris,
High Hat

WEIGHT: 13.2 lbs.

SIZE: 26¾" x 11¾" (L x W)

PHOTOMETRICS CHART

(Performance data for this unit are measured at peak focus using FEL - 1KW. lamp.)

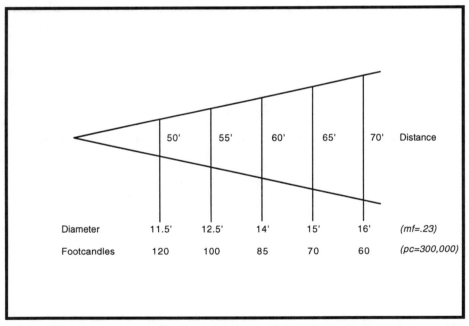

	50'	55'	60'	65'	70'	Distance
Diameter	11.5'	12.5'	14'	15'	16'	*(mf=.23)*
Footcandles	120	100	85	70	60	*(pc=300,000)*

* In the 1960s, Strand Century used model number 2215 for a 3½" x 5" ERS. See page 143.

STRAND LIGHTING
Model No. 2220, "Leko"

BEAM ANGLE: 17° peak, 13° cosine

FIELD ANGLE: 19° peak, 20° cosine

LAMP BASE TYPE: Medium 2-Pin

LAMP MOUNT: Axial

STANDARD LAMP: FEL - 1 KW.

OTHER LAMPS: EHD - 500 W. *(cf=.36)*
                EHG - 750 W. *(cf=.54)*

INTEGRAL PATTERN SLOT: Yes

ACCESSORIES AVAILABLE:
Color Frame (7½" sq.), Pattern Holder, Drop-in Iris, High Hat

WEIGHT: 12.8 lbs.

SIZE: 22¼" x 11¾" (L x W)

PHOTOMETRICS CHART
(Performance data for this unit are measured at peak focus using FEL - 1KW. lamp.)

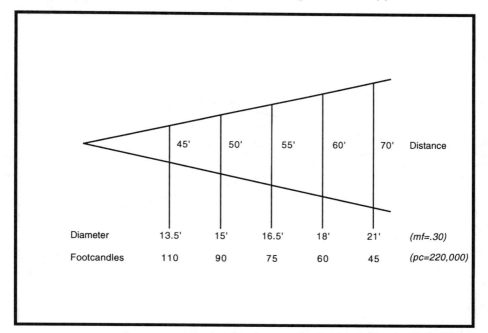

Distance	45'	50'	55'	60'	70'	
Diameter	13.5'	15'	16.5'	18'	21'	*(mf=.30)*
Footcandles	110	90	75	60	45	*(pc=220,000)*

263

STRAND LIGHTING
Model No. 2275, "Quartet 25"

BEAM ANGLE: 11°

FIELD ANGLE: 19.4°

LAMP BASE TYPE: GY 9.5

LAMP MOUNT: Axial

STANDARD LAMP: FRE - 650 W.

OTHER LAMPS: FRG - 500 W. *(cf=.87)*

INTEGRAL PATTERN SLOT: Yes

ACCESSORIES AVAILABLE:
Color Frame, Pattern Holder, Iris

WEIGHT: 9.7 lbs.

SIZE: 16¹¹⁄₁₆" x 9⅝" x 11¼" (L x W x H)

PHOTOMETRICS CHART
(Performance data for this unit are measured using FRE - 650 W. lamp.)

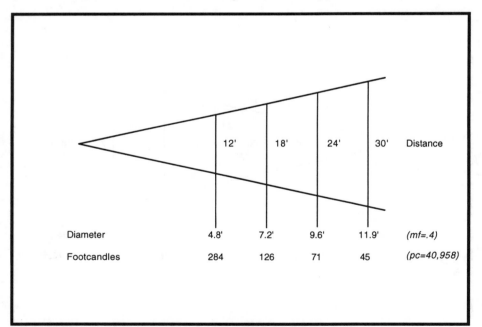

	12'	18'	24'	30'	Distance
Diameter	4.8'	7.2'	9.6'	11.9'	*(mf=.4)*
Footcandles	284	126	71	45	*(pc=40,958)*

* Manufacturer's catalog does not specify the size or focal length of this lens system.

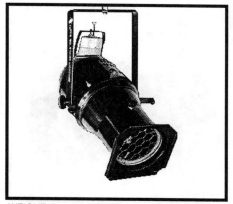

TIMES SQUARE LIGHTING
Model No. 6E6x16

BEAM ANGLE: (not specified in mfg. catalog)

FIELD ANGLE: 17° *

LAMP BASE TYPE: Medium Prefocus

LAMP MOUNT: Off Axis

STANDARD LAMP: EGG - 750 W.

OTHER LAMPS: EGE - 500 W. *(cf=.66)*
EGJ - 1 KW. *(cf=1.75)*

INTEGRAL PATTERN SLOT: Yes

ACCESSORIES AVAILABLE:
Color Frame (7½" sq.), Color Wheel, Pattern Holder

WEIGHT: (not specified in manufacturer's catalog)

SIZE: (not specified in manufacturer's catalog)

PHOTOMETRICS CHART
(Information not provided in manufacturer's catalog about lamp used to measure performance data.)

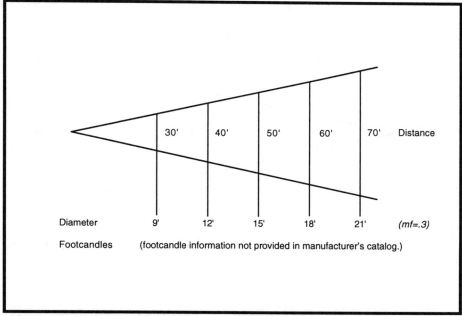

	30'	40'	50'	60'	70'	Distance
Diameter	9'	12'	15'	18'	21'	*(mf=.3)*
Footcandles	(footcandle information not provided in manufacturer's catalog.)					

* Beam and field angles not specified in manufacturer's catalog. This field angle is calculated from given
distance and field diameter data. See Introduction for more information about calculating photometric data.

TIMES SQUARE LIGHTING
Model No. Q6x16

BEAM ANGLE: (not specified in mfg. catalog)

FIELD ANGLE: 18°

LAMP MOUNT: Axial

LAMP BASE TYPE: Medium 2-Pin

STANDARD LAMP: FEL - 1 KW.

OTHER LAMPS: EHD - 500 W. *(cf=.39)*
EHG - 750 W. *(cf=.56)*

INTEGRAL PATTERN SLOT: Yes

ACCESSORIES AVAILABLE:
Color Frame, Pattern Holder, Snoot, Hand,
Motorized and Indexing Color Wheels, Iris Kit

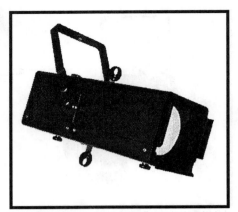

WEIGHT: (not specified in manufacturer's catalog)

SIZE: (not specified in manufacturer's catalog)

PHOTOMETRICS CHART
(Performance data for this unit are measured using FEL - 1 KW. lamp.)

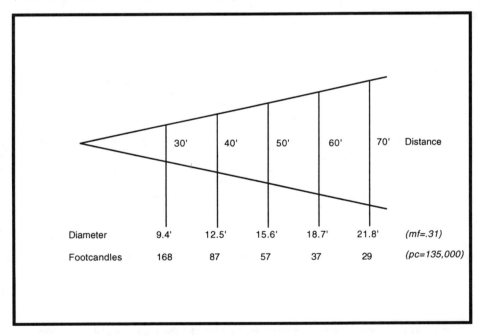

Distance	30'	40'	50'	60'	70'	
Diameter	9.4'	12.5'	15.6'	18.7'	21.8'	*(mf=.31)*
Footcandles	168	87	57	37	29	*(pc=135,000)*

ALTMAN STAGE LIGHTING
Model No. 366*

BEAM ANGLE: (not specified in mfg. catalog)

FIELD ANGLE: 24° **

LAMP BASE TYPE: Medium Prefocus

LAMP MOUNT: Off Axis

STANDARD LAMP: 750T12/9 - 750 W.

OTHER LAMPS: (none specified in mfg. catalog)

INTEGRAL PATTERN SLOT: Yes

WEIGHT: (not specified in manufacturer's catalog)

SIZE: (not specified in manufacturer's catalog)

ACCESSORIES AVAILABLE:
Color Frame (10" sq.), Pattern Holder, Iris Kit Available

PHOTOMETRICS CHART
(Performance data for this unit are measured using a 750T12/9 - 750 W. lamp.)

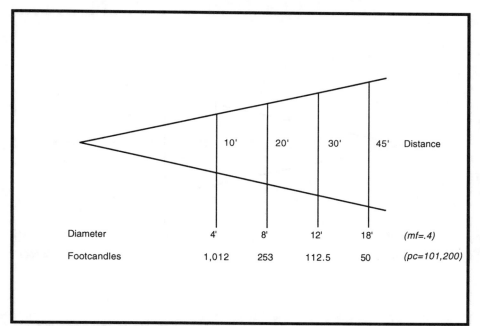

	10'	20'	30'	45'	Distance
Diameter	4'	8'	12'	18'	(mf=.4)
Footcandles	1,012	253	112.5	50	(pc=101,200)

* There are actually two Altman 8x8 instruments bearing the model number 366. This page describes the older version, manufactured in the '70s.

** Beam and field angles not specified in manufacturer's catalog. This field angle is calculated from given distance and field diameter data. See Introduction for more information about calculating photometric data.

ALTMAN STAGE LIGHTING
Model No. 367

BEAM ANGLE: (not specified in mfg. catalog)

FIELD ANGLE: 25° *

LAMP BASE TYPE: Mogul Bipost

LAMP MOUNT: Off Axis

STANDARD LAMP: T24 - 1 KW. or 2 KW.**

OTHER LAMPS: 1500T24/6**

INTEGRAL PATTERN SLOT: No

ACCESSORIES AVAILABLE:
Color Frame (10" sq.)

WEIGHT: (not specified in manufacturer's catalog)

SIZE: (not specified in manufacturer's catalog)

PHOTOMETRICS CHART
(Performance data for this unit are measured using an unspecified 2 KW. lamp.)

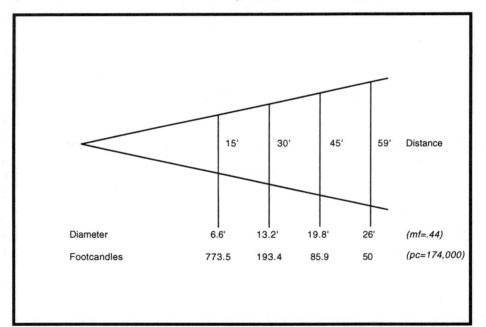

	15'	30'	45'	59'	Distance
Diameter	6.6'	13.2'	19.8'	26'	(mf=.44)
Footcandles	773.5	193.4	85.9	50	(pc=174,000)

* Beam and field angles not specified in manufacturer's catalog. This field angle is calculated from given distance and field diameter data. See Introduction for more information about calculating photometric data.
** Description of standard lamp in Altman catalog is not specific. Without more information about the standard lamp, it is not possible to determine a correction factor *(cf)* for the 1500 W. lamp.

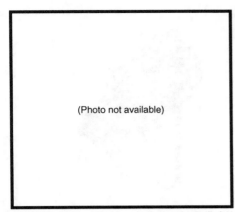

(Photo not available)

CAPITOL STAGE LIGHTING
Model Nos. 1059*

BEAM ANGLE: (not specified in mfg. catalog)

FIELD ANGLE: 25.5° **

LAMP BASE TYPE: Medium Prefocus

LAMP MOUNT: Off Axis

STANDARD LAMP: Q-T16CL - 1 KW.

OTHER LAMPS: (none specified in mfg. catalog)

INTEGRAL PATTERN SLOT: No*

WEIGHT: (not specified in manufacturer's catalog)

SIZE: 28" (L)

ACCESSORIES AVAILABLE:
Color Frame, Snoot

PHOTOMETRICS CHART
(Performance data for this unit are measured using QT6/CL - 1 KW. lamp.)

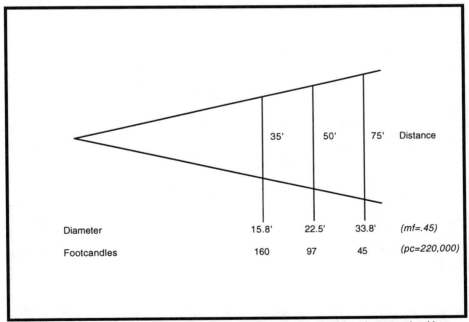

	35'	50'	75'	Distance
Diameter	15.8'	22.5'	33.8'	*(mf=.45)*
Footcandles	160	97	45	*(pc=220,000)*

* Instruments in this series are equipped as follows - 1059 has shutters, 10519 has shutters and an iris, 1059T has shutters and a template slot.
** Beam and field angles not specified in manufacturer's catalog. This field angle is calculated from given distance and field diameter data. See Introduction for more information about calculating photometric data.

CENTURY LIGHTING
Model Nos. 1562 (w/ shutters)
1563 (w/ iris, handle)

BEAM ANGLE: (not specified in manufacturer's catalog)

FIELD ANGLE: 24°

LAMP BASE TYPE: Mogul Bipost

LAMP MOUNT: Off Axis

STANDARD LAMP: 3M/T32/2 C13D*

OTHER LAMPS: (not specified in mfg. catalog)

INTEGRAL PATTERN SLOT: No

ACCESSORIES AVAILABLE:
Color Frame (10" sq.)

WEIGHT: (not specified in manufacturer's catalog)

SIZE: (not specified in manufacturer's catalog)

PHOTOMETRICS CHART
(Performance data for this unit are measured using 3M/T32/2 C13D lamp.)

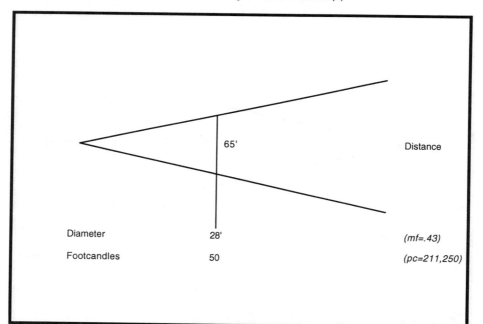

	65'	Distance
Diameter	28'	(mf=.43)
Footcandles	50	(pc=211,250)

* This is an obsolete lamp but it is listed as the standard lamp since the manufacturer uses this lamp to describe the photometric performance.

CENTURY LIGHTING
Model Nos. 1567 (w/ shutters)
1568 (w/ iris, handle)

BEAM ANGLE: 18°

FIELD ANGLE: 23°

LAMP BASE TYPE: Mogul Bipost

LAMP MOUNT: Off Axis

STANDARD LAMP: 2M/T30/1 - 2 KW.*

OTHER LAMPS: 1M/T24/5 -   1 KW.
1500 T24/6 - 1.5 KW.

INTEGRAL PATTERN SLOT: Yes

ACCESSORIES AVAILABLE:
Color Frame (10" sq.), Pattern Holder, Snoot

WEIGHT: (not specified in manufacturer's catalog)

SIZE: 33" x 20" x 29" (L x W x H)

PHOTOMETRICS CHART
(Performance data for this unit are measured using 2M/30/1 - 2 KW. lamp.)

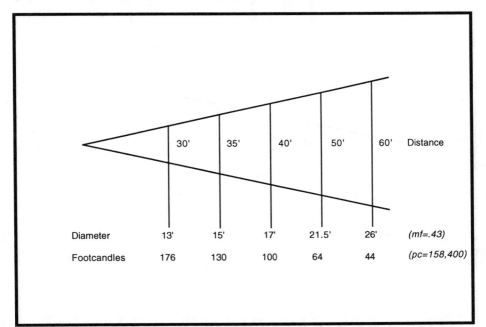

	30'	35'	40'	50'	60'	Distance
Diameter	13'	15'	17'	21.5'	26'	*(mf=.43)*
Footcandles	176	130	100	64	44	*(pc=158,400)*

* This is an obsolete lamp but it is listed as the standard lamp since the manufacturer uses this lamp to describe the photometric performance. Correction factor *(cf)* information is not available for the other lamps because no initial lumens data is available for the obsolete lamp.

CENTURY LIGHTING
Model No. 1575*

BEAM ANGLE: 16°

FIELD ANGLE: 20°

LAMP BASE TYPE: Medium Prefocus

LAMP MOUNT: Off Axis

STANDARD LAMP: 750T12/9 - 750 W.

OTHER LAMPS: 500 12/8 -   500 W. *(cf=.53)*
              1M/T 12/2 - 1 KW. *(cf=1.47)*

INTEGRAL PATTERN SLOT: Yes (no. 1575T only*)

ACCESSORIES AVAILABLE:
Color Frame (10" sq.), Pattern Holder, Snoot

WEIGHT: (not specified in manufacturer's catalog)
SIZE: 29⅝" x 15" x 31⅞" (L x W x H)

PHOTOMETRICS CHART
(Performance data for this unit are measured using 750T12/9 - 750 W. lamp.)

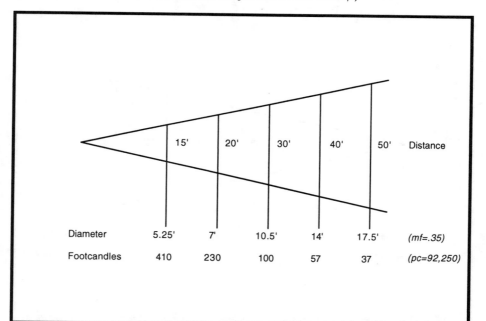

	15'	20'	30'	40'	50'	Distance
Diameter	5.25'	7'	10.5'	14'	17.5'	*(mf=.35)*
Footcandles	410	230	100	57	37	*(pc=92,250)*

* Instruments in this series are equipped as follows - 1572 has an iris and handle, 1573 has a permanent template holder, 1575 has four shutters, and 1575T has four shutters and a template slot.

CENTURY STRAND
Model Nos. 2347 (w/ shutters)
2348 (w/ iris)

BEAM ANGLE: 12.5°

FIELD ANGLE: 22.5°

LAMP BASE TYPE: Medium Prefocus

LAMP MOUNT: Axial

STANDARD LAMP: EGJ - 1 KW.

OTHER LAMPS: EGE - 500 W. *(cf=.38)*
EGG - 750 W. *(cf=.57)*

INTEGRAL PATTERN SLOT: Yes (w/ shutters only)

WEIGHT: (not specified in manufacturer's catalog)

SIZE: 34½" x 12" x 27½" (L x W x H)

ACCESSORIES AVAILABLE:
Color Frame, Pattern Holder, Snoot

PHOTOMETRICS CHART
(Performance data for this unit are measured using EGJ - 1 KW. lamp.)

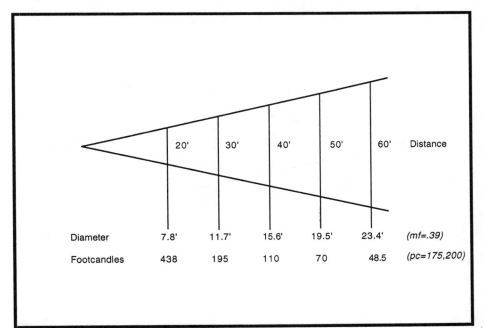

	20'	30'	40'	50'	60'	Distance
Diameter	7.8'	11.7'	15.6'	19.5'	23.4'	*(mf=.39)*
Footcandles	438	195	110	70	48.5	*(pc=175,200)*

273

ELECTRO CONTROLS
Model No. 3365*

BEAM ANGLE: (not specified in mfg. catalog)

FIELD ANGLE: 24.8°

LAMP BASE TYPE: Mini-Can Screw

LAMP MOUNT: Axial

STANDARD LAMP: EGJ - 1 KW.

OTHER LAMPS: EGE - 500 W. *(cf=.38)*
EGG - 750 W. *(cf=.57)*

INTEGRAL PATTERN SLOT: Yes

ACCESSORIES AVAILABLE:
Color Frame (7½" sq.), Iris (as an option)

WEIGHT: (not specified in manufacturer's catalog)

SIZE: (not specified in manufacturer's catalog)

PHOTOMETRICS CHART
(Performance data for this unit are measured using EGJ - 1 KW. lamp.)

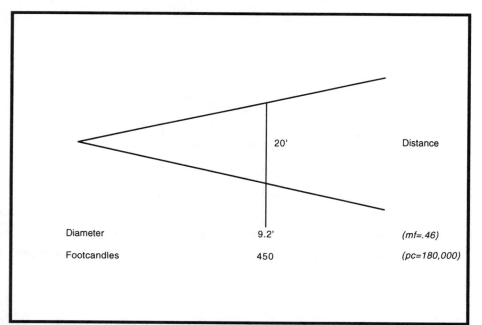

	20'	Distance
Diameter	9.2'	*(mf=.46)*
Footcandles	450	*(pc=180,000)*

* This model number, 3365, can accept 6 different lens configurations: 1 6x6 step lens (this page), 2 6x9 plano lenses, 2 6x12 lenses, 2 6x16 plano lenses, 2 6x20 plano lenses, and 1 6x8 step lens. See Index for the page numbers of other versions of model 3365.

WEIGHT: (not specified in manufacturer's catalog)

SIZE: 24" x 12" x 17" (L x W x H)

ELECTRO CONTROLS
Model No. 3366*

BEAM ANGLE: 9.3°

FIELD ANGLE: 24.8°

LAMP BASE TYPE: Medium Prefocus

LAMP MOUNT: Axial

STANDARD LAMP: EGJ - 1 KW.

OTHER LAMPS: EGE - 500 W. *(cf=.38)*
EGG - 750 W. *(cf=.57)*

INTEGRAL PATTERN SLOT: No

ACCESSORIES AVAILABLE:
Color Frame (7½" sq.), Iris (as an option)

PHOTOMETRICS CHART
(Performance data for this unit are measured using EGJ - 1 KW. lamp.)

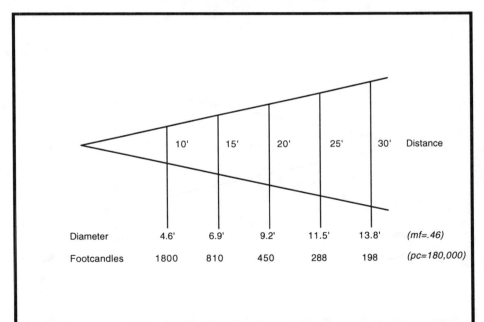

	10'	15'	20'	25'	30'	Distance
Diameter	4.6'	6.9'	9.2'	11.5'	13.8'	*(mf=.46)*
Footcandles	1800	810	450	288	198	*(pc=180,000)*

* This model number, 3366, can accept 3 different lens configurations - two 6x9 plano lenses, two 6x12 plano lenses, and one 6x6 step lens (this page). See Index for the page numbers of other versions of model 3366.

ELECTRO CONTROLS
Model No. 3386A/8

BEAM ANGLE: 13°

FIELD ANGLE: 23°

LAMP BASE TYPE: Medium Prefocus

LAMP MOUNT: Off Axis

STANDARD LAMP: EGJ - 1 KW.

OTHER LAMPS: EGE - 500 W. *(cf=.38)*
EGG - 750 W. *(cf=.57)*

INTEGRAL PATTERN SLOT: Yes

ACCESSORIES AVAILABLE:
Color Frame (10" sq.), Iris

WEIGHT: (not specified in manufacturer's catalog)
SIZE: 28¼" x 15½" x 20¾" (L x W x H)

PHOTOMETRICS CHART
(Performance data for this unit are measured using EGJ - 1 KW. lamp.)

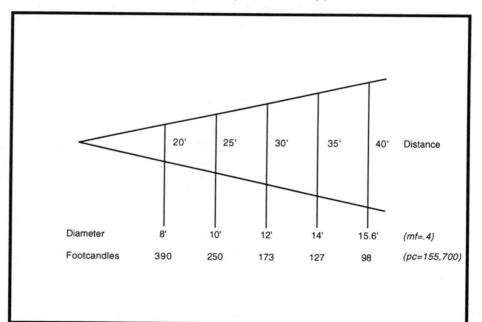

Distance	20'	25'	30'	35'	40'	
Diameter	8'	10'	12'	14'	15.6'	*(mf=.4)*
Footcandles	390	250	173	127	98	*(pc=155,700)*

WEIGHT: 41 lbs.
SIZE: 40" x 22" x 31" (L x W x H)

KLIEGL BROS.
Model No. 1374

BEAM ANGLE: (not specified in mfg. catalog)

FIELD ANGLE: 27° *

LAMP BASE TYPE: Mogul Bipost

LAMP MOUNT: Off Axis

STANDARD LAMP: 3M/T32/2 - 3 KW.

OTHER LAMPS: (none specified in mfg. catalog)

INTEGRAL PATTERN SLOT: No

ACCESSORIES AVAILABLE:
Color Frame (11" sq.)

PHOTOMETRICS CHART
(Performance data for this lamp are measured using 3M/T32/2 - 3 KW. lamp)

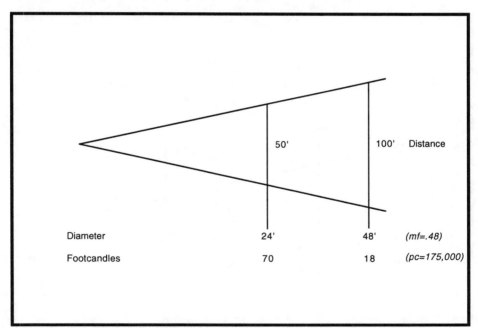

	50'	100'	Distance
Diameter	24'	48'	*(mf=.48)*
Footcandles	70	18	*(pc=175,000)*

* Beam and field angles not specified in manufacturer's catalog. This field angle is calculated from given distance and field diameter data. See Introduction for more information about calculating photometric data.

ALTMAN STAGE LIGHTING
Model No. 1KL6-12

BEAM ANGLE: 10° (peak),12° (flat field)

FIELD ANGLE: 12°

LAMP BASE TYPE: Medium 2-Pin

LAMP MOUNT: Axial

STANDARD LAMP: FEL - 1 KW.

OTHER LAMPS: EHD - 500 W. *(cf=.36)*
　　　　　　　EHG - 750 W. *(cf=.55)*

INTEGRAL PATTERN SLOT: Yes

ACCESSORIES AVAILABLE:
Color Frame, Pattern Holder, Snoot, Motorized &
Non-motorized Color Wheels, Iris

WEIGHT: (not specified in manufacturer's catalog)

SIZE: 25⅜" x 17" x 19" (L x W x H)

PHOTOMETRICS CHART
(Performance data for this unit are measured at peak focus using FEL - 1 KW. lamp.)

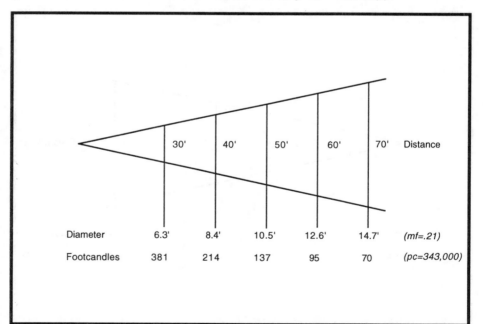

	30'	40'	50'	60'	70'	Distance
Diameter	6.3'	8.4'	10.5'	12.6'	14.7'	*(mf=.21)*
Footcandles	381	214	137	95	70	*(pc=343,000)*

ALTMAN STAGE LIGHTING
Model No. 1KL8-10*

BEAM ANGLE: 7.5° (peak), 12° (flat field)

FIELD ANGLE: 12°

LAMP BASE TYPE: Medium 2-Pin

LAMP MOUNT: Axial

STANDARD LAMP: FEL - 1 KW.

OTHER LAMPS: EHD - 500 W. *(cf=.36)*
EHG - 750 W. *(cf=.55)*

INTEGRAL PATTERN SLOT: Yes

ACCESSORIES AVAILABLE:
Color Frame (8" sq.), Pattern Holder, Snoot, Iris

WEIGHT: (not specified in manufacturer's catalog)

SIZE: 28⅜" x 17" x 19" (L x W x H)

PHOTOMETRICS CHART
(Performance data for this unit are measured at peak focus using FEL - 1 KW. lamp.)

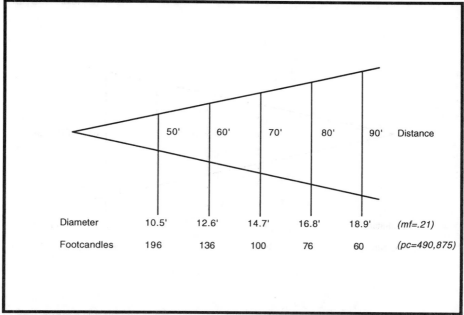

	50'	60'	70'	80'	90'	Distance
Diameter	10.5'	12.6'	14.7'	16.8'	18.9'	*(mf=.21)*
Footcandles	196	136	100	76	60	*(pc=490,875)*

* With optional iris installed, the model number is 1KL8-I-10.

279

ALTMAN STAGE LIGHTING
Model No. 360Q 6x22

BEAM ANGLE: 8°

FIELD ANGLE: 11°

LAMP BASE TYPE: Medium 2-Pin

LAMP MOUNT: Axial

STANDARD LAMP: EHF - 750 W.

OTHER LAMPS: EHD - 500 W. *(cf=.5)*
EHG - 750 W. *(cf=.75)*

INTEGRAL PATTERN SLOT: Yes

ACCESSORIES AVAILABLE:
Color Frame (7½" sq.), Pattern Holder, Snoot,
Motorized & Non-motorized Color Wheels, Iris

WEIGHT: (not specified in manufacturer's catalog)
SIZE: 30½" x 13" x 16¼" (L x W x H)

PHOTOMETRICS CHART
(Performance data for this unit are measured using EHF - 750 W. lamp.)

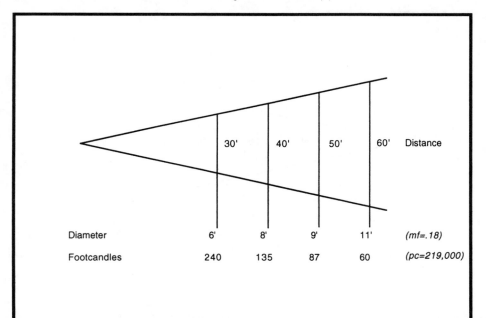

	30'	40'	50'	60'	Distance
Diameter	6'	8'	9'	11'	*(mf=.18)*
Footcandles	240	135	87	60	*(pc=219,000)*

ALTMAN STAGE LIGHTING
Model No. 365/8x11

BEAM ANGLE: 9°

FIELD ANGLE: 14°

LAMP BASE TYPE: Medium Prefocus

LAMP MOUNT: Off Axis

STANDARD LAMP: DNT - 750 W.

OTHER LAMPS: DEB - 500 W. *(cf=.53)*
EGE - 500 W. *(cf=.6)*
EGG - 750 W. *(cf=.9)*

INTEGRAL PATTERN SLOT: Yes

ACCESSORIES AVAILABLE:
Color Frame (10" sq.), Pattern Holder, Snoot,
Motorized & Non-motorized Color Wheels

WEIGHT: (not specified in manufacturer's catalog)

SIZE: 21" x 13¼" x 18½" (L x W x H)

PHOTOMETRICS CHART

(Performance data for this unit are calculated from manufacturer's field angle and peak candela data using DNT - 750 W. lamp.)

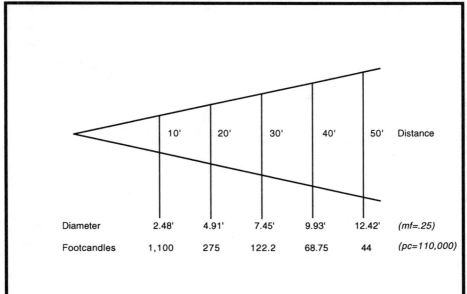

	10'	20'	30'	40'	50'	Distance
Diameter	2.48'	4.91'	7.45'	9.93'	12.42'	*(mf=.25)*
Footcandles	1,100	275	122.2	68.75	44	*(pc=110,000)*

ALTMAN STAGE LIGHTING
Model No. 365/8

BEAM ANGLE: 10°

FIELD ANGLE: 18°

LAMP BASE TYPE: Medium Prefocus

LAMP MOUNT: Off Axis

STANDARD LAMP: DNT - 750 W.

OTHER LAMPS: DEB - 500 W. *(cf=.53)*
EGE - 500 W. *(cf=.6)*
EGG - 750 W. *(cf=.9)*

INTEGRAL PATTERN SLOT: Yes

ACCESSORIES AVAILABLE:
Color Frame (10" sq.), Pattern Holder, Snoot,
Motorized & Non-motorized Color Wheels

WEIGHT: (not specified in manufacturer's catalog)

SIZE: 21" x 13¼" x 18½" (L x W x H)

PHOTOMETRICS CHART

(Performance data for this unit are measured using DNT - 750 W. lamp.)

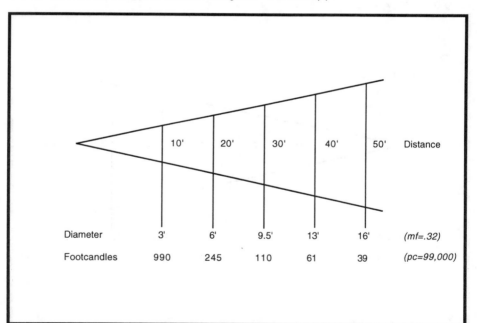

	10'	20'	30'	40'	50'	Distance
Diameter	3'	6'	9.5'	13'	16'	*(mf=.32)*
Footcandles	990	245	110	61	39	*(pc=99,000)*

ALTMAN STAGE LIGHTING
Model No. 366*

BEAM ANGLE: 8°

FIELD ANGLE: 19°

LAMP BASE TYPE: Medium Prefocus

LAMP MOUNT: Off Axis

STANDARD LAMP: DNT - 750 W.

OTHER LAMPS: DEB - 500 W. *(cf=.53)*
EGE - 500 W. *(cf=.6)*
EGG - 750 W. *(cf=.9)*
EGM - 1 KW. *(cf=1.3)*

INTEGRAL PATTERN SLOT: Yes

ACCESSORIES AVAILABLE:
Color Frame (10" sq.), Pattern Holder, Snoot, Motorized & Non-motorized Color Wheels

WEIGHT: (not specified in manufacturer's catalog)

SIZE: 27⅞" x 14¾" x 21½" (L x W x H)

PHOTOMETRICS CHART

(Performance data for this unit are measured using DNT - 750 W. lamp.)

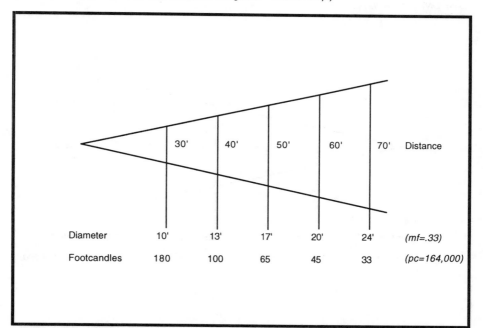

	30'	40'	50'	60'	70'	Distance
Diameter	10'	13'	17'	20'	24'	*(mf=.33)*
Footcandles	180	100	65	45	33	*(pc=164,000)*

* There are actually two Altman 8x8 instruments bearing the model number 366. This page describes the unit currently being manufactured.

ALTMAN STAGE LIGHTING
Model No. 366/8x11

BEAM ANGLE: 7°

FIELD ANGLE: 14°

LAMP BASE TYPE: Medium Prefocus

LAMP MOUNT: Off Axis

STANDARD LAMP: DNT - 750 W.

OTHER LAMPS: DEB - 500 W. *(cf=.65)*
EGE - 500 W. *(cf=.61)*
EGG - 750 W. *(cf=.92)*
EGM - 1 KW. *(cf=1.62)*

INTEGRAL PATTERN SLOT: Yes

ACCESSORIES AVAILABLE:
Color Frame (10" sq.), Pattern Holder, Snoot,
Motorized & Non-motorized Color Wheels

WEIGHT: (not specified in manufacturer's catalog)

SIZE: 27⅛" x 14¾" x 21½" (L x W x H)

PHOTOMETRICS CHART

(Performance data for this unit are calculated from manufacturer's field angle and peak candela data using DNT - 750 W. lamp.)

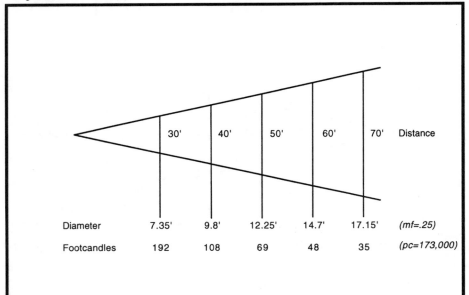

	30'	40'	50'	60'	70'	Distance
Diameter	7.35'	9.8'	12.25'	14.7'	17.15'	*(mf=.25)*
Footcandles	192	108	69	48	35	*(pc=173,000)*

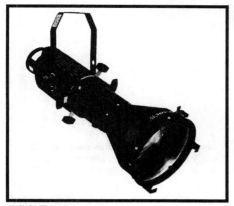

ALTMAN
Model No. S6-10, "Shakespeare"

BEAM ANGLE: 7° (cosine)

FIELD ANGLE: 10°

LAMP BASE TYPE: Medium 2-Pin

LAMP MOUNT: Axial

STANDARD LAMP: HX600 - 575 W.

OTHER LAMPS: EHG - 750 W. *(cf=.93)*
EHD - 500 W. *(cf-.63)*

INTEGRAL PATTERN SLOT: Yes

WEIGHT: 30 lbs.

SIZE: 34" x 12" x 17½" (L x W x H) (without c-clamp)

ACCESSORIES AVAILABLE:
Color Frame (12" sq.), Snoot, Pattern Holder

PHOTOMETRICS CHART
(Performance data for this unit are measured at cosine focus using HX600 - 575 W. lamp.)

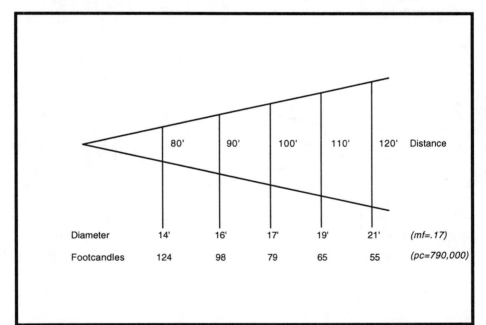

	80'	90'	100'	110'	120'	Distance
Diameter	14'	16'	17'	19'	21'	*(mf=.17)*
Footcandles	124	98	79	65	55	*(pc=790,000)*

ALTMAN
Model No. S6-12, "Shakespeare"

BEAM ANGLE: 7° (cosine)

FIELD ANGLE: 12°

LAMP BASE TYPE: Medium 2-Pin

LAMP MOUNT: Axial

STANDARD LAMP: HX600 - 575 W.

OTHER LAMPS: EHG - 750 W. *(cf=.93)*
EHD - 500 W. *(cf-.63)*

INTEGRAL PATTERN SLOT: Yes

ACCESSORIES AVAILABLE:
Color Frame, Snoot, Pattern Holder

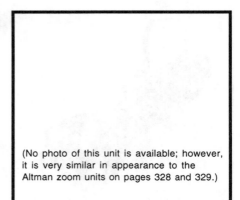

(No photo of this unit is available; however, it is very similar in appearance to the Altman zoom units on pages 328 and 329.)

WEIGHT: 30 lbs.

SIZE: 33³⁄₁₆" x 12" x 17½" (L x W x H)

PHOTOMETRICS CHART

(Performance data for this unit are measured at cosine focus using HX600 - 575 W. lamp.)

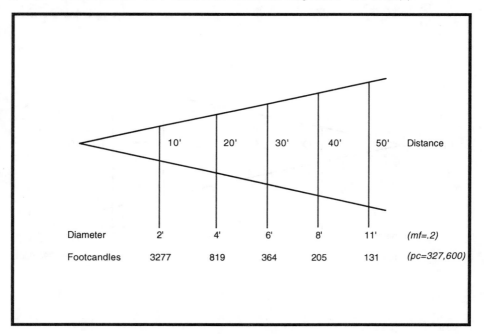

	10'	20'	30'	40'	50'	Distance
Diameter	2'	4'	6'	8'	11'	*(mf=.2)*
Footcandles	3277	819	364	205	131	*(pc=327,600)*

BERKEY COLORTRAN
Model No. 212-082*

BEAM ANGLE: 7.5° (peak & flat field)

FIELD ANGLE: 15° (peak & flat field)

LAMP BASE TYPE: Medium Prefocus

LAMP MOUNT: Axial

STANDARD LAMP: EGJ - 1 KW.

OTHER LAMPS: EGE - 500 W. *(cf=.39)*
EGG - 750 W. *(cf=.57)*

INTEGRAL PATTERN SLOT: Yes

WEIGHT: 37.8 lbs.

SIZE: 42½" x 17" x 37" (L x W x H)

ACCESSORIES AVAILABLE:
Color Frame, Pattern Holder

PHOTOMETRICS CHART
(Performance data for this unit are measured at peak focus using an EGJ - 1 KW. lamp.)

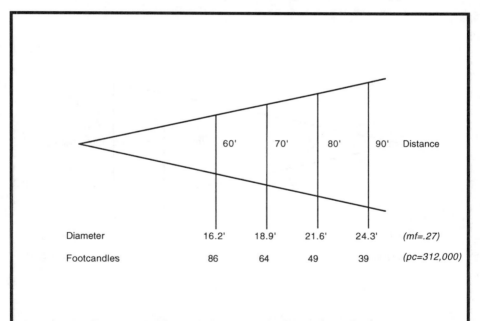

	60'	70'	80'	90'	Distance
Diameter	16.2'	18.9'	21.6'	24.3'	*(mf=.27)*
Footcandles	86	64	49	39	*(pc=312,000)*

* This model can be identified with model numbers -082, -085, -086, or -087 depending on the wire leads and connectors supplied.

BERKEY COLORTRAN
Model No. 213-102*

BEAM ANGLE: 5.8°

FIELD ANGLE: 10°

LAMP BASE TYPE: Medium 2-Pin

LAMP MOUNT: Axial

STANDARD LAMP: FEL - 1 KW.

OTHER LAMPS: EHD - 500 W. *(cf=.39)*
EHG - 750 W. *(cf=.56)*

INTEGRAL PATTERN SLOT: Yes

ACCESSORIES AVAILABLE:
Color Frame, Pattern Holder, Iris Kit

WEIGHT: 21.1 lbs.

SIZE: 30½" x 15" x 24½"

PHOTOMETRICS CHART
(Performance data for this unit are measured at peak focus using FEL - 1 KW. lamp.)

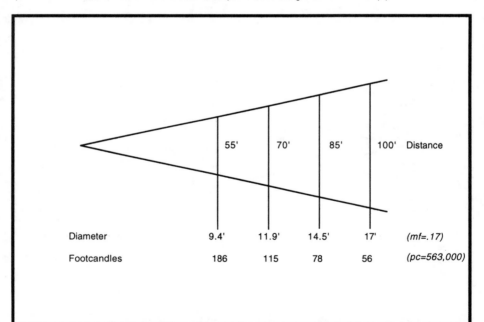

	55'	70'	85'	100'	Distance
Diameter	9.4'	11.9'	14.5'	17'	*(mf=.17)*
Footcandles	186	115	78	56	*(pc=563,000)*

* This model can be identified with model numbers -102, -105, -106, or -107 depending on the wire leads and connectors supplied.

WEIGHT: (not specified in manufacturer's catalog)

SIZE: 33¼" x 15½" x 35" (L x W x H)

CENTURY LIGHTING
Model Nos. 1521 (w/ shutters)
1522 (w/ iris, handle)

BEAM ANGLE: 8°

FIELD ANGLE: 14°

LAMP BASE TYPE: Medium Prefocus

LAMP MOUNT: Off Axis

STANDARD LAMP: 750T12/9 - 750 W.

OTHER LAMPS: 500 T12/8 or 9 - 500 W. *(cf=.53)*
1M/T12/2 - 1 KW. *(cf=1.47)*

INTEGRAL PATTERN SLOT: Yes

ACCESSORIES AVAILABLE:
Color Frame, Pattern Holder

PHOTOMETRICS CHART
(Performance data for this unit are measured using 750T12/9 - 750 W. lamp.)

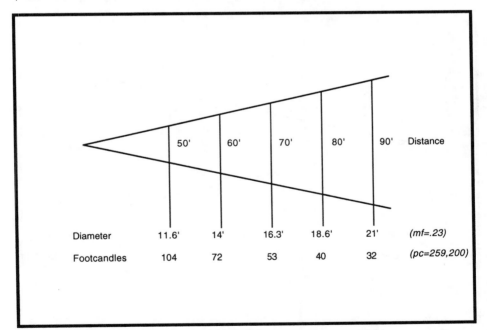

	50'	60'	70'	80'	90'	Distance
Diameter	11.6'	14'	16.3'	18.6'	21'	*(mf=.23)*
Footcandles	104	72	53	40	32	*(pc=259,200)*

CENTURY LIGHTING
Model Nos. 1551 (w/ shutters)
1552 (w/ iris, handle)

BEAM ANGLE: (not specified in manufacturer's catalog)

FIELD ANGLE: 16°

LAMP BASE TYPE: Mogul Bipost

LAMP MOUNT: Off Axis

STANDARD LAMP: 2M/T30/1 C13D*

OTHER LAMPS: (not specified in mfg. catalog)

INTEGRAL PATTERN SLOT: No

ACCESSORIES AVAILABLE:
Color Frame

WEIGHT: (not specified in manufacturer's catalog)

SIZE: (not specified in manufacturer's catalog)

PHOTOMETRICS CHART
(Performance data for this unit are measured using 2M/T30/1 C13D lamp.)

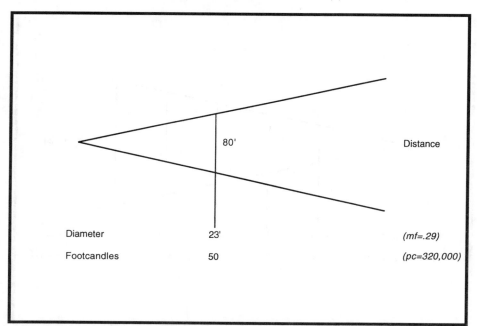

	80'	Distance
Diameter	23'	(mf=.29)
Footcandles	50	(pc=320,000)

* This is an obsolete lamp but it is listed as the standard lamp since the manufacturer uses this lamp to describe the photometric performance.

CENTURY LIGHTING
Model Nos. 1553 (w/ shutters)
1554 (w/ iris, handle)

BEAM ANGLE: (not specified in manufacturer's catalog)

FIELD ANGLE: 16°

LAMP BASE TYPE: Mogul Bipost

LAMP MOUNT: Off Axis

STANDARD LAMP: 3M/T32/2 C13D*

OTHER LAMPS: (not specified in mfg. catalog)

INTEGRAL PATTERN SLOT: No

ACCESSORIES AVAILABLE:
Color Frame

WEIGHT: (not specified in manufacturer's catalog)

SIZE: (not specified in manufacturer's catalog)

PHOTOMETRICS CHART
(Performance data for this unit are measured using 3M/T32/2 C13D lamp.)

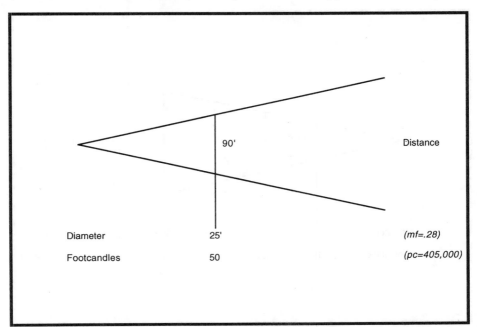

	90'	Distance
Diameter	25'	(mf=.28)
Footcandles	50	(pc=405,000)

* This is an obsolete lamp but it is listed as the standard lamp since the manufacturer uses this lamp to describe the photometric performance.

CENTURY STRAND
Model Nos.2457 (w/ shutters)
2458 (w/ iris, handle)

BEAM ANGLE: 7.5°

FIELD ANGLE: 15°

LAMP BASE TYPE: Medium Prefocus

LAMP MOUNT: Axial

STANDARD LAMP: EGJ - 1 KW.

OTHER LAMPS: EGE - 500 W. *(cf=.38)*
EGG - 750 W. *(cf=.57)*

INTEGRAL PATTERN SLOT: Yes

ACCESSORIES AVAILABLE:
Color Frame, Pattern Holder, Snoot

WEIGHT: (not specified in manufacturer's catalog)

SIZE: 38" x 12" x 21½" (L x W x H)

PHOTOMETRICS CHART
(Performance data for this unit are measured using EGJ - 1 KW. lamp.)

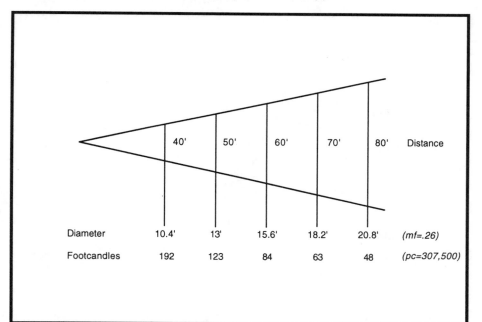

	40'	50'	60'	70'	80'	Distance
Diameter	10.4'	13'	15.6'	18.2'	20.8'	*(mf=.26)*
Footcandles	192	123	84	63	48	*(pc=307,500)*

CENTURY STRAND
Model Nos.2567 (w/ shutters)
2568 (w/ iris, handle)

BEAM ANGLE: 7.5°

FIELD ANGLE: 10.5°

LAMP BASE TYPE: Medium Prefocus

LAMP MOUNT: Axial

STANDARD LAMP: EGJ - 1 KW.

OTHER LAMPS: EGE - 500 W. *(cf=.38)*
EGG - 750 W. *(cf=.57)*

INTEGRAL PATTERN SLOT: Yes (2567 only)

ACCESSORIES AVAILABLE:
Color Frame, Pattern Holder, Snoot

WEIGHT: (not specified in manufacturer's catalog)

SIZE: 38" x 12" x 27½" (L x W x H)

PHOTOMETRICS CHART
(Performance data for this unit are measured using EGJ - 1 KW. lamp.)

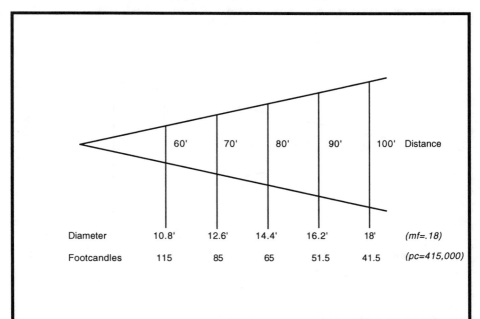

	60'	70'	80'	90'	100'	Distance
Diameter	10.8'	12.6'	14.4'	16.2'	18'	*(mf=.18)*
Footcandles	115	85	65	51.5	41.5	*(pc=415,000)*

CENTURY STRAND
Model Nos. 2667 (w/ shutters)
2668 (w/ iris, handle)

BEAM ANGLE: 7.5°

FIELD ANGLE: 10.5°

LAMP BASE TYPE: Medium Prefocus

LAMP MOUNT: Axial

STANDARD LAMP: EGJ - 1 KW.

OTHER LAMPS: EGE - 500 W. *(cf=.38)*
EGG - 750 W. *(cf=.57)*

INTEGRAL PATTERN SLOT: Yes (2667 only)

ACCESSORIES AVAILABLE:
Color Frame, Pattern Holder, Snoot

WEIGHT: (not specified in manufacturer's catalog)
SIZE: 40½" x 14" x 28½" (L x W x H)

PHOTOMETRICS CHART
(Performance data for this unit are measured using EGJ - 1 KW. lamp.)

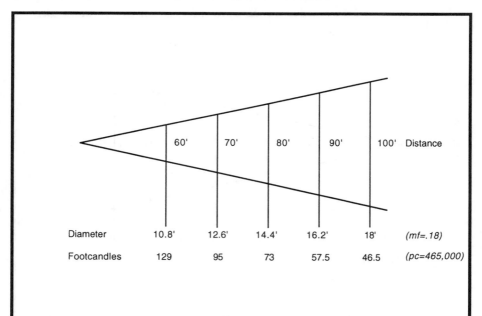

	60'	70'	80'	90'	100'	Distance
Diameter	10.8'	12.6'	14.4'	16.2'	18'	*(mf=.18)*
Footcandles	129	95	73	57.5	46.5	*(pc=465,000)*

WEIGHT: (not specified in manufacturer's catalog)

SIZE: 32" x 15½" x 20¾" (L x W x H)

ELECTRO CONTROLS
Model No. 3386A/11

BEAM ANGLE: 11°

FIELD ANGLE: 15°

LAMP BASE TYPE: Medium Prefocus

LAMP MOUNT: Off Axis

STANDARD LAMP: EGJ - 1 KW.

OTHER LAMPS: EGE - 500 W. *(cf=.38)*
EGG - 750 W. *(cf=.57)*

INTEGRAL PATTERN SLOT: Yes

ACCESSORIES AVAILABLE:
Color Frame (10" sq.), Iris

PHOTOMETRICS CHART
(Performance data for this unit are measured using EGJ - 1 KW. lamp.)

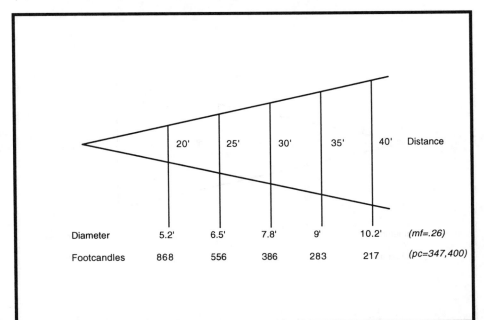

	20'	25'	30'	35'	40'	Distance
Diameter	5.2'	6.5'	7.8'	9'	10.2'	*(mf=.26)*
Footcandles	868	556	386	283	217	*(pc=347,400)*

295

ELECTRONIC THEATRE CONTROLS (ETC)
Model No. 410, "Source Four"

BEAM ANGLE: 6.9°

FIELD ANGLE: 10.3°

LAMP BASE TYPE: Medium 2-Pin

LAMP MOUNT: Axial

STANDARD LAMP: HPL 575/115 - 575 W.

OTHER LAMPS: HPL 550/77 - 550W.*

INTEGRAL PATTERN SLOT: Yes

ACCESSORIES AVAILABLE:
Color Frame (12" sq.), Pattern Holder, Donut, Snoot

WEIGHT: 17.5 lbs.

SIZE: 29½" x 12½" x 20⅝" (L x W x H)

PHOTOMETRICS CHART

(Performance data for this unit are measured at cosine focus using HPL 575/110 lamp.)

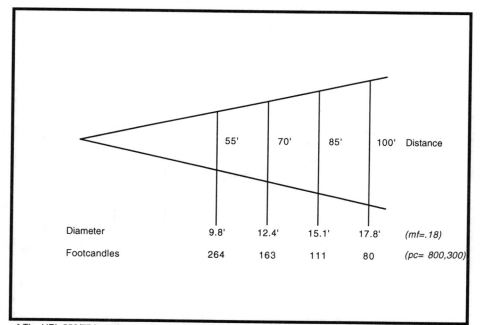

	55'	70'	85'	100'	Distance
Diameter	9.8'	12.4'	15.1'	17.8'	(mf=.18)
Footcandles	264	163	111	80	(pc= 800,300)

* The HPL 550/77 lamp is used with an ETC multiplexing system, used to allow multiple units to share a common circuit, with separate dimming of each.

KLIEGL BROS.
Model No. 1165/8

BEAM ANGLE: (not specified in mfg. catalog)

FIELD ANGLE: 14° *

LAMP BASE TYPE: Medium Bipost

LAMP MOUNT: Off Axis

STANDARD LAMP: 500/T14/8 - 500 W.

OTHER LAMPS: 500/T14/7 - 500 W. *(cf=1.0)*
750/T14 - 750 W. *(cf=1.89)*

INTEGRAL PATTERN SLOT: No

ACCESSORIES AVAILABLE:
Color Frame

WEIGHT: 15 lbs.

SIZE: (not specified in manufacturer's catalog)

PHOTOMETRICS CHART
(Performance data for this unit are measured using 500/T14/8 - 500 W. lamp.)

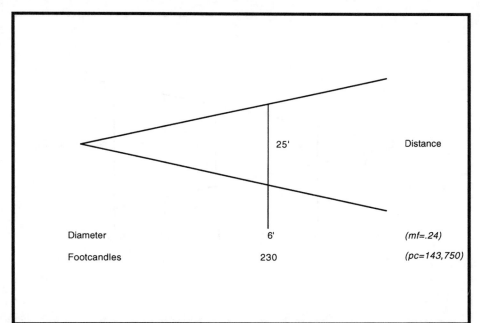

	25'	Distance
Diameter	6'	*(mf=.24)*
Footcandles	230	*(pc=143,750)*

* Beam and field angles not specified in manufacturer's catalog. This field angle is calculated from given distance and field diameter data. See Introduction for more information about calculating photometric data.

KLIEGL BROS.
Model No. 1355/8*

BEAM ANGLE: (not specified in mfg. catalog)

FIELD ANGLE: 12°

LAMP BASE TYPE: Medium 2-Pin

LAMP MOUNT: Off Axis

STANDARD LAMP: EHG - 750 W.

OTHER LAMPS: EHD - 500 W. *(cf=.69)*
EHF - 750 W. *(cf=1.32)*

INTEGRAL PATTERN SLOT: Yes

ACCESSORIES AVAILABLE:
Color Frame (10½" sq.), Pattern Holder, Snoot

WEIGHT: 16 lbs.
SIZE: 26½" x 13½" x 19½" (L x W x H)

PHOTOMETRICS CHART
(Performance data for this unit are measured using EHG - 750 W. lamp.)

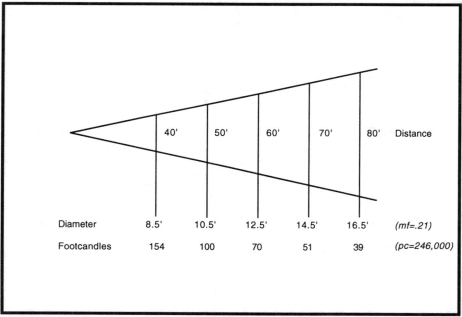

	40'	50'	60'	70'	80'	Distance
Diameter	8.5'	10.5'	12.5'	14.5'	16.5'	*(mf=.21)*
Footcandles	154	100	70	51	39	*(pc=246,000)*

* With optional iris installed, the model number is 1355I/8.

KLIEGL BROS.
Model No. 1357/8*

BEAM ANGLE: (not specified in mfg. catalog)

FIELD ANGLE: 13° **

LAMP BASE TYPE: Recessed Single-Contact

LAMP MOUNT: Off Axis

STANDARD LAMP: DWT - 1 KW.

OTHER LAMPS: FER - 1 KW. *(cf=1.18)*

INTEGRAL PATTERN SLOT: Yes

WEIGHT: 24 lbs.

SIZE: 31½" x 16" x 27¼" (L x W x H)

ACCESSORIES AVAILABLE:
Color Frame (10" sq.), Pattern Holder

PHOTOMETRICS CHART
(Performance data for this unit are measured using DWT - 1 KW. lamp.)

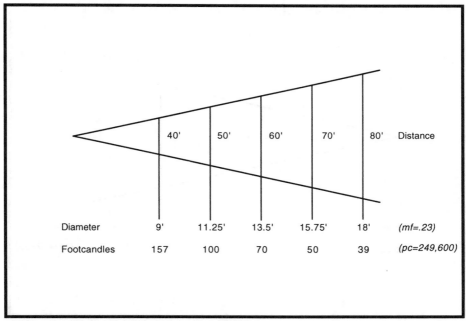

	40'	50'	60'	70'	80'	Distance
Diameter	9'	11.25'	13.5'	15.75'	18'	*(mf=.23)*
Footcandles	157	100	70	50	39	*(pc=249,600)*

* With optional iris installed, the model number is 1357I/8.

KLIEGL BROS.
Model No. 1358/10*

BEAM ANGLE: (not specified in mfg. catalog)

FIELD ANGLE: 12.5°

LAMP BASE TYPE: Mogul Bipost

LAMP MOUNT: Off Axis

STANDARD LAMP: BWA - 2 KW.

OTHER LAMPS: CYX - 2 KW. *(cf=1.0)*

INTEGRAL PATTERN SLOT: Yes

ACCESSORIES AVAILABLE:
Color Frame, Pattern Holder

WEIGHT: 33 lbs.

SIZE: 28½" x 18" x 26" (L x W x H)

PHOTOMETRICS CHART
(Performance data for this unit are measured using BWA - 2 KW. lamp.)

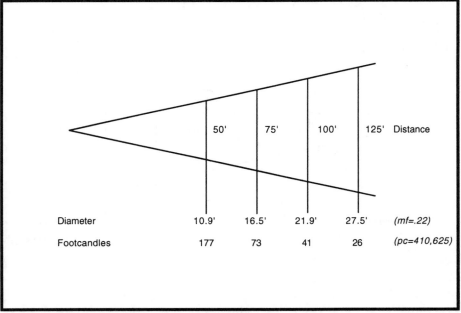

	50'	75'	100'	125'	Distance
Diameter	10.9'	16.5'	21.9'	27.5'	*(mf=.22)*
Footcandles	177	73	41	26	*(pc=410,625)*

* With optional iris installed, the model number is 1358I/10.

KLIEGL BROS.
Model No. 1358/12*

BEAM ANGLE: (not specified in mfg. catalog)

FIELD ANGLE: 12°

LAMP BASE TYPE: Mogul Bipost

LAMP MOUNT: Off Axis

STANDARD LAMP: BWA - 2 KW.

OTHER LAMPS: CYX - 2 KW. *(cf=1.0)*

INTEGRAL PATTERN SLOT: Yes

WEIGHT: 34 lbs.

SIZE: 28½" x 18" x 26" (L x W x H)

ACCESSORIES AVAILABLE:
Color Frame, Pattern Holder

PHOTOMETRICS CHART
(Performance data for this unit are measured using BWA - 2 KW. lamp.)

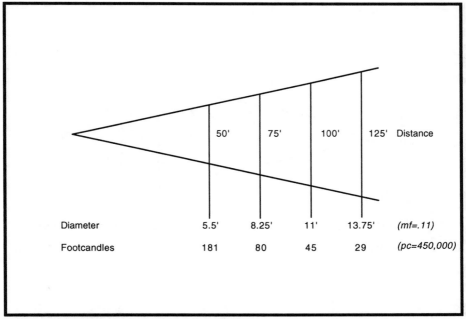

	50'	75'	100'	125'	Distance
Diameter	5.5'	8.25'	11'	13.75'	*(mf=.11)*
Footcandles	181	80	45	29	*(pc=450,000)*

* With optional iris installed, the model number is 1358I/10.

KLIEGL BROS.
Model No. 1358/8*

BEAM ANGLE: (not specified in mfg. catalog)

FIELD ANGLE: 19°

LAMP BASE TYPE: Mogul Bipost

LAMP MOUNT: Off Axis

STANDARD LAMP: BWA - 2 KW.

OTHER LAMPS: CYX - 2 KW. *(cf=1.0)*

INTEGRAL PATTERN SLOT: Yes

ACCESSORIES AVAILABLE:
Color Frame (10" sq.), Pattern Holder

WEIGHT: (not specified in manufacturer's catalog)

SIZE: 23½" x 18" x 26" (L x W x H)

PHOTOMETRICS CHART
(Performance data for this unit are measured using BWA - 2 KW. lamp.)

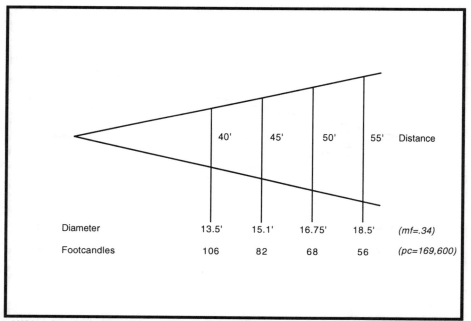

	40'	45'	50'	55'	Distance
Diameter	13.5'	15.1'	16.75'	18.5'	*(mf=.34)*
Footcandles	106	82	68	56	*(pc=169,600)*

* With optional iris installed, the model number is 1358I/8.

WEIGHT: 11 lbs.

SIZE: (not specified in manufacturer's catalog)

KLIEGL BROS.
Model No. 1365/8

BEAM ANGLE: (not specified in mfg. catalog)

FIELD ANGLE: 13° *

LAMP BASE TYPE: Medium Bipost

LAMP MOUNT: Off Axis

STANDARD LAMP: 500/T14/8 - 500 W.

OTHER LAMPS: 500/T14/7 - 500 W. *(cf=1.0)*
750/T14 - 750 W. *(cf=1.89)*

INTEGRAL PATTERN SLOT: No

ACCESSORIES AVAILABLE:
Color Frame

PHOTOMETRICS CHART
(Performance data for this unit are measured using 500/T14/8 - 500 W. lamp.)

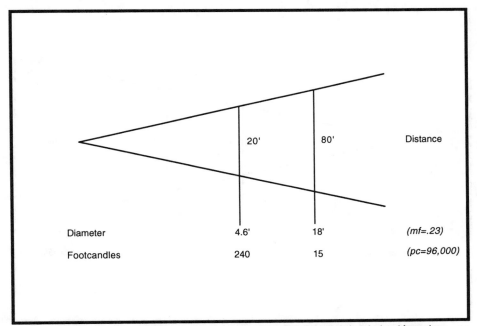

	20'	80'	Distance
Diameter	4.6'	18'	*(mf=.23)*
Footcandles	240	15	*(pc=96,000)*

* Beam and field angles not specified in manufacturer's catalog. This field angle is calculated from given distance and field diameter data. See Introduction for more information about calculating photometric data.

KLIEGL BROS.
Model No. 1373

BEAM ANGLE: (not specified in mfg. catalog)

FIELD ANGLE: 14° *

LAMP BASE TYPE: Mogul Bipost

LAMP MOUNT: Off Axis

STANDARD LAMP: 3M/T32/2 - 3 KW.**

OTHER LAMPS: (none specified in mfg. catalog)

INTEGRAL PATTERN SLOT: No

ACCESSORIES AVAILABLE:
Color Frame, Motorized & Non-motorized Color Wheels

WEIGHT: 45 lbs.

SIZE: (not specified in manufacturer's catalog)

PHOTOMETRICS CHART
(Performance data for this unit are measured using 3M/T32/2 - 3 KW. lamp.)

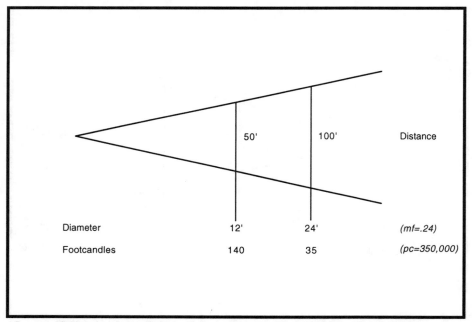

	50'	100'	Distance
Diameter	12'	24'	*(mf=.24)*
Footcandles	140	35	*(pc=350,000)*

* Beam and field angles not specified in manufacturer's catalog. This field angle is calculated from given distance and field diameter data. See Introduction for more information about calculating photometric data.
** This is an obsolete lamp but it is listed as the standard lamp since the manufacturer uses this lamp to describe the photometric performance.

KLIEGL BROS.
Model No. 1375

BEAM ANGLE: (not specified in mfg. catalog)

FIELD ANGLE: 18° *

LAMP BASE TYPE: Mogul Bipost

LAMP MOUNT: Off Axis

STANDARD LAMP: 3M/T32/2 - 3 KW.**

OTHER LAMPS: (none specified in mfg. catalog)

INTEGRAL PATTERN SLOT: No

ACCESSORIES AVAILABLE:
Color Frame, Motorized & Non-motorized Color Wheels

WEIGHT: 42 lbs.

SIZE: (not specified in manufacturer's catalog)

PHOTOMETRICS CHART
(Performance data for this unit are measured using 3M/T32/2 - 3 KW. lamp.)

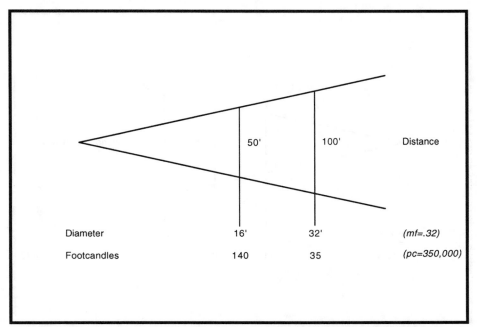

	50'	100'	Distance
Diameter	16'	32'	(mf=.32)
Footcandles	140	35	(pc=350,000)

* Beam and field angles not specified in manufacturer's catalog. This field angle is calculated from given distance and field diameter data. See Introduction for more information about calculating photometric data.
** This is an obsolete lamp but it is listed as the standard lamp since the manufacturer uses this lamp to describe the photometric performance.

305

KLIEGL BROS.
Model No. 1550/8*

BEAM ANGLE: (not specified in mfg. catalog)

FIELD ANGLE: 10° (cosine)

LAMP BASE TYPE: Medium 2-Pin

LAMP MOUNT: Axial

STANDARD LAMP: FEL - 1 KW.

OTHER LAMPS: EHD - 500 W. *(cf=.39)*
                 EHG - 750 W. *(cf=.56)*

INTEGRAL PATTERN SLOT: Yes

ACCESSORIES AVAILABLE:
Color Frame, Pattern Holder, Drop-in Iris

WEIGHT: (not specified in manufacturer's catalog)

SIZE: 36" x 18" (L x H)

PHOTOMETRICS CHART
(Performance data for this unit are measured using FEL - 1 KW. lamp.)

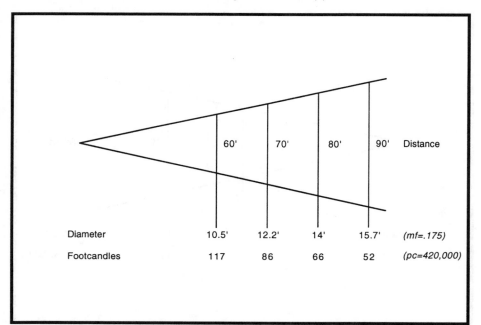

	60'	70'	80'	90'	Distance
Diameter	10.5'	12.2'	14'	15.7'	*(mf=.175)*
Footcandles	117	86	66	52	*(pc=420,000)*

* Lens barrels in this series of instruments are color-coded to indentify the field angle. This 10° model is color-coded with maroon.

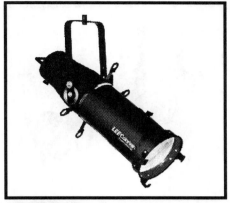

LEE COLORTRAN
Model No. 650-072*

BEAM ANGLE: 5.8° (peak), 6.7° (cosine)

FIELD ANGLE: 10°

LAMP BASE TYPE: Medium 2-Pin

LAMP MOUNT: Axial

STANDARD LAMP: FEL - 1 KW.

OTHER LAMPS: EHD - 500 W. *(cf=.39)*
EHG - 750 W. *(cf=.56)*

INTEGRAL PATTERN SLOT: Yes

WEIGHT: 20.9 lbs.

SIZE: 29½" x 15" x 25½" (L x W x H)

ACCESSORIES AVAILABLE:
Color Frame, Pattern Holder, Iris Kit

PHOTOMETRICS CHART
(Performance data for this unit are measured at peak focus using FEL - 1 KW. lamp.)

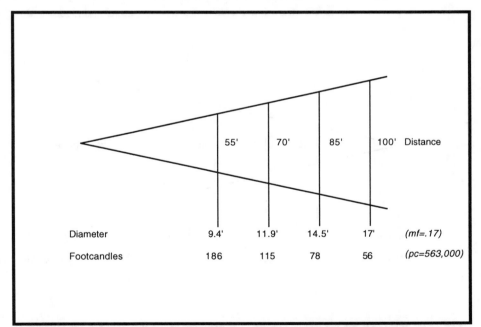

	55'	70'	85'	100'	Distance
Diameter	9.4'	11.9'	14.5'	17'	*(mf=.17)*
Footcandles	186	115	78	56	*(pc=563,000)*

* This model can be identified with model numbers -072, -074, -075, or -076 depending on the wire leads and connectors supplied.

LIGHTING & ELECTRONICS
Model No. AQ61-22

BEAM ANGLE: 7°

FIELD ANGLE: 11°

LAMP BASE TYPE: Medium 2-Pin

LAMP MOUNT: Axial

STANDARD LAMP: FEL - 1 KW.

OTHER LAMPS: EHG - 750 W. *(cf=.56)*
             Hx600 - 600 W. *(cf=.93)*

INTEGRAL PATTERN SLOT: Yes

ACCESSORIES AVAILABLE:
Color Frame (7½" sq.), Pattern Holder, Iris

WEIGHT: (not specified in manufacturer's catalog)
SIZE: 29" (L)

PHOTOMETRICS CHART
(Performance data for this unit are measured using FEL - 1 KW. lamp.)

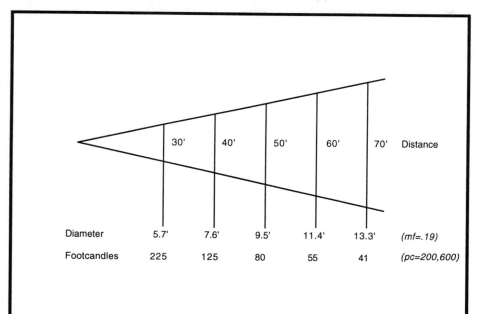

	30'	40'	50'	60'	70'	Distance
Diameter	5.7'	7.6'	9.5'	11.4'	13.3'	*(mf=.19)*
Footcandles	225	125	80	55	41	*(pc=200,600)*

WEIGHT: 17 lbs.

SIZE: 29" x 12" x 24" (L x W x H)

STRAND CENTURY
Model Nos.2113 (w/ shutters)
2114 (plus iris)

BEAM ANGLE: 7° (peak), 9° (cosine)

FIELD ANGLE: 12° (peak)

LAMP BASE TYPE: Medium 2-Pin

LAMP MOUNT: Axial

STANDARD LAMP: FEL - 1 KW.

OTHER LAMPS: EHD - 500 W. *(cf=.39)*
EHG - 750 W. *(cf=.56)*

INTEGRAL PATTERN SLOT: Yes

ACCESSORIES AVAILABLE:
Color Frame, Pattern Holder, Snoot, Iris Kit

PHOTOMETRICS CHART

(Performance data for this unit are measured at peak focus using FEL - 1 KW. lamp.)

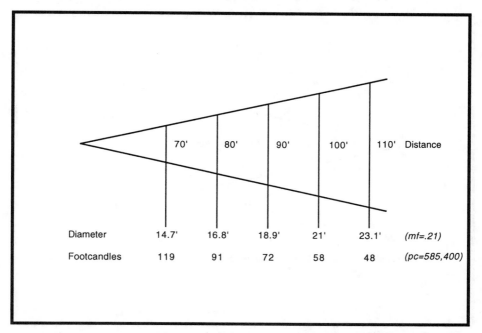

	70'	80'	90'	100'	110'	Distance
Diameter	14.7'	16.8'	18.9'	21'	23.1'	*(mf=.21)*
Footcandles	119	91	72	58	48	*(pc=585,400)*

TIMES SQUARE LIGHTING
Model No. 8EPC

BEAM ANGLE: (not specified in mfg. catalog)

FIELD ANGLE: 19° *

LAMP BASE TYPE: Medium Prefocus

LAMP MOUNT: Off Axis

STANDARD LAMP: 750T12/9 - 750 W.

OTHER LAMPS: 500T12/9 -   500 W. *(cf=.65)*
　　　　　　　1000T12/2 - 1 KW. *(cf=1.47)*

INTEGRAL PATTERN SLOT: Yes

ACCESSORIES AVAILABLE:
Color Frame (10" sq.), Pattern Holder

WEIGHT: (not specified in manufacturer's catalog)

SIZE: 23" (L)

PHOTOMETRICS CHART

(Information about lamp used to measure performance data is not provided in manufacturer's catalog.)

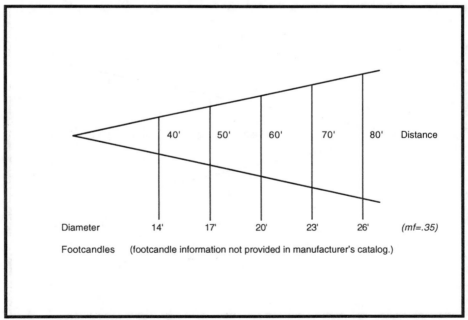

	40'	50'	60'	70'	80'	Distance
Diameter	14'	17'	20'	23'	26'	*(mf=.35)*
Footcandles	(footcandle information not provided in manufacturer's catalog.)					

* Beam and field angles not specified in manufacturer's catalog. This field angle is calculated from given distance and field diameter data. See Introduction for more information about calculating photometric data.

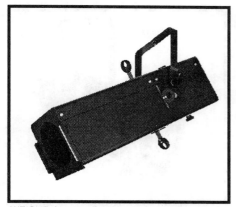

TIMES SQUARE LIGHTING
Model No. Q6x22

BEAM ANGLE: (not specified in mfg. catalog)

FIELD ANGLE: 10°

LAMP BASE TYPE: Medium 2-Pin

LAMP MOUNT: Axial

STANDARD LAMP: FEL - 1 KW.

OTHER LAMPS: EHD - 500 W. *(cf=.39)*
EHG - 750 W. *(cf=.56)*

INTEGRAL PATTERN SLOT: Yes

ACCESSORIES AVAILABLE:
Color Frame, Pattern Holder, Snoot,
Hand/Motorized/Indexing Color Wheels, Iris Kit

WEIGHT: (not specified in manufacturer's catalog)

SIZE: (not specified in manufacturer's catalog)

PHOTOMETRICS CHART
(Performance data for this unit are measured using FEL - 1 KW. lamp.)

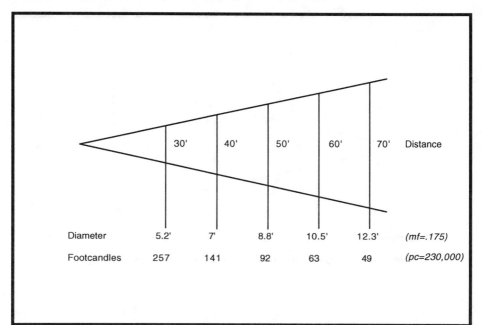

	30'	40'	50'	60'	70'	Distance
Diameter	5.2'	7'	8.8'	10.5'	12.3'	*(mf=.175)*
Footcandles	257	141	92	63	49	*(pc=230,000)*

TIMES SQUARE LIGHTING
Model No. Q8EPC

BEAM ANGLE: (not specified in mfg. catalog)

FIELD ANGLE: 15° *

LAMP BASE TYPE: Medium 2-Pin

LAMP MOUNT: Axial

STANDARD LAMP: EHG - 750 W.

OTHER LAMPS: EHD - 500 W. *(cf=.69)*
FEL - 1 KW. *(cf=1.79)*

INTEGRAL PATTERN SLOT: Yes

ACCESSORIES AVAILABLE:
Color Frame, Pattern Holder

WEIGHT: 16 lbs.

SIZE: (not specified in manufacturer's catalog)

PHOTOMETRICS CHART

(Information about lamp used to measure performance data is not provided in manufacturer's catalog.)

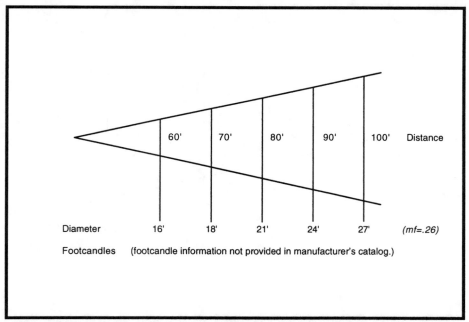

Distance	60'	70'	80'	90'	100'
Diameter	16'	18'	21'	24'	27'	*(mf=.26)*
Footcandles	(footcandle information not provided in manufacturer's catalog.)

* Beam and field angles not specified in manufacturer's catalog. This field angle is calculated from given distance and field diameter data. See Introduction for more information about calculating photometric data.

WEIGHT: (not specified in manufacturer's catalog)

SIZE: 35½" x 17" x 24" (L x W x H)

ALTMAN STAGE LIGHTING
Model No. 1KL10-5*

BEAM ANGLE: 5° (peak), 6.9° (flat field)

FIELD ANGLE: 6.9°

LAMP BASE TYPE: Medium 2-Pin

LAMP MOUNT: Axial

STANDARD LAMP: FEL - 1 KW.

OTHER LAMPS: EHD - 500 W. *(cf=.36)*
EHG - 750 W. *(cf=.55)*

INTEGRAL PATTERN SLOT: Yes

ACCESSORIES AVAILABLE:
Color Frame, Pattern Holder, Snoot

PHOTOMETRICS CHART
(Performance data for this unit are measured at peak focus using FEL - 1 KW. lamp.)

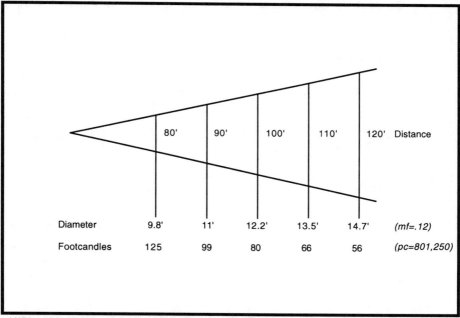

	80'	90'	100'	110'	120'	Distance
Diameter	9.8'	11'	12.2'	13.5'	14.7'	*(mf=.12)*
Footcandles	125	99	80	66	56	*(pc=801,250)*

* With optional iris installed, the model number is 1KL10-I-5.

313

ALTMAN
Model No. S6-5, "Shakespeare"

BEAM ANGLE: 5° (cosine)

FIELD ANGLE: 7°

LAMP BASE TYPE: Medium 2-Pin

LAMP MOUNT: Axial

STANDARD LAMP: HX600 - 575 W.

OTHER LAMPS: EHG - 750 W. *(cf=.93)*
EHD - 500 W. *(cf-.63)*

INTEGRAL PATTERN SLOT: Yes

ACCESSORIES AVAILABLE:
Color Frame (14" sq.), Snoot, Pattern Holder

WEIGHT: 30 lbs.

SIZE: 42½" x 14" x 19½" (L x W x H) (without c-clamp)

PHOTOMETRICS CHART
(Performance data for this unit are measured at cosine focus using HX600 - 575 W. lamp.)

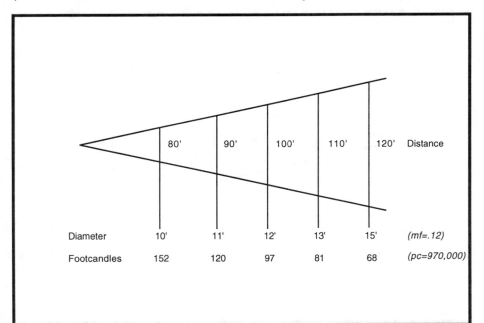

	80'	90'	100'	110'	120'	Distance
Diameter	10'	11'	12'	13'	15'	*(mf=.12)*
Footcandles	152	120	97	81	68	*(pc=970,000)*

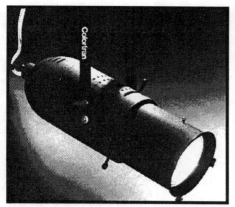

WEIGHT: 37.8 lbs.

SIZE: 42½" x 17" x 37" (L x W x H)

BERKEY COLORTRAN
Model No. 212-092*

BEAM ANGLE: 6.5°

FIELD ANGLE: 9°

LAMP BASE TYPE: Medium Prefocus

LAMP MOUNT: Axial

STANDARD LAMP: EGJ - 1 KW.

OTHER LAMPS: EGE - 500 W. *(cf=.40)*
EGG - 750 W. *(cf=.61)*

INTEGRAL PATTERN SLOT: Yes

ACCESSORIES AVAILABLE:
Color Frame, Pattern Holder

PHOTOMETRICS CHART
(Performance data for this unit are measured at peak focus using EGJ - 1 KW. lamp.)

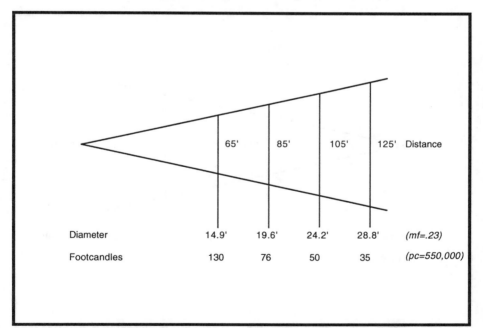

	65'	85'	105'	125'	Distance
Diameter	14.9'	19.6'	24.2'	28.8'	*(mf=.23)*
Footcandles	130	76	50	35	*(pc=550,000)*

* This model can be identified with model numbers -092, -095, -096, or -097 depending on the wire leads and connectors supplied.

BERKEY COLORTRAN
Model No. 213-112*

BEAM ANGLE: 3.3°

FIELD ANGLE: 5°

LAMP BASE TYPE: Medium 2-Pin

LAMP MOUNT: Axial

STANDARD LAMP: FEL - 1 KW.

OTHER LAMPS: EHD - 500 W. *(cf=.39)*
EHG - 750 W. *(cf=.56)*

INTEGRAL PATTERN SLOT: Yes

ACCESSORIES AVAILABLE:
Color Frame, Pattern Holder, Iris Kit

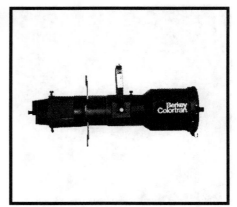

WEIGHT: 30.3 lbs.

SIZE: 41½" x 15" x 37½" (L x W x H)

PHOTOMETRICS CHART
(Performance data for this unit are measured at peak focus using FEL - 1 KW. lamp.)

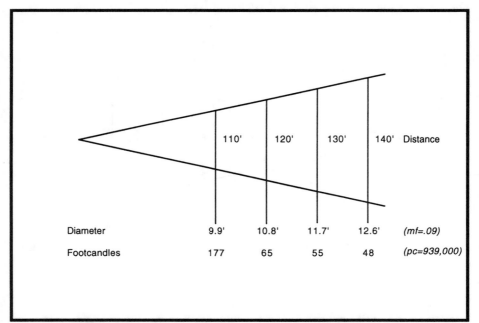

	110'	120'	130'	140'	Distance
Diameter	9.9'	10.8'	11.7'	12.6'	*(mf=.09)*
Footcandles	177	65	55	48	*(pc=939,000)*

* This model can be identified with model numbers -112, -115, -116, or -117 depending on the wire leads and connectors supplied.

CCT LIGHTING
Model No. Z00FV

BEAM ANGLE: 4.5°

FIELD ANGLE: 6.5°

LAMP BASE TYPE: Medium 2-Pin

LAMP MOUNT: Axial

STANDARD LAMP: FEL - 1 KW.

OTHER LAMPS: EHD - 500 W. *(cf= .35)*
EHG - 750 W. *(cf= .55)*

INTEGRAL PATTERN SLOT: Yes

ACCESSORIES AVAILABLE:
Color Frame (12" sq.), Drop-in Iris, Rotatable Gobo Holder

·WEIGHT: 38 lbs.

SIZE: 35" x 12½" x 16½" (L x W x H)

PHOTOMETRICS CHART

(Performance data for this unit are measured at peak focus using FEL 1 KW. lamp.)

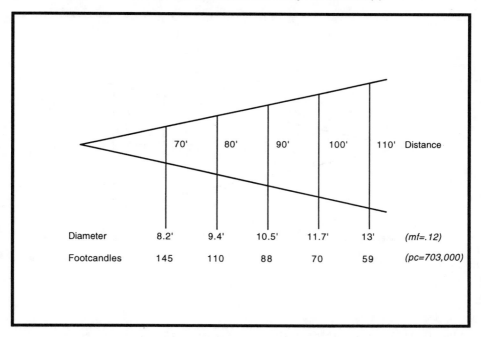

	70'	80'	90'	100'	110'	Distance
Diameter	8.2'	9.4'	10.5'	11.7'	13'	*(mf=.12)*
Footcandles	145	110	88	70	59	*(pc=703,000)*

317

ELECTRONIC THEATRE CONTROLS (ETC)
Model No. 405, "Source Four"

BEAM ANGLE: 4.6°

FIELD ANGLE: 6.8°

LAMP BASE TYPE: Medium 2-Pin

LAMP MOUNT: Axial

STANDARD LAMP: HPL 575/115 - 575 W.

OTHER LAMPS: HPL 550/77 - 550W.*

INTEGRAL PATTERN SLOT: Yes

WEIGHT: 21.3 lbs.

SIZE: 37⅞"" x 14¼" x 21⅜" (L x W x H)

ACCESSORIES AVAILABLE:
Color Frame (14" sq.), Pattern Holder, Donut, Snoot

PHOTOMETRICS CHART

(Performance data for this unit are measured at cosine focus using HPL 575/110 lamp.)

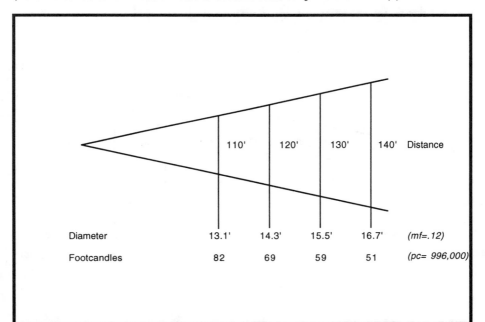

	110'	120'	130'	140'	Distance
Diameter	13.1'	14.3'	15.5'	16.7'	*(mf=.12)*
Footcandles	82	69	59	51	*(pc= 996,000)*

* The HPL 550/77 lamp is used with an ETC multiplexing system, used to allow multiple units to share a common circuit, with separate dimming of each.

KLIEGL BROS.
Model No. 1357/10*

BEAM ANGLE: (not specified in mfg. catalog)

FIELD ANGLE: 9°

LAMP BASE TYPE: Recessed Single-Contact

LAMP MOUNT: Off Axis

STANDARD LAMP: DWT - 1 KW.

OTHER LAMPS: FER -1 KW. *(cf=1.18)*

INTEGRAL PATTERN SLOT: Yes

WEIGHT: 28 lbs.

SIZE: 35½" x 16" x 27¼" (L x W x H)

ACCESSORIES AVAILABLE:
Color Frame (10" sq.), Pattern Holder

PHOTOMETRICS CHART
(Performance data for this unit are measured using DWT - 1 KW. lamp.)

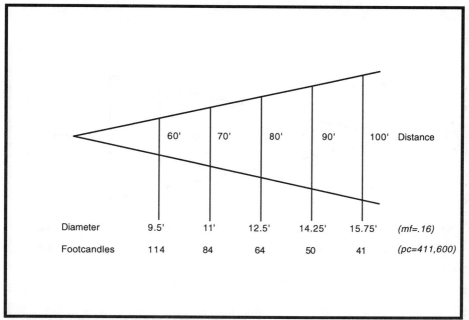

	60'	70'	80'	90'	100'	Distance
Diameter	9.5'	11'	12.5'	14.25'	15.75'	*(mf=.16)*
Footcandles	114	84	64	50	41	*(pc=411,600)*

* With optional iris installed, the model number is 1357I/10

KLIEGL BROS.
Model No. 1357/12*

BEAM ANGLE: (not specified in mfg. catalog)

FIELD ANGLE: 9°

LAMP BASE TYPE: Recessed Single-Contact

LAMP MOUNT: Off Axis

STANDARD LAMP: DWT - 1 KW.

OTHER LAMPS: FER - 1 KW. *(cf=1.18)*

INTEGRAL PATTERN SLOT: Yes

ACCESSORIES AVAILABLE:
Color Frame, Pattern Holder

WEIGHT: 28 lbs.

SIZE: 37" x 16" x 27¼" (L x W x H)

PHOTOMETRICS CHART
(Performance data for this unit are measured using DWT - 1 KW. lamp.)

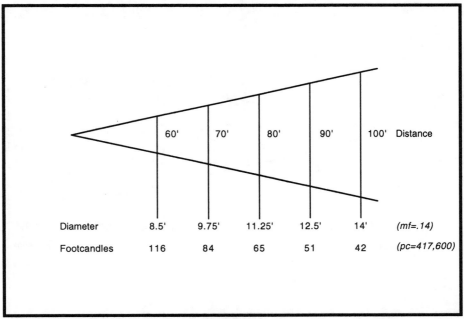

	Distance	60'	70'	80'	90'	100'	
Diameter		8.5'	9.75'	11.25'	12.5'	14'	*(mf=.14)*
Footcandles		116	84	65	51	42	*(pc=417,600)*

* With optional iris installed, the model number is 1357I/12.

320

LEE COLORTRAN
Model No. 650-082*

BEAM ANGLE: 3.3°

FIELD ANGLE: 5°

LAMP BASE TYPE: Medium 2-Pin

LAMP MOUNT: Axial

STANDARD LAMP: FEL - 1 KW.

OTHER LAMPS: EHD - 500 W. *(cf=.39)*
EHG - 750 W. *(cf=.56)*

INTEGRAL PATTERN SLOT: Yes

ACCESSORIES AVAILABLE:
Color Frame, Pattern Holder, Iris Kit

WEIGHT: 30.1 lbs.

SIZE: 40½" x 15" x 25½" (L x W x H)

PHOTOMETRICS CHART
(Performance data for this unit are measured at peak focus using FEL - 1 KW. lamp.)

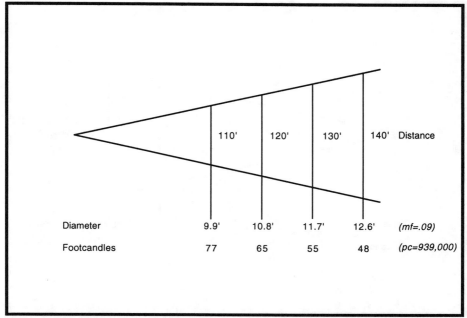

	110'	120'	130'	140'	Distance
Diameter	9.9'	10.8'	11.7'	12.6'	*(mf=.09)*
Footcandles	77	65	55	48	*(pc=939,000)*

* This model can be identified with model numbers -082, -084, -085, or -086 depending on the wire leads and connectors supplied.

STRAND CENTURY
Model Nos. 2123 (w/ shutters)
2124 (plus iris)

BEAM ANGLE: 6° (peak and cosine)

FIELD ANGLE: 9°

LAMP BASE TYPE: Medium 2-Pin

LAMP MOUNT: Axial

STANDARD LAMP: FEL - 1 KW.

OTHER LAMPS: EHD - 500 W. *(cf=.39)*
EHG - 750 W. *(cf=.56)*

INTEGRAL PATTERN SLOT: Yes

ACCESSORIES AVAILABLE:
Color Frame, Pattern Holder, Snoot

WEIGHT: 24.5 lbs.

SIZE: 36" x 12" x 25" (L x W x H)

PHOTOMETRICS CHART
(Performance data for this unit are measured at peak focus using FEL - 1 KW. lamp.)

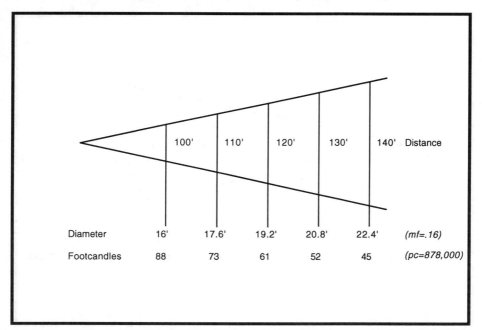

	100'	110'	120'	130'	140' Distance
Diameter	16'	17.6'	19.2'	20.8'	22.4' *(mf=.16)*
Footcandles	88	73	61	52	45 *(pc=878,000)*

ALTMAN STAGE LIGHTING
Model No. 1KL6-2040Z

BEAM ANGLE, NARROW: 9.5° (peak), 15° (flat field)
BEAM ANGLE, WIDE: 19° (peak), 31.5° (flat field)
FIELD ANGLE, NARROW: 20°
FIELD ANGLE, WIDE: 40°
LAMP BASE TYPE: Medium 2-Pin
LAMP MOUNT: Axial
STANDARD LAMP: FEL - 1 KW.
OTHER LAMPS: EHD - 500 W. *(cf=.36)*
EHG - 750 W. *(cf=.55)*
INTEGRAL PATTERN SLOT: Yes

ACCESSORIES AVAILABLE:
Color Frame, Pattern Holder, Snoot, Motorized & Non-motorized Color Wheels

WEIGHT: (not specified in manufacturer's catalog)

SIZE: 26⅞" x 17" x 19" (L x W x H)

PHOTOMETRICS CHART

(Performance data for this unit are measured at peak focus using FEL - 1 KW. lamp.)

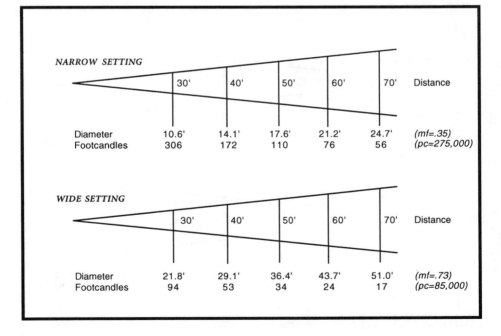

NARROW SETTING

	30'	40'	50'	60'	70'	Distance
Diameter	10.6'	14.1'	17.6'	21.2'	24.7'	*(mf=.35)*
Footcandles	306	172	110	76	56	*(pc=275,000)*

WIDE SETTING

	30'	40'	50'	60'	70'	Distance
Diameter	21.8'	29.1'	36.4'	43.7'	51.0'	*(mf=.73)*
Footcandles	94	53	34	24	17	*(pc=85,000)*

ALTMAN STAGE LIGHTING
Model No. 1KL8-1424Z*

BEAM ANGLE, NARROW: 8° (peak), 14° (flat field)

BEAM ANGLE, WIDE: 14° (peak), 24° (flat field)

FIELD ANGLE, NARROW: 14° (peak & flat field)

FIELD ANGLE, WIDE: 24° (peak & flat field)

LAMP BASE TYPE: Medium 2-Pin

LAMP MOUNT: Axial

STANDARD LAMP: FEL - 1KW.

OTHER LAMPS: EHD - 500 W. *(cf=.36)*
EHG - 750 W. *(cf=.55)*

INTEGRAL PATTERN SLOT: Yes

ACCESSORIES AVAILABLE:
Color Frame, Pattern Holder, Snoot, Iris

WEIGHT: (not specified in manufacturer's catalog)

SIZE: 32" x 10½" x 17" (L x W x H)

PHOTOMETRICS CHART

(Performance data for this unit are measured at peak focus using FEL - 1 KW. lamp.)

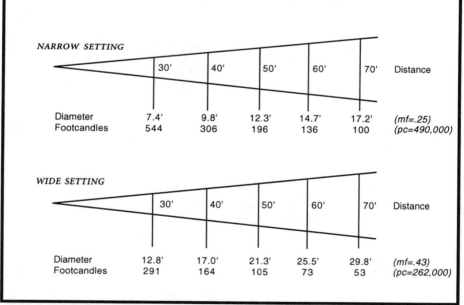

NARROW SETTING

	30'	40'	50'	60'	70'	Distance
Diameter	7.4'	9.8'	12.3'	14.7'	17.2'	*(mf=.25)*
Footcandles	544	306	196	136	100	*(pc=490,000)*

WIDE SETTING

	30'	40'	50'	60'	70'	Distance
Diameter	12.8'	17.0'	21.3'	25.5'	29.8'	*(mf=.43)*
Footcandles	291	164	105	73	53	*(pc=262,000)*

* With optional iris installed, the model number is 1KL8-I-1424Z.

WEIGHT: (not specified in manufacturer's catalog)

SIZE: 24" x 11" x 10¼" (L x W x H)

ALTMAN STAGE LIGHTING
Model Nos.4.5-1530Z-MC (Mini-Can)
4.5-1530Z-MT (Med. 2-Pin)

BEAM ANGLE, NARROW: 8° (peak), 13° (flat field)
BEAM ANGLE, WIDE: 13° (peak), 18° (flat field)
FIELD ANGLE, NARROW: 15° (peak & flat field)
FIELD ANGLE, WIDE: 30° (peak & flat field)
LAMP BASE TYPE: Mini-Can Screw or Medium 2-Pin
LAMP MOUNT: Axial
STANDARD LAMP: EHF - 750 W.
OTHER LAMPS: EHT - 250 W. (MC) *(cf=.24)*
EVR - 500 W. (MC) *(cf=.50)*
EHD - 500 W. (M2P) *(cf=.50)*
EHG - 750 W. (M2P) *(cf=.75)*
INTEGRAL PATTERN SLOT: Yes
ACCESSORIES AVAILABLE:
Color Frame, Pattern Holder, Snoot, Aperture Gobos

PHOTOMETRICS CHART
(Performance data for this unit are measured at peak focus using EHF - 750 W. lamp.)

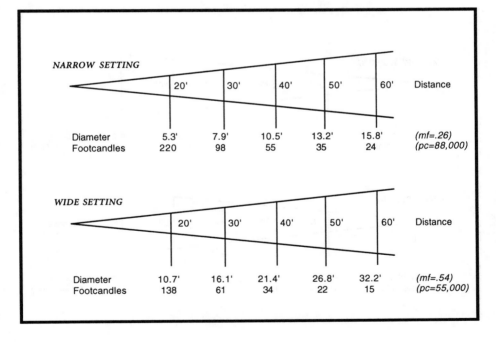

NARROW SETTING

	20'	30'	40'	50'	60'	Distance
Diameter	5.3'	7.9'	10.5'	13.2'	15.8'	*(mf=.26)*
Footcandles	220	98	55	35	24	*(pc=88,000)*

WIDE SETTING

	20'	30'	40'	50'	60'	Distance
Diameter	10.7'	16.1'	21.4'	26.8'	32.2'	*(mf=.54)*
Footcandles	138	61	34	22	15	*(pc=55,000)*

ALTMAN STAGE LIGHTING
Model Nos. 4.5-2550Z-MC (Mini-Can)
4.5-2550Z-MT (Med 2-Pin)

BEAM ANGLE, NARROW: 17° (peak), 22° (flat field)
BEAM ANGLE, WIDE: 22° (peak), 25° (flat field)
FIELD ANGLE, NARROW: 25° (peak & flat field)
FIELD ANGLE, WIDE: 50° (peak & flat field)
LAMP BASE TYPE: Mini-Can Screw or Medium 2-Pin
LAMP MOUNT: Axial
STANDARD LAMP: EHF - 750 W.

OTHER LAMPS: EHT - 250 W. (MC) *(cf=.24)*
EVR - 500 W. (MC) *(cf=.50)*
EHD - 500 W. (M2P) *(cf=.50)*
EHG - 750 W. (M2P) *(cf=.275)*

INTEGRAL PATTERN SLOT: Yes
ACCESSORIES AVAILABLE:
Color Frame, Pattern Holder, Snoot, Aperture Gobos

WEIGHT: (not specified in manufacturer's catalog)
SIZE: 19⅛" x 11" x 10¼" (L x W x H)

PHOTOMETRICS CHART

(Performance data for this unit are measured at peak focus using EHF - 750 W. lamp.)

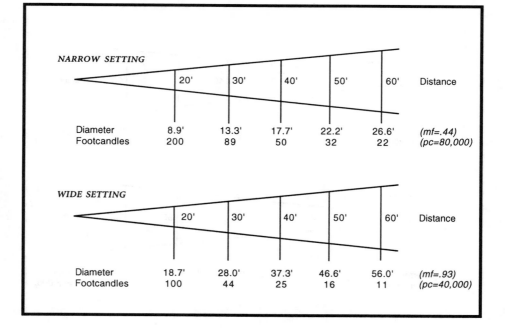

NARROW SETTING

	20'	30'	40'	50'	60'	Distance
Diameter	8.9'	13.3'	17.7'	22.2'	26.6'	*(mf=.44)*
Footcandles	200	89	50	32	22	*(pc=80,000)*

WIDE SETTING

	20'	30'	40'	50'	60'	Distance
Diameter	18.7'	28.0'	37.3'	46.6'	56.0'	*(mf=.93)*
Footcandles	100	44	25	16	11	*(pc=40,000)*

WEIGHT: (not specified in manufacturer's catalog)

SIZE: 19⅜" x 11" x 10¼" (L x W x H)

ALTMAN STAGE LIGHTING
Model Nos. 4.5-3060Z-MC (Mini-Can)
4.5-3060Z-MT (Med. 2-Pin)

BEAM ANGLE, NARROW: 17° (peak), 25° (flat field)
BEAM ANGLE, WIDE: 23° (peak), 50° (flat field)
FIELD ANGLE, NARROW: 30° (peak & flat field)
FIELD ANGLE, WIDE: 60° (peak & flat field)
LAMP BASE TYPE: Mini-Can Screw or Medium 2-Pin
LAMP MOUNT: Axial
STANDARD LAMP: EHF - 750 W.
OTHER LAMPS: EHT - 250 W. (MC) *(cf=.24)*
EVR - 500 W. (MC) *(cf=.50)*
EHD - 500 W. (M2P) *(cf=.50)*
EHG - 750 W. (M2P) *(cf=.275)*
INTEGRAL PATTERN SLOT: Yes
ACCESSORIES AVAILABLE:
Color Frame, Pattern Holder, Snoot, Aperture Gobos

PHOTOMETRICS CHART
(Performance data for this unit are measured at peak focus using EHF - 750 W. lamp.)

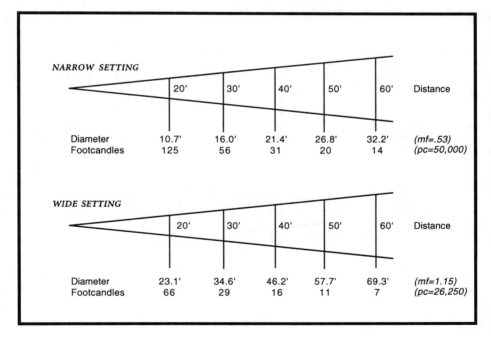

NARROW SETTING

	20'	30'	40'	50'	60'	Distance
Diameter	10.7'	16.0'	21.4'	26.8'	32.2'	*(mf=.53)*
Footcandles	125	56	31	20	14	*(pc=50,000)*

WIDE SETTING

	20'	30'	40'	50'	60'	Distance
Diameter	23.1'	34.6'	46.2'	57.7'	69.3'	*(mf=1.15)*
Footcandles	66	29	16	11	7	*(pc=26,250)*

ALTMAN
Model No. S6-15-35, "Shakespeare"

BEAM ANGLE, NARROW: 9° (cosine)

BEAM ANGLE, WIDE: 15° (cosine)

FIELD ANGLE, NARROW: 15°

FIELD ANGLE, WIDE: 35°

LAMP BASE TYPE: Medium 2-Pin

LAMP MOUNT: Axial

STANDARD LAMP: HX600 - 575 W.

OTHER LAMPS: EHG - 750 W. *(cf=.93)*
                EHD - 500 W. *(cf-.63)*

INTEGRAL PATTERN SLOT: Yes

WEIGHT: 30 lbs.

SIZE: 33³⁄₁₆" x 12" x 17½" (L x W x H)

ACCESSORIES AVAILABLE:
Color Frame, Snoot, Pattern Holder

PHOTOMETRICS CHART

(Performance data for this unit are measured at cosine focus using HX600 - 575 W. lamp.)

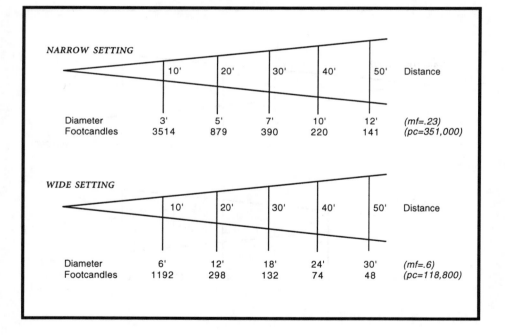

NARROW SETTING

	10'	20'	30'	40'	50'	Distance
Diameter	3'	5'	7'	10'	12'	*(mf=.23)*
Footcandles	3514	879	390	220	141	*(pc=351,000)*

WIDE SETTING

	10'	20'	30'	40'	50'	Distance
Diameter	6'	12'	18'	24'	30'	*(mf=.6)*
Footcandles	1192	298	132	74	48	*(pc=118,800)*

ALTMAN
Model No. S6-30-55, "Shakespeare"

BEAM ANGLE, NARROW: 13° (cosine)

BEAM ANGLE, WIDE: 15° (cosine)

FIELD ANGLE, NARROW: 30°

FIELD ANGLE, WIDE: 55°

LAMP BASE TYPE: Medium 2-Pin

LAMP MOUNT: Axial

STANDARD LAMP: HX600 - 575 W.

OTHER LAMPS: EHG - 750 W. *(cf=.93)*
EHD - 500 W. *(cf-.63)*

INTEGRAL PATTERN SLOT: Yes

ACCESSORIES AVAILABLE:
Color Frame, Snoot, Pattern Holder

WEIGHT: 30 lbs.

SIZE: 33³⁄₁₆" x 12" x 17½" (L x W x H)

PHOTOMETRICS CHART

(Performance data for this unit are measured at peak using HX600 - 575 W. lamp.)

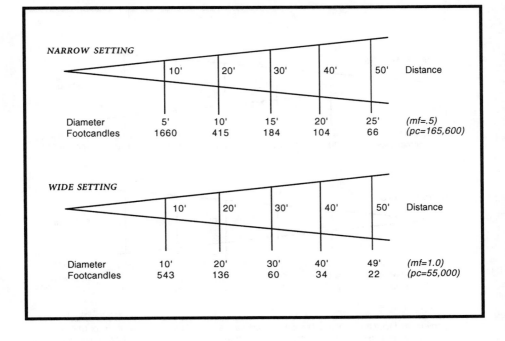

NARROW SETTING

	10'	20'	30'	40'	50'	Distance
Diameter	5'	10'	15'	20'	25'	*(mf=.5)*
Footcandles	1660	415	184	104	66	*(pc=165,600)*

WIDE SETTING

	10'	20'	30'	40'	50'	Distance
Diameter	10'	20'	30'	40'	49'	*(mf=1.0)*
Footcandles	543	136	60	34	22	*(pc=55,000)*

AVAB AMERICA
Model No. HPZ 112

BEAM ANGLE, NARROW: (not specified in mfg. catalog)

BEAM ANGLE, WIDE: (not specified in mfg. catalog)

FIELD ANGLE, NARROW: 12°

FIELD ANGLE, WIDE: 27°

LAMP BASE TYPE: Medium Bipost

LAMP MOUNT: Off Axis

STANDARD LAMP: EGT - 1 KW.

OTHER LAMPS: (not specified in mfg. catalog)

INTEGRAL PATTERN SLOT: Yes

WEIGHT: (not specified in manufacturer's catalog)

SIZE: 28" x 11⅞" x 16⅞" (L x W x H)

ACCESSORIES AVAILABLE:
Optional Angled Yoke Available

PHOTOMETRICS CHART

(Performance data for this unit are measured using EGT - 1KW. lamp.)

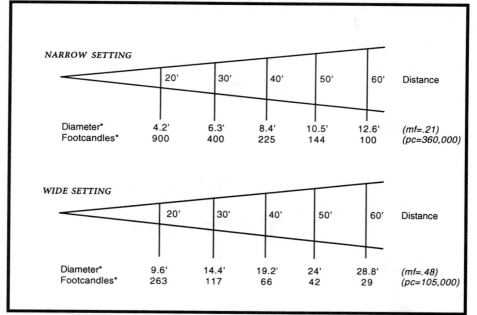

NARROW SETTING

	20'	30'	40'	50'	60'	Distance
Diameter*	4.2'	6.3'	8.4'	10.5'	12.6'	(mf=.21)
Footcandles*	900	400	225	144	100	(pc=360,000)

WIDE SETTING

	20'	30'	40'	50'	60'	Distance
Diameter*	9.6'	14.4'	19.2'	24'	28.8'	(mf=.48)
Footcandles*	263	117	66	42	29	(pc=105,000)

* The diameter and footcandle information in this chart was calculated using multiplying factor (*mf*) and peak candela (*pc*) data given in catalog. See Introduction for more information about calculating photometric data.

AVAB AMERICA
Model No. HPZ 115

BEAM ANGLE, NARROW: (not specified in mfg. catalog)

BEAM ANGLE, WIDE: (not specified in mfg. catalog)

FIELD ANGLE, NARROW: 15°

FIELD ANGLE, WIDE: 38°

LAMP BASE TYPE: Medium Bipost

LAMP MOUNT: Off Axis

STANDARD LAMP: EGT - 1 KW.

OTHER LAMPS: (not specified in mfg. catalog)

INTEGRAL PATTERN SLOT: Yes

WEIGHT: (not specified in manufacturer's catalog)

SIZE: 25⅜" x 11⅞" x 16⅞" (L x W x H)

ACCESSORIES AVAILABLE:
Optional Angled Yoke Available

PHOTOMETRICS CHART
(Performance data for this unit are measured using EGT - 1KW. lamp.)

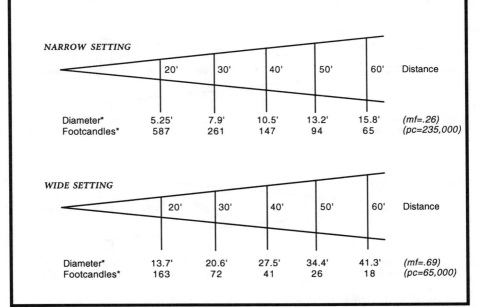

NARROW SETTING

	20'	30'	40'	50'	60'	Distance
Diameter*	5.25'	7.9'	10.5'	13.2'	15.8'	(mf=.26)
Footcandles*	587	261	147	94	65	(pc=235,000)

WIDE SETTING

	20'	30'	40'	50'	60'	Distance
Diameter*	13.7'	20.6'	27.5'	34.4'	41.3'	(mf=.69)
Footcandles*	163	72	41	26	18	(pc=65,000)

* The diameter and footcandle information in this chart was calculated using multiplying factor (mf) and peak candela (pc) data given in catalog. See Introduction for more information about calculating photometric data.

AVAB AMERICA
Model No. HPZ 211

BEAM ANGLE, NARROW: (not specified in mfg. catalog)

BEAM ANGLE, WIDE: (not specified in mfg. catalog)

FIELD ANGLE, NARROW: 11°

FIELD ANGLE, WIDE: 27°

LAMP BASE TYPE: Medium Bipost

LAMP MOUNT: Off Axis

STANDARD LAMP: EGT - 1 KW.

OTHER LAMPS: (not specified in mfg. catalog)

INTEGRAL PATTERN SLOT: Yes

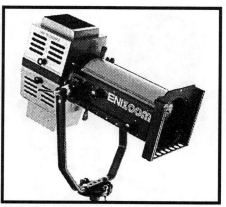

WEIGHT: (not specified in manufacturer's catalog)

SIZE: 33¼" x 17" x 25⅜" (L x W x H)

ACCESSORIES AVAILABLE:
Optional Angled Yoke Available

PHOTOMETRICS CHART

(Performance data for this unit are measured using EGT - 1KW. lamp.)

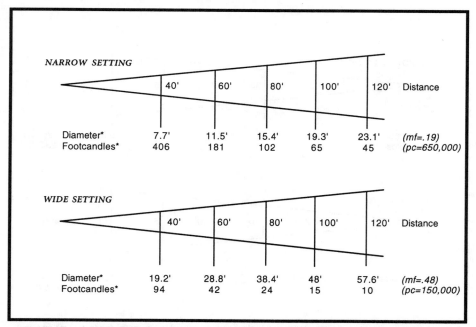

NARROW SETTING

	40'	60'	80'	100'	120'	Distance
Diameter*	7.7'	11.5'	15.4'	19.3'	23.1'	(mf=.19)
Footcandles*	406	181	102	65	45	(pc=650,000)

WIDE SETTING

	40'	60'	80'	100'	120'	Distance
Diameter*	19.2'	28.8'	38.4'	48'	57.6'	(mf=.48)
Footcandles*	94	42	24	15	10	(pc=150,000)

* The diameter and footcandle information in this chart was calculated using multiplying factor (*mf*) and peak candela (*pc*) data given in catalog. See Introduction for more information about calculating photometric data.

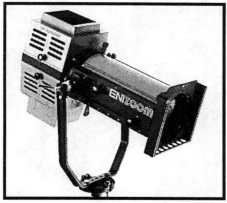

WEIGHT: (not specified in manufacturer's catalog)
SIZE: 31" x 17" x 25⅜" (L x W x H)

AVAB AMERICA
Model No. HPZ 215

BEAM ANGLE, NARROW: (not specified in mfg. catalog)
BEAM ANGLE, WIDE: (not specified in mfg. catalog)
FIELD ANGLE, NARROW: 15°
FIELD ANGLE, WIDE: 40°
LAMP BASE TYPE: Medium Bipost
LAMP MOUNT: Off Axis
STANDARD LAMP: EGT - 1 KW.
OTHER LAMPS: (not specified in mfg. catalog)
INTEGRAL PATTERN SLOT: Yes

ACCESSORIES AVAILABLE:
Optional Angled Yoke Available

PHOTOMETRICS CHART
(Performance data for this unit are measured using EGT - 1KW. lamp.)

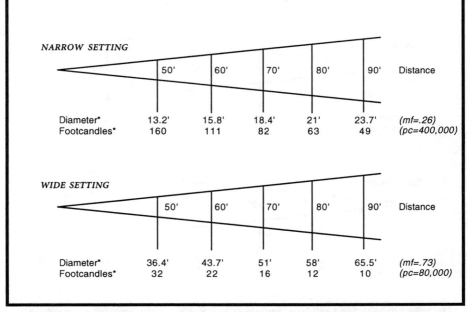

NARROW SETTING

	50'	60'	70'	80'	90'	Distance
Diameter*	13.2'	15.8'	18.4'	21'	23.7'	(mf=.26)
Footcandles*	160	111	82	63	49	(pc=400,000)

WIDE SETTING

	50'	60'	70'	80'	90'	Distance
Diameter*	36.4'	43.7'	51'	58'	65.5'	(mf=.73)
Footcandles*	32	22	16	12	10	(pc=80,000)

* The diameter and footcandle information in this chart was calculated using multiplying factor (*mf*) and peak candela (*pc*) data given in catalog. See Introduction for more information about calculating photometric data.

BERKEY COLORTRAN
Model No. 213-002*

BEAM ANGLE, NARROW: 14°

BEAM ANGLE, WIDE: 21°

FIELD ANGLE, NARROW: 24°

FIELD ANGLE, WIDE: 43°

LAMP BASE TYPE: Medium 2-Pin

LAMP MOUNT: Axial

STANDARD LAMP: FEL - 1 KW.

OTHER LAMPS: EHD - 500 W. *(cf=.39)*
EHG - 750 W. *(cf=.56)*

INTEGRAL PATTERN SLOT: Yes

ACCESSORIES AVAILABLE:
Color Frame, Pattern Holder

WEIGHT: 34 lbs.

SIZE: 25¾" x 14½" x 27½" (L x W x H)

PHOTOMETRICS CHART

(Performance data for this unit are measured using FEL - 1 KW. lamp.)

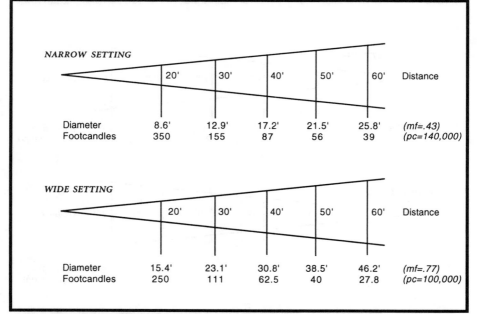

NARROW SETTING

	20'	30'	40'	50'	60'	Distance
Diameter	8.6'	12.9'	17.2'	21.5'	25.8'	*(mf=.43)*
Footcandles	350	155	87	56	39	*(pc=140,000)*

WIDE SETTING

	20'	30'	40'	50'	60'	Distance
Diameter	15.4'	23.1'	30.8'	38.5'	46.2'	*(mf=.77)*
Footcandles	250	111	62.5	40	27.8	*(pc=100,000)*

* This model can be identified with model numbers -002, -005, -006, or -007 depending on the wire leads and connectors supplied.

WEIGHT: 18.25 lbs.

SIZE: 26⅝" x 14½" x 24" (L x W x H)

BERKEY COLORTRAN
Model No. 213-162*

BEAM ANGLE, NARROW: 7.5°

BEAM ANGLE, WIDE: 14°

FIELD ANGLE, NARROW: 15°

FIELD ANGLE, WIDE: 35°

LAMP BASE TYPE: Medium 2-Pin

LAMP MOUNT: Axial

STANDARD LAMP: FEL - 1 KW.

OTHER LAMPS: EHD - 500 W. *(cf=.30)*
                  EHG - 750 W. *(cf=.60)*
                  FEP - 1 KW. *(cf=.70)*

INTEGRAL PATTERN SLOT: Yes

ACCESSORIES AVAILABLE:
Color Frame (7½" sq.), Pattern Holder, Iris Kit

PHOTOMETRICS CHART

(Performance data for this unit are measured using FEL - 1 KW. lamp.)

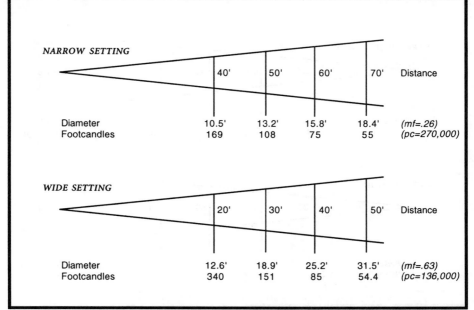

NARROW SETTING

	40'	50'	60'	70'	Distance
Diameter	10.5'	13.2'	15.8'	18.4'	*(mf=.26)*
Footcandles	169	108	75	55	*(pc=270,000)*

WIDE SETTING

	20'	30'	40'	50'	Distance
Diameter	12.6'	18.9'	25.2'	31.5'	*(mf=.63)*
Footcandles	340	151	85	54.4	*(pc=136,000)*

* This model can be identified with model numbers -162, -165, -166, or -167 depending on the wire leads and connectors supplied.

CCT LIGHTING
Model No. Z00FS

BEAM ANGLE, NARROW: 6° (peak) 8° (flat field)

BEAM ANGLE, WIDE: 9° (peak) 20° (flat field)

FIELD ANGLE, NARROW: 10° (peak) 20° (flat field)

FIELD ANGLE, WIDE: 22° (peak + flat field)

LAMP BASE TYPE: Medium 2-Pin

LAMP MOUNT: Axial

STANDARD LAMP: FEL - 1 KW.

OTHER LAMPS: EHD - 500 W. *(cf=.36)*
EHG - 750 W. *(cf=.55)*

INTEGRAL PATTERN SLOT: Yes

WEIGHT: 38 lbs.

SIZE: 35" x 12½" x 16½" (L x W x H)

ACCESSORIES AVAILABLE:
Color Frame, Drop-in Iris, Pattern Holder, Rotatable Pattern Holder

PHOTOMETRICS CHART

(Performance data for this unit are measured at peak focus using FEL - 1KW. lamp.)

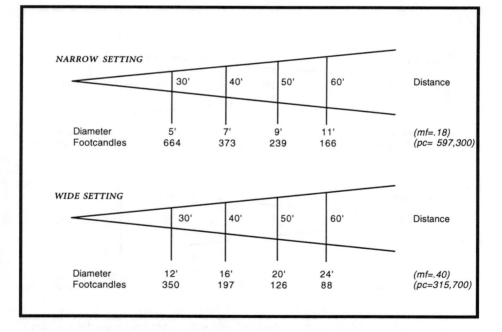

NARROW SETTING

	30'	40'	50'	60'	Distance
Diameter	5'	7'	9'	11'	*(mf=.18)*
Footcandles	664	373	239	166	*(pc= 597,300)*

WIDE SETTING

	30'	40'	50'	60'	Distance
Diameter	12'	16'	20'	24'	*(mf=.40)*
Footcandles	350	197	126	88	*(pc=315,700)*

WEIGHT: 25 lbs.

SIZE: 24½" x 10" x 15½" (L x W x H)

CCT LIGHTING
Model No. Z00FW

BEAM ANGLE, NARROW: 11.3° (peak) 18° (flat field)

BEAM ANGLE, WIDE: 19.3° (peak) 42.3° (flat field)

FIELD ANGLE, NARROW: 19° (peak + flat field)

FIELD ANGLE, WIDE: 45° (peak + flat field)

LAMP BASE TYPE: Medium 2-Pin

LAMP MOUNT: Axial

STANDARD LAMP: FEL - 1 KW.

OTHER LAMPS: EHD - 500 W. *(cf=.36)*
EHG - 750 W. *(cf=.55)*

INTEGRAL PATTERN SLOT: Yes

ACCESSORIES AVAILABLE:
Color Frame, Drop-in Iris, Pattern Holder, Rotatable
Pattern Holder

PHOTOMETRICS CHART

(Performance data for this unit are measured at peak focus using FEL - 1KW. lamp.)

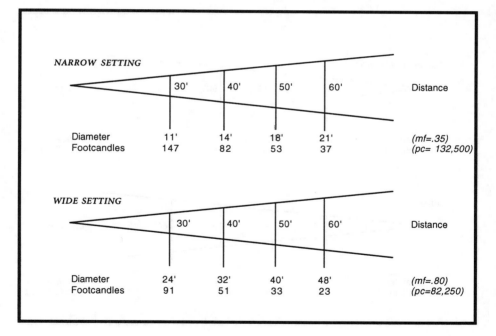

NARROW SETTING

	30'	40'	50'	60'	Distance
Diameter	11'	14'	18'	21'	*(mf=.35)*
Footcandles	147	82	53	37	*(pc= 132,500)*

WIDE SETTING

	30'	40'	50'	60'	Distance
Diameter	24'	32'	40'	48'	*(mf=.80)*
Footcandles	91	51	33	23	*(pc=82,250)*

CCT LIGHTING
Model No. Z00FX

BEAM ANGLE, NARROW: 8.5° (peak) 13° (flat field)

BEAM ANGLE, WIDE: 10.5° (peak) 31.5° (flat field)

FIELD ANGLE, NARROW: 15° (peak + flat field)

FIELD ANGLE, WIDE: 34° (peak + flat field)

LAMP BASE TYPE: Medium 2-Pin

LAMP MOUNT: Axial

STANDARD LAMP: FEL - 1 KW.

OTHER LAMPS: EHD - 500 W. *(cf=.36)*
                EHG - 750 W. *(cf=.55)*

INTEGRAL PATTERN SLOT: Yes

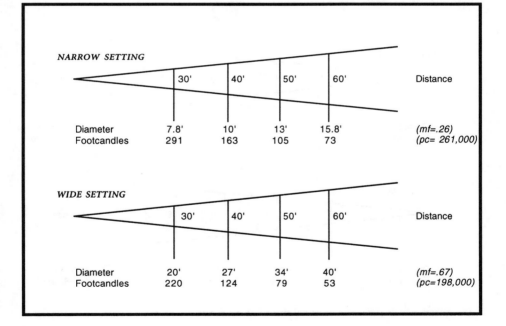

WEIGHT: 26 lbs.

SIZE: 28½" x 10" x 15½" (L x W x H)

ACCESSORIES AVAILABLE:
Color Frame, Drop-in Iris, Pattern Holder, Rotatable
Pattern Holder

PHOTOMETRICS CHART

(Performance data for this unit are measured at peak focus using FEL - 1KW. lamp.)

NARROW SETTING

	30'	40'	50'	60'	Distance
Diameter	7.8'	10'	13'	15.8'	*(mf=.26)*
Footcandles	291	163	105	73	*(pc= 261,000)*

WIDE SETTING

	30'	40'	50'	60'	Distance
Diameter	20'	27'	34'	40'	*(mf=.67)*
Footcandles	220	124	79	53	*(pc=198,000)*

WEIGHT: 27 lbs.

SIZE: 31" x 10" x 15½" (L x W x H)

CCT LIGHTING
Model No. Z00FY

BEAM ANGLE, NARROW: 7.3° (peak) 10.3° (flat field)

BEAM ANGLE, WIDE: 11° (peak) 26° (flat field)

FIELD ANGLE, NARROW: 11° (peak + flat field)

FIELD ANGLE, WIDE: 28° (peak + flat field)

LAMP BASE TYPE: Medium 2-Pin

LAMP MOUNT: Axial

STANDARD LAMP: FEL - 1 KW.

OTHER LAMPS: EHD - 500 W. *(cf=.36)*
EHG - 750 W. *(cf=.55)*

INTEGRAL PATTERN SLOT: Yes

ACCESSORIES AVAILABLE:
Color Frame, Drop-in Iris, Pattern Holder, Rotatable Pattern Holder

PHOTOMETRICS CHART

(Performance data for this unit are measured at peak focus using FEL - 1KW. lamp.)

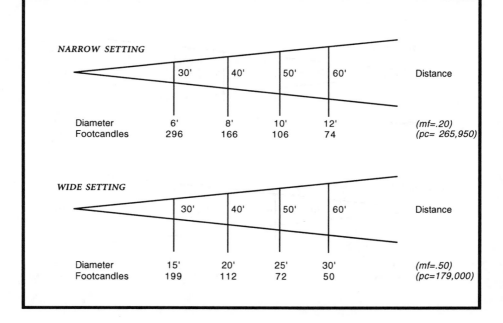

NARROW SETTING

	30'	40'	50'	60'	Distance
Diameter	6'	8'	10'	12'	*(mf=.20)*
Footcandles	296	166	106	74	*(pc= 265,950)*

WIDE SETTING

	30'	40'	50'	60'	Distance
Diameter	15'	20'	25'	30'	*(mf=.50)*
Footcandles	199	112	72	50	*(pc=179,000)*

CCT LIGHTING
Model No. Z00TS

BEAM ANGLE, NARROW: 6° (peak) 8° (flat field)

BEAM ANGLE, WIDE: 9° (peak) 20° (flat field)

FIELD ANGLE, NARROW: 10° (peak) 20° (flat field)

FIELD ANGLE, WIDE: 22° (peak + flat field)

LAMP BASE TYPE: Medium 2-Pin

LAMP MOUNT: Axial

STANDARD LAMP: FEL - 1 KW.

OTHER LAMPS: EHD - 500 W. *(cf=.36)*
EHG - 750 W. *(cf=.55)*

INTEGRAL PATTERN SLOT: Yes

WEIGHT: 38 lbs.

SIZE: 35" x 12½" x 16½" (L x W x H)

ACCESSORIES AVAILABLE:
Color Frame (12" sq.), Drop-in Iris, Pattern Holder,
Rotatable Pattern Holder

PHOTOMETRICS CHART

(Performance data for this unit are measured at peak focus using FEL - 1KW. lamp.)

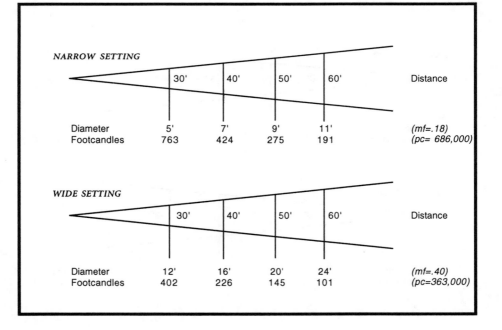

NARROW SETTING

	30'	40'	50'	60'	Distance
Diameter	5'	7'	9'	11'	*(mf=.18)*
Footcandles	763	424	275	191	*(pc= 686,000)*

WIDE SETTING

	30'	40'	50'	60'	Distance
Diameter	12'	16'	20'	24'	*(mf=.40)*
Footcandles	402	226	145	101	*(pc=363,000)*

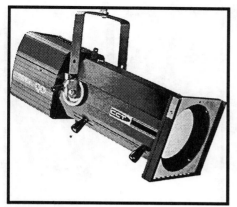

WEIGHT: 25 lbs.

SIZE: 24½" x 10" x 15½" (L x W x H)

CCT LIGHTING
Model No. Z00TW

BEAM ANGLE, NARROW: 11.3° (peak) 18° (flat field)

BEAM ANGLE, WIDE: 19.3° (peak) 42.3° (flat field)

FIELD ANGLE, NARROW: 19° (peak + flat field)

FIELD ANGLE, WIDE: 45° (peak + flat field)

LAMP BASE TYPE: Medium 2-Pin

LAMP MOUNT: Axial

STANDARD LAMP: FEL - 1 KW.

OTHER LAMPS: EHD - 500 W. *(cf=.36)*
EHG - 750 W. *(cf=.55)*

INTEGRAL PATTERN SLOT: Yes

ACCESSORIES AVAILABLE:
Color Frame (10" sq.), Drop-in Iris, Pattern Holder, Rotatable Pattern Holder

PHOTOMETRICS CHART

(Performance data for this unit are measured at peak focus using FEL - 1KW. lamp.)

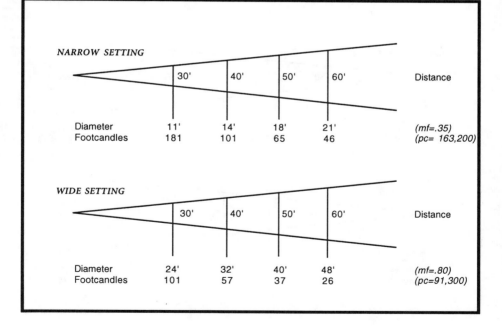

NARROW SETTING

	30'	40'	50'	60'	Distance
Diameter	11'	14'	18'	21'	*(mf=.35)*
Footcandles	181	101	65	46	*(pc= 163,200)*

WIDE SETTING

	30'	40'	50'	60'	Distance
Diameter	24'	32'	40'	48'	*(mf=.80)*
Footcandles	101	57	37	26	*(pc=91,300)*

CCT LIGHTING
Model No. Z00TX

BEAM ANGLE, NARROW: 8.5° (peak) 13° (flat field)

BEAM ANGLE, WIDE: 10.5° (peak) 31.5° (flat field)

FIELD ANGLE, NARROW: 15° (peak + flat field)

FIELD ANGLE, WIDE: 34° (peak + flat field)

LAMP BASE TYPE: Medium 2-Pin

LAMP MOUNT: Axial

STANDARD LAMP: FEL - 1 KW.

OTHER LAMPS: EHD - 500 W. *(cf=.36)*
EHG - 750 W. *(cf=.55)*

INTEGRAL PATTERN SLOT: Yes

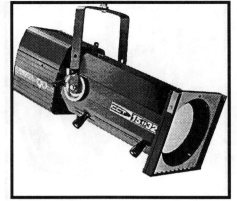

WEIGHT: 26 lbs.

SIZE: 28½" x 10" x 15½" (L x W x H)

ACCESSORIES AVAILABLE:
Color Frame (10" sq.), Drop-in Iris, Pattern Holder,
Rotatable Pattern Holder

PHOTOMETRICS CHART

(Performance data for this unit are measured at peak focus using FEL - 1KW. lamp.)

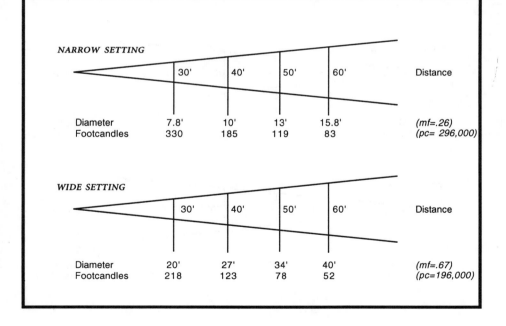

NARROW SETTING

	30'	40'	50'	60'	Distance
Diameter	7.8'	10'	13'	15.8'	*(mf=.26)*
Footcandles	330	185	119	83	*(pc= 296,000)*

WIDE SETTING

	30'	40'	50'	60'	Distance
Diameter	20'	27'	34'	40'	*(mf=.67)*
Footcandles	218	123	78	52	*(pc=196,000)*

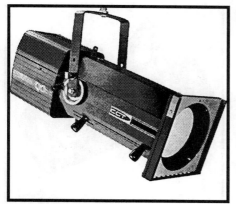

WEIGHT: 27 lbs.

SIZE: 31" x 10" x 15½" (L x W x H)

CCT LIGHTING
Model No. Z00TY

BEAM ANGLE, NARROW: 7.3° (peak) 10.3° (flat field)

BEAM ANGLE, WIDE: 11° (peak) 26° (flat field)

FIELD ANGLE, NARROW: 11° (peak + flat field)

FIELD ANGLE, WIDE: 28° (peak + flat field)

LAMP BASE TYPE: Medium 2-Pin

LAMP MOUNT: Axial

STANDARD LAMP: FEL - 1 KW.

OTHER LAMPS: EHD - 500 W. *(cf=.36)*
EHG - 750 W. *(cf=.55)*

INTEGRAL PATTERN SLOT: Yes

ACCESSORIES AVAILABLE:
Color Frame (10" sq.), Drop-in Iris, Pattern Holder, Rotatable Pattern Holder

PHOTOMETRICS CHART

(Performance data for this unit are measured at peak focus using FEL - 1KW. lamp.)

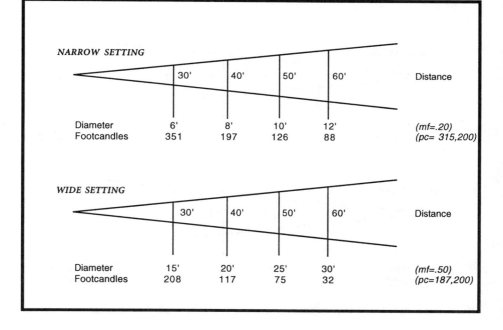

NARROW SETTING

	30'	40'	50'	60'	Distance
Diameter	6'	8'	10'	12'	*(mf=.20)*
Footcandles	351	197	126	88	*(pc= 315,200)*

WIDE SETTING

	30'	40'	50'	60'	Distance
Diameter	15'	20'	25'	30'	*(mf=.50)*
Footcandles	208	117	75	32	*(pc=187,200)*

CCT LIGHTING
Model No. Z0602

BEAM ANGLE, NARROW:11.5° (peak) 16.5° (flat field)

BEAM ANGLE, WIDE: 17° (peak) 35° (flat field)

FIELD ANGLE, NARROW: 21° (peak + flat field)

FIELD ANGLE, WIDE: 36° (peak + flat field)

LAMP BASE TYPE: GY9.5

LAMP MOUNT: Axial

STANDARD LAMP: FMR - 600 W.

OTHER LAMPS: (not specified in mfg. catalog)

INTEGRAL PATTERN SLOT: Yes

WEIGHT: 8 lbs.

SIZE: 17" x 12" (L x W)

ACCESSORIES AVAILABLE:
Color Frame, Drop-in Iris, Rotatable Pattern Holder

PHOTOMETRICS CHART
(Performance data for this unit are measured at peak focus using FMR - 600W. lamp.)

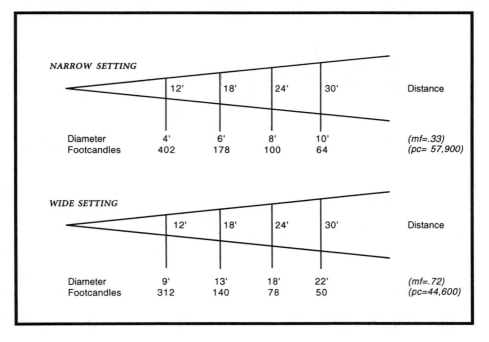

NARROW SETTING

	12'	18'	24'	30'	Distance
Diameter	4'	6'	8'	10'	*(mf=.33)*
Footcandles	402	178	100	64	*(pc= 57,900)*

WIDE SETTING

	12'	18'	24'	30'	Distance
Diameter	9'	13'	18'	22'	*(mf=.72)*
Footcandles	312	140	78	50	*(pc=44,600)*

CCT LIGHTING
Model No. Z0603

BEAM ANGLE, NARROW:12° (peak) 20° (flat field)

BEAM ANGLE, WIDE: 21° (peak) 31° (flat field)

FIELD ANGLE, NARROW: 30° (peak + flat field)

FIELD ANGLE, WIDE: 45° (peak + flat field)

LAMP BASE TYPE: GY9.5

LAMP MOUNT: Axial

STANDARD LAMP: FMR - 600 W.

OTHER LAMPS: (not specified in mfg. catalog)

INTEGRAL PATTERN SLOT: Yes

WEIGHT: 8 lbs.

SIZE: 14" x 12½" (L x W)

ACCESSORIES AVAILABLE:
Color Frame (5" sq.), Drop-in Iris, Rotatable Pattern Holder

PHOTOMETRICS CHART

(Performance data for this unit are measured at peak focus using FMR - 600W. lamp.)

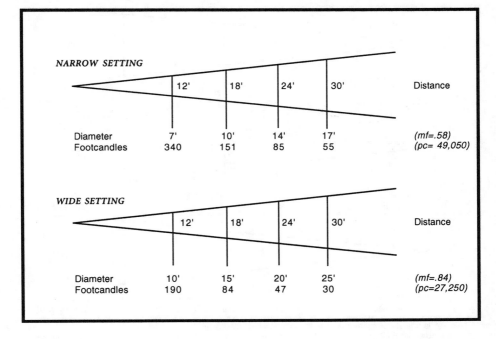

NARROW SETTING

	12'	18'	24'	30'	Distance
Diameter	7'	10'	14'	17'	*(mf=.58)*
Footcandles	340	151	85	55	*(pc= 49,050)*

WIDE SETTING

	12'	18'	24'	30'	Distance
Diameter	10'	15'	20'	25'	*(mf=.84)*
Footcandles	190	84	47	30	*(pc=27,250)*

COLORTRAN
Model No. 213-152**

BEAM ANGLES: 11°, 22°, 22°

FIELD ANGLES: 30°, 40°, 50°

LAMP BASE TYPE: Mini-Can Screw

LAMP MOUNT: Axial

STANDARD LAMP: EVR - 500 W.

OTHER LAMPS: EHT -   250 W. *(cf=.48)*
                    Q400 CL/MC - 400 W. *(cf=.79)*

INTEGRAL PATTERN SLOT: Yes

ACCESSORIES AVAILABLE:
Color Frame, Pattern Holder, Iris Assembly

WEIGHT: 8.75 lbs.

SIZE: 19" x 12⅞" x 20⅛" (L x W x H)

PHOTOMETRICS CHART

(Performance data for this unit are measured at peak focus using EVR - 500 W. lamp.)

30° POSITION

	15'	20'	25'	30'	Distance
Diameter	8'	10.7'	13.4'	16.1'	*(mf=.54)*
Footcandles	258	145	93	64	*(pc=58,000)*

40° POSITION

	10'	15'	20'	25'	Distance
Diameter	7.3'	10.9'	14.6'	18.2'	*(mf=.73)*
Footcandles	316	140	79	51	*(pc=31,600)*

50° POSITION

	7.5'	10'	12.5'	15'	Distance
Diameter	7'	9.3'	11.7'	14'	*(mf=.93)*
Footcandles	346	195	125	86	*(pc=19,476)*

* This unit is adjustable by rearranging the lenses within the barrel.
** This model can be identified with model numbers -152, -155, -156, or -157 depending on the wire leads and connectors supplied.

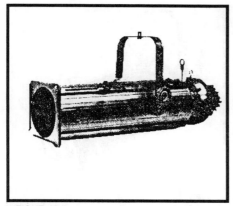

ELECTRO CONTROLS
Model No. 7331A*

BEAM ANGLE, NARROW: 5.4°

BEAM ANGLE, WIDE: 7.5°

FIELD ANGLE, NARROW: 10.9°

FIELD ANGLE, WIDE: 15.4°

LAMP BASE TYPE: Medium Prefocus

LAMP MOUNT: Axial

STANDARD LAMP: EGJ - 1 KW.

OTHER LAMPS: EGE - 500 W. *(cf=.40)*
EGG - 750 W. *(cf=.59)*
EGM - 1 KW. *(cf=.84)*

INTEGRAL PATTERN SLOT: Yes

ACCESSORIES AVAILABLE:
Color Frame (7½" sq.), Pattern Holder, Snoot, Iris

WEIGHT: 33 lbs.

SIZE: 40" x 15½" x 26¼" (L x W x H)

PHOTOMETRICS CHART
(Performance data for this unit are measured using EGJ - 1 KW. lamp.)

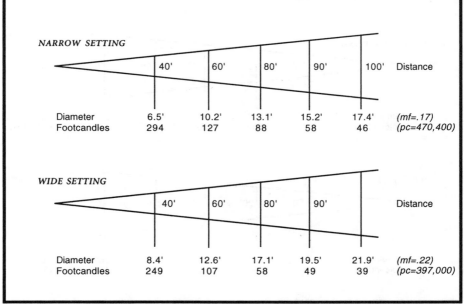

NARROW SETTING

	40'	60'	80'	90'	100'	Distance
Diameter	6.5'	10.2'	13.1'	15.2'	17.4'	*(mf=.17)*
Footcandles	294	127	88	58	46	*(pc=470,400)*

WIDE SETTING

	40'	60'	80'	90'		Distance
Diameter	8.4'	12.6'	17.1'	19.5'	21.9'	*(mf=.22)*
Footcandles	249	107	58	49	39	*(pc=397,000)*

* With optional iris installed, the model number is 7331AI.

ELECTRO CONTROLS
Model No. 7367A*

BEAM ANGLE, NARROW: 12°

BEAM ANGLE, WIDE: 18.6°

FIELD ANGLE, NARROW: 23°

FIELD ANGLE, WIDE: 33.4°

LAMP BASE TYPE: Medium Prefocus

LAMP MOUNT: Axial

STANDARD LAMP: EGJ - 1 KW

OTHER LAMPS: EGE - 500 W. *(cf=.40)*
EGG - 750 W. *(cf=.59)*
EGM -1 KW. *(cf=.84)*

INTEGRAL PATTERN SLOT: Yes

ACCESSORIES AVAILABLE:
Color Frame (7½" sq.), Pattern Holder, Snoot, Iris

WEIGHT: 22.5 lbs.

SIZE: 25¹¹⁄₁₆" x 15½" x 26⁵⁄₁₆" (L x W x H)

PHOTOMETRICS CHART
(Performance data for this unit are measured using EGJ - 1 KW. lamp.)

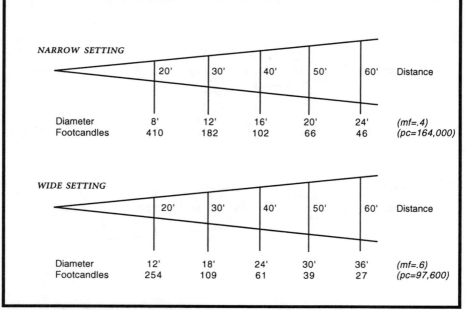

NARROW SETTING

	20'	30'	40'	50'	60'	Distance
Diameter	8'	12'	16'	20'	24'	*(mf=.4)*
Footcandles	410	182	102	66	46	*(pc=164,000)*

WIDE SETTING

	20'	30'	40'	50'	60'	Distance
Diameter	12'	18'	24'	30'	36'	*(mf=.6)*
Footcandles	254	109	61	39	27	*(pc=97,600)*

* With optional iris installed, instead of shutters, the model number is 7367AI.

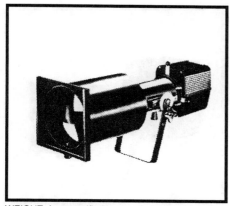

KLIEGL/CCT
Model No. 1010, "Aimslite"

BEAM ANGLE, NARROW: 5° (peak), 9° (flat field)

BEAM ANGLE, WIDE: 8° (peak), 20° (flat field)

FIELD ANGLE, NARROW: 9° * (peak & flat field)

FIELD ANGLE, WIDE: 14° * (peak), 22° flat field)

LAMP BASE TYPE: Medium 2-Pin

LAMP MOUNT: Axial

STANDARD LAMP: FEL - 1 KW.

OTHER LAMPS: EHD - 500 W. *(cf=.39)*
EHG - 750 W. *(cf=.56)*

INTEGRAL PATTERN SLOT: Yes

ACCESSORIES AVAILABLE:
Color Frame, Pattern Holder, Remote Motorized
4-color Changer

WEIGHT: (not specified in manufacturer's catalog)

SIZE: (not specified in manufacturer's catalog)

PHOTOMETRICS CHART
(Performance data for this unit are measured at peak focus using FEL - 1 KW. lamp.)

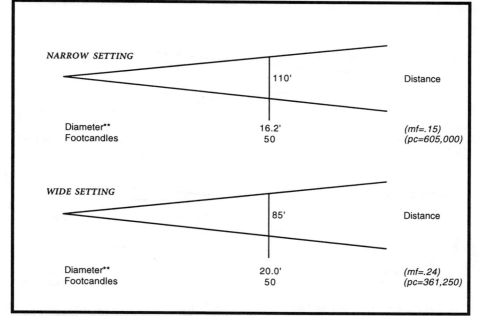

NARROW SETTING

110' — Distance

| Diameter** | 16.2' | *(mf=.15)* |
| Footcandles | 50 | *(pc=605,000)* |

WIDE SETTING

85' — Distance

| Diameter** | 20.0' | *(mf=.24)* |
| Footcandles | 50 | *(pc=361,250)* |

* Manufacturer describes this angle as "cut-off angle" which may be larger than the actual field angle.
** Manufacturer describes this diameter as "field diameter."

KLIEGL/CCT
Model No. 1015, "Aimslite"

BEAM ANGLE, NARROW: 7° (peak), 8° (flat field)

BEAM ANGLE, WIDE: 10° (peak), 16° (flat field)

FIELD ANGLE, NARROW: 11° *(peak), 12° (flat field)

FIELD ANGLE, WIDE: 16° *(peak), 24° (flat field)

LAMP BASE TYPE: Medium 2-Pin

LAMP MOUNT: Axial

STANDARD LAMP: FEL - 1 KW.

OTHER LAMPS: EHD - 500 W. *(cf=.39)*
　　　　　　　　 EHG - 750 W. *(cf=.56)*

INTEGRAL PATTERN SLOT: Yes

ACCESSORIES AVAILABLE:
Color Frame, Pattern Holder, Remote Motorized
4-color Changer

WEIGHT: (not specified in manufacturer's catalog)

SIZE: (not specified in manufacturer's catalog)

PHOTOMETRICS CHART
(Performance data for this unit are measured at peak focus using FEL - 1 KW. lamp.)

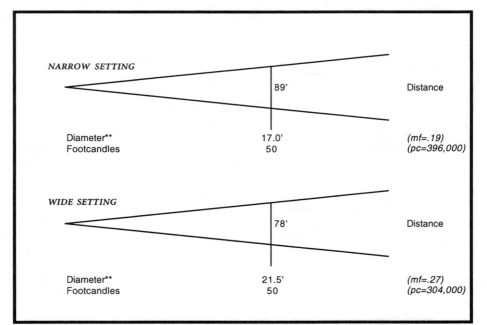

NARROW SETTING

89'　　　　　　　Distance

| Diameter** | 17.0' | *(mf=.19)* |
| Footcandles | 50 | *(pc=396,000)* |

WIDE SETTING

78'　　　　　　　Distance

| Diameter** | 21.5' | *(mf=.27)* |
| Footcandles | 50 | *(pc=304,000)* |

* Manufacturer describes this angle as "cut-off angle" which may be larger than the actual field angle.

** Manufacturer describes this diameter as "field diameter."

KLIEGL/CCT
Model No. 1030, "Aimslite"

BEAM ANGLE, NARROW: 12° (peak), 18° (flat field)

BEAM ANGLE, WIDE: 15° (peak), 40° (flat field)

FIELD ANGLE, NARROW: 21° * (peak & flat field)

FIELD ANGLE, WIDE: 40° * (peak & flat field)

LAMP BASE TYPE: Medium 2-Pin

LAMP MOUNT: Axial

STANDARD LAMP: FEL - 1 KW.

OTHER LAMPS: EHD - 500 W. *(cf=.39)*
EHG - 750 W. *(cf=.56)*

INTEGRAL PATTERN SLOT: Yes

ACCESSORIES AVAILABLE:
Color Frame, Pattern Holder, Remote Motorized
4-color Changer

WEIGHT: (not specified in manufacturer's catalog)

SIZE: (not specified in manufacturer's catalog)

PHOTOMETRICS CHART
(Performance data for this unit are measured at peak focus using FEL - 1 KW. lamp.)

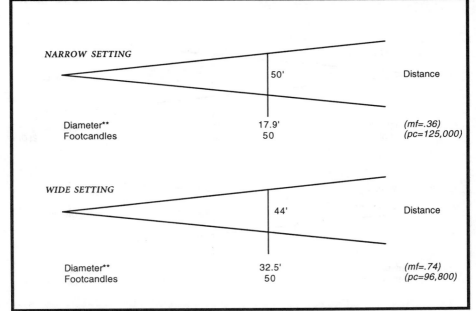

NARROW SETTING

	50'	Distance
Diameter**	17.9'	*(mf=.36)*
Footcandles	50	*(pc=125,000)*

WIDE SETTING

	44'	Distance
Diameter**	32.5'	*(mf=.74)*
Footcandles	50	*(pc=96,800)*

* Manufacturer describes this angle as "cut-off angle" which may be larger than the actual field angle.

** Manufacturer describes this diameter as "field diameter."

KLIEGL/CCT
Model No. 1040, "Aimslite"

BEAM ANGLE, NARROW: 15° (peak), 16° (flat field)

BEAM ANGLE, WIDE: 16° (peak), 20° (flat field)

FIELD ANGLE, NARROW: 27° * (peak & flat field)

FIELD ANGLE, WIDE: 44° * (peak & flat field)

LAMP BASE TYPE: Medium 2-Pin

LAMP MOUNT: Axial

STANDARD LAMP: FEL - 1 KW.

OTHER LAMPS: EHD - 500 W. *(cf=.39)*
EHG - 750 W. *(cf=.56)*

INTEGRAL PATTERN SLOT: Yes

ACCESSORIES AVAILABLE:
Color Frame, Pattern Holder, Remote Motorized
4-color Changer

WEIGHT: (not specified in manufacturer's catalog)

SIZE: (not specified in manufacturer's catalog)

PHOTOMETRICS CHART
(Performance data for this unit are measured at peak focus using FEL - 1 KW. lamp.)

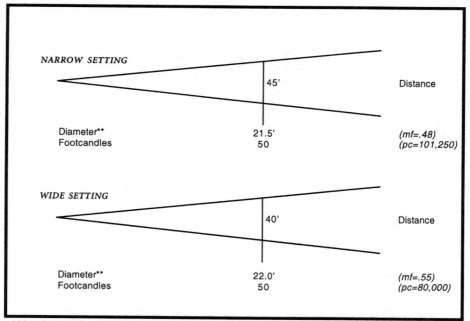

NARROW SETTING

| | 45' | Distance |

| Diameter** | 21.5' | *(mf=.48)* |
| Footcandles | 50 | *(pc=101,250)* |

WIDE SETTING

| | 40' | Distance |

| Diameter** | 22.0' | *(mf=.55)* |
| Footcandles | 50 | *(pc=80,000)* |

* Manufacturer describes this angle as "cut-off angle" which may be larger than the actual field angle.
** Manufacturer describes this diameter as "field diameter."

WEIGHT: 36.7 lbs.

SIZE: 35⅝" x 12⅝" x 17" (L x W x H)

KLIEGL/CCT
Model No. 1990, "Silhouette"*

BEAM ANGLE, NARROW: 6°

BEAM ANGLE, WIDE: 9°

FIELD ANGLE, NARROW: 10°

FIELD ANGLE, WIDE: 22°

LAMP BASE TYPE: Medium 2-Pin

LAMP MOUNT: Axial

STANDARD LAMP: FEL - 1 KW.

OTHER LAMPS: EHD - 500 W. *(cf=.39)*
EHG - 750 W. *(cf=.56)*

INTEGRAL PATTERN SLOT: Yes

ACCESSORIES AVAILABLE:
Color Frame (12½" sq.), Pattern Holder, Remote 4-Color Changer, 2 KW Base-Down Lamp House, Drop-in or Bolt-in Iris

PHOTOMETRICS CHART

(Performance data for this unit are measured at peak focus using FEL - 1 KW. lamp.)

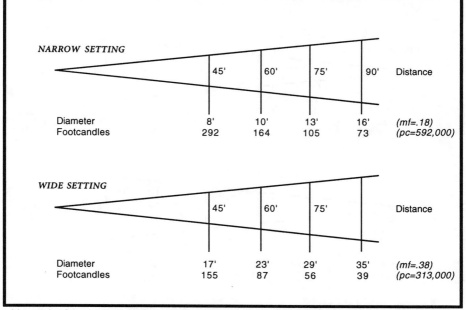

NARROW SETTING					
	45'	60'	75'	90'	Distance
Diameter	8'	10'	13'	16'	*(mf=.18)*
Footcandles	292	164	105	73	*(pc=592,000)*

WIDE SETTING					
	45'	60'	75'		Distance
Diameter	17'	23'	29'	35'	*(mf=.38)*
Footcandles	155	87	56	39	*(pc=313,000)*

* Lens tubes/barrel assemblies in the 1990 series are interchangeable. See photometrics for other lens configurations in the series on the following pages.

KLIEGL/CCT
Model No. 1991, "Silhouette"*

BEAM ANGLE, NARROW: 7°

BEAM ANGLE, WIDE: 11°

FIELD ANGLE, NARROW: 12°

FIELD ANGLE, WIDE: 26°

LAMP BASE TYPE: Medium 2-Pin

LAMP MOUNT: Axial

STANDARD LAMP: FEL - 1 KW.

OTHER LAMPS: EHD - 500 W. *(cf=.39)*
EHG - 750 W. *(cf=.56)*

INTEGRAL PATTERN SLOT: Yes

ACCESSORIES AVAILABLE:
Color Frame (10" sq.), Pattern Holder, Remote 4-Color
Changer, 2 KW Base-Down Lamp House, Drop-in or
Bolt-in Iris

WEIGHT: 27.9 lbs.

SIZE: 29⅝" x 10¼" x 17" (L x W x H)

PHOTOMETRICS CHART

(Performance data for this unit are measured at peak focus using FEL - 1 KW. lamp.)

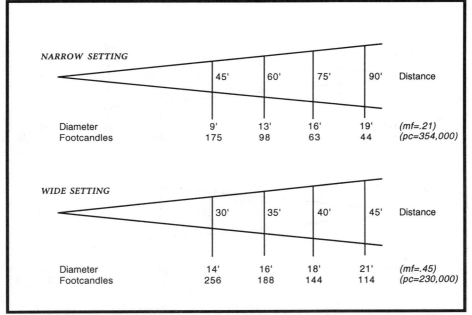

NARROW SETTING

	45'	60'	75'	90'	Distance
Diameter	9'	13'	16'	19'	*(mf=.21)*
Footcandles	175	98	63	44	*(pc=354,000)*

WIDE SETTING

	30'	35'	40'	45'	Distance
Diameter	14'	16'	18'	21'	*(mf=.45)*
Footcandles	256	188	144	114	*(pc=230,000)*

* Lens tubes/barrel assemblies in the 1990 series are interchangeable. See photometrics for other lens
configurations in the series on previous and following pages.

WEIGHT: 27.9 lbs.

SIZE: 27⅜" x 10¼" x 17" (L x W x H)

KLIEGL/CCT
Model No. 1992, "Silhouette"*

BEAM ANGLE, NARROW: 10°

BEAM ANGLE, WIDE: 12°

FIELD ANGLE, NARROW: 20°

FIELD ANGLE, WIDE: 31°

LAMP BASE TYPE: Medium 2-Pin

LAMP MOUNT: Axial

STANDARD LAMP: FEL - 1 KW.

OTHER LAMPS: EHD - 500 W. *(cf=.39)*
EHG - 750 W. *(cf=.56)*

INTEGRAL PATTERN SLOT: Yes

ACCESSORIES AVAILABLE:
Color Frame (10" sq.), Pattern Holder, Remote 4-Color Changer, 2 KW Base-Down Lamp House, Drop-in or Bolt-in Iris

PHOTOMETRICS CHART

(Performance data for this unit are measured at peak focus using FEL - 1 KW. lamp.)

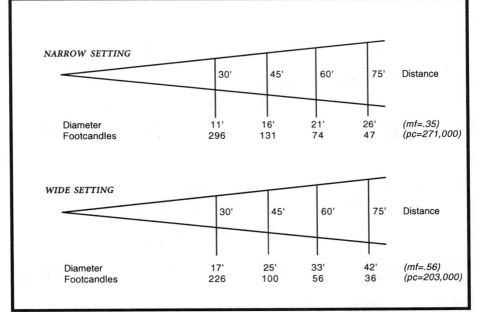

NARROW SETTING

	30'	45'	60'	75'	Distance
Diameter	11'	16'	21'	26'	*(mf=.35)*
Footcandles	296	131	74	47	*(pc=271,000)*

WIDE SETTING

	30'	45'	60'	75'	Distance
Diameter	17'	25'	33'	42'	*(mf=.56)*
Footcandles	226	100	56	36	*(pc=203,000)*

* Lens tubes/barrel assemblies in the 1990 series are interchangeable. See photometrics for other lens configurations in the series on previous and following pages.

KLIEGL/CCT
Model No. 1993, "Silhouette"*

BEAM ANGLE, NARROW: 12°

BEAM ANGLE, WIDE: 15°

FIELD ANGLE, NARROW: 22°

FIELD ANGLE, WIDE: 38°

LAMP BASE TYPE: Medium 2-Pin

LAMP MOUNT: Axial

STANDARD LAMP: FEL - 1 KW.

OTHER LAMPS: EHD - 500 W. *(cf=.39)*
EHG - 750 W. *(cf=.56)*

INTEGRAL PATTERN SLOT: Yes

WEIGHT: 20 lbs.

SIZE: 21¾" x 8⅜" x 17" (L x W x H)

ACCESSORIES AVAILABLE:
Color Frame (7½" sq.), Pattern Holder, Remote 4-Color Changer, 2 KW Base-Down Lamp House, Drop-in or Bolt-in Iris

PHOTOMETRICS CHART

(Performance data for this unit are measured at peak focus using FEL - 1 KW. lamp.)

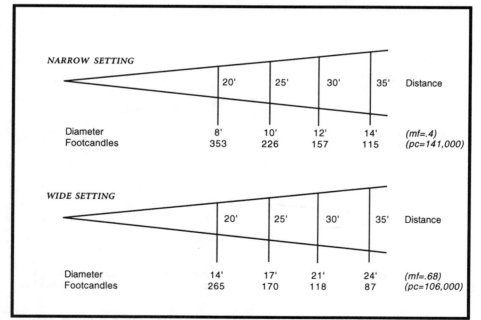

NARROW SETTING

	20'	25'	30'	35'	Distance
Diameter	8'	10'	12'	14'	*(mf=.4)*
Footcandles	353	226	157	115	*(pc=141,000)*

WIDE SETTING

	20'	25'	30'	35'	Distance
Diameter	14'	17'	21'	24'	*(mf=.68)*
Footcandles	265	170	118	87	*(pc=106,000)*

* Lens tube/barrel assemblies in the 1990 series are interchangeable. See photometrics for other lens configurations in the series on previous and following pages.

WEIGHT: 20.5 lbs.

SIZE: 21¾" x 8⅝" x 17" (L x W x H)

KLIEGL/CCT
Model No. 1994, "Silhouette"*

BEAM ANGLE, NARROW: 15°

BEAM ANGLE, WIDE: 20°

FIELD ANGLE, NARROW: 31°

FIELD ANGLE, WIDE: 47°

LAMP BASE TYPE: Medium 2-Pin

LAMP MOUNT: Axial

STANDARD LAMP: FEL - 1 KW.

OTHER LAMPS: EHD - 500 W. *(cf=.39)*
EHG - 750 W. *(cf=.56)*

INTEGRAL PATTERN SLOT: Yes

ACCESSORIES AVAILABLE:
Color Frame (7½" sq.), Pattern Holder, Remote 4-Color Changer, 2 KW Base-Down Lamp House, Drop-in or Bolt-in Iris

PHOTOMETRICS CHART
(Performance data for this unit are measured at peak focus using FEL - 1 KW. lamp.)

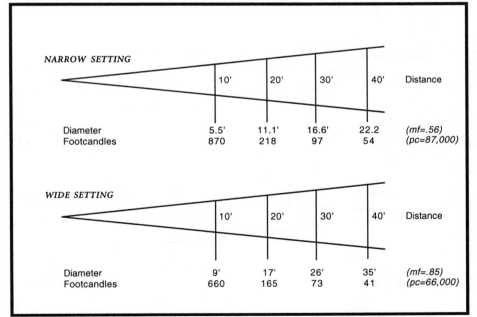

NARROW SETTING

	10'	20'	30'	40'	Distance
Diameter	5.5'	11.1'	16.6'	22.2	*(mf=.56)*
Footcandles	870	218	97	54	*(pc=87,000)*

WIDE SETTING

	10'	20'	30'	40'	Distance
Diameter	9'	17'	26'	35'	*(mf=.85)*
Footcandles	660	165	73	41	*(pc=66,000)*

* Lens tube/barrel assemblies in the 1990 series are interchangeable. See photometrics for other lens configurations in the series on previous pages.

KLIEGL/CCT
Model No. 2010, "Aimslite"

BEAM ANGLE, NARROW: 8° (flat field)

BEAM ANGLE, WIDE: 14° (flat field)

FIELD ANGLE, NARROW: 10° * (flat field)

FIELD ANGLE, WIDE: 20° * (flat field)

LAMP BASE TYPE: Mogul Bipost

LAMP MOUNT: Axial

STANDARD LAMP: BWA - 2 KW.

OTHER LAMPS: (none specified in mfg. catalog)

INTEGRAL PATTERN SLOT: Yes

ACCESSORIES AVAILABLE:
Color Frame, Pattern Holder, Remote Motorized
4-color Changer

WEIGHT: (not specified in manufacturer's catalog)

SIZE: (not specified in manufacturer's catalog)

PHOTOMETRICS CHART
(Performance data for this unit are measured at flat field focus using BWA - 2 KW. lamp.)

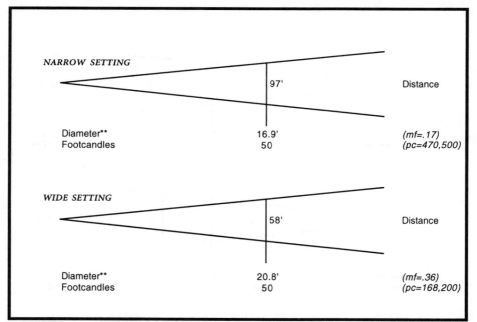

NARROW SETTING

	97'	Distance
Diameter**	16.9'	(mf=.17)
Footcandles	50	(pc=470,500)

WIDE SETTING

	58'	Distance
Diameter**	20.8'	(mf=.36)
Footcandles	50	(pc=168,200)

* Manufacturer describes this angle as "cut-off angle" which may be larger than the actual field angle.
** Manufacturer describes this diameter as "field diameter."

KLIEGL/CCT
Model No. 2015, "Aimslite"

BEAM ANGLE, NARROW: 10° (flat field)

BEAM ANGLE, WIDE: 15° (flat field)

FIELD ANGLE, NARROW: 13° * (flat field)

FIELD ANGLE, WIDE: 27° * (flat field)

LAMP BASE TYPE: Mogul Bipost

LAMP MOUNT: Axial

STANDARD LAMP: BWA - 2 KW.

OTHER LAMPS: (none specified in mfg. catalog)

INTEGRAL PATTERN SLOT: Yes

WEIGHT: (not specified in manufacturer's catalog)

SIZE: (not specified in manufacturer's catalog)

ACCESSORIES AVAILABLE:
Color Frame, Pattern Holder, Remote Motorized
4-color Changer

PHOTOMETRICS CHART

(Performance data for this unit are measured at flat field focus using BWA - 2 KW. lamp.)

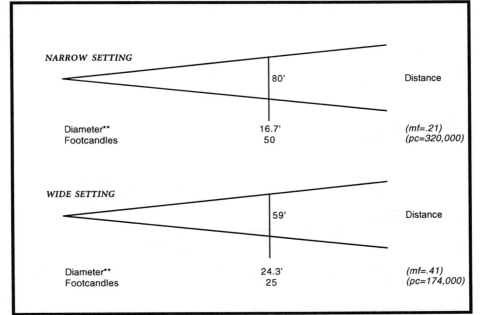

NARROW SETTING		
	80'	Distance
Diameter**	16.7'	(mf=.21)
Footcandles	50	(pc=320,000)

WIDE SETTING		
	59'	Distance
Diameter**	24.3'	(mf=.41)
Footcandles	25	(pc=174,000)

* Manufacturer describes this angle as "cut-off angle" which may be larger than the actual field angle.
** Manufacturer describes this diameter as "field diameter."

KLIEGL/CCT
Model No. 2030, "Aimslite"

BEAM ANGLE, NARROW: 14° (flat field)

BEAM ANGLE, WIDE: 31° (flat field)

FIELD ANGLE, NARROW: 21° * (flat field)

FIELD ANGLE, WIDE: 44° * (flat field)

LAMP BASE TYPE: Mogul Bipost

LAMP MOUNT: Axial

STANDARD LAMP: BWA - 2 KW.

OTHER LAMPS: (none specified in mfg. catalog)

INTEGRAL PATTERN SLOT: Yes

ACCESSORIES AVAILABLE:
Color Frame, Pattern Holder, Remote Motorized
4-color Changer

WEIGHT: (not specified in manufacturer's catalog)

SIZE: (not specified in manufacturer's catalog)

PHOTOMETRICS CHART
(Performance data for this unit are measured at flat field focus using BWA - 2 KW. lamp.)

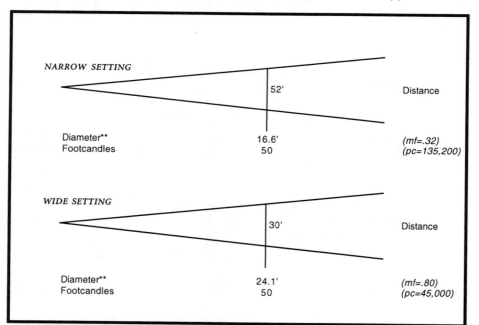

NARROW SETTING

	52'	Distance
Diameter**	16.6'	(mf=.32)
Footcandles	50	(pc=135,200)

WIDE SETTING

	30'	Distance
Diameter**	24.1'	(mf=.80)
Footcandles	50	(pc=45,000)

* Manufacturer describes this angle as "cut-off angle" which may be larger than the actual field angle.
** Manufacturer describes this diameter as "field diameter."

WEIGHT: 14.9 lbs.

SIZE: 15½" x 14" x 22" (L x W x H)

LEE COLORTRAN
Model No. 213-302*

BEAM ANGLE, NARROW: 17° (peak), 26° (cosine)

BEAM ANGLE, WIDE: 27° (peak), 43° (cosine)

FIELD ANGLE, NARROW: 40° (peak & cosine)

FIELD ANGLE, WIDE: 65° (peak & cosine)

LAMP BASE TYPE: Medium 2-Pin

LAMP MOUNT: Axial

STANDARD LAMP: FMR - 600 W.

OTHER LAMPS: FNA - 300 W. *(cf=.46)*

INTEGRAL PATTERN SLOT: Yes

ACCESSORIES AVAILABLE:
Color Frame, Pattern Holder, Iris Kit

PHOTOMETRICS CHART
(Performance data for this unit are measured at cosine focus using FMR - 600 W. lamp.)

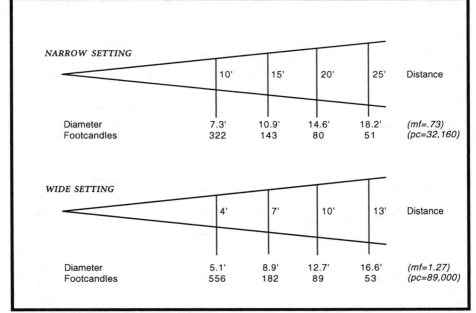

	10'	15'	20'	25'	Distance
NARROW SETTING					
Diameter	7.3'	10.9'	14.6'	18.2'	*(mf=.73)*
Footcandles	322	143	80	51	*(pc=32,160)*

	4'	7'	10'	13'	Distance
WIDE SETTING					
Diameter	5.1'	8.9'	12.7'	16.6'	*(mf=1.27)*
Footcandles	556	182	89	53	*(pc=89,000)*

* This model can be identified with model numbers -302, -305, -306, or -307 depending on the wire leads and connectors supplied.

LEE COLORTRAN
Model No. 213-312*

BEAM ANGLE, NARROW: 9.5° (peak), 16.7° (cosine)
BEAM ANGLE, WIDE: 18° (peak), 33.3° (cosine)
FIELD ANGLE, NARROW: 25° (peak & cosine)
FIELD ANGLE, WIDE: 50° (peak & cosine)
LAMP BASE TYPE: Medium 2-Pin

LAMP MOUNT: Axial

STANDARD LAMP: FMR - 600 W.

OTHER LAMPS: FNA - 300 W. *(cf=.46)*

INTEGRAL PATTERN SLOT: Yes

ACCESSORIES AVAILABLE:
Color Frame, Pattern Holder, Iris Kit

WEIGHT: 15.5 lbs.
SIZE: 18" x 14" x 22" (L x W x H)

PHOTOMETRICS CHART
(Performance data for this unit are measured at cosine focus using FMR - 600 W. lamp.)

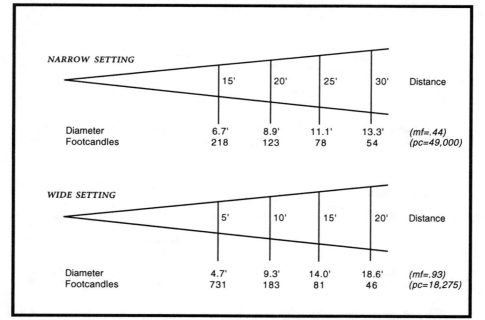

NARROW SETTING

	15'	20'	25'	30'	Distance
Diameter	6.7'	8.9'	11.1'	13.3'	*(mf=.44)*
Footcandles	218	123	78	54	*(pc=49,000)*

WIDE SETTING

	5'	10'	15'	20'	Distance
Diameter	4.7'	9.3'	14.0'	18.6'	*(mf=.93)*
Footcandles	731	183	81	46	*(pc=18,275)*

* This model can be identified with model numbers -312, -315, -316, or -317 depending on the wire leads
and connectors supplied.

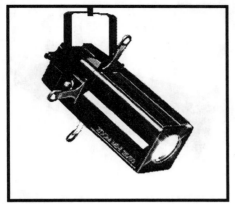

LEE COLORTRAN
Model No. 213-322*

BEAM ANGLE, NARROW: 9° (peak), 10° (cosine)
BEAM ANGLE, WIDE: 11° (peak), 20° (cosine)
FIELD ANGLE, NARROW: 15° (peak & cosine)
FIELD ANGLE, WIDE: 30° (peak & cosine)
LAMP BASE TYPE: Medium 2-Pin
LAMP MOUNT: Axial
STANDARD LAMP: FMR - 600 W.
OTHER LAMPS: FNA - 300 W. *(cf=.46)*
INTEGRAL PATTERN SLOT: Yes

WEIGHT: 16 lbs.
SIZE: 21" x 14" x 22" (L x W x H)

ACCESSORIES AVAILABLE:
Color Frame, Pattern Holder, Iris Kit

PHOTOMETRICS CHART

(Performance data for this unit are measured at cosine focus using FMR - 600 W. lamp.)

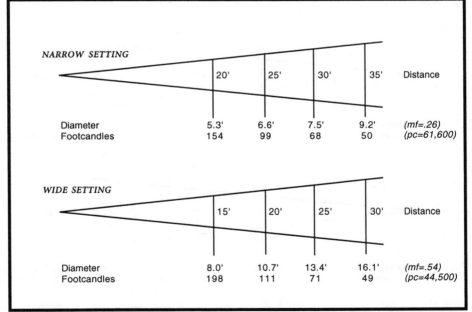

NARROW SETTING

	20'	25'	30'	35'	Distance
Diameter	5.3'	6.6'	7.5'	9.2'	*(mf=.26)*
Footcandles	154	99	68	50	*(pc=61,600)*

WIDE SETTING

	15'	20'	25'	30'	Distance
Diameter	8.0'	10.7'	13.4'	16.1'	*(mf=.54)*
Footcandles	198	111	71	49	*(pc=44,500)*

* This model can be identified with model numbers -322, -325, -326, or -327 depending on the wire leads and connectors supplied.

LIGHTING & ELECTRONICS
Model No. AQ61-ZM

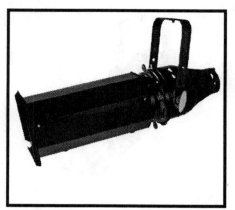

BEAM ANGLE, NARROW: 9° (peak), 13° (cosine)
BEAM ANGLE, WIDE: 18° (peak), 26° (cosine)
FIELD ANGLE, NARROW: 20° (peak & cosine)
FIELD ANGLE, WIDE: 40° (peak & cosine)
LAMP BASE TYPE: Medium 2-Pin
LAMP MOUNT: Axial
STANDARD LAMP: EHG - 750 W.
OTHER LAMPS: EHD - 500 W. *(cf=.68*)*
　　　　　　　 FLK - 575 W. *(cf=1.93*)*
INTEGRAL PATTERN SLOT: Yes
ACCESSORIES AVAILABLE:
Color Frame, Pattern Holder, Iris Kit (factory installed)

WEIGHT: 19 lbs.
SIZE: 25⅝" x 7⅝" x 7⅝" (L x W x H)

PHOTOMETRICS CHART
(Performance data for this unit are measured at cosine focus using EHG - 750 W. lamp.)

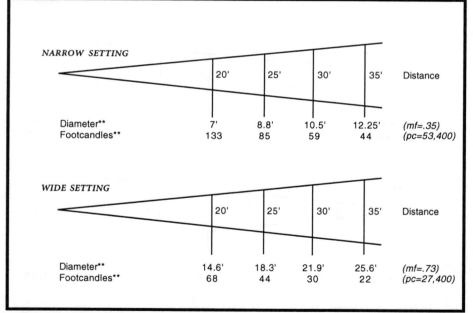

NARROW SETTING

	20'	25'	30'	35'	Distance
Diameter**	7'	8.8'	10.5'	12.25'	(mf=.35)
Footcandles**	133	85	59	44	(pc=53,400)

WIDE SETTING

	20'	25'	30'	35'	Distance
Diameter**	14.6'	18.3'	21.9'	25.6'	(mf=.73)
Footcandles**	68	44	30	22	(pc=27,400)

* This lamp multiplying factor (mf) is based on measured output for this particular instrument, not the initial lumens of the lamp itself.
** The diameter and footcandle information in this chart was calculated using multiplying factor (mf) and peak candela (pc) data given in catalog. See Introduction for more information about calculating photometric data.

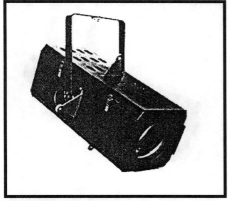

LITTLE STAGE LIGHTING
Model No. E509

BEAM ANGLE, NARROW: 12.5°

BEAM ANGLE, WIDE: 15.1°

FIELD ANGLE, NARROW: 21.5°

FIELD ANGLE, WIDE: 36°

LAMP BASE TYPE: Medium 2-Pin

LAMP MOUNT: Axial

STANDARD LAMP: FEL - 1 KW.

OTHER LAMPS: (none specified in mfg. catalog)

INTEGRAL PATTERN SLOT: Yes

ACCESSORIES AVAILABLE:
Color Frame (7½" sq.), Iris

WEIGHT: (not specified in manufacturer's catalog)

SIZE: 21" x 14" x 20½" (L x W x H)

PHOTOMETRICS CHART
(Performance data for this unit are measured using FEL - 1 KW. lamp.)

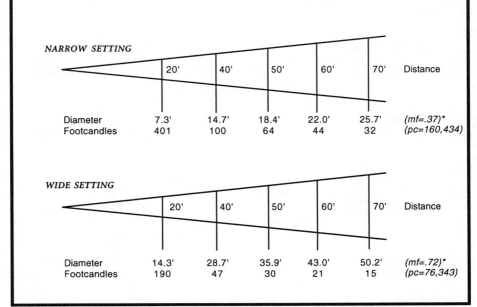

NARROW SETTING

	20'	40'	50'	60'	70'	Distance
Diameter	7.3'	14.7'	18.4'	22.0'	25.7'	(mf=.37)*
Footcandles	401	100	64	44	32	(pc=160,434)

WIDE SETTING

	20'	40'	50'	60'	70'	Distance
Diameter	14.3'	28.7'	35.9'	43.0'	50.2'	(mf=.72)*
Footcandles	190	47	30	21	15	(pc=76,343)

* Manufacturer's catalog gives multiplying factor *(mf)* data for "cut-off beam spread": narrow=.388 and wide=.788.

STRAND CENTURY
Model No. 2205

BEAM ANGLE, NARROW: 16° (cosine)

BEAM ANGLE, WIDE: 20° (cosine)

FIELD ANGLE, NARROW: 25° (cosine)

FIELD ANGLE, WIDE: 50° (cosine)

LAMP BASE TYPE: Mini-Can Screw

LAMP MOUNT: Axial

STANDARD LAMP: EVR - 500 W.

OTHER LAMPS: Q150CL/MC - 150 W. *(cf=.27)*
EHT - 250 W. *(cf=.48)*
Q400CL/MC - 400 W. *(cf=.79*

INTEGRAL PATTERN SLOT: Yes

ACCESSORIES AVAILABLE:
Color Frame, Pattern Holder, Snoot, Iris Kit

WEIGHT: 15 lbs.

SIZE: 20½" x 8¾" x 14¾" (L x W x H)

PHOTOMETRICS CHART

(Performance data for this unit are measured at cosine focus using EVR - 500 W. lamp.)

NARROW SETTING

	20'	25'	30'	35'	Distance
Diameter	8.8'	11.0'	13.2'	15.4'	*(mf=.44)*
Footcandles	65	42	29	21	*(pc=26,000)*

WIDE SETTING

	7.5'	10'	12.5'	15'	Distance
Diameter	7.0'	9.3'	11.6'	14.0'	*(mf=.93)*
Footcandles	250	140	90	62	*(pc=14,000)*

STRAND LIGHTING
Model No. 12011 (OPTIQUE 8/17)

BEAM ANGLE, NARROW: 6.6°

BEAM ANGLE, WIDE: 17°

FIELD ANGLE, NARROW: 6.9°

FIELD ANGLE, WIDE: 17°

LAMP BASE TYPE: GX9.5

LAMP MOUNT: Off Axis

STANDARD LAMP: T29-1200 W. *

OTHER LAMPS: (not specified in mfg. catalog)

INTEGRAL PATTERN SLOT: Yes

WEIGHT: N/A

SIZE: 25½" x 11¾" x 16⅛" (L x W x H)

ACCESSORIES AVAILABLE:
Color Frame (7⁵⁄₁₆" sq.), Iris, Color Scroller, Pattern Holder

PHOTOMETRICS CHART
(Performance data for this unit are measured at peak focus using T-29 - 1200W. lamp.)

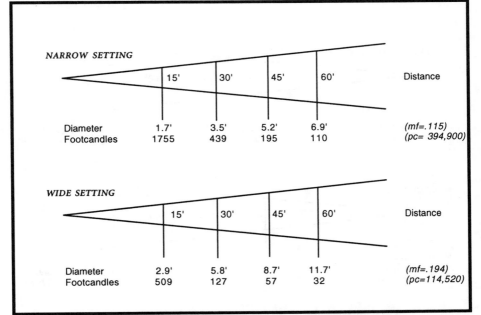

NARROW SETTING

	15'	30'	45'	60'	Distance
Diameter	1.7'	3.5'	5.2'	6.9'	*(mf=.115)*
Footcandles	1755	439	195	110	*(pc= 394,900)*

WIDE SETTING

	15'	30'	45'	60'	Distance
Diameter	2.9'	5.8'	8.7'	11.7'	*(mf=.194)*
Footcandles	509	127	57	32	*(pc=114,520)*

* "T-29" is an L. I. F. (Lighting Industries Federation, a UK trade association) code for the lamp.
Further information was unavailable—the lamp may be purchased from Strand Lighting.

STRAND LIGHTING
Model No. 12021(OPTIQUE 15/42)

BEAM ANGLE, NARROW: 10.7°

BEAM ANGLE, WIDE: 22.3°

FIELD ANGLE, NARROW: 17.7°

FIELD ANGLE, WIDE: 41.2°

LAMP BASE TYPE: GX9.5

LAMP MOUNT: Off Axis

STANDARD LAMP: T29-1200 W. *

OTHER LAMPS: (not specified in mfg. catalog)

INTEGRAL PATTERN SLOT: Yes

WEIGHT: N/A

SIZE: 25½" x 11¾" x 16⅛" (L x W x H)

ACCESSORIES AVAILABLE:
Color Frame (7⁵⁄₁₆" sq.), Iris, Color Scroller, Pattern Holder

PHOTOMETRICS CHART
(Performance data for this unit are measured at peak focus using T-29 - 1200W. lamp.)

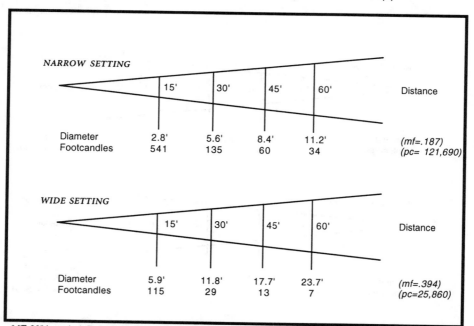

NARROW SETTING

	15'	30'	45'	60'	Distance
Diameter	2.8'	5.6'	8.4'	11.2'	*(mf=.187)*
Footcandles	541	135	60	34	*(pc= 121,690)*

WIDE SETTING

	15'	30'	45'	60'	Distance
Diameter	5.9'	11.8'	17.7'	23.7'	*(mf=.394)*
Footcandles	115	29	13	7	*(pc=25,860)*

* "T-29" is an L. I. F. (Lighting Industries Federation, a UK trade association) code for the lamp.
Further information was unavailable—the lamp may be purchased from Strand Lighting.

WEIGHT: 23.7 lbs.

SIZE: 27" x 12¼" x 21¾" (L x W x H)

STRAND LIGHTING
Model No. 2206

BEAM ANGLE, NARROW: 10° (cosine)

BEAM ANGLE, WIDE: 24° (cosine)

FIELD ANGLE, NARROW: 15° * (cosine)

FIELD ANGLE, WIDE: 40° * (cosine)

LAMP BASE TYPE: Medium 2-Pin

LAMP MOUNT: Axial

STANDARD LAMP: FEL - 1 KW.

OTHER LAMPS: EHD - 500 W. (cf=.36)
                 EHG - 750 W. (cf=.54)

INTEGRAL PATTERN SLOT: Yes

ACCESSORIES AVAILABLE:
Color Frame, Pattern Holder, Snoot, Iris Kit

PHOTOMETRICS CHART

(Performance data for this unit are measured at cosine focus using FEL - 1 KW. lamp.)

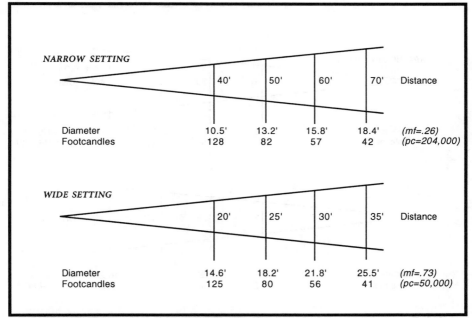

NARROW SETTING

	40'	50'	60'	70'	Distance
Diameter	10.5'	13.2'	15.8'	18.4'	(mf=.26)
Footcandles	128	82	57	42	(pc=204,000)

WIDE SETTING

	20'	25'	30'	35'	Distance
Diameter	14.6'	18.2'	21.8'	25.5'	(mf=.73)
Footcandles	125	80	56	41	(pc=50,000)

* Manafacturer's catalog describes this angle as "cut-off angle."

ERS — Zoom

STRAND LIGHTING
Model No. 2271, "Cantata"*

BEAM ANGLE, NARROW: 10.9°

BEAM ANGLE, WIDE: 13.5°

FIELD ANGLE, NARROW: 11° **

FIELD ANGLE, WIDE: 26° **

LAMP BASE TYPE: Medium 2-Pin

LAMP MOUNT: Axial

STANDARD LAMP: FEL - 1 KW.

OTHER LAMPS: EHD - 500 W. *(cf=.39)*
EHG - 750 W. *(cf=.54)*

INTEGRAL PATTERN SLOT: Yes

ACCESSORIES AVAILABLE:
Color Frame, Pattern Holder, Iris

WEIGHT: 28.2 lbs.

SIZE: 36⅛" x 11¾" x 16½" (L x W x H)

PHOTOMETRICS CHART
(Performance data for this unit are measured using FEL - 1 KW. lamp.)

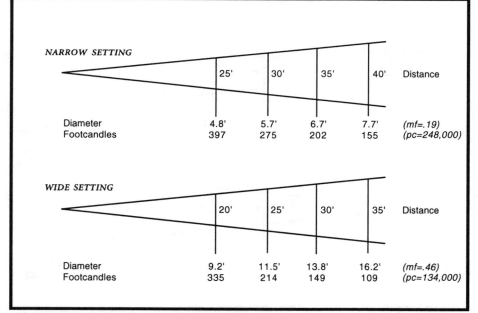

NARROW SETTING

	25'	30'	35'	40'	Distance
Diameter	4.8'	5.7'	6.7'	7.7'	*(mf=.19)*
Footcandles	397	275	202	155	*(pc=248,000)*

WIDE SETTING

	20'	25'	30'	35'	Distance
Diameter	9.2'	11.5'	13.8'	16.2'	*(mf=.46)*
Footcandles	335	214	149	109	*(pc=134,000)*

* Lens tubes/barrel assemblies in the Cantata series are interchangeable. See photometrics for other lens configurations in the series on the following pages.
** Manufacturer's catalog describes this angle as "cut-off angle."

STRAND LIGHTING
Model No. 2272, "Cantata"*

BEAM ANGLE, NARROW: 12.5°

BEAM ANGLE, WIDE: 17.1°

FIELD ANGLE, NARROW: 18° **

FIELD ANGLE, WIDE: 32° **

LAMP BASE TYPE: Medium 2-Pin

LAMP MOUNT: Axial

STANDARD LAMP: FEL - 1 KW.

OTHER LAMPS: EHD - 500 W. *(cf=.39)*
EHG - 750 W. *(cf=.54)*

INTEGRAL PATTERN SLOT: Yes

ACCESSORIES AVAILABLE:
Color Frame, Pattern Holder, Iris

WEIGHT: 26.4 lbs.

SIZE: 30⅜" x 11¾" x 16½" (L x W x H)

PHOTOMETRICS CHART

(Performance data for this unit are measured using FEL - 1 KW. lamp.)

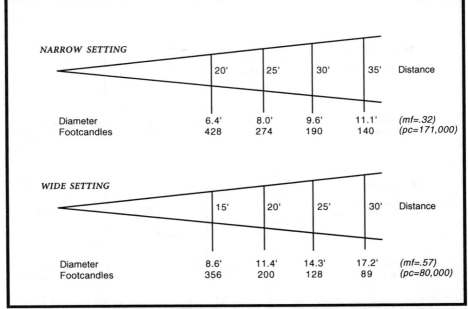

NARROW SETTING

	20'	25'	30'	35'	Distance
Diameter	6.4'	8.0'	9.6'	11.1'	*(mf=.32)*
Footcandles	428	274	190	140	*(pc=171,000)*

WIDE SETTING

	15'	20'	25'	30'	Distance
Diameter	8.6'	11.4'	14.3'	17.2'	*(mf=.57)*
Footcandles	356	200	128	89	*(pc=80,000)*

* Lens tubes/barrel assemblies in the Cantata series are interchangeable. See photometrics for other lens configurations in the series on previous and following pages.
** Manufacturer's catalog describes this angle as "cut-off angle."

ERS — Zoom

STRAND LIGHTING
Model No. 2273, "Cantata"*

BEAM ANGLE, NARROW: 18.3°

BEAM ANGLE, WIDE: 23.4°

FIELD ANGLE, NARROW: 26° **

FIELD ANGLE, WIDE: 44° **

LAMP BASE TYPE: Medium 2-Pin

LAMP MOUNT: Axial

STANDARD LAMP: FEL - 1 KW.

OTHER LAMPS: EHD - 500 W. *(cf=.39)*
EHG - 750 W. *(cf=.54)*

INTEGRAL PATTERN SLOT: Yes

ACCESSORIES AVAILABLE:
Color Frame, Pattern Holder, Iris

WEIGHT: 24.2 lbs.

SIZE: 26¼" x 11¾" x 16½" (L x W x H)

PHOTOMETRICS CHART
(Performance data for this unit are measured using FEL - 1 KW. lamp.)

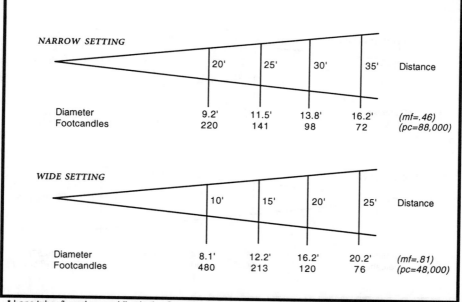

NARROW SETTING

	20'	25'	30'	35'	Distance
Diameter	9.2'	11.5'	13.8'	16.2'	*(mf=.46)*
Footcandles	220	141	98	72	*(pc=88,000)*

WIDE SETTING

	10'	15'	20'	25'	Distance
Diameter	8.1'	12.2'	16.2'	20.2'	*(mf=.81)*
Footcandles	480	213	120	76	*(pc=48,000)*

* Lens tubes/barrel assemblies in the Cantata series are interchangeable. See photometrics for other lens configurations in the series on the previous pages.
** Manufacturer's catalog describes this angle as "cut-off angle."

WEIGHT: 10.5 lbs.

SIZE: 18¹¹⁄₁₆" x 9⅝" x 11¼" (L x W x H)

STRAND LIGHTING
Model No. 2274, "Quartet 22/40"

BEAM ANGLE, NARROW: 11.3°

BEAM ANGLE, WIDE: 12.8°

FIELD ANGLE, NARROW: 24.4°

FIELD ANGLE, WIDE: 30.6°

LAMP BASE TYPE: GY9.5

LAMP MOUNT: Axial

STANDARD LAMP: FRE - 650 W.

OTHER LAMPS: (none specified in mfg. catalog)

INTEGRAL PATTERN SLOT: Yes

ACCESSORIES AVAILABLE:
Color Frame, Pattern Holder, Iris

PHOTOMETRICS CHART

(Performance data for this unit are measured using FRE - 650 W. lamp.)

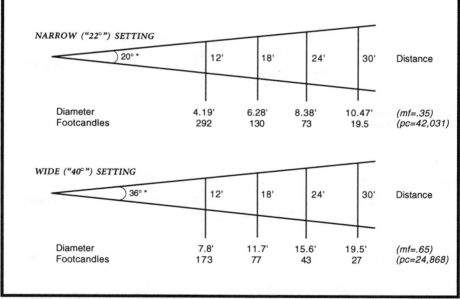

NARROW ("22°") SETTING

	12'	18'	24'	30'	Distance
Diameter	4.19'	6.28'	8.38'	10.47'	(mf=.35)
Footcandles	292	130	73	19.5	(pc=42,031)

20° *

WIDE ("40°") SETTING

	12'	18'	24'	30'	Distance
Diameter	7.8'	11.7'	15.6'	19.5'	(mf=.65)
Footcandles	173	77	43	27	(pc=24,868)

36° *

* The given distance and diameter data, described in the catalog as "narrowest" and "widest," yield angles of 20° and 36° rather than the given "field angles" of 24.4° and 30.6°. See Introduction for more information about calculating photometric data.

TIMES SQUARE LIGHTING
Model No. MR75Z

BEAM ANGLES: (not specified in mfg. catalog)

FIELD ANGLE, NARROW: 29° *

FIELD ANGLE, WIDE: 42° **

LAMP BASE TYPE: 2-Pin (GX5.3)

LAMP MOUNT: Axial

STANDARD LAMP: EYJ (MR-16) - 75 W. (12 V.)

OTHER LAMPS: EXZ (MR-16) - 50 W. (12 V.) *(cf=.65)*

INTEGRAL PATTERN SLOT: Yes

ACCESSORIES AVAILABLE:
Color Frame, Pattern Holder, "Circular Reducer"
Aperture Pattern

WEIGHT: (not specified in manufacturer's catalog)

SIZE: 8¾" x 7⅛" x 10½" (L x W x H)

PHOTOMETRICS CHART

(Performance data for this unit are measured using EYJ - 75 W. (12V.) lamp.)

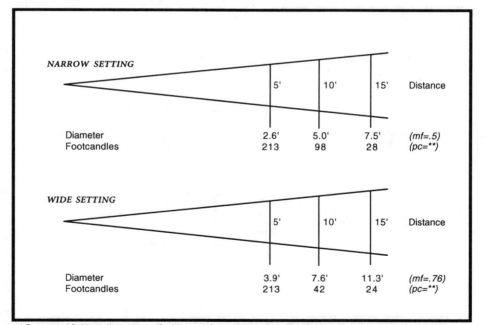

NARROW SETTING

	5'	10'	15'	Distance
Diameter	2.6'	5.0'	7.5'	(mf=.5)
Footcandles	213	98	28	(pc=**)

WIDE SETTING

	5'	10'	15'	Distance
Diameter	3.9'	7.6'	11.3'	(mf=.76)
Footcandles	213	42	24	(pc=**)

* Beam and field angles not specified in manufacturer's catalog. This field angle is calculated from given distance and field diameter data. See Introduction for more information about calculating photometric data.
** Manufacturer's footcandle data is not consistent enough to calculate a reliable peak candela figure for all throw distances.

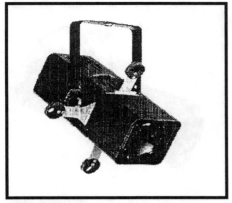

TIMES SQUARE LIGHTING
Model No. Q325

BEAM ANGLES: (not specified in mfg. catalog)

FIELD ANGLES: 30° - 50° **

LAMP BASE TYPE: Mini-Can Screw

LAMP MOUNT: Axial

STANDARD LAMP: EHV - 325 W.

OTHER LAMPS: EHT -   250 W. *(cf=.64)*
               Q400 CL/MC - 400 W. *(cf=1.06)*

INTEGRAL PATTERN SLOT: Yes

WEIGHT: 5 lbs.

SIZE: 12" x 10" x 10¾" (L x W x H)

ACCESSORIES AVAILABLE:
Color Frame, Pattern Holder, Adaptor for use with
tracklight systems

PHOTOMETRICS CHART
(Performance data for this unit are measured using EHV - 325 W. lamp.)

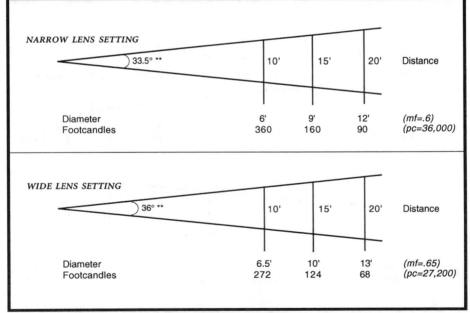

NARROW LENS SETTING

33.5° **   10'   15'   20'   Distance

		10'	15'	20'	
Diameter		6'	9'	12'	*(mf=.6)*
Footcandles		360	160	90	*(pc=36,000)*

WIDE LENS SETTING

36° **   10'   15'   20'   Distance

		10'	15'	20'	
Diameter		6.5'	10'	13'	*(mf=.65)*
Footcandles		272	124	68	*(pc=27,200)*

* This unit is adjustable by rearranging the lenses within the barrel.
** Beam and field angles not specified in manufacturer's catalog. "Beam spread" described as 30° to 50°.
Given distance and diameter data yield angles of 33.5° for narrow lens setting and 36° for wide lens setting.
See Introduction for more information on calculating photometric data.

TIMES SQUARE LIGHTING
Model No. Q3Z

BEAM ANGLES: (not specified in mfg. catalog)

FIELD ANGLE, NARROW: 25°

FIELD ANGLE, WIDE: 40°

LAMP BASE TYPE: Mini-Can Screw

LAMP MOUNT: Axial

STANDARD LAMP: EVR - 500 W.

OTHER LAMPS: Q150CL/MC - 150 W. *(cf=.27)*
EHT - 250 W. *(cf=.48)*
Q325CL/MC - 325 W. *(cf=.75)*
Q400CL/MC - 400 W. *(cf=.79)*

INTEGRAL PATTERN SLOT: Yes

ACCESSORIES AVAILABLE:
Color Frame, Pattern Holder, Iris Kit

WEIGHT: (not specified in manufacturer's catalog)

SIZE: 16" x 12½" x 13" (L x W x H)

PHOTOMETRICS CHART

(Performance data for this unit are measured using EVR - 500 W. lamp.)

NARROW SETTING

	20'	25'	30'	35'	Distance
Diameter	8.8'	11'	13.2'	15.4'	*(mf=.44)*
Footcandles	83	53	37	27	*(pc=33,200)*

WIDE SETTING

	10'	15'	20'	25'	Distance
Diameter	7.3'	11'	14.6'	18.3'	*(mf=.73)*
Footcandles	268	119	67	43	*(pc=26,800)*

TIMES SQUARE LIGHTING
Model No. Q4W

BEAM ANGLES: (not specified in mfg. catalog)

FIELD ANGLES: 32°, 40°, 48° **

LAMP BASE TYPE: Mini-Can Screw

LAMP MOUNT: Axial

STANDARD LAMP: EVR - 500 W.

OTHER LAMPS: EHT - 250 W. *(cf=.48)*
EHV - 325 W. *(cf=.75)*
EYT - 750 W. *(cf=1.78)*

INTEGRAL PATTERN SLOT: Yes

ACCESSORIES AVAILABLE:
Color Frame, Pattern Holder, Iris, Adaptors for use
with track lighting systems.

WEIGHT: 11 lbs.

SIZE: 15¾" x 10" (L x H)

PHOTOMETRICS CHART

(Performance data for this unit are measured using EVR - 500 W. lamp.)

NARROW SETTING — 25° **

	10'	15'	20'	25'	Distance
Diameter	4.5'	7'	9'	11'	*(mf=.45)*
Footcandles	402	179	101	64	*(pc=40,400)*

MEDIUM SETTING — 39° **

	10'	15'	20'	25'	Distance
Diameter	7'	11'	14.5'	18'	*(mf=.7)*
Footcandles	295	131	74	47	*(pc=29,500)*

WIDE SETTING — 44° **

	10'	15'	20'	25'	Distance
Diameter	8'	12'	16'	20'	*(mf=.8)*
Footcandles	260	116	65	42	*(pc=26,000)*

* This unit is adjustable by rearranging the lenses within the barrel.

** The given diameter and distance data, defined in catalog as "narrow position," "medium position," and "wide position," yield angles of approximately 25°, 39° and 44° rather than the given field angles of 32°, 40° and 48°. See Introduction for more information about calculating photometric data.

TIMES SQUARE LIGHTING
Model No. Q4Z

BEAM ANGLES: (not specified in mfg. catalog)

FIELD ANGLE, NARROW: 30° *

FIELD ANGLE, WIDE: 50° *

LAMP BASE TYPE: Mini-Can Screw**

LAMP MOUNT: Axial

STANDARD LAMP: EVR - 500 W.

OTHER LAMPS: EHT -   250 W. *(cf=.48)*
         Q325CL/MC - 325 W. *(cf=.75)*
         Q400CL/MC - 400 W. *(cf=.79)*

INTEGRAL PATTERN SLOT: Yes

ACCESSORIES AVAILABLE:
Color Frame, Pattern Holder, Iris Kit

WEIGHT: (not specified in manufacturer's catalog)

SIZE: 16" x 12½" x 13" (L x W x H)

PHOTOMETRICS CHART
(Performance data for this unit are measured using EVR - 500 W. lamp.)

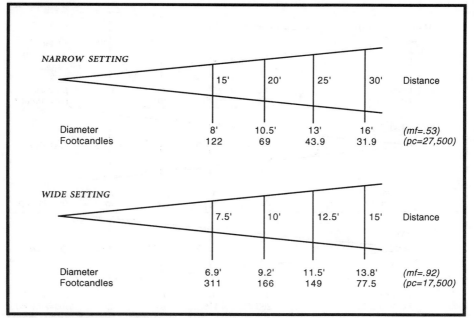

NARROW SETTING

	15'	20'	25'	30'	Distance
Diameter	8'	10.5'	13'	16'	*(mf=.53)*
Footcandles	122	69	43.9	31.9	*(pc=27,500)*

WIDE SETTING

	7.5'	10'	12.5'	15'	Distance
Diameter	6.9'	9.2'	11.5'	13.8'	*(mf=.92)*
Footcandles	311	166	149	77.5	*(pc=17,500)*

* Catalog describes this angle as "beam spread."
** Instrument can be ordered with a Medium 2-Pin lamp socket.

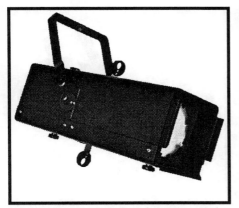

TIMES SQUARE LIGHTING
Model No. Q6V

BEAM ANGLES: 30°, 40°, 50°

FIELD ANGLES: (not specified in mfg. catalog)

LAMP BASE TYPE: Medium 2-Pin

LAMP MOUNT: Axial

STANDARD LAMP: FEL - 1 KW.

OTHER LAMPS: EHD - 500 W. *(cf=.39)*
EHG - 750 W. *(cf=.56)*

INTEGRAL PATTERN SLOT: Yes

ACCESSORIES AVAILABLE:
Color Frame, Pattern Holder, Snoot, Drop-in Iris

WEIGHT: (not specified in manufacturer's catalog)

SIZE: (not specified in manufacturer's catalog)

PHOTOMETRICS CHART
(Performance data for this unit are measured using FEL - 1 KW. lamp.)

30° POSITION

	30'	40'	50'	60'	Distance
Diameter (beam angle)	16'	21'	27'	32'	*(mf=.54)*
Footcandles	215	120	76	52	*(pc=192,000)*

40° POSITION

	30'	40'	50'	60'	Distance
Diameter (beam angle)	22'	24'	36'	44'	*(mf=.72)*
Footcandles	162	92	58	39	*(pc=145,000)*

50° POSITION

	30'	40'	50'	60'	Distance
Diameter (beam angle)	28'	37'	47'	56'	*(mf=.94)*
Footcandles	107	59	37	25	*(pc=92,500)*

* This unit is adjustable by rearranging the lenses within the barrel.

TIMES SQUARE LIGHTING
Model No. Q6Z

BEAM ANGLES: (not specified in mfg. catalog)

FIELD ANGLE, NARROW: 20°

FIELD ANGLE, WIDE: 40°

LAMP BASE TYPE: Medium 2-Pin

LAMP MOUNT: Axial

STANDARD LAMP: FEL - 1 KW.

OTHER LAMPS: EHD - 500 W. *(cf=.39)*
             EHG - 750 W. *(cf=.54)*

INTEGRAL PATTERN SLOT: Yes

ACCESSORIES AVAILABLE:
Color Frame, Pattern Holder, Iris

WEIGHT: 20 lbs.

SIZE: 25½" x 15" x 19¼" (L x W x H)

PHOTOMETRICS CHART

(Performance data for this unit are measured using FEL - 1 KW. lamp.)

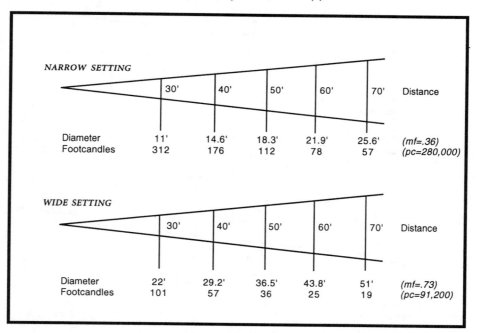

NARROW SETTING

	30'	40'	50'	60'	70'	Distance
Diameter	11'	14.6'	18.3'	21.9'	25.6'	*(mf=.36)*
Footcandles	312	176	112	78	57	*(pc=280,000)*

WIDE SETTING

	30'	40'	50'	60'	70'	Distance
Diameter	22'	29.2'	36.5'	43.8'	51'	*(mf=.73)*
Footcandles	101	57	36	25	19	*(pc=91,200)*

HIGH END SYSTEMS
Model No. Studio Color - 575M*

SPOT BEAM SPREAD: 8° **

FLOOD BEAM SPREAD: 17° **

BEAM IRIS RANGE: (no iris in this unit)

PAN/TILT RANGE: 370° pan, 240° tilt

ACCURACY:1.44° /step (LWR protocol), 0.93° (DMX)

GOBOS: (not specified in manufacturer's catalog)

COLOR: CMY color mixing, plus one 6-color wheel
(includes one color correction filter)

WEIGHT: 68 lbs.

SIZE: 12" x 19" x 25¹⁄₁₆" (L x W x H)

DATA CONNECTORS – 1: ground
(3-pin XLR)   2: DMX data, negative (black)
                       3: DMX data, positive (red)

DIFFUSION: two 90° rotating lenticular frost,
with both in, beam spread is 22°

LAMPS: MSR 575/2 (Philips) - 575 W.
MSD 575 (Phillips) - 575 W. *(cf=1.0)*

PHOTOMETRICS CHART
(Performance data for this unit are measured using MSR 575/2 - 575 W. lamp)

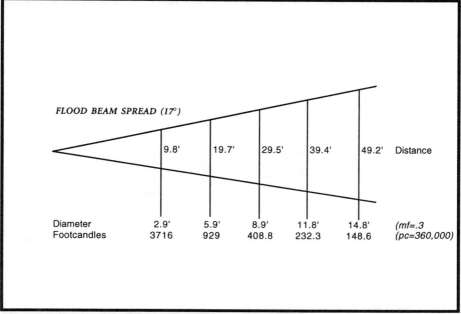

FLOOD BEAM SPREAD (17°)

	9.8'	19.7'	29.5'	39.4'	49.2'	Distance
Diameter	2.9'	5.9'	8.9'	11.8'	14.8'	*(mf=.3*
Footcandles	3716	929	408.8	232.3	148.6	*(pc=360,000)*

* Version "M" has a magnetic ballast and accepts only 208–250 VAC, 50/60 Hz.
**Accessory effects lenses are also available (VNSP, NSP, MFL, WFL, and VWFL). No
photometrics data were available in manufacturer's catalog.

381

HIGH END SYSTEMS
Model No. Studio Color - 575S*

SPOT BEAM SPREAD: 8° **

FLOOD BEAM SPREAD: 17° **

BEAM IRIS RANGE: (no iris in this unit)

PAN/TILT RANGE: 370° pan, 240° tilt

ACCURACY:1.44° per step (LWR protocol), 0.93° (DMX)

GOBOS: (not specified in manufacturer's catalog)

COLOR: CMY color mixing, plus one 6-color wheel
      (includes one color correction filter)

DIFFUSION: two 90° rotating lenticular frost,
      with both in, beam spread is 22°

LAMPS: MSR 575/2 (Philips) - 575 W.
      MSD 575 (Phillips) - 575 W. *(cf=1.0)*

WEIGHT: 57 lbs.

SIZE: 12" x 19" x 25¹¹⁄₁₆" (L x W x H)

DATA CONNECTORS – 1: ground
(3-pin XLR)   2: DMX data, negative (black)
               3: DMX data, positive (red)

PHOTOMETRICS CHART

(Performance data for this unit are measured using MSR 575/2 - 575 W. lamp)

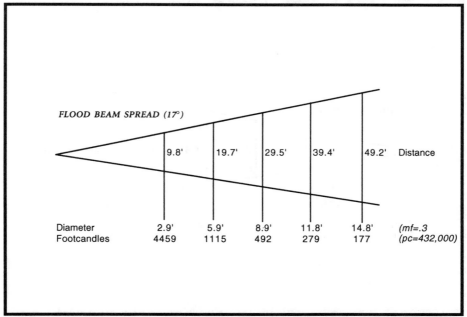

FLOOD BEAM SPREAD (17°)

		Distance
9.8'	19.7'	29.5'	39.4'	49.2'

Diameter	2.9'	5.9'	8.9'	11.8'	14.8'	*(mf=.3*
Footcandles	4459	1115	492	279	177	*(pc=432,000)*

* Version "S" has an electronic ballast which automatically accommodates voltages between 100 and 250 VAC, 50/60 Hz.

**Accessory effects lenses are also available (VNSP, NSP, MFL, WFL, and VWFL). No photometrics data were available in manufacturer's catalog.

MARTIN PROFESSIONAL A/S
Model No. MAC 600

BEAM SPREAD: 25° field *

BEAM IRIS RANGE: *

PAN/TILT RANGE: 440° pan, 306° tilt

ACCURACY: 1.7° per step (Martin) , 1.2° (DMX)

GOBOS: (not available on this unit)

COLOR: CMY color mixing, plus one 4-color wheel
(includes one color correction filter)

DIFFUSION: 2 combinable beam shaping
modules + frost effect

LAMP: MSR 575 (Philips) - 575 W.

WEIGHT: 68 lbs.

SIZE: 17¾" x 13¾" x 26" (L x W x H)

DATA CONNECTORS –1: ground
(3-pin XLR)   2: DMX data, positive (red)
                 3: DMX data, negative (black)

PHOTOMETRICS CHART
(Performance data for this unit are measured using MSR 575 - 575 W. lamp.)

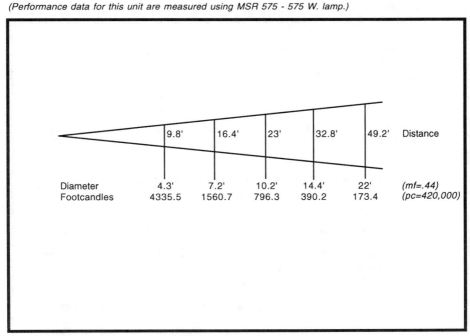

	9.8'	16.4'	23'	32.8'	49.2'	Distance
Diameter	4.3'	7.2'	10.2'	14.4'	22'	(mf=.44)
Footcandles	4335.5	1560.7	796.3	390.2	173.4	(pc=420,000)

* This unit has a fixed-focus fresnel lens, and it doesn't have an iris.

383

MARTIN PROFESSIONAL A/S
Model No. MAC 1200

FLOOD BEAM SPREAD: 32° *

BEAM IRIS RANGE: *

PAN/TILT RANGE: 440° pan, 306° tilt

ACCURACY: 0.13° per step (Martin) , .007° (DMX)

GOBOS: 5 aperture gobos* + 1 gobo

COLOR: CMY color mixing, plus one 4-color wheel
      (includes one color correction filter)

DIFFUSION: variable*

LAMP: MSR 1200 (Philips) - 1.2 KW.

WEIGHT: 104.5 lbs.

SIZE: 22¹¹⁄₁₆" x 18⅞" x 39½" (L x W x H)

DATA CONNECTORS −1: ground
(3-pin XLR)   2: DMX data, positive (red)
              3: DMX data, negative (black)

PHOTOMETRICS CHART
(Performance data for this unit are measured using MSR 1200 - 1.2 KW. lamp.)

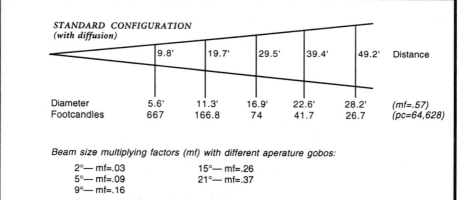

STANDARD CONFIGURATION
(with diffusion)

Distance	9.8'	19.7'	29.5'	39.4'	49.2'
Diameter	5.6'	11.3'	16.9'	22.6'	28.2'	*(mf=.57)*
Footcandles	667	166.8	74	41.7	26.7	*(pc=64,628)*

Beam size multiplying factors (mf) with different aperture gobos:

    2°— mf=.03   15°— mf=.26
    5°— mf=.09   21°— mf=.37
    9°— mf=.16

    Intensity data for aperature configurations not given in manufacturer's catalog.

Intensity data using different combinations of the diffuser filter and the condenser lens:

    − diffuser / − condenser pc= 69,296
    − diffuser / + condenser pc= 102,825
    + diffuser / + condenser pc= 88,357

    Beam size data for these configurations not given in manufacturer's catalog.

* This unit has five aperature gobos—2°, 5°, 9°, 15° and 21°—which are used to modify the beam size. It also has a system to smooth the hot spot which uses a diffusion filter and condenser lens in different combinations. The standard configuration, with a beam size of 32°, includes the diffuser filter. See photometrics chart for intensity data for other configurations.

VARI-LITE, INC.
Model No. VL2C

SPOT BEAM SPREAD: 7° beam, 15° field (peak)

FLOOD BEAM SPREAD: 9° beam, 20° field (peak)

BEAM IRIS RANGE: 4° to 22° *

PAN/TILT RANGE: 360° pan, 270° tilt

ACCURACY: 0.3° per step

GOBOS: 9 positions

COLOR: special 2-wheel mixing system (120+ colors),
plus 1 color correction filter

DIFFUSION:**

LAMP: HTI 600 W S/E (OSRAM) - 600 W.

WEIGHT: 58 lbs.***

SIZE: 17½" x 17½" x 25¾" (L x W x H)

DATA CONNECTORS – Vari*Lite protocol only

PHOTOMETRICS CHART
(Performance data for this unit are measured at peak focus using HTI 600 W S/E lamp.)

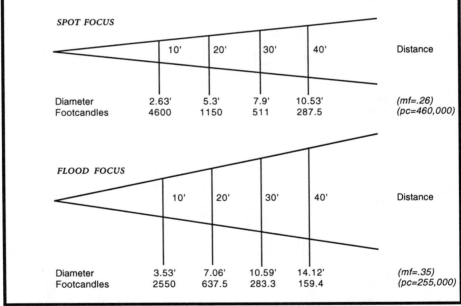

SPOT FOCUS

	10'	20'	30'	40'	Distance
Diameter	2.63'	5.3'	7.9'	10.53'	*(mf=.26)*
Footcandles	4600	1150	511	287.5	*(pc=460,000)*

FLOOD FOCUS

	10'	20'	30'	40'	Distance
Diameter	3.53'	7.06'	10.59'	14.12'	*(mf=.35)*
Footcandles	2550	637.5	283.3	159.4	*(pc=255,000)*

* With lamp focused for flat field performance, this unit can achieve a 26° beam spread, which the iris restricts to 22°.

** This unit uses a variable focus lens to "diffuse" the edges of its light beam.

*** The weight of a required outboard "repeater" unit is not included in this weight.

385

MOVING HEAD UNITS

VARI-LITE, INC.
Model No. VL4

SPOT BEAM SPREAD: 4° beam, 12° field

FLOOD BEAM SPREAD: 9° beam, 18° field

BEAM IRIS RANGE: *

PAN/TILT RANGE: 360° pan, 270° tilt

ACCURACY: 0.3° per step (VL protocol), 3.6° (DMX)

GOBOS: (no gobos in this unit)

COLOR: CMA (cyan/magenta/amber) color mixing,
    plus one color correction filter

DIFFUSION: textured glass panels

LAMP: HTI 400 W S/E (OSRAM) - 400 W.

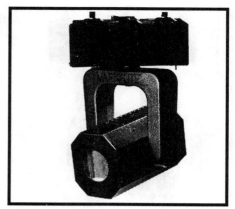

WEIGHT: 38 lbs.**

SIZE: 14" x 14" x 20⅜" (L x W x H)

DATA CONNECTORS – Vari*Lite protocol only

PHOTOMETRICS CHART
(Performance data for this unit are measured using HTI 400 W. S/E lamp.)

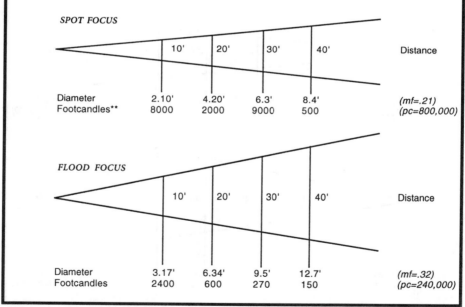

SPOT FOCUS

	10'	20'	30'	40'	Distance
Diameter	2.10'	4.20'	6.3'	8.4'	(mf=.21)
Footcandles**	8000	2000	9000	500	(pc=800,000)

FLOOD FOCUS

	10'	20'	30'	40'	Distance
Diameter	3.17'	6.34'	9.5'	12.7'	(mf=.32)
Footcandles	2400	600	270	150	(pc=240,000)

* Iris in this unit is out of the focal plane and is used for dimming. Adjustable lenses provide beam shaping.
** The weight of a required outboard "repeater" unit is not included in this weight.

386

WEIGHT: 29 lbs.**

SIZE: 17" x 16" x 26¾" (L x W x H)

VARI-LITE, INC.
Model No. VL5Arc

SPOT BEAM SPREAD: 5° beam, 13° field *

FLOOD BEAM SPREAD: 17.5° beam, 34° field *

BEAM IRIS RANGE: 5° to 34°

PAN/TILT RANGE: 360° pan, 270° tilt

ACCURACY: 0.3° per step (VL protocol), 3.6° (DMX)

GOBOS: (no gobos in this unit)

COLOR: CMA (cyan/magenta/amber) color mixing

DIFFUSION: variable liquid lens *

LAMP: MSR 575 (Philips) - 575 W.

DATA CONNECTORS –1: ground
(5-pin XLR)   2: DMX data, negative (black)
    3: DMX data, positive (red)
    4: unused
    5: unused

PHOTOMETRICS CHART

(Performance data for this unit are measured at field angle using MSR 575 - 575 W. lamp.)

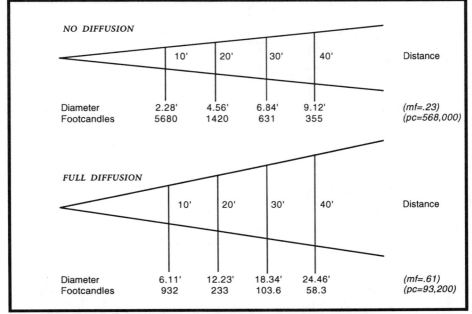

NO DIFFUSION

	10'	20'	30'	40'	Distance
Diameter	2.28'	4.56'	6.84'	9.12'	(mf=.23)
Footcandles	5680	1420	631	355	(pc=568,000)

FULL DIFFUSION

	10'	20'	30'	40'	Distance
Diameter	6.11'	12.23'	18.34'	24.46'	(mf=.61)
Footcandles	932	233	103.6	58.3	(pc=93,200)

* "Vari*Beam" fluid-filled membrane is used to diffuse the beam size from "spot" to "flood."

** The weight of a required outboard "repeater" unit is not included in this weight.

387

MOVING HEAD UNITS

VARI-LITE, INC.
Model No. VL5/5B* (plain lens)

SPOT BEAM SPREAD: 6° beam, 11° field

FLOOD BEAM SPREAD: 11° beam, 41° field (diffusion)

BEAM IRIS RANGE: (no iris on this unit)

PAN/TILT RANGE: 360° pan, 270° tilt

ACCURACY: 0.3° per step (VL protocol), 3.6° (DMX)

GOBOS: (no gobos on this unit)

COLOR: CMA (cyan/magenta/amber) color mixing

DIFFUSION: glass diffuser panels

LAMPS: MSR 1200 (Philips) - 1.2 KW.

WEIGHT: 25 lbs.**

SIZE: 16¼" x 16" x 26¾" (L x W x H)

DATA CONNECTORS – 1: ground
(5-pin XLR) 2: DMX data, negative (black)
3: DMX data, positive (red)
4: unused
5: unused

PHOTOMETRICS CHART

(Performance data for this unit are measured using MSR 1200 - 1.2 KW. lamp.)

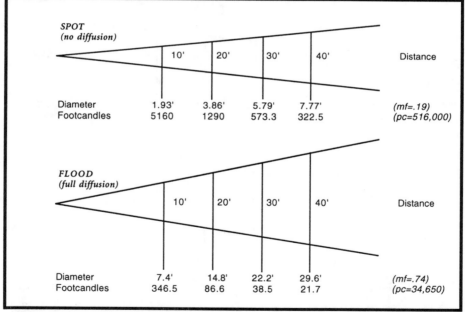

SPOT (no diffusion)

	10'	20'	30'	40'	Distance
Diameter	1.93'	3.86'	5.79'	7.77'	(mf=.19)
Footcandles	5160	1290	573.3	322.5	(pc=516,000)

FLOOD (full diffusion)

	10'	20'	30'	40'	Distance
Diameter	7.4'	14.8'	22.2'	29.6'	(mf=.74)
Footcandles	346.5	86.6	38.5	21.7	(pc=34,650)

* VL5 contains the standard palette of dichroics; VL5B contains the "theatre palette" of less saturated colors.
** The weight of a required outboard "repeater" unit is not included in this weight.

WEIGHT: 25 lbs.**

SIZE: 16¼" x 16" x 26¾" (L x W x H)

VARI-LITE, INC.
Model No. VL5/5B* (narrow spot lens)

SPOT BEAM SPREAD: 8° beam, 15° field

FLOOD BEAM SPREAD: 20° beam, 42° field (diffusion)

BEAM IRIS RANGE: (no iris on this unit)

PAN/TILT RANGE: 360° pan, 270° tilt

ACCURACY: 0.3° per step (VL protocol), 3.6° (DMX)

GOBOS: (no gobos on this unit)

COLOR: CMA (cyan/magenta/amber) color mixing

DIFFUSION: glass diffuser panels

LAMPS: MSR 1200 (Philips) - 1.2 KW.

DATA CONNECTORS – 1: ground
(5-pin XLR)   2: DMX data, negative (black)
                   3: DMX data, positive (red)
                   4: unused
                   5: unused

PHOTOMETRICS CHART

(Performance data for this unit are measured using MSR 1200 - 1.2 KW. lamp.)

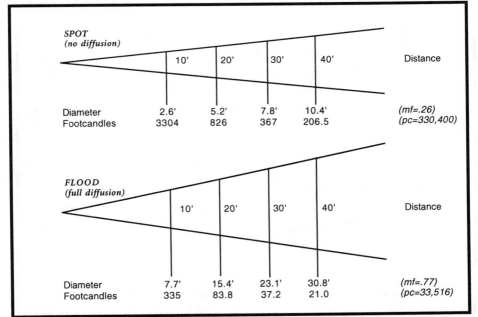

SPOT (no diffusion)

	10'	20'	30'	40'	Distance
Diameter	2.6'	5.2'	7.8'	10.4'	*(mf=.26)*
Footcandles	3304	826	367	206.5	*(pc=330,400)*

FLOOD (full diffusion)

	10'	20'	30'	40'	Distance
Diameter	7.7'	15.4'	23.1'	30.8'	*(mf=.77)*
Footcandles	335	83.8	37.2	21.0	*(pc=33,516)*

* VL5 contains the standard palette of dichroics; VL5B contains the "theatre palette" of less saturated colors.
** The weight of a required outboard "repeater" unit is not included in this weight.

389

VARI-LITE, INC.
Model No. VL5/5B* (narrow flood lens)

SPOT BEAM SPREAD: 21° x 8° beam, 31° x 18° field

FLOOD BEAM SPREAD: 27° x 21° beam, 51° x 43° field

BEAM IRIS RANGE: (no iris on this unit)

PAN/TILT RANGE: 360° pan, 270° tilt

ACCURACY: 0.3° per step (VL protocol), 3.6° (DMX)

GOBOS: (no gobos on this unit)

COLOR: CMA (cyan/magenta/amber) color mixing

DIFFUSION: glass diffuser panels

LAMPS: MSR 1200 (Philips) - 1.2 KW.

WEIGHT: 25 lbs.**

SIZE: 16¼" x 16" x 26¾" (L x W x H)

DATA CONNECTORS – 1: ground
(5-pin XLR)   2: DMX data, negative (black)
                       3: DMX data, positive (red)
                       4: unused
                       5: unused

PHOTOMETRICS CHART

(Performance data for this unit are measured using MSR 1200 - 1.2 KW. lamp.)

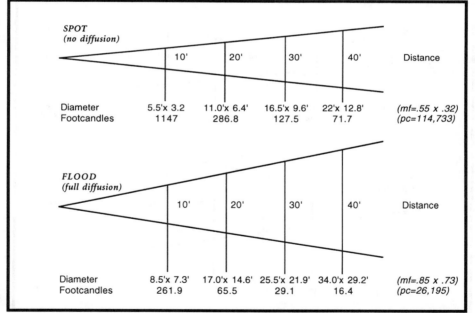

SPOT
(no diffusion)

	10'	20'	30'	40'	Distance
Diameter	5.5'x 3.2	11.0'x 6.4'	16.5'x 9.6'	22'x 12.8'	*(mf=.55 x .32)*
Footcandles	1147	286.8	127.5	71.7	*(pc=114,733)*

FLOOD
(full diffusion)

	10'	20'	30'	40'	Distance
Diameter	8.5'x 7.3'	17.0'x 14.6'	25.5'x 21.9'	34.0'x 29.2'	*(mf=.85 x .73)*
Footcandles	261.9	65.5	29.1	16.4	*(pc=26,195)*

* VL5 contains the standard palette of dichroics; VL5B contains the "theatre palette" of less saturated colors.
** The weight of a required outboard "repeater" unit is not included in this weight.

390

VARI-LITE, INC.
Model No. VL5/5B* (med. flood lens)

SPOT BEAM SPREAD: 19° x 11° beam, 34° x 19° field

FLOOD BEAM SPREAD: 25° x 20° beam, 51° x 43° field

BEAM IRIS RANGE: (no iris on this unit)

PAN/TILT RANGE: 360° pan, 270° tilt

ACCURACY: 0.3° per step (VL protocol), 3.6° (DMX)

GOBOS: (no gobos on this unit)

COLOR: CMA (cyan/magenta/amber) color mixing

DIFFUSION: glass diffuser panels

LAMPS: MSR 1200 (Philips) - 1.2 KW.

DATA CONNECTORS – 1: ground
(5-pin XLR)   2: DMX data, negative (black)
                      3: DMX data, positive (red)
                      4: unused
                      5: unused

WEIGHT: 25 lbs.**

SIZE: 16¼" x 16" x 26¾" (L x W x H)

PHOTOMETRICS CHART

(Performance data for this unit are measured using MSR 1200 - 1.2 KW. lamp.)

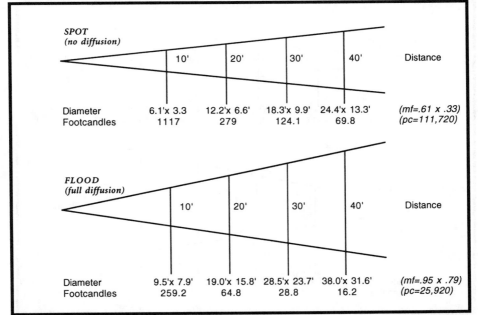

SPOT
(no diffusion)

	10'	20'	30'	40'	Distance
Diameter	6.1'x 3.3	12.2'x 6.6'	18.3'x 9.9'	24.4'x 13.3'	*(mf=.61 x .33)*
Footcandles	1117	279	124.1	69.8	*(pc=111,720)*

FLOOD
(full diffusion)

	10'	20'	30'	40'	Distance
Diameter	9.5'x 7.9'	19.0'x 15.8'	28.5'x 23.7'	38.0'x 31.6'	*(mf=.95 x .79)*
Footcandles	259.2	64.8	28.8	16.2	*(pc=25,920)*

* VL5 contains the standard palette of dichroics; VL5B contains the "theatre palette" of less saturated colors.

** The weight of a required outboard "repeater" unit is not included in this weight.

VARI-LITE, INC.
Model No. VL5/5B* (wide flood lens)

SPOT BEAM SPREAD: 49° x 19° beam, 61° x 29° field

FLOOD BEAM SPREAD: 41° x 24° beam, 72° x 48° field

BEAM IRIS RANGE: (no iris on this unit)

PAN/TILT RANGE: 360° pan, 270° tilt

ACCURACY: 0.3° per step (VL protocol), 3.6° (DMX)

GOBOS: (no gobos on this unit)

COLOR: CMA (cyan/magenta/amber) color mixing

DIFFUSION: glass diffuser panels

LAMPS: MSR 1200 (Philips) - 1.2 KW.

WEIGHT: 25 lbs.**

SIZE: 16¼" x 16" x 26¾" (L x W x H)

DATA CONNECTORS – 1: ground
(5-pin XLR)   2: DMX data, negative (black)
         3: DMX data, positive (red)
         4: unused
         5: unused

PHOTOMETRICS CHART

(Performance data for this unit are measured using MSR 1200 - 1.2 KW. lamp.)

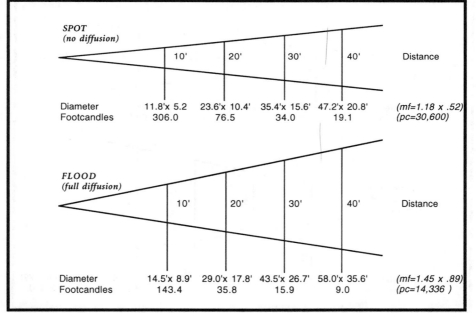

SPOT
(no diffusion)

	10'	20'	30'	40'	Distance
Diameter	11.8'x 5.2	23.6'x 10.4'	35.4'x 15.6'	47.2'x 20.8'	*(mf=1.18 x .52)*
Footcandles	306.0	76.5	34.0	19.1	*(pc=30,600)*

FLOOD
(full diffusion)

	10'	20'	30'	40'	Distance
Diameter	14.5'x 8.9'	29.0'x 17.8'	43.5'x 26.7'	58.0'x 35.6'	*(mf=1.45 x .89)*
Footcandles	143.4	35.8	15.9	9.0	*(pc=14,336)*

* VL5 contains the standard palette of dichroics; VL5B contains the "theatre palette" of less saturated colors.
** The weight of a required outboard "repeater" unit is not included in this weight.

392

WEIGHT: 22 lbs.****

SIZE: 16½" x 16½" x 24¹¹⁄₁₆" (L x W x H)

VARI-LITE, INC.
Model No. VL6 (NFOV)*

SPOT BEAM SPREAD: 6° beam, 15° field

FLOOD BEAM SPREAD: 8° beam, 17° field

BEAM IRIS RANGE: (not specified in mfg. catalog)

PAN/TILT RANGE: 360° pan, 270° tilt

ACCURACY: 0.3° per step (VL protocol), 3.6° (DMX)

GOBOS: 11 positions **

COLOR: one 11-color wheel **

DIFFUSION: ***

LAMP: MSR 400SA (Philips) - 400W.

DATA CONNECTORS – 1: ground
(5-pin XLR)   2: DMX data, negative (black)
                   3: DMX data, positive (red)
                   4: unused
                   5: unused

PHOTOMETRICS CHART

(Performance data for this unit are measured at peak focus using MSR 400SA - 400 W. lamp.)

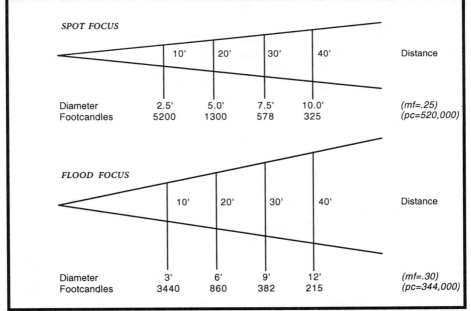

SPOT FOCUS

	10'	20'	30'	40'	Distance
Diameter	2.5'	5.0'	7.5'	10.0'	(mf=.25)
Footcandles	5200	1300	578	325	(pc=520,000)

FLOOD FOCUS

	10'	20'	30'	40'	Distance
Diameter	3'	6'	9'	12'	(mf=.30)
Footcandles	3440	860	382	215	(pc=344,000)

* VL6s can accept three lenses. This page describes the unit with the narrow field of view lens (NFOV).
** This unit uses two identical wheels which may be filled with color, gobos or any combination.
*** This unit uses a variable focus lens to "diffuse" the edges of its light beam.
**** The weight of a required outboard "repeater" unit is not included in this weight.

MOVING HEAD UNITS

VARI-LITE, INC.
Model No. VL6 (MFOV)*

SPOT BEAM SPREAD: 8° beam, 21° field

FLOOD BEAM SPREAD: 10° beam, 23° field

BEAM IRIS RANGE: (not specified in mfg. catalog)

PAN/TILT RANGE: 360° pan, 270° tilt

ACCURACY: 0.3° per step (VL protocol), 3.6° (DMX)

GOBOS: 11 positions **

COLOR: one 11-color wheel **

DIFFUSION: ***

LAMP: MSR 400SA (Philips) - 400W.

WEIGHT: 22 lbs.****

SIZE: 16½" x 16½" x 24¹¹⁄₁₆" (L x W x H)

DATA CONNECTORS – 1: ground
(5-pin XLR)   2: DMX data, negative (black)
                      3: DMX data, positive (red)
                      4: unused
                      5: unused

PHOTOMETRICS CHART

(Performance data for this unit are measured at peak focus using MSR 400SA - 400 W. lamp.)

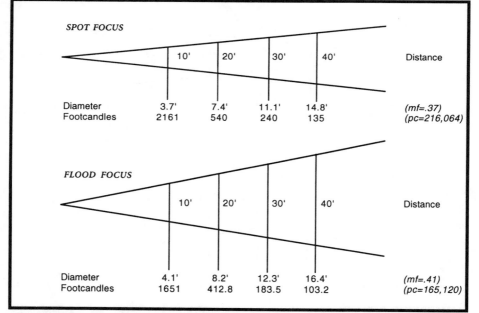

SPOT FOCUS

	10'	20'	30'	40'	Distance
Diameter	3.7'	7.4'	11.1'	14.8'	*(mf=.37)*
Footcandles	2161	540	240	135	*(pc=216,064)*

FLOOD FOCUS

	10'	20'	30'	40'	Distance
Diameter	4.1'	8.2'	12.3'	16.4'	*(mf=.41)*
Footcandles	1651	412.8	183.5	103.2	*(pc=165,120)*

* VL6s can accept three lenses. This page describes the unit with the medium field of view lens (MFOV).

** This unit uses two identical wheels which may be filled with color, gobos or any combination.

*** This unit uses a variable focus lens to "diffuse" the edges of its light beam.

**** The weight of a required outboard "repeater" unit is not included in this weight.

394

WEIGHT: 22 lbs.****

SIZE: 16½" x 16½" x 24¹¹⁄₁₆" (L x W x H)

VARI-LITE, INC.
Model No. VL6 (WFOV)*

SPOT BEAM SPREAD: 12° beam, 35° field

FLOOD BEAM SPREAD: 15° beam, 36° field

BEAM IRIS RANGE: (not specified in mfg. catalog)

PAN/TILT RANGE: 360° pan, 270° tilt

ACCURACY: 0.3° per step (VL protocol), 3.6° (DMX)

GOBOS: 11 positions **

COLOR: one 11-color wheel **

DIFFUSION: ***

LAMP: MSR 400SA (Philips) - 400W.

DATA CONNECTORS – 1: ground
(5-pin XLR)   2: DMX data, negative (black)
    3: DMX data, positive (red)
    4: unused
    5: unused

PHOTOMETRICS CHART

(Performance data for this unit are measured at peak focus using MSR 400SA - 400 W. lamp.)

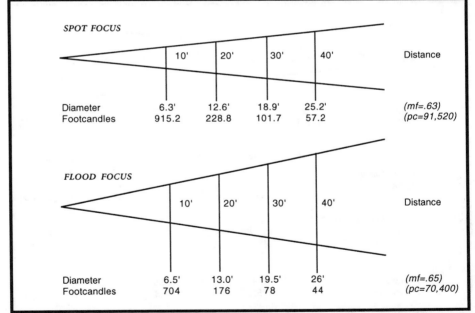

SPOT FOCUS

	10'	20'	30'	40'	Distance
Diameter	6.3'	12.6'	18.9'	25.2'	(mf=.63)
Footcandles	915.2	228.8	101.7	57.2	(pc=91,520)

FLOOD FOCUS

	10'	20'	30'	40'	Distance
Diameter	6.5'	13.0'	19.5'	26'	(mf=.65)
Footcandles	704	176	78	44	(pc=70,400)

* VL6s can accept three lenses. This page describes the unit with the wide field of view lens (WFOV).

** This unit uses two identical wheels which may be filled with color, gobos or any combination.

*** This unit uses a variable focus lens to "diffuse" the edges of its light beam.

**** The weight of a required outboard "repeater" unit is not included in this weight.

395

CLAY PAKY
Model No. C11066 "Golden Scan 2"

SPOT BEAM SPREAD: (not specified in mfg. catalog)

FLOOD BEAM SPREAD: 11.74° field *

BEAM IRIS RANGE: (not specified in mfg. catalog)

PAN/TILT RANGE: 150° pan, 110° tilt

ACCURACY: 0.59° per step (CP protocol), 0.43° (DMX)

GOBOS: 8 (2 wheels x 4 patterns)

COLOR: one 7-color wheel

DIFFUSION: (not specified in mfg. catalog)

LAMP: HMI 1200 W/GS

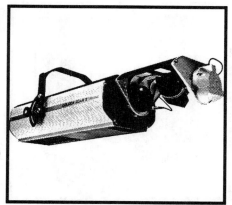

WEIGHT: 98.1 lbs.

SIZE: 44⅞" x 16⅞" x 17⅞" (L x W x H)

DATA CONNECTORS – 1: ground
(5-pin XLR)   2: DMX data, negative (black)
                      3: DMX data, positive (red)
                      4: unused
                      5: **

PHOTOMETRICS CHART
(Performance data for this unit are measured using HMI 1200 W/GS lamp.)

1:2.5 / 250mm lens (11.74°)*

	16.4'	32.8'	49.2'	65.6'	82'	Distance
Diameter	2.85'	6.56'	9.91'	13.48'	16.86'	*(mf=.2)*
Footcandles	1486.4	371.6	165.2	92.9	59.5	*(pc=399,782)*

	16.4'	32.8'	49.2'	65.6'	82'	Distance

1:3.5 / 200mm (14.9°)*

	16.4'	32.8'	49.2'	65.6'	82'	Distance
Diameter	4.28'	8.5'	12.86'	17.1'	21.4'	*(mf=.26)*
Footcandles	929	232.3	103.2	58.1	37.2	*(pc=249,864)*

1:1.8 / 170mm (17.96°)*

	16.4'	32.8'	49.2'	65.6'	82'	
Diameter	5.18'	10.37'	15.55'	20.73'	25.59'	*(mf=.316)*
Footcandles	844	211	93.8	52.7	33.8	*(pc=227,008)*

1:3 / 150mm (18.7°)*

	16.4'	32.8'	49.2'	65.6'	82'	
Diameter	5.4'	10.8'	16.23'	21.65'	27.06'	*(mf=.33)*
Footcandles	501.3	125.3	55.7	31.3	20	*(pc=134,833)*

* Beam and field angles not specified in manufacturer's catalog. This field angle is calculated from given distance and field diameter data. Also, this unit accepts three optional lenses as described in the photometrics chart above.

** Some units of this model source +25V DC on pin 5 (the table printed on the fixture near the DMX connector will label pin 5, "Is 18-25VDC"). This high voltage on pin 5 does not conform to the USITT DMX512/1990 standard and may cause damage to other equipment on a DMX network.

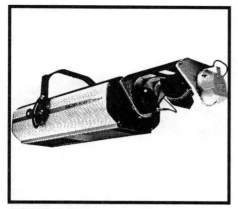

WEIGHT: 70.5 lbs.

SIZE: 40" x 16⅞" x 17⅞" (L x W x H)

CLAY PAKY
Model No. C11067 "Golden Scan 2"

SPOT BEAM SPREAD: (not specified in mfg. catalog)

FLOOD BEAM SPREAD: 11.74° field *

BEAM IRIS RANGE: (not specified in mfg. catalog)

PAN/TILT RANGE: 150° pan, 110° tilt

ACCURACY: 0.59° per step (CP protocol), 0.43° (DMX)

GOBOS: 8 (2 wheels x 4 patterns)

COLOR: one 7-color wheel

DIFFUSION: (not specified in mfg. catalog)

LAMP: HMI 575 W/GS

DATA CONNECTORS – 1: ground
(5-pin XLR) 2: DMX data, negative (black)
3: DMX data, positive (red)
4: unused
5: **

PHOTOMETRICS CHART

(Performance data for this unit are measured using HMI 575 W/GS lamp.)

1:2.5 / 250mm lens (11.74°)*

	16.4'	32.8'	49.2'	65.6'	82'	Distance
Diameter	2.85'	6.56'	9.91'	13.48'	16.86'	(mf=.2)
Footcandles	929	232.3	103.2	58	37	(pc=249,864)

	16.4'	32.8'	49.2'	65.6'	82'	Distance
1:3.5 / 200mm (14.9°)*						
Diameter	4.28'	8.5'	12.86'	17.1'	21.4'	(mf=.26)
Footcandles	557.4	139.4	61.9	34.8	22.3	(pc=149,918)
1:1.8 / 170mm (17.96°)*						
Diameter	5.18'	10.37'	15.55'	20.73'	25.59'	(mf=.316)
Footcandles	341.7	85.4	38	21.4	13.7	(pc=91,912)
1:3 / 150mm (18.7°)*						
Diameter	5.4'	10.8'	16.23'	21.65'	27.06'	(mf=.33)
Footcandles	232.3	58.1	25.8	14.5	9.3	(pc=62,477)

* Beam and field angles not specified in manufacturer's catalog. This field angle is calculated from given distance and field diameter data. Also, this unit accepts three optional lenses as described in the photometrics chart above.

** Some units of this model source +25V DC on pin 5 (the table printed on the fixture near the DMX connector will label pin 5, "Is 18-25VDC"). This high voltage on pin 5 does not conform to the USITT DMX512/1990 standard and may cause damage to other equipment on a DMX network.

CLAY PAKY
Model No. C11068* "Golden Scan 3"

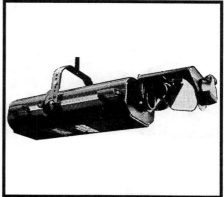

SPOT BEAM SPREAD: (not specified in mfg. catalog)

FLOOD BEAM SPREAD: 11° **

BEAM IRIS RANGE: (not specified in mfg. catalog)

PAN/TILT RANGE: 150° pan, 110° tilt

ACCURACY: 0.6° per step (pan), 0.4° per step (tilt)

GOBOS: 4 positions, and 4-facet prism

COLOR: one 7-color wheel,
       plus 2 color correction filters

DIFFUSION: frost filter

LAMP: HMI 575 - 575 W.

WEIGHT: 71.9 lbs. (with mirror head)

SIZE: 43½" x 17⅛" x 14⅛" (L x W x H)

DATA CONNECTORS – 1: ground
(5-pin XLR)   2: DMX data, negative (black)
                   3: DMX data, positive (red)
                   4: unused
                   5: ***

PHOTOMETRICS CHART
(Performance data for this unit are measured using HMI 575 - 575 W. lamp)

*1:2.5 / 250mm lens (11°)***

	16.4'	32.8'	49.2'	65.6'	82'	Distance
Diameter	3.28'	6.56'	9.84'	13.12'	16.4'	*(mf=.2)*
Footcandles	989	247	109.9	61.8	39.5	*(pc=265,948)*
Wide-Angle Option (16°)**						
Diameter	4.6'	9.19'	13.8'	18.37'	23'	*(mf=.28)*
Footcandles	468	117	52	29.25	18.72	*(pc=125,873)*

*1:3.3 / 300mm lens (9.5°)***

	16.4'	32.8'	49.2'	65.6'	82'	Distance
Diameter	2.69'	5.38'	8.07'	10.76'	13.55'	*(mf=.16)*
Footcandles	1235	308.8	137.2	77.2	49.4	*(pc=332,219)*
Wide-Angle Option (13°)**						
Diameter	3.77'	7.55'	11.15'	15.09'	19.03'	*(mf=.23)*
Footcandles	535.5	133.87	59.5	33.5	21.4	*(pc=144,028)*

* Also available in touring version (TV) with features such as a lamp hour meter, graduated scale on mounting bracket and mirror head, side handles, etc.

** Beam and field angles not specified in manufacturer's catalog. This angle is calculated from given distance and diameter data. Also, this unit accepts an optional, manually-operated, "flip-in," wide-angle lens, as well as an optional narrow-focus lens.
*** Some units of this model source +25V DC on pin 5 (the table printed on the fixture near the DMX connector will label pin 5, "Is 18-25VDC"). This high voltage on pin 5 does not conform to the USITT DMX512/1990 standard and may cause damage to other equipment on a DMX network.

CLAY PAKY
Model No. C11069* "Golden Scan 3"

SPOT BEAM SPREAD: (not specified in mfg. catalog)

FLOOD BEAM SPREAD: 11° **

BEAM IRIS RANGE: (not specified in mfg. catalog)

PAN/TILT RANGE: 150° pan, 110° tilt

ACCURACY: 0.6° per step (pan), 0.4° per step (tilt)

GOBOS: 4 positions, and 4-facet prism

COLOR: one 7-color wheel,
      plus 2 color correction filters

DIFFUSION: frost filter

LAMP: HMI 1200 - 1200 W.

DATA CONNECTORS – 1: ground
(5-pin XLR)   2: DMX data, negative (black)
                  3: DMX data, positive (red)
                  4: unused
                  5: ***

WEIGHT: 71.9 lbs. (with mirror head)

SIZE: 48½" x 17⅛" x 14⅛" (L x W x H)

PHOTOMETRICS CHART

(Performance data for this unit are measured using HMI 1200 - 1200 W. lamp)

1:2.5 / 250mm lens (11°)**

	16.4'	32.8'	49.2'	65.6'	82'	Distance
Diameter	3.28'	6.56'	9.84'	13.12'	16.4'	(mf=.2)
Footcandles	2384	596	264.9	149	95.4	(pc=641,201)
Wide-Angle Option (16°)						
Diameter	4.6'	9.19'	13.8'	18.37'	23'	(mf=.28)
Footcandles	1044	261	116	65.25	41.76	(pc=280,794)

1:3.3 / 300mm lens (9.5°)**

	16.4'	32.8'	49.2'	65.6'	82'	Distance
Diameter	2.69'	5.38'	8.07'	10.76'	13.55'	(mf=.16)
Footcandles	2973	743.2	330.3	185.8	119	(pc=799,564)
Wide-Angle Option (13°)						
Diameter	3.77'	7.55'	11.15'	15.09'	19.03'	(mf=.23)
Footcandles	1170	292.5	130	73.1	46.8	(pc=314,683)

* Also available in touring version (TV) with features such as a lamp hour meter, graduated
scale on mounting bracket and mirror head, side handles, etc.
** Beam and field angles not specified in manufacturer's catalog. This angle is calculated
from given distance and diameter data. Also, this unit accepts an optional,
manually-operated, "flip-in," wide-angle lens, as well as an optional narrow-focus lens.
*** Some units of this model source +25V DC on pin 5 (the table printed on the fixture near
the DMX connector will label pin 5, "Is 18-25VDC"). This high voltage on pin 5 does not
conform to the USITT DMX512/1990 standard and may cause damage to other equipment
on a DMX network.

CLAY PAKY
Model No. C11150 "Golden Scan HPE"

SPOT BEAM SPREAD: 16° w/optional lens adapter*

FLOOD BEAM SPREAD: 24.5°*

BEAM IRIS RANGE: (not specified in mfg. catalog)

PAN/TILT RANGE: 150° pan, 110° tilt

ACCURACY: 0.59° per step (CP protocol), 0.43° (DMX)

GOBOS: 4 rotating slots

COLOR: two 7-color wheels,
plus 1 color filter on frost/effects wheel

DIFFUSION: 2 frost filters: +¾° and +8°
5 prisms

LAMP: HMI 1200 W/GS - 1.2 kW.

WEIGHT: 93.03 lbs. (w/ mirror head)

SIZE: 48½" x 17⅛" x 8⅞" (L x W x H)

DATA CONNECTORS – 1: ground
(5-pin XLR)**   2: DMX data, negative (black)
3: DMX data, positive (red)
4: unused
5: ***

PHOTOMETRICS CHART

(Performance data for this unit are measured using HMI 1200 W/GS - 1.2 kW. lamp)

STANDARD LENS (24.5°)

	16.4'	32.8'	49.2'	65.6'	82'	Distance
Diameter	7.2'	14.4'	21.7'	28.9'	36'	(mf=.44)
Footcandles	502	125	56	31	20	(pc=135,556)
With Supplementary Lens (16°)*						
Diameter	4.6'	9.2'	13.8'	17.1'	23'	(mf=.28)
Footcandles	873	218.3	97	54.6	34.9	(pc=234,802)

OPTIONAL LENS (13.5°)*

	16.4'	32.8'	49.2'	65.6'	82'	Distance
Diameter	3.9'	7.8'	11.7'	15.6'	19.5'	(mf=.24)
Footcandles	1096	274	122	69	44	(pc=295,318)
With Supplementary Lens (9°)*						
Diameter	2.6'	5.2'	7.9'	10.5'	13.1'	(mf=.16)
Footcandles	1626	406	181	95	65	(pc=438,136)

* Beam and field angles not specified in manufacturer's catalog. This angle is calculated from given distance and diameter data. Also, this unit accepts an optional, manually-operated, "flip-in" lens as well as an optional narrow-focus lens.

** This unit can also be controlled by the RS 232 protocol, and by 0-10V analog.

*** Some units of this model source +25V DC on pin 5 (the table printed on the fixture near the DMX connector will label pin 5, "Is 18-25VDC"). This high voltage on pin 5 does not conform to the USITT DMX512/1990 standard and may cause damage to other equipment on a DMX network.

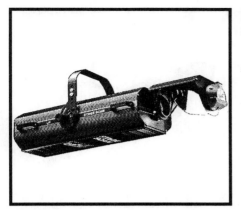

CLAY PAKY
Model No. C11155* "Stage Scan"

SPOT BEAM SPREAD: (not specified in mfg. catalog)

FLOOD BEAM SPREAD: 25° **

BEAM IRIS RANGE: (not specified in mfg. catalog)

PAN/TILT RANGE: 150° pan, 110° tilt

ACCURACY: 0.58° per step (CP protocol), 0.43° (DMX)

GOBOS: 4 positions

COLOR: RGB color mixing

DIFFUSION: 3 frost filters, 1 fixed prism,
          4 rotating prisms

LAMP: HMI 1200 W/GS - 1.2 kW.

DATA CONNECTORS – 1: ground
(5-pin XLR)   2: DMX data, negative (black)
                   3: DMX data, positive (red)
                   4: unused
                   5: ***

WEIGHT: 98.8 lbs.

SIZE: 48⅜" x 17⅛" x 17⅛" (L x W x H)

PHOTOMETRICS CHART

(Performance data for this unit are measured at peak focus using HMI 1200 W/GS - 1.2 kW. lamp)

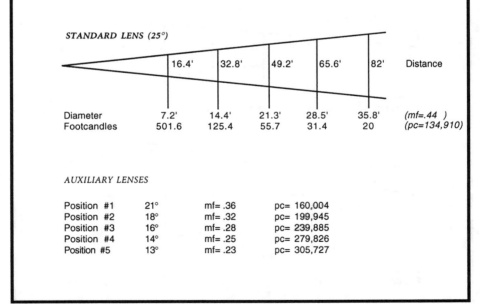

STANDARD LENS (25°)

	16.4'	32.8'	49.2'	65.6'	82'	Distance
Diameter	7.2'	14.4'	21.3'	28.5'	35.8'	*(mf=.44)*
Footcandles	501.6	125.4	55.7	31.4	20	*(pc=134,910)*

AUXILIARY LENSES

Position #1	21°	mf= .36	pc= 160,004
Position #2	18°	mf= .32	pc= 199,945
Position #3	16°	mf= .28	pc= 239,885
Position #4	14°	mf= .25	pc= 279,826
Position #5	13°	mf= .23	pc= 305,727

* This unit operates on 200-240 VAC, 50/60 Hz.
** Five channel-controlled auxiliary lenses (on one wheel) can be used to narrow the beam spread. Diameter multiplying factors (mf) and peak candela multipliers (pc) are noted in the photometrics chart above.
*** Some units of this model source +25V DC on pin 5 (the table printed on the fixture near the DMX connector will label pin 5, "Is 18-25VDC"). This high voltage on pin 5 does not conform to the USITT DMX512/1990 standard and may cause damage to other equipment on a DMX network.

MOVING MIRROR UNITS

CLAY PAKY
Model No. C11200* "Super Scan Zoom"

SPOT BEAM SPREAD: 8°

FLOOD BEAM SPREAD: 16°

BEAM IRIS RANGE: .5° to 16°

PAN/TILT RANGE: 150° pan, 110° tilt

ACCURACY: 0.59° per step (CP protocol), 0.43° (DMX)

GOBOS: 4 rotating slots

COLOR: RGB color mixing, plus one 7-color wheel
(includes 2 color correction filters)

DIFFUSION: 2 frost filters: +1° and +9.5°

LAMP: HMI 1200 W/GS - 1.2 kW.

WEIGHT: 105.16 lbs.

SIZE: 57⅞" x 17" x 14⅛" (L x W x H)

DATA CONNECTORS – 1: ground
(5-pin XLR)**   2: DMX data, negative (black)
3: DMX data, positive (red)
4: unused
5: ***

PHOTOMETRICS CHART
(Performance data for this unit are measured using HMI 1200 W/GS - 1.2 kW. lamp)

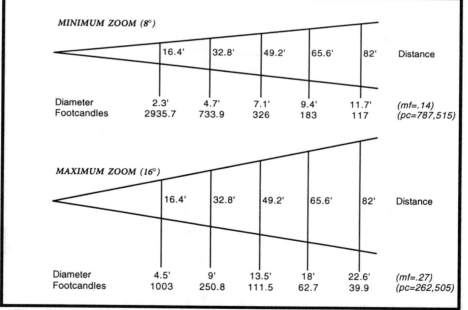

* This unit operates on 200-240 VAC, 50/60 Hz.

** This unit can also be controlled by the RS232 protocol, and by 0-10V analog.

*** Some units of this model source +25V DC on pin 5 (the table printed on the fixture near the DMX connector will label pin 5, "Is 18-25VDC"). This high voltage on pin 5 does not conform to the USITT DMX512/1990 standard and may cause damage to other equipment on a DMX network.

HIGH END SYSTEMS
Model No. Cyberlight* (narrow)

SPOT BEAM SPREAD: 8.5° (narrow zoom position)

FLOOD BEAM SPREAD: 12.5° (wide zoom position)**

BEAM IRIS RANGE: (no iris in this unit)

PAN/TILT RANGE: 170° pan, 110° tilt

ACCURACY:LWR protocol:
0.002° /step (pan), 0.001°/step (tilt)
DMX protocol:
0.66° (pan), 1.7° (tilt)

GOBOS: 8 slots

COLOR: CMY color mixing, plus one 7-color wheel,
plus 3 color correction filters

DIFFUSION: frost flags

LAMP: MSR 1200 (Philips) - 1.2 kW.

WEIGHT: 100.5 lbs.

SIZE: 44" x 16" x 17⅜" (L x W x H)

DATA CONNECTORS – 1: ground
(3-pin XLR)   2: DMX data, negative (black)
                     3: DMX data, positive (red)

PHOTOMETRICS CHART
(Performance data for this unit are measured using MSR 1200 - 1.2 kW. lamp)

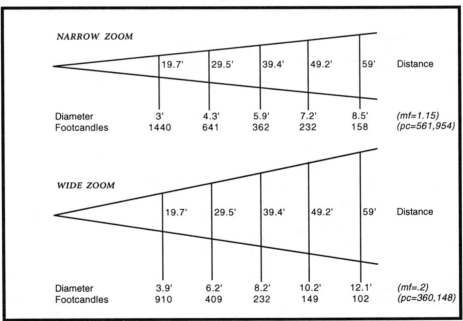

NARROW ZOOM

	19.7'	29.5'	39.4'	49.2'	59'	Distance
Diameter	3'	4.3'	5.9'	7.2'	8.5'	(mf=1.15)
Footcandles	1440	641	362	232	158	(pc=561,954)

WIDE ZOOM

	19.7'	29.5'	39.4'	49.2'	59'	Distance
Diameter	3.9'	6.2'	8.2'	10.2'	12.1'	(mf=.2)
Footcandles	910	409	232	149	102	(pc=360,148)

* Unit can also be purchased as a "Cyberlight CX" or "Cyberlight SV." The CX uses a manually adjustable zoom and the SV is 75% quieter.
** This page describes the unit with optional narrow angle lens. The effects wheel includes a supplementary lens, but manufacturer's catalog did not have photometrics data for it.

HIGH END SYSTEMS
Model No. Cyberlight* (standard)

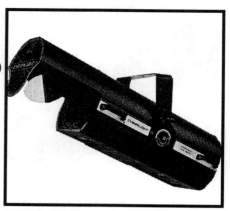

SPOT BEAM SPREAD: 12.5° (narrow zoom position)

FLOOD BEAM SPREAD: 26° (wide zoom position)**

BEAM IRIS RANGE: (no iris in this unit)

PAN/TILT RANGE: 170° pan, 110° tilt

ACCURACY: LWR protocol:
  0.002°/step (pan), 0.001°/step (tilt)
  DMX protocol:
  0.66° (pan), 1.7° (tilt)

GOBOS: 8 slots

COLOR: CMY color mixing, plus one 7-color wheel,
  plus 3 color correction filters

DIFFUSION: frost flags

LAMP: MSR 1200 (Philips) - 1.2 kW.

WEIGHT: 100.5 lbs.

SIZE: 44" x 16" x 17⅜" (L x W x H)

DATA CONNECTORS – 1: ground
(3-pin XLR)   2: DMX data, negative (black)
                  3: DMX data, positive (red)

PHOTOMETRICS CHART

(Performance data for this unit are measured using MSR 1200 - 1.2 kW. lamp)

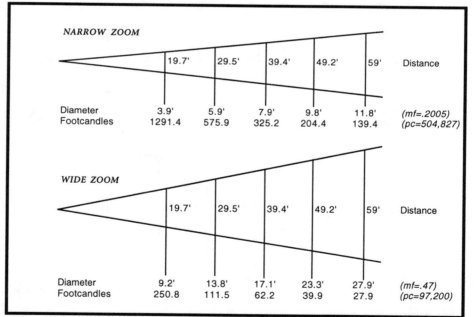

NARROW ZOOM

	19.7'	29.5'	39.4'	49.2'	59'	Distance
Diameter	3.9'	5.9'	7.9'	9.8'	11.8'	(mf=.2005)
Footcandles	1291.4	575.9	325.2	204.4	139.4	(pc=504,827)

WIDE ZOOM

	19.7'	29.5'	39.4'	49.2'	59'	Distance
Diameter	9.2'	13.8'	17.1'	23.3'	27.9'	(mf=.47)
Footcandles	250.8	111.5	62.2	39.9	27.9	(pc=97,200)

* Unit can also be purchased as a "Cyberlight CX" or "Cyberlight SV." The CX uses a manually adjustable zoom and the SV is 75% quieter.

** This page describes the unit with the standard lens. The effects wheel includes a supplementary lens, but manufacturer's catalog did not have photometrics data for it.

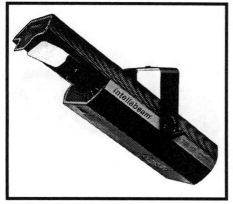

HIGH END SYSTEMS
Model No. Intellabeam 700HX

BEAM SPREAD: 12.5° *

BEAM IRIS RANGE: (no iris in this unit)

PAN/TILT RANGE: 170° pan, 110° tilt

ACCURACY: .66° per step (LWR protocol), 0.43° (DMX)

GOBOS: 11 slots

COLOR: two 11-color wheels (one has bi-colored filters)

DIFFUSION: (not specified in mfg. catalog)

LAMP: MSR 700 (Philips) -700 W.

WEIGHT: 66 lbs.

SIZE: 36½" x 16" x 13¾" (L x W x H)

DATA CONNECTORS – 1: ground
(3-pin XLR)   2: DMX data, negative (black)
                    3: DMX data, positive (red)

PHOTOMETRICS CHART

(Performance data for this unit are measured using MSR 700 - 700 W. lamp)

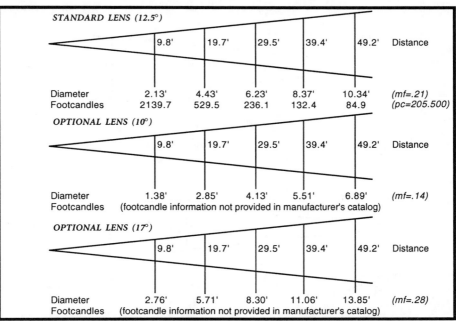

STANDARD LENS (12.5°)

	9.8'	19.7'	29.5'	39.4'	49.2'	Distance
Diameter	2.13'	4.43'	6.23'	8.37'	10.34'	(mf=.21)
Footcandles	2139.7	529.5	236.1	132.4	84.9	(pc=205.500)

OPTIONAL LENS (10°)

	9.8'	19.7'	29.5'	39.4'	49.2'	Distance
Diameter	1.38'	2.85'	4.13'	5.51'	6.89'	(mf=.14)
Footcandles	(footcandle information not provided in manufacturer's catalog)					

OPTIONAL LENS (17°)

	9.8'	19.7'	29.5'	39.4'	49.2'	Distance
Diameter	2.76'	5.71'	8.30'	11.06'	13.85'	(mf=.28)
Footcandles	(footcandle information not provided in manufacturer's catalog)					

* The standard configuration of this unit is with the 12.5° lens. Optional lenses include a 10° (nominal) and a 17° (nominal) lens.

MARTIN PROFESSIONAL A/S
Model No. PAL 1200

SPOT BEAM SPREAD: 15° (narrow zoom position)

FLOOD BEAM SPREAD: 26° (wide zoom position)

BEAM IRIS RANGE: (no iris in this unit)

PAN/TILT RANGE: 287° pan, 85° tilt

ACCURACY: 1.1° per step (Martin), .34° (DMX)

GOBOS: 5 slots, plus wheel with 4-leaf framing shutters

COLOR: CMY color mixing, plus one 4-color wheel
(includes 1 color correction filter)

DIFFUSION: 1 frost (variable), approx. + 9.5°
when fully engaged

LAMP: MSR 1200 (Philips) - 1.2 KW.

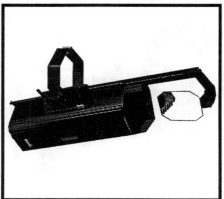

WEIGHT: 135.6 lbs.

SIZE: 52⅜" x 17³⁄₁₆" x 26¼" (L x W x H)

DATA CONNECTORS –1: ground
(3-pin XLR)   2: DMX data, positive (red)
    3: DMX data, negative (black)

PHOTOMETRICS CHART
(Performance data for this unit are measured using MSR 1200 - 1.2 KW. lamp.)

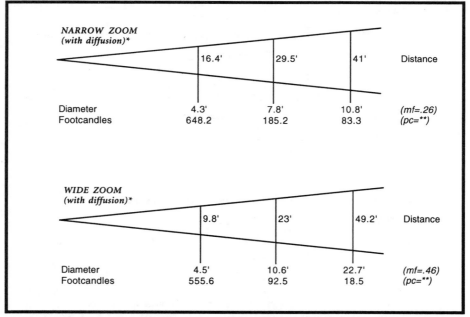

NARROW ZOOM (with diffusion)*

	16.4'	29.5'	41'	Distance
Diameter	4.3'	7.8'	10.8'	(mf=.26)
Footcandles	648.2	185.2	83.3	(pc=**)

WIDE ZOOM (with diffusion)*

	9.8'	23'	49.2'	Distance
Diameter	4.5'	10.6'	22.7'	(mf=.46)
Footcandles	555.6	92.5	18.5	(pc=**)

* This unit has a system to smooth the hot spot which uses a diffusion filter and condenser lens in different combinations (see also model "Mac 1200").

** Footcandle data given in manufacturer's catalog is not consistent enough to calculate a reliable peak candela (pc) figure for all throw distances.

406

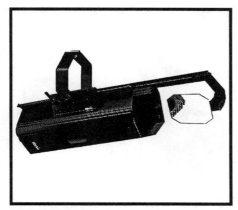

MARTIN PROFESSIONAL A/S
Model No. PAL 1200-E*

SPOT BEAM SPREAD: 15° (narrow zoom position)

FLOOD BEAM SPREAD: 26° (wide zoom position)

BEAM IRIS RANGE: .75° to 26°

PAN/TILT RANGE: 287° pan, 85° tilt

ACCURACY: 1.1° per step (Martin), .34° (DMX)

GOBOS: 10 slots on two wheels, including 1 rotating
prism, plus wheel with 4-leaf framing shutters

COLOR: CMY color mixing, plus one 4-color wheel
(includes 1 color correction filter)

DIFFUSION: 1 frost (variable), approx. + 9.5°
when fully engaged

LAMP: MSR 1200 (Philips) - 1.2 KW.

WEIGHT: 135.6 lbs.

SIZE: 52⅜" x 17³⁄₁₆" x 26¼" (L x W x H)

DATA CONNECTORS –1: ground
(3-pin XLR)   2: DMX data, positive (red)
3: DMX data, negative (black)

PHOTOMETRICS CHART
(Performance data for this unit are measured using MSR 1200 - 1.2 KW. lamp.)

NARROW ZOOM
(with diffusion)*

	16.4'	29.5'	41'	Distance
Diameter	4.3'	7.8'	10.8'	(mf=.26)
Footcandles	648.2	185.2	83.3	(pc=***)

WIDE ZOOM
(with diffusion)*

	9.8'	23'	49.2'	Distance
Diameter	4.5'	10.6'	22.7'	(mf=.46)
Footcandles	555.6	92.5	18.5	(pc=***)

* The only difference between this unit and the PAL 1200 is the iris and additional gobo wheels.

** This unit has a system to smooth the hot spot which uses a diffusion filter and condenser lens in different combinations (see also model "Mac 1200").

*** Footcandle data given in manufacturer's catalog is not consistent enough to calculate a reliable peak candela (pc) figure for all throw distances.

MARTIN PROFESSIONAL A/S
Model No. Roboscan Pro 218

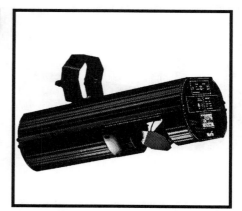

BEAM SPREAD: 11.5° *

BEAM IRIS RANGE: (no iris in this unit)

PAN/TILT RANGE: 180° pan, 90° tilt

ACCURACY: .70° (pan), .35° (tilt) (DMX)

GOBOS: 17 slots

COLOR: one 17-color wheel
        (includes 2 multi-color filters)

DIFFUSION: none

LAMP: MSD 200 (Philips) - 200 W.

WEIGHT: 30.8 lbs.

SIZE: 21" x 11" x 7½" (L x W x H)

DATA CONNECTORS – 1: ground
(3-pin XLR - Martin   2: DMX data, positive (red)
Interface required)   3: DMX data, negative (black)

PHOTOMETRICS CHART

(Performance data for this unit are measured using MSD 200 - 200 W. lamp.)

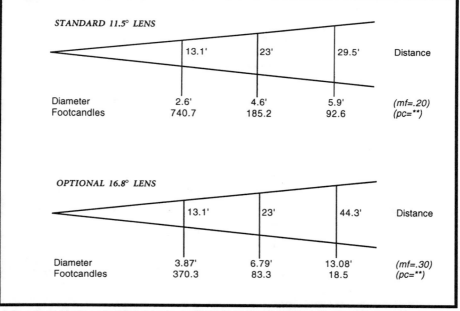

STANDARD 11.5° LENS

	13.1'	23'	29.5'	Distance
Diameter	2.6'	4.6'	5.9'	(mf=.20)
Footcandles	740.7	185.2	92.6	(pc=**)

OPTIONAL 16.8° LENS

	13.1'	23'	44.3'	Distance
Diameter	3.87'	6.79'	13.08'	(mf=.30)
Footcandles	370.3	83.3	18.5	(pc=**)

* An optional 16.8° lens for this unit is also available and is listed in the photometrics chart.
** Footcandle data given in manufacturer's catalog is not consistent enough to calculate a reliable
      peak candela (pc) figure for all throw distances.

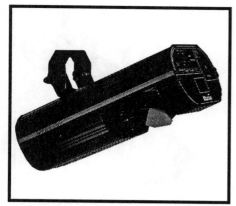

MARTIN PROFESSIONAL A/S
Model No. Roboscan Pro 518

BEAM SPREAD: 12.5°*

BEAM IRIS RANGE: (no iris in this unit)

PAN/TILT RANGE: 176° pan, 85° tilt

ACCURACY: .028° (pan), .051° (tilt) (Martin protocol)

GOBOS: 5 slots

COLOR: one 17-color wheel (includes 2 multi-color filters), plus 1 color correction filter

DIFFUSION: frost filter, prism

LAMP: MSD 200 (Philips) - 200 W.

WEIGHT: 31 lbs.

SIZE: 20⅞" x 11" x 7⁵⁄₁₆" (L x W x H)

DATA CONNECTORS –1: ground
(3-pin XLR - Martin   2: DMX data, positive (red)
Interface required)   3: DMX data, negative (black)

PHOTOMETRICS CHART

(Performance data for this unit are measured using MSD 200 - 200 W. lamp.)

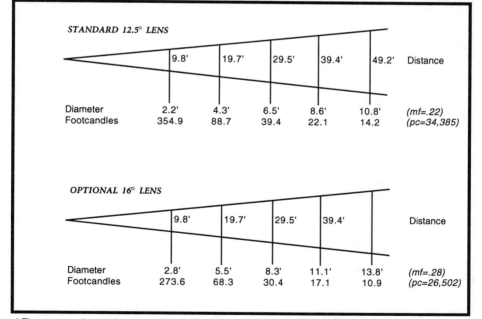

STANDARD 12.5° LENS

	9.8'	19.7'	29.5'	39.4'	49.2'	Distance
Diameter	2.2'	4.3'	6.5'	8.6'	10.8'	*(mf=.22)*
Footcandles	354.9	88.7	39.4	22.1	14.2	*(pc=34,385)*

OPTIONAL 16° LENS

	9.8'	19.7'	29.5'	39.4'		Distance
Diameter	2.8'	5.5'	8.3'	11.1'	13.8'	*(mf=.28)*
Footcandles	273.6	68.3	30.4	17.1	10.9	*(pc=26,502)*

* There are two lenses listed here: the standard 12.5° lens, and the optional 16° lens.

MARTIN PROFESSIONAL A/S
Model: Roboscan Pro 1220 CMYR

SPOT BEAM SPREAD: (not specified in mfg. catalog)

FLOOD BEAM SPREAD: 18.5°

BEAM IRIS RANGE: .925° to 18.5°

PAN/TILT RANGE: 176° pan, 85° tilt

ACCURACY: .028° (pan), .056° (tilt) (Martin protocol)

GOBOS: 13 slots

COLOR: CMY color mixing, plus two 9-color wheels

DIFFUSION: frost filter, 3 prisms

LAMP: MSR 1200 (Philips) - 1.2 KW.
        HTI 1200 (OSRAM) - 1.2 KW. *(cf= 1.0)*

WEIGHT:117 lbs.*

SIZE: 43⅝" x 12⅜" x 11¼" (L x W x H)

DATA CONNECTORS – 1: ground
(3-pin XLR - Martin   2: DMX data, positive (red)
Interface required)   3: DMX data, negative (black)

PHOTOMETRICS CHART
(Performance data for this unit are measured using MSR 1200 - 1.2 KW. lamp.)

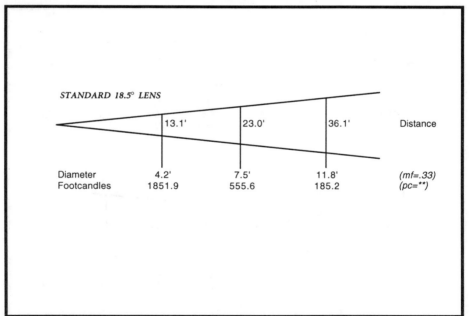

STANDARD 18.5° LENS

	13.1'	23.0'	36.1'	Distance
Diameter	4.2'	7.5'	11.8'	*(mf=.33)*
Footcandles	1851.9	555.6	185.2	*(pc=**)*

* Unit is also available in a "Studio" (ST) configuration, which uses an outboard ballast (28 lbs.). Unit weight drops to 103 lbs.

** Footcandle data given in manufacturer's catalog is not consistent enough to calculate a reliable peak candela (pc) figure for all throw distances.

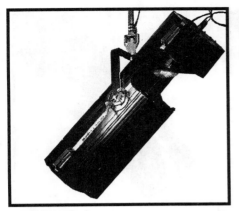

MARTIN PROFESSIONAL A/S
Model: Roboscan Pro 1220 RPR

SPOT BEAM SPREAD: (not specified in mfg. catalog)

FLOOD BEAM SPREAD: 12.5°

BEAM IRIS RANGE: .625° to 13.5°

PAN/TILT RANGE: 176° pan, 85° tilt

ACCURACY: .028° (pan), .056° (tilt) (Martin protocol)

GOBOS: 13 slots

COLOR: two 9-color wheels

DIFFUSION: 4 rotating prisms

LAMP: MSR 1200 (Philips) - 1.2 KW.
        HTI 1200 (OSRAM) - 1.2 KW. *(cf= 1.0)*

WEIGHT:117 lbs.*

SIZE: 43⅝" x 12⅜" x 11¼" (L x W x H)

DATA CONNECTORS –1: ground
(3-pin XLR - Martin   2: DMX data, positive (red)
Interface required)   3: DMX data, negative (black)

PHOTOMETRICS CHART
(Performance data for this unit are measured using MSR 1200 - 1.2 KW. lamp.)

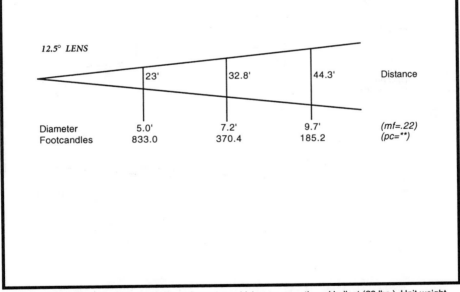

12.5° LENS

	23'	32.8'	44.3'	Distance
Diameter	5.0'	7.2'	9.7'	*(mf=.22)*
Footcandles	833.0	370.4	185.2	*(pc=**)*

* Unit is also available in a "Studio" (ST) configuration, which uses an outboard ballast (28 lbs.). Unit weight drops to 103 lbs.
** Footcandle data given in manufacturer's catalog is not consistent enough to calculate a reliable peak candela (pc) figure for all throw distances.

MARTIN PROFESSIONAL A/S
Model: Roboscan Pro 1220 XR

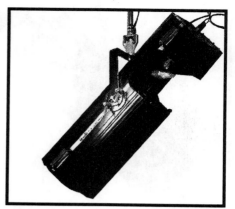

SPOT BEAM SPREAD: (not specified in mfg. catalog)

FLOOD BEAM SPREAD: 18.5°*

BEAM IRIS RANGE: .925° to 18.5°

PAN/TILT RANGE: 176° pan, 85° tilt

ACCURACY: .028° (pan), .056° (tilt) (Martin protocol)

GOBOS: 13 slots

COLOR: one 9-color wheel, plus 1 color correction filter on effects wheel

DIFFUSION: frost filter, 3 prisms

LAMP: MSR 1200 (Philips) - 1.2 KW.
HTI 1200 (OSRAM) - 1.2 KW. *(cf= 1.0)*

WEIGHT: 117 lbs.**

SIZE: 43⅝" x 12⅜" x 11¼" (L x W x H)

DATA CONNECTORS – 1: ground
(3-pin XLR - Martin   2: DMX data, positive (red)
Interface required)   3: DMX data, negative (black)

PHOTOMETRICS CHART

(Performance data for this unit are measured using MSR 1200 - 1.2 KW. lamp.)

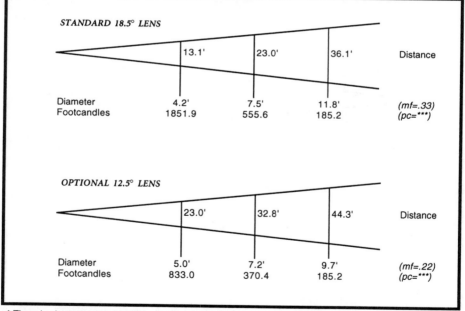

STANDARD 18.5° LENS

	13.1'	23.0'	36.1'	Distance
Diameter	4.2'	7.5'	11.8'	*(mf=.33)*
Footcandles	1851.9	555.6	185.2	*(pc=***)*

OPTIONAL 12.5° LENS

	23.0'	32.8'	44.3'	Distance
Diameter	5.0'	7.2'	9.7'	*(mf=.22)*
Footcandles	833.0	370.4	185.2	*(pc=***)*

* There is also an optional 12.5° lens, which is listed in the photometrics chart above.
** Unit is also available in a "Studio" (ST) configuration, which uses an outboard ballast (28 lbs.).

412

*** Footcandle data given in manufacturer's catalog is not consistent enough to calculate a reliable peak candela (pc) figure for all throw distances.

ALTMAN STAGE LIGHTING
Model No. 660

SPOT BEAM SPREAD: 8° beam, 18° field

FLOOD BEAM SPREAD: 8° beam, 32° field

LAMP BASE TYPE: Medium Prefocus

REFLECTOR DIAMETER: 10"

STANDARD LAMP: BTN - 750 W.

OTHER LAMPS: 250 T20/47 - 250 W. *(cf=.27)*
DNW - 500 W. *(cf=.59)*
BTP - 750 W. *(cf=1.2)*
1 MT20P/SP - 1 KW. *(cf=1.4)*
BTL - 750 W. *(cf=.65)*
BTR - 1 KW. *(cf=1.6)*

ACCESSORIES AVAILABLE:
Color Frame

WEIGHT: (not specified in manufacturer's catalog)

SIZE: 11½" x 13" x 16" (L x W x H)

PHOTOMETRICS CHART
(Performance data for this unit are measured using BTN - 750 W. lamp.)

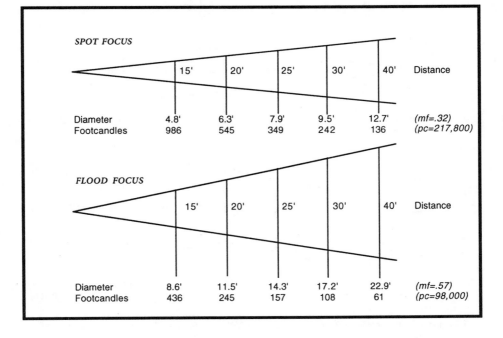

SPOT FOCUS

	15'	20'	25'	30'	40'	Distance
Diameter	4.8'	6.3'	7.9'	9.5'	12.7'	*(mf=.32)*
Footcandles	986	545	349	242	136	*(pc=217,800)*

FLOOD FOCUS

	15'	20'	25'	30'	40'	Distance
Diameter	8.6'	11.5'	14.3'	17.2'	22.9'	*(mf=.57)*
Footcandles	436	245	157	108	61	*(pc=98,000)*

BERKEY COLORTRAN
Model No. 216-072*

SPOT BEAM SPREAD: 5.7° beam, 11° field

FLOOD BEAM SPREAD: 14° beam, 20° field

LAMP BASE TYPE: Medium Prefocus

REFLECTOR DIAMETER: 10"

STANDARD LAMP: BTL - 500 W.

OTHER LAMPS: BTP - 750 W. *(cf=1.8)*
BTR - 1 KW. *(cf=2.4)*

ACCESSORIES AVAILABLE:
Color Frame

WEIGHT: 10 lbs.

SIZE: 11½" x 14¼" x 21⅝" (L x W x H)

PHOTOMETRICS CHART
(Performance data for this unit are measured using BTL - 500 W. lamp.)

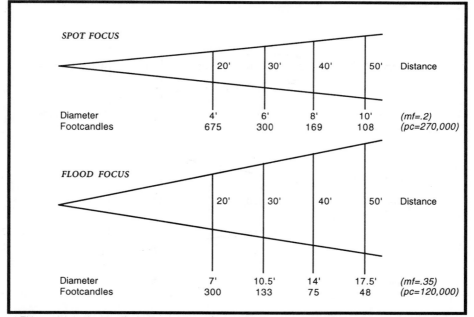

SPOT FOCUS

		20'	30'	40'	50'	Distance
Diameter		4'	6'	8'	10'	*(mf=.2)*
Footcandles		675	300	169	108	*(pc=270,000)*

FLOOD FOCUS

		20'	30'	40'	50'	Distance
Diameter		7'	10.5'	14'	17.5'	*(mf=.35)*
Footcandles		300	133	75	48	*(pc=120,000)*

* This model can be identified with model numbers -072, -075, -076, or -077 depending on the wire leads and connectors supplied.

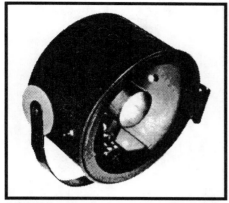

CENTURY LIGHTING
Model No. 1515

SPOT BEAM SPREAD: 15° field*

FLOOD BEAM SPREAD: 25° field*

LAMP BASE TYPE: Medium Prefocus

REFLECTOR DIAMETER: 10"

STANDARD LAMP: 750T20/C13 - 750 W.

OTHER LAMPS: 500T20/C13 -   500 W. *(cf=.59)*
500T20/C13D - 500 W. *(cf=.65)*

WEIGHT: (not specified in manufacturer's catalog)

SIZE: 9¾" x 15" x 19" (L x W x H)

ACCESSORIES AVAILABLE:
Color Frame

PHOTOMETRICS CHART
(Performance data for this unit are measured using 750T20/C13 - 750 W. lamp.)

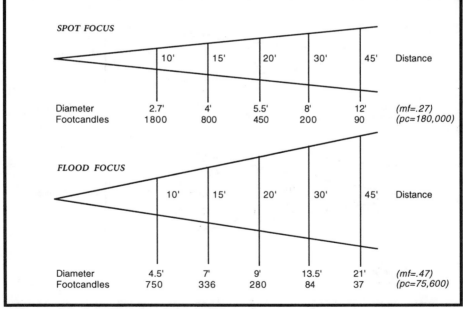

SPOT FOCUS

	10'	15'	20'	30'	45'	Distance
Diameter	2.7'	4'	5.5'	8'	12'	*(mf=.27)*
Footcandles	1800	800	450	200	90	*(pc=180,000)*

FLOOD FOCUS

	10'	15'	20'	30'	45'	Distance
Diameter	4.5'	7'	9'	13.5'	21'	*(mf=.47)*
Footcandles	750	336	280	84	37	*(pc=75,600)*

* Beam angles not specified in manufacturer's catalog.

BEAM PROJECTOR

CENTURY STRAND
Model No. 4121

SPOT BEAM SPREAD: 15° field*

FLOOD BEAM SPREAD: 25° field*

LAMP BASE TYPE: Medium Prefocus

REFLECTOR DIAMETER: 10"

STANDARD LAMP: BTR - 1 KW.

OTHER LAMPS: BTL - 500 W. *(cf=.39)*
BTN - 750 W. *(cf=.62)*
500T20/64 - 500 W. *(cf=.35)*
750T20P/SP - 750 W. *(cf=.60)*

ACCESSORIES AVAILABLE:
Color Frame

WEIGHT: (not specified in manufacturer's catalog)

SIZE: 11" x 12" x 15¾" (L x W x H)

PHOTOMETRICS CHART
(Performance data for this unit are measured using BTR - 1 KW. lamp.)

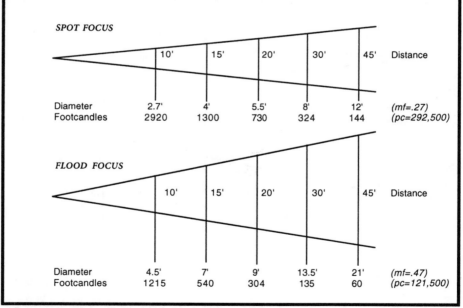

SPOT FOCUS

	10'	15'	20'	30'	45'	Distance
Diameter	2.7'	4'	5.5'	8'	12'	*(mf=.27)*
Footcandles	2920	1300	730	324	144	*(pc=292,500)*

FLOOD FOCUS

	10'	15'	20'	30'	45'	Distance
Diameter	4.5'	7'	9'	13.5'	21'	*(mf=.47)*
Footcandles	1215	540	304	135	60	*(pc=121,500)*

* Beam angles not specified in manufacturer's catalog.

ELECTRO CONTROLS
Model No. 7701A

SPOT BEAM SPREAD: 9.5° beam, 22.5° field

FLOOD BEAM SPREAD: 37° beam, 47° field

LAMP BASE TYPE: Medium 2-Pin

REFLECTOR DIAMETER: 10"

STANDARD LAMP: FEL - 1 KW.

OTHER LAMPS: EHD - 500 W. *(cf=.39)*
EHG - 750 W. *(cf=.56)*

WEIGHT: (not specified in manufacturer's catalog)

SIZE: 11⅜" x 14½" (L x W)

ACCESSORIES AVAILABLE:
Color Frame (12⅛" sq.)

PHOTOMETRICS CHART
(Performance data for this unit are measured using FEL - 1 KW. lamp.)

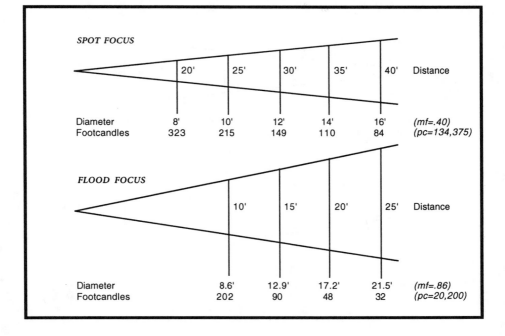

SPOT FOCUS

	20'	25'	30'	35'	40'	Distance
Diameter	8'	10'	12'	14'	16'	*(mf=.40)*
Footcandles	323	215	149	110	84	*(pc=134,375)*

FLOOD FOCUS

	10'	15'	20'	25'	Distance
Diameter	8.6'	12.9'	17.2'	21.5'	*(mf=.86)*
Footcandles	202	90	48	32	*(pc=20,200)*

KLIEGL BROS.
Model No. 1143

SPOT BEAM SPREAD: 13° field*

FLOOD BEAM SPREAD: 20° field*

LAMP BASE TYPE: Mogul Prefocus

REFLECTOR DIAMETER: 15"

STANDARD LAMP: CWZ - 1.5 KW.

OTHER LAMPS: BVT/BVV - 1 KW. *(cf=.64)*
             BVW -   2 KW. *(cf=1.53)*

ACCESSORIES AVAILABLE:
Color Frame

WEIGHT: (not specified in manufacturer's catalog)

SIZE: 11½" x 15" x 22¾" (L x W x H)

PHOTOMETRICS CHART
(Performance data for this unit are measured using CWZ - 1500 W. lamp.)

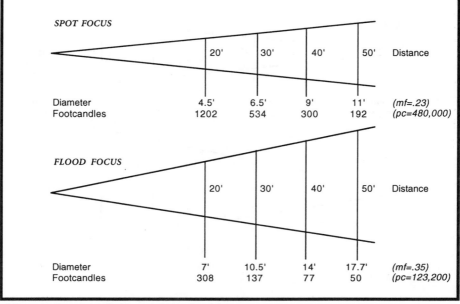

SPOT FOCUS

	20'	30'	40'	50'	Distance
Diameter	4.5'	6.5'	9'	11'	*(mf=.23)*
Footcandles	1202	534	300	192	*(pc=480,000)*

FLOOD FOCUS

	20'	30'	40'	50'	Distance
Diameter	7'	10.5'	14'	17.7'	*(mf=.35)*
Footcandles	308	137	77	50	*(pc=123,200)*

* Beam angles not specified in manufacturer's catalog.

KLIEGL BROS.
Model No. 3440

SPOT BEAM SPREAD: 10° field*

FLOOD BEAM SPREAD: 22° field*

LAMP BASE TYPE: Medium Bipost

REFLECTOR DIAMETER: 11"

STANDARD LAMP: EGT - 1 KW.

OTHER LAMPS: EGN - 500 W. *(cf=.46)*
EGR - 750 W. *(cf=.74)*

WEIGHT: (not specified in manufacturer's catalog)

SIZE: 10¾" x 13" x 17¼" (L x W x H)

ACCESSORIES AVAILABLE:
Color Frame

PHOTOMETRICS CHART
(Performance data for this unit are measured using EGT - 1 KW. lamp.)

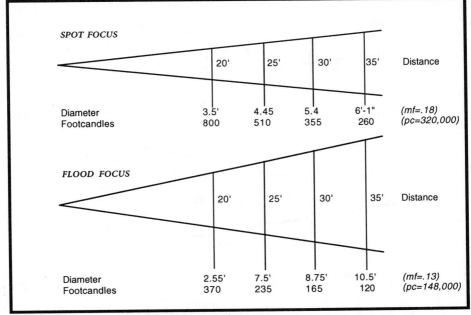

SPOT FOCUS

	20'	25'	30'	35'	Distance
Diameter	3.5'	4.45	5.4	6'-1"	*(mf=.18)*
Footcandles	800	510	355	260	*(pc=320,000)*

FLOOD FOCUS

	20'	25'	30'	35'	Distance
Diameter	2.55'	7.5'	8.75'	10.5'	*(mf=.13)*
Footcandles	370	235	165	120	*(pc=148,000)*

* Beam angles not specified in manufacturer's catalog.

LIGHTING & ELECTRONICS
Model No. 63-32

SPOT BEAM SPREAD: 17° (beam)*

FLOOD BEAM SPREAD: 23° (beam)*

LAMP BASE TYPE: Medium Prefocus

STANDARD LAMP: BFE - 750 W.

OTHER LAMPS: DNW - 500 W. *(cf=.59)*

ACCESSORIES AVAILABLE:
Color Frame

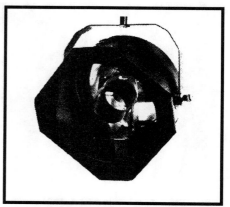

WEIGHT: (not specified in manufacturer's catalog)

SIZE: 9¾" x 14" (L x W)

PHOTOMETRICS CHART
(Performance data for this unit are measured using BFE - 750 W. lamp.)

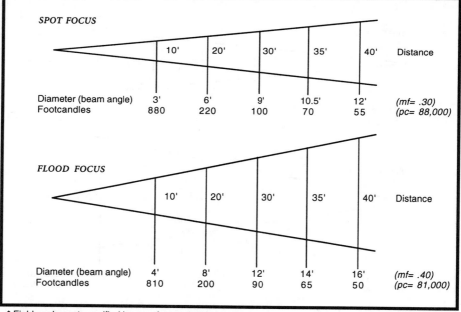

SPOT FOCUS

	10'	20'	30'	35'	40'	Distance
Diameter (beam angle)	3'	6'	9'	10.5'	12'	*(mf= .30)*
Footcandles	880	220	100	70	55	*(pc= 88,000)*

FLOOD FOCUS

	10'	20'	30'	35'	40'	Distance
Diameter (beam angle)	4'	8'	12'	14'	16'	*(mf= .40)*
Footcandles	810	200	90	65	50	*(pc= 81,000)*

* Field angles not specified in manufacturer's catalog.

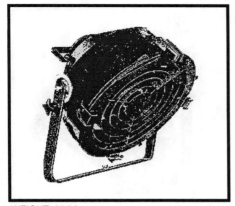

LUDWIG PANI
Model No. P1001

SPOT BEAM SPREAD: 5° beam, 8° field

FLOOD BEAM SPREAD: 7.5° beam, 12° field

LAMP BASE TYPE: K39d

REFLECTOR DIAMETER: 14.17"

STANDARD LAMP: 1 KW/24V, K39d, mirror domed*

OTHER LAMPS: *

ACCESSORIES AVAILABLE:
Color Frame, 220V. - 24V. transformer

WEIGHT: 25.2 kg.

SIZE: 385mm x 500mm x 480mm (L x W x H)

PHOTOMETRICS CHART
(Performance data for this unit are measured using 1 KW/24V, K39d, mirror domed lamp.)

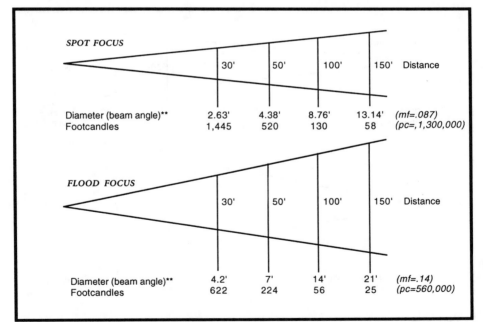

SPOT FOCUS

	30'	50'	100'	150'	Distance
Diameter (beam angle)**	2.63'	4.38'	8.76'	13.14'	(mf=.087)
Footcandles	1,445	520	130	58	(pc=,1,300,000)

FLOOD FOCUS

	30'	50'	100'	150'	Distance
Diameter (beam angle)**	4.2'	7'	14'	21'	(mf=.14)
Footcandles	622	224	56	25	(pc=560,000)

* Catalog specifies three lamps from different manufacturers: Radium 578 K,kv 21202419, Osram 555622 Kku, and Philips 7064 K/02. All three are 1 KW/24V, K39d, mirror domed, with 23,000 initial lumens and 100 hrs. life.

** Photometric data in manufacturer's catalog is given in metric units, and was converted to English units for this book. Also, these diameter and footcandle figures were calculated from beam angles and peak candela data given in catalog. See Introduction for more information about calculating photometric data.

BEAM PROJECTOR

LUDWIG PANI
Model No. P250

SPOT BEAM SPREAD: 6° beam, 10.5° field

FLOOD BEAM SPREAD: 14° beam, 23.5° field

LAMP BASE TYPE: E27

REFLECTOR DIAMETER: 7.09"

STANDARD LAMP: 250 W/24V, E27, mirror domed*

OTHER LAMPS: *

ACCESSORIES AVAILABLE:
Color Frame (185mm sq.), 220V. - 24V.
transformer

WEIGHT: 6.12 kg.

SIZE: 350mm x 210mm x 370mm (L x W x H)

PHOTOMETRICS CHART

(Performance data for this unit are measured using 250 W/24V, E27, mirror domed lamp.)

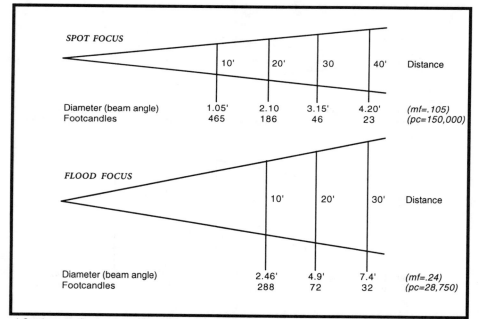

SPOT FOCUS

	10'	20'	30	40'	Distance
Diameter (beam angle)	1.05'	2.10	3.15'	4.20'	(mf=.105)
Footcandles	465	186	46	23	(pc=150,000)

FLOOD FOCUS

	10'	20'	30'	Distance
Diameter (beam angle)	2.46'	4.9'	7.4'	(mf=.24)
Footcandles	288	72	32	(pc=28,750)

* Catalog specifies two lamps from different manufacturers: Radium 564 A,kv 21207704 and Osram 555440
Aku. Both are 250 W/24V, E27, mirror domed, with 5,500 initial lumens and 100 hrs. life.
** Photometric data in manufacturer's catalog is given in metric units, and was converted to English units for
this book. Also, these diameter and footcandle figures were calculated from beam angles and peak candela
data given in catalog. See Introduction for more information about calculating photometric data.

LUDWIG PANI
Model No. P500

SPOT BEAM SPREAD: 5° beam, 9° field

FLOOD BEAM SPREAD: 10° beam, 13.5° field

LAMP BASE TYPE: E40

REFLECTOR DIAMETER: 9.84"

STANDARD LAMP: 500 W/24V, E40, mirror domed*

OTHER LAMPS: *

ACCESSORIES AVAILABLE:
Color Frame (275mm sq.), 220V. - 24V. transformer

WEIGHT: 13.3 kg.

SIZE: 370mm x 360mm x 390mm (L x W x H)

PHOTOMETRICS CHART
(Performance data for this unit are measured using 500 W/24V, E40, mirror domed lamp.)

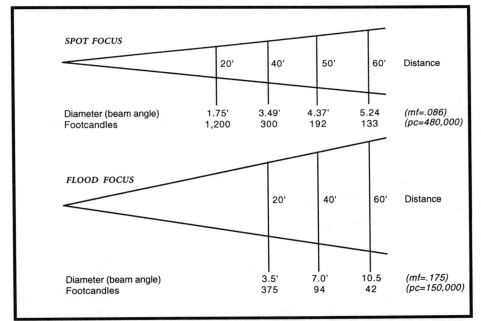

SPOT FOCUS	20'	40'	50'	60'	Distance
Diameter (beam angle)	1.75'	3.49'	4.37'	5.24	*(mf=.086)*
Footcandles	1,200	300	192	133	*(pc=480,000)*

FLOOD FOCUS	20'	40'	60'	Distance
Diameter (beam angle)	3.5'	7.0'	10.5	*(mf=.175)*
Footcandles	375	94	42	*(pc=150,000)*

* Catalog specifies three lamps from different manufacturers: Radium 570 B,kv 21201618, Osram 555529 Dku, and Phillips 162 G/02. All three are 500 W/24V, E40, mirror domed, with 12,600 initial lumens and 100 hrs. life.

** Photometric data in manufacturer's catalog is given in metric units, and was converted to English units for this book. Also, these diameter and footcandle figures were calculated from beam angles and peak candela data given in catalog. See Introduction for more information about calculating photometric data.

BEAM PROJECTOR

STRAND CENTURY
Model No. 4122

SPOT BEAM SPREAD: 15° field*

FLOOD BEAM SPREAD: 25° field*

LAMP BASE TYPE: Medium Prefocus

REFLECTOR DIAMETER: 10"

STANDARD LAMP: BTR - 1 KW.

OTHER LAMPS: BTL -   500 W. *(cf=.39)*
           BTN -   750 W. *(cf=.62)*
           500T20/64 -   500 W. *(cf=.35)*
           750T20P/SP - 750 W. *(cf=.6)*

ACCESSORIES AVAILABLE:
Color Frame

WEIGHT: (not specified in manufacturer's catalog)

SIZE: 10⅜" x 14⅜" (L x W)

PHOTOMETRICS CHART
(Performance data for this unit are measured using BTR - 1 KW. lamp.)

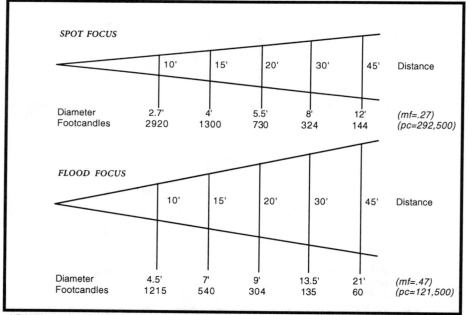

SPOT FOCUS

	10'	15'	20'	30'	45'	Distance
Diameter	2.7'	4'	5.5'	8'	12'	*(mf=.27)*
Footcandles	2920	1300	730	324	144	*(pc=292,500)*

FLOOD FOCUS

	10'	15'	20'	30'	45'	Distance
Diameter	4.5'	7'	9'	13.5'	21'	*(mf=.47)*
Footcandles	1215	540	304	135	60	*(pc=121,500)*

* Beam angles not specified in manufacturer's catalog.

D 424

STRAND CENTURY
Model No. 4125

SPOT BEAM SPREAD: 9° beam, 16° field

FLOOD BEAM SPREAD: 22° beam, 30° field

LAMP BASE TYPE: Mogul Prefocus

REFLECTOR DIAMETER: 15"

STANDARD LAMP: BVW - 2 KW.

OTHER LAMPS: BVT - 1 KW. *(cf=.42)*

ACCESSORIES AVAILABLE:
Color Frame

WEIGHT: (not specified in manufacturer's catalog)

SIZE: 14½" x 18⅛" x 23" (L x W x H)

PHOTOMETRICS CHART
(Performance data for this unit are measured using BVW - 2 KW. lamp.)

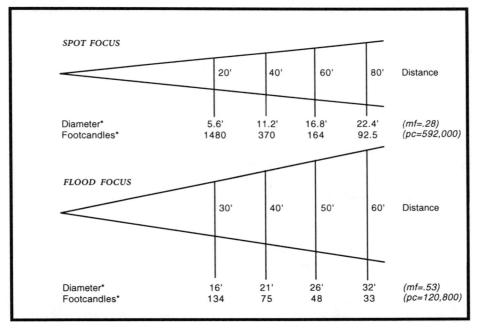

SPOT FOCUS

	20'	40'	60'	80'	Distance
Diameter*	5.6'	11.2'	16.8'	22.4'	*(mf=.28)*
Footcandles*	1480	370	164	92.5	*(pc=592,000)*

FLOOD FOCUS

	30'	40'	50'	60'	Distance
Diameter*	16'	21'	26'	32'	*(mf=.53)*
Footcandles*	134	75	48	33	*(pc=120,800)*

* The diameter and footcandle information in this chart was calculated using multiplying factor *(mf)* and peak candela *(pc)* data given in catalog. See Introduction for more information about calculating photometric data.

BEAM PROJECTOR

STRAND LIGHTING
Model No. 13011, "Beamlite 500"

BEAM SPREAD*: 4.2° beam, 6.9° field

LAMP BASE TYPE: E40**

REFLECTOR DIAMETER: 9½"

STANDARD LAMP: E40 Base, 500 W., 24V.***

ACCESSORIES AVAILABLE:
Color Frame (10¹³⁄₁₆" sq.)

WEIGHT: 28.6 lbs.

SIZE: 16½" x 13¹³⁄₁₆" x 15¼" (L x W x H)

PHOTOMETRICS CHART
(Performance data for this unit are measured using 500 W., 24V lamp.)

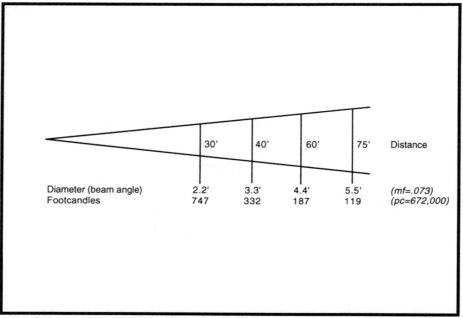

	30'	40'	60'	75'	Distance
Diameter (beam angle)	2.2'	3.3'	4.4'	5.5'	*(mf=.073)*
Footcandles	747	332	187	119	*(pc=672,000)*

* Manufacturer's catalog does not indicate that this beam projector is adjustable for "spot" and "flood" focus.
** This is most likely a mogul screw base.
*** Further information unavailable. A 24V. toroidal transformer is integral in this unit.

STRAND LIGHTING
Model No. 13021, "Beamlite

BEAM SPREAD*: 4.7° beam, 7° field

LAMP BASE TYPE: K39d**

REFLECTOR DIAMETER: 13"

STANDARD LAMP: K39d Base, 1 KW., 24V.***

ACCESSORIES AVAILABLE:
Color Frame (14⁵⁄₁₆" sq.)

WEIGHT: 37.5 lbs.

SIZE: 18¹⁵⁄₁₆" x 14¹³⁄₁₆" x 8¾" (L x W x H)

PHOTOMETRICS CHART
(Performance data for this unit are measured using 1 KW., 24V lamp.)

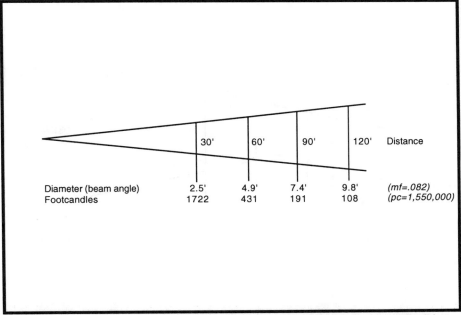

	30'	60'	90'	120'	Distance
Diameter (beam angle)	2.5'	4.9'	7.4'	9.8'	(mf=.082)
Footcandles	1722	431	191	108	(pc=1,550,000)

* Manufacturer's catalog does not indicate that this beam projector is adjustable for "spot" and "flood" focus.
** This base is a clamp with leads going to a terminal block.
*** Further information unavailable a 24V. toroidal transformer is integral in this unit.

STRAND CENTURY
Model No. 4202

BEAM ANGLE: 90°

FIELD ANGLE: 105°

FOCUSABLE: No

LAMP BASE TYPE: Medium Screw

STANDARD LAMP: 200A/CL - 200 W.

OTHER LAMPS: 150A/CL - 150 W. *(cf=.76)*

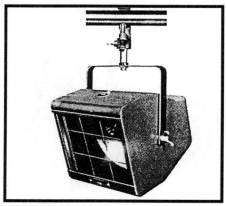

WEIGHT: 8 lbs.

ACCESSORIES AVAILABLE:
Color Frame

SIZE: 9½" x 12¾" x 11½" (L x W x H)

PHOTOMETRICS CHART
(Performance data for this unit are measured using 200A/CL - 200 W. lamp.)

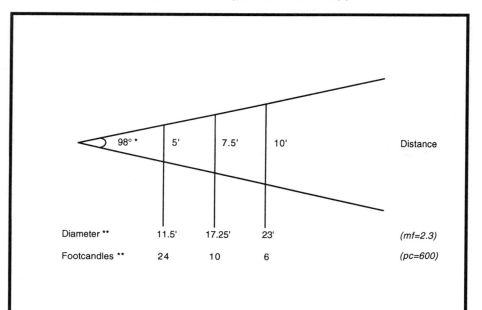

	98° *	5'	7.5'	10'	Distance
Diameter **		11.5'	17.25'	23'	*(mf=2.3)*
Footcandles **		24	10	6	*(pc=600)*

* The given multiplying factor *(mf)* for this instrument yields an angle of 98° rather than the stated "field angle" of 105°.

** The diameter and footcandle information in this chart was calculated using multiplying factor *(mf)* and peak candela *(pc)* data given in catalog. See Introduction for more information about calculating photometric data.

ALTMAN STAGE LIGHTING
Model No. 153

BEAM ANGLE: 57°

FIELD ANGLE: 103°

FOCUSABLE: No

LAMP BASE TYPE: Medium Screw

STANDARD LAMP: 400G/FL - 400 W.

OTHER LAMPS: 150P25/2 - 150 W. *(cf=.3)*
250G/FL - 250 W. *(cf=.52)*
250G/SP - 250 W. *(cf=.64)*
400G/SP - 400 W. *(1.22)*

WEIGHT: (not specified in manufacturer's catalog)

SIZE: 8" x 10¾" x 14⅛" (L x W x H)

ACCESSORIES AVAILABLE:
Color Frame

PHOTOMETRICS CHART

(Performance data for this unit are measured using 400G/FL - 400 W. lamp.)

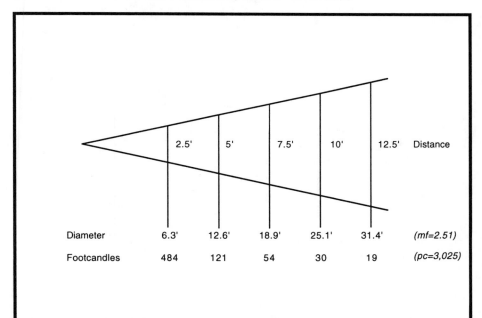

	2.5'	5'	7.5'	10'	12.5'	Distance
Diameter	6.3'	12.6'	18.9'	25.1'	31.4'	*(mf=2.51)*
Footcandles	484	121	54	30	19	*(pc=3,025)*

SCOOP — 10"

LIGHTING & ELECTRONICS
Model No. 63-10

BEAM ANGLE: 75°

FIELD ANGLE: 108°

FOCUSABLE: No

LAMP BASE TYPE: Medium Screw

STANDARD LAMP: 400G/FL - 400 W.

OTHER LAMPS: 250G/SP - 250 W. *(cf=.64)*
250G/FL - 250 W. *(cf=.52)*
400G/SP - 400 W. *(cf=1.22)*

ACCESSORIES AVAILABLE:
Color Frame

WEIGHT: 5 lbs.

SIZE: 7" x 11" x 11" (L x W x H)

PHOTOMETRICS CHART

(Performance data for this unit are measured using 400G/FL - 400 W. lamp.)

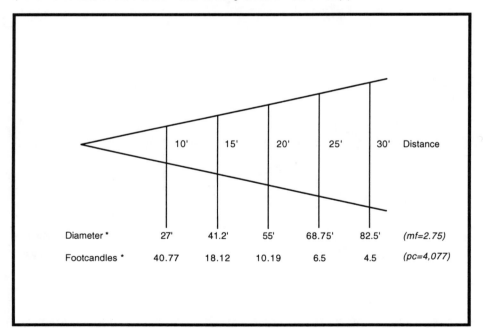

	10'	15'	20'	25'	30'	Distance
Diameter *	27'	41.2'	55'	68.75'	82.5'	*(mf=2.75)*
Footcandles *	40.77	18.12	10.19	6.5	4.5	*(pc=4,077)*

* The diameter and footcandle information in this chart was calculated using multiplying factor *(mf)* and peak candela *(pc)* data given in catalog. See Introduction for more information about calculating photometric data.

430

STRAND CENTURY
Model No. 4205

BEAM ANGLE: 80°

FIELD ANGLE: 105°

FOCUSABLE: No

LAMP BASE TYPE: Mogul Screw

STANDARD LAMP: PS35 - 500 W.

OTHER LAMPS: PS35 - 300 W. *(cf=.54)*

WEIGHT: 12 lbs.

SIZE: 14½" x 14½" x 15¾" (L x W x H)

ACCESSORIES AVAILABLE:
Color Frame

PHOTOMETRICS CHART
(Performance data for this unit are measured using PS35 - 500 W. lamp.)

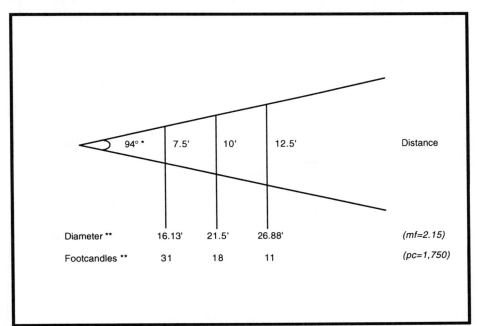

	94° *	7.5'	10'	12.5'	Distance
Diameter **		16.13'	21.5'	26.88'	*(mf=2.15)*
Footcandles **		31	18	11	*(pc=1,750)*

* The given multiplying factor *(mf)* for this instrument yields an angle of 94° rather than the stated "field angle" of 105°.

** The diameter and footcandle information in this chart was calculated using multiplying factor *(mf)* and peak candela *(pc)* data given in catalog. See Introduction for more information about calculating photometric data.

431

ALTMAN STAGE LIGHTING
Model No. 154

BEAM ANGLE: 47°

FIELD ANGLE: 90°

FOCUSABLE: No

LAMP BASE TYPE: Mogul Screw

STANDARD LAMP: 1000 IF - 1 KW.

OTHER LAMPS: 300IF - 300 W. *(cf=.25)*
500IF - 500 W. *(cf=.45)*
750IF - 750 W. *(cf=.72)*
DSE - 1 KW. *(cf=1.14)*

WEIGHT: (not specified in manufacturer's catalog)

SIZE: 13½" x 16" x 15⅛" (L x W x H)

ACCESSORIES AVAILABLE:
Color Frame

PHOTOMETRICS CHART
(Performance data for this unit are measured using 1000 IF - 1KW. lamp.)

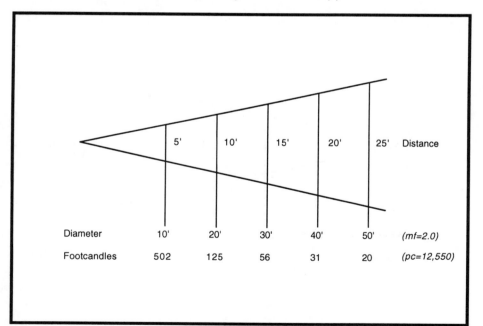

	5'	10'	15'	20'	25'	Distance
Diameter	10'	20'	30'	40'	50'	*(mf=2.0)*
Footcandles	502	125	56	31	20	*(pc=12,550)*

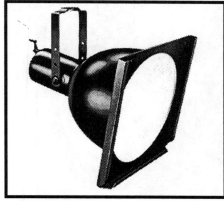

ALTMAN STAGE LIGHTING
Model No. 160

SPOT BEAM SPREAD: 32° beam, 84° field

FLOOD BEAM SPREAD: 70° beam, 108° field

FOCUSABLE: Yes

LAMP BASE TYPE: Medium Prefocus

STANDARD LAMP: EGK - 1 KW.

OTHER LAMPS: EGE - 500 W. *(cf=.4)*
EGG - 750 W. *(cf=.6)*
EGJ - 1 KW. *(cf=1.0)*

WEIGHT: (not specified in manufacturer's catalog)

SIZE: 19½" x 16" x 17" (L x W x H)

ACCESSORIES AVAILABLE:
Color Frame

PHOTOMETRICS CHART
(Performance data for this unit are measured using EGK - 1 KW. lamp.)

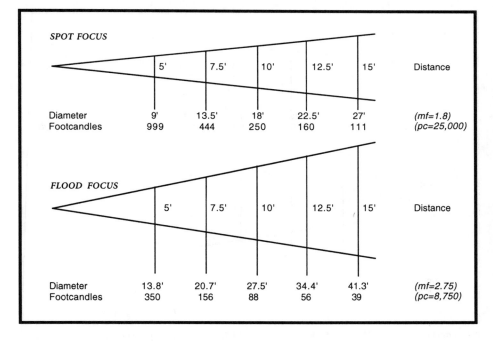

SPOT FOCUS

	5'	7.5'	10'	12.5'	15'	Distance
Diameter	9'	13.5'	18'	22.5'	27'	*(mf=1.8)*
Footcandles	999	444	250	160	111	*(pc=25,000)*

FLOOD FOCUS

	5'	7.5'	10'	12.5'	15'	Distance
Diameter	13.8'	20.7'	27.5'	34.4'	41.3'	*(mf=2.75)*
Footcandles	350	156	88	56	39	*(pc=8,750)*

SCOOP — 14"

BERKEY COLORTRAN
Model No. 216-052 *

BEAM ANGLE, SPOT: 33°

BEAM ANGLE, FLOOD: 74°

FIELD ANGLE, SPOT: 73°

FIELD ANGLE, FLOOD: 104°

FOCUSABLE: Yes

LAMP BASE TYPE: Medium Prefocus

STANDARD LAMP: EGJ - 1 KW.

OTHER LAMPS: EGG - 750 W. *(cf=.61)*

WEIGHT: 9.8 lbs.

SIZE: 20" x 16⁵⁄₁₆" x 23¹³⁄₁₆" (L x W x H)

ACCESSORIES AVAILABLE:
Color Frame

PHOTOMETRICS CHART
(Performance data for this unit are measured using EGJ - 1 KW. lamp.)

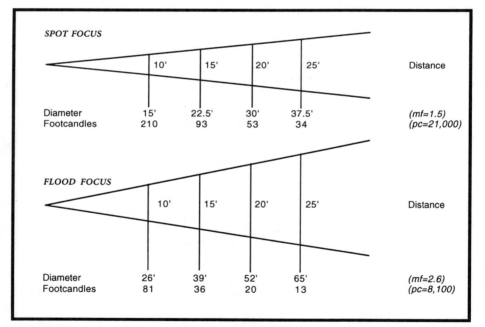

SPOT FOCUS

	10'	15'	20'	25'	Distance
Diameter	15'	22.5'	30'	37.5'	*(mf=1.5)*
Footcandles	210	93	53	34	*(pc=21,000)*

FLOOD FOCUS

	10'	15'	20'	25'	Distance
Diameter	26'	39'	52'	65'	*(mf=2.6)*
Footcandles	81	36	20	13	*(pc=8,100)*

* This model can be identified with model number -052, -055, -056, or -057 depending on the wire leads and connectors supplied.

CENTURY LIGHTING
Model No. 1313

BEAM ANGLE: 64°

FIELD ANGLE: 110°

FOCUSABLE: No

LAMP BASE TYPE: Medium Prefocus

STANDARD LAMP: Q1000T12/4 - 1 KW.*

OTHER LAMPS: Q500T12 - 500 W.
Q750T12 - 750 W.

WEIGHT: (not specified in manufacturer's catalog)

SIZE: 13¾" x 16" x 20¼" (L x W x H)

ACCESSORIES AVAILABLE:
Color Frame

PHOTOMETRICS CHART

(Performance data for this unit are measured using Q1000T12/4 - 1 KW. lamp.)

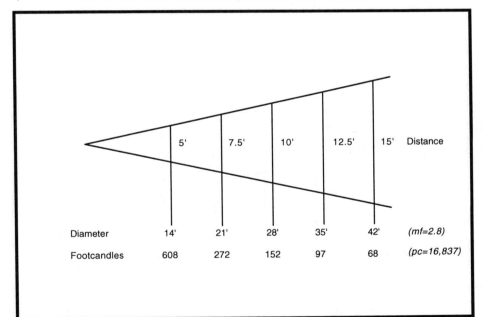

	5'	7.5'	10'	12.5'	15'	Distance
Diameter	14'	21'	28'	35'	42'	*(mf=2.8)*
Footcandles	608	272	152	97	68	*(pc=16,837)*

* This is an obsolete lamp but it is listed as the standard lamp since the manufacturer uses this lamp to describe the photometric performance. Correction factor *(cf)* information is not available for the other lamps because no initial lumens data is available for the obsolete lamp.

SCOOP — 14"

CENTURY STRAND
Model No. 4271

BEAM ANGLE: 88°

FIELD ANGLE: 130° *

FOCUSABLE: No

LAMP BASE TYPE: Medium Prefocus

STANDARD LAMP: EGK - 1 KW.

OTHER LAMPS: Q1000/T12/4 - 1 KW.**

ACCESSORIES AVAILABLE:
Color Frame

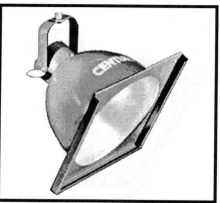

WEIGHT: (not specified in manufacturer's catalog)

SIZE: 13¾" x 16" x 20¼" (L x W x H)

PHOTOMETRICS CHART

(Performance data for this unit are measured using EGK - 1 KW. lamp.)

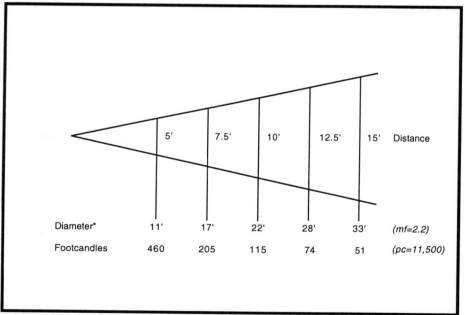

	5'	7.5'	10'	12.5'	15'	Distance
Diameter*	11'	17'	22'	28'	33'	(mf=2.2)
Footcandles	460	205	115	74	51	(pc=11,500)

* There seems to be an inconsistency in the data given in the manufacturer's catalog. The distance and diameter data, which is described as representing the field angle, yield an angle of 95.5°, not the stated field angle of 130°.

** Unable to identify lamp sufficiently to provide correction factor *(cf)* information.

436

ELECTRO CONTROLS
Model No. 35145

SPOT BEAM SPREAD: 45° beam, 80° field

FLOOD BEAM SPREAD: 115° beam, 150° field

FOCUSABLE: Yes

LAMP BASE TYPE: Medium 2 Pin

STANDARD LAMP: FEL - 1 KW.

OTHER LAMPS: EHD - 500 W. *(cf=.39)*
EHG - 750 W. *(cf=.56)*
EHH - 1 KW. *(cf=.76)*

WEIGHT: (not specified in manufacturer's catalog)

SIZE: 20" x 16" x 18¼" (L x W x H)

ACCESSORIES AVAILABLE:
Color Frame (15¾" x 16")

PHOTOMETRICS CHART
(Performance data for this unit are measured using FEL - 1 KW. lamp.)

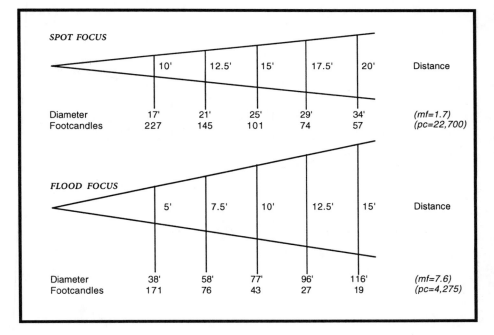

SPOT FOCUS

	10'	12.5'	15'	17.5'	20'	Distance
Diameter	17'	21'	25'	29'	34'	*(mf=1.7)*
Footcandles	227	145	101	74	57	*(pc=22,700)*

FLOOD FOCUS

	5'	7.5'	10'	12.5'	15'	Distance
Diameter	38'	58'	77'	96'	116'	*(mf=7.6)*
Footcandles	171	76	43	27	19	*(pc=4,275)*

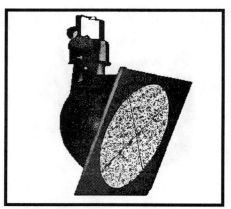

SCOOP — 14"

ELECTRO CONTROLS
Model No. 7514A

BEAM ANGLE: 80°

FIELD ANGLE: 120°

FOCUSABLE: No

LAMP BASE TYPE: Mogul Screw

STANDARD LAMP: 500 IF - 500 W.

OTHER LAMPS: (none specified in mfg. catalog)

ACCESSORIES AVAILABLE:
Color Frame (15¾" x 16")

WEIGHT: (not specified in manufacturer's catalog)

SIZE: 16⅛" x 20" (W x H)

PHOTOMETRICS CHART

(Performance data for this unit are measured using 500IF - 500 W. lamp.)

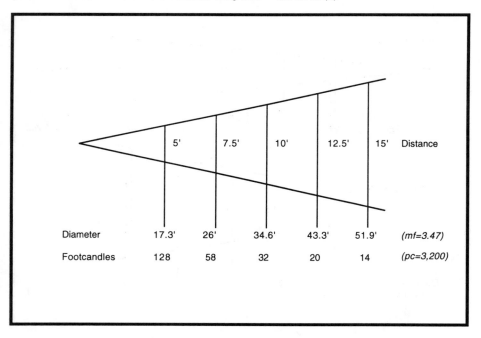

	5'	7.5'	10'	12.5'	15'	Distance
Diameter	17.3'	26'	34.6'	43.3'	51.9'	*(mf=3.47)*
Footcandles	128	58	32	20	14	*(pc=3,200)*

ELECTRO CONTROLS
Model No. 7515A

SPOT BEAM SPREAD: 45° beam, 80° field

FLOOD BEAM SPREAD: 115° beam, 150° field

FOCUSABLE: Yes

LAMP BASE TYPE: Medium Prefocus

STANDARD LAMP: EGJ - 1 KW.

OTHER LAMPS: EGE - 500 W. *(cf=.38)*
EGG - 750 W. *(cf=.57)*

WEIGHT: (not specified in manufacturer's catalog)

SIZE: 20" x 16" x 18¼" (L x W x H)

ACCESSORIES AVAILABLE:
Color Frame (15¾" x 16")

PHOTOMETRICS CHART

(Performance data for this unit are measured using EGJ - 1 KW. lamp.)

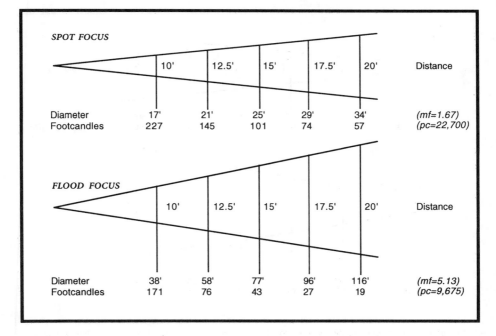

SPOT FOCUS

	10'	12.5'	15'	17.5'	20'	Distance
Diameter	17'	21'	25'	29'	34'	*(mf=1.67)*
Footcandles	227	145	101	74	57	*(pc=22,700)*

FLOOD FOCUS

	10'	12.5'	15'	17.5'	20'	Distance
Diameter	38'	58'	77'	96'	116'	*(mf=5.13)*
Footcandles	171	76	43	27	19	*(pc=9,675)*

LIGHTING & ELECTRONICS
Model No. 63-14

SPOT BEAM SPREAD: 34° beam, 90° field

FLOOD BEAM SPREAD: 68° beam, 105° field

FOCUSABLE: Yes

LAMP BASE TYPE: Medium Prefocus

STANDARD LAMP: EGJ - 1 KW.

OTHER LAMPS: EGE - 500 W. *(cf=.38)*
EGG - 750 W. *(cf=.57)*

ACCESSORIES AVAILABLE:
Color Frame

WEIGHT: 14 lbs.

SIZE: 19" x 16" x 16" (L x W x H)

PHOTOMETRICS CHART

(Performance data for this unit are measured using EGJ - 1 KW. lamp.)

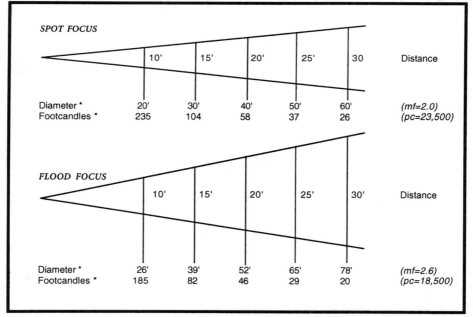

SPOT FOCUS

	10'	15'	20'	25'	30	Distance
Diameter *	20'	30'	40'	50'	60'	*(mf=2.0)*
Footcandles *	235	104	58	37	26	*(pc=23,500)*

FLOOD FOCUS

	10'	15'	20'	25'	30'	Distance
Diameter *	26'	39'	52'	65'	78'	*(mf=2.6)*
Footcandles *	185	82	46	29	20	*(pc=18,500)*

* The diameter and footcandle information in this chart was calculated using multiplying factor *(mf)* and peak candela *(pc)* data given in catalog. See Introduction for more information about calculating photometric data.

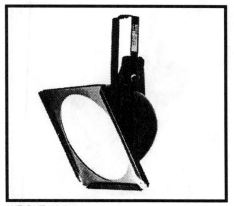

LIGHTING & ELECTRONICS
Model No. 63-15

BEAM ANGLE: 85°

FIELD ANGLE: 125°

FOCUSABLE: No

LAMP BASE TYPE: Mogul Screw

STANDARD LAMP: DKX - 1.5 KW.

OTHER LAMPS: DSE - 1 KW. *(cf=.68)*
DSF - 1.5 KW. *(cf=1.0)*

WEIGHT: 13 lbs.

SIZE: 16" x 16" x 14" (L x W x H)

ACCESSORIES AVAILABLE:
Color Frame

PHOTOMETRICS CHART
(Performance data for this unit are measured using DKX - 1.5 KW. lamp.)

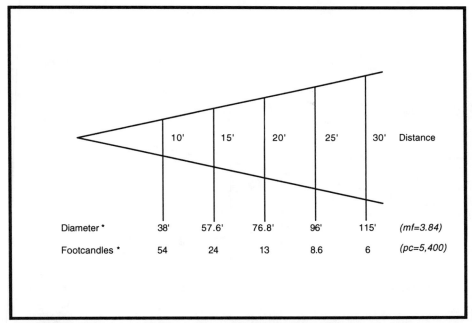

	10'	15'	20'	25'	30'	Distance
Diameter *	38'	57.6'	76.8'	96'	115'	*(mf=3.84)*
Footcandles *	54	24	13	8.6	6	*(pc=5,400)*

* The diameter and footcandle information in this chart was calculated using multiplying factor *(mf)* and peak candela *(pc)* data given in catalog. See Introduction for more information about calculating photometric data.

SCOOP — 14"

STRAND CENTURY
Model No. 4291

SPOT BEAM SPREAD: 56° beam, 100° field

FLOOD BEAM SPREAD: 100° beam, 150° field

FOCUSABLE: Yes

LAMP BASE TYPE: Medium Prefocus

STANDARD LAMP: EGK - 1 KW.

OTHER LAMPS: EGE - 500 W. *(cf=.41)*
EGG - 750 W. *(cf=.61)*
EGJ - 1 KW. *(cf=1.0)*

ACCESSORIES AVAILABLE:
Color Frame

WEIGHT: 11 lbs.
SIZE: 20" x 16" x 17⅞" (L x W x H)

PHOTOMETRICS CHART
(Performance data for this unit are measured using EGK - 1 KW. lamp.)

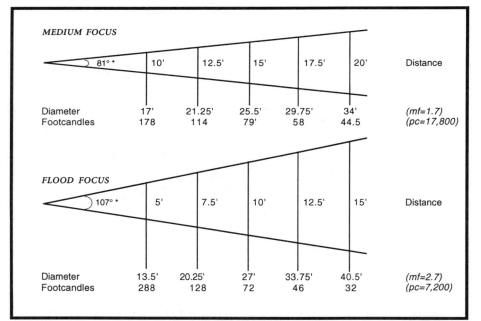

MEDIUM FOCUS

	81° *	10'	12.5'	15'	17.5'	20'	Distance
Diameter		17'	21.25'	25.5'	29.75'	34'	(mf=1.7)
Footcandles		178	114	79'	58	44.5	(pc=17,800)

FLOOD FOCUS

	107° *	5'	7.5'	10'	12.5'	15'	Distance
Diameter		13.5'	20.25'	27'	33.75'	40.5'	(mf=2.7)
Footcandles		288	128	72	46	32	(pc=7,200)

* The given distance and diameter data, described in the catalog as "Medium Flood" and "Flood," yield angles of 81° and 107° rather than the given "field angles" of 100° at Spot Focus and 150° at Flood Focus. See Introduction for more information about calculating photometric data.

442

TIMES SQUARE LIGHTING
Model No. Q114PC

SPOT BEAM SPREAD: 81° field*

FLOOD BEAM SPREAD: 108° field*

FOCUSABLE: Yes

LAMP BASE TYPE: Medium Prefocus

STANDARD LAMP: EGJ - 1 KW.

OTHER LAMPS: EGE - 500 W. *(cf=.41)*
EGG - 750 W. *(cf=.61)*

ACCESSORIES AVAILABLE:
Color Frame

WEIGHT: (not specified in manufacturer's catalog)

SIZE: 17" x 14" x 19" (L x W x H)

PHOTOMETRICS CHART
(Performance data for this unit are measured using EGJ - 1 KW. lamp.)

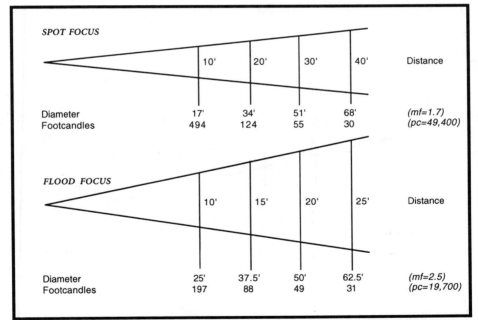

SPOT FOCUS

	10'	20'	30'	40'	Distance
Diameter	17'	34'	51'	68'	*(mf=1.7)*
Footcandles	494	124	55	30	*(pc=49,400)*

FLOOD FOCUS

	10'	15'	20'	25'	Distance
Diameter	25'	37.5'	50'	62.5'	*(mf=2.5)*
Footcandles	197	88	49	31	*(pc=19,700)*

* Beam and field angles not specified in manufacturer's catalog. This field angle is calculated from given distance and field diameter data. See Introduction for more information about calculating photometric data.

TIMES SQUARE LIGHTING
Model No. Q145

BEAM ANGLE: (not specified in mfg. catalog)

FIELD ANGLE: 90° *

FOCUSABLE: No

LAMP BASE TYPE: Mogul Screw

STANDARD LAMP: 1000 IF - 1 KW.

OTHER LAMPS: 300/300 IF - 300 W. *(cf=.25)*
500/500 IF - 500 W. *(cf=.45)*
750/750 IF - 750 W. *(cf=.72)*
DSE - 1 KW. *(cf=1.14)*

ACCESSORIES AVAILABLE:
Color Frame

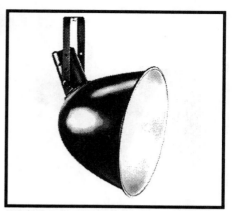

WEIGHT: (not specified in manufacturer's catalog)

SIZE: (not specified in manufacturer's catalog)

PHOTOMETRICS CHART
(Performance data for this unit are measured using 1000 IF - 1KW. lamp.)

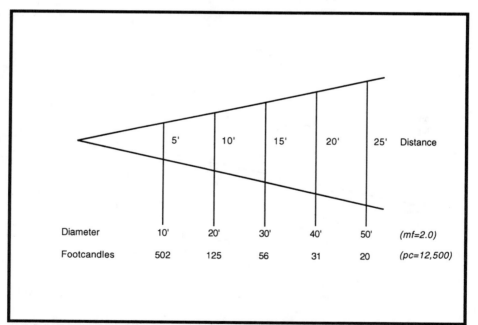

	5'	10'	15'	20'	25'	Distance
Diameter	10'	20'	30'	40'	50'	*(mf=2.0)*
Footcandles	502	125	56	31	20	*(pc=12,500)*

* Beam and field angles not specified in manufacturer's catalog. This field angle is calculated from given distance and field diameter data. See Introduction for more information about calculating photometric data.

ALTMAN STAGE LIGHTING
Model No. 161

BEAM ANGLE: 43°

FIELD ANGLE: 90°

FOCUSABLE: No

LAMP BASE TYPE: Recessed Single-Contact

STANDARD LAMP: FCM - 1 KW.

OTHER LAMPS: Q500T3 - 500 W. *(cf=.40)*
FDN - 500 W. *(cf=.47)*
EMD - 750 W. *(cf=.71)*
FHM - 1 KW. *(cf=.98)*

WEIGHT: (not specified in manufacturer's catalog)

SIZE: 14" x 16½" x 15⅛" (L x W x H)

ACCESSORIES AVAILABLE:
Color Frame

PHOTOMETRICS CHART
(Performance data for this unit are measured using FCM - 1KW. lamp.)

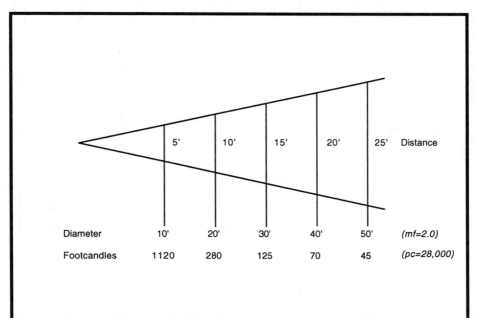

	5'	10'	15'	20'	25'	Distance
Diameter	10'	20'	30'	40'	50'	*(mf=2.0)*
Footcandles	1120	280	125	70	45	*(pc=28,000)*

ALTMAN STAGE LIGHTING
Model No. 261

SPOT BEAM SPREAD: 12° beam, 54° field

FLOOD BEAM SPREAD: 46° beam, 100° field

FOCUSABLE: Yes

LAMP BASE TYPE: Mogul Screw

STANDARD LAMP: BWF - 2 KW.

OTHER LAMPS: 1500Q/CL/48 -1500 W. *(cf=.58)*
BWG -   2 KW. *(cf=.97)*

ACCESSORIES AVAILABLE:
Color Frame

WEIGHT: (not specified in manufacturer's catalog)

SIZE: 17¾" x 16⅞" x 15¾" (L x W x H)

PHOTOMETRICS CHART
(Performance data for this unit are measured using BWF - 2 KW. lamp.)

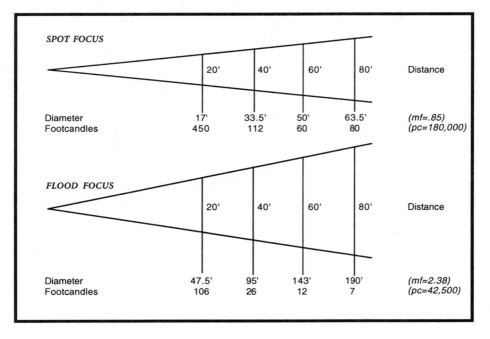

SPOT FOCUS

	20'	40'	60'	80'	Distance
Diameter	17'	33.5'	50'	63.5'	*(mf=.85)*
Footcandles	450	112	60	80	*(pc=180,000)*

FLOOD FOCUS

	20'	40'	60'	80'	Distance
Diameter	47.5'	95'	143'	190'	*(mf=2.38)*
Footcandles	106	26	12	7	*(pc=42,500)*

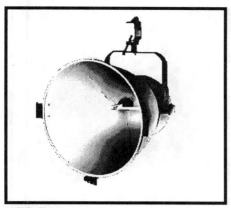

KLIEGL BROS.
Model No. 3451

BEAM ANGLE: (not specified in mfg. catalog)

FIELD ANGLE: 102°

FOCUSABLE: No

LAMP BASE TYPE: Recessed Single-Contact

STANDARD LAMP: FHM - 1 KW.

OTHER LAMPS: EHZ - 300 W. *(cf=.22)*
FCZ - 500 W. *(cf=.34)*

WEIGHT: 8 lbs.

SIZE: 15½" x 18½" x 23½" (L x W x H)

ACCESSORIES AVAILABLE:
Color Frame

PHOTOMETRICS CHART
(Performance data for this unit are measured using FHM - 1 KW. lamp.)

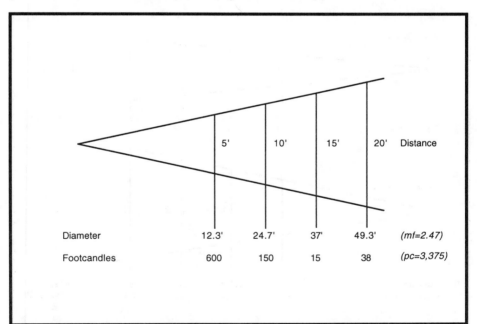

	5'	10'	15'	20'	Distance
Diameter	12.3'	24.7'	37'	49.3'	*(mf=2.47)*
Footcandles	600	150	15	38	*(pc=3,375)*

447

KLIEGL BROS.
Model No. 3452

SPOT BEAM SPREAD: 65° field*

FLOOD BEAM SPREAD: 102° field*

FOCUSABLE: Yes

LAMP BASE TYPE: Recessed Single-Contact

STANDARD LAMP: FHM - 1 KW.

OTHER LAMPS: EHZ - 300 W. *(cf=.22)*
FCZ - 500 W. *(cf=.34)*

ACCESSORIES AVAILABLE:
Color Frame

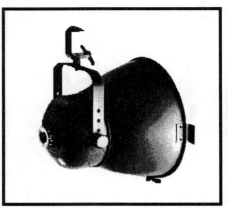

WEIGHT: 8 lbs.

SIZE: 15½" x 18½" x 23½" (L x W x H)

PHOTOMETRICS CHART
(Performance data for this unit are measured using FHM - 1 KW. lamp.)

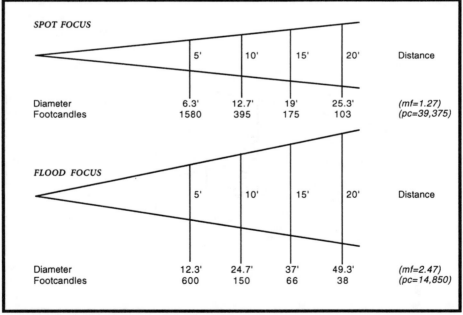

SPOT FOCUS

	5'	10'	15'	20'	Distance
Diameter	6.3'	12.7'	19'	25.3'	*(mf=1.27)*
Footcandles	1580	395	175	103	*(pc=39,375)*

FLOOD FOCUS

	5'	10'	15'	20'	Distance
Diameter	12.3'	24.7'	37'	49.3'	*(mf=2.47)*
Footcandles	600	150	66	38	*(pc=14,850)*

* Beam angles not specified in manufacturer's catalog.

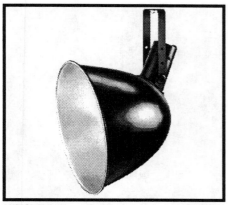

ALTMAN STAGE LIGHTING
Model No. 155

BEAM ANGLE: 60°

FIELD ANGLE: 106°

FOCUSABLE: No

LAMP BASE TYPE: Mogul Screw

STANDARD LAMP: DKX - 1500 W.

OTHER LAMPS: 500IF - 500 W. *(cf=.31)*
750IF - 750 W. *(cf=.50)*
1000IF - 1 KW. *(cf=.70)*
DSE - 1 KW. *(cf=.79)*

ACCESSORIES AVAILABLE:
Color Frame

WEIGHT: (not specified in manufacturer's catalog)

SIZE: 14½" x 18½" x 24½" (L x W x H)

PHOTOMETRICS CHART
(Performance data for this unit are measured using DKX - 1500 W. lamp.)

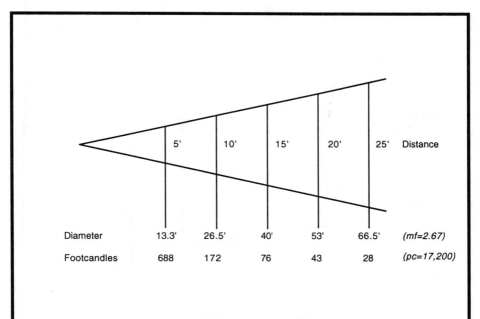

	5'	10'	15'	20'	25'	Distance
Diameter	13.3'	26.5'	40'	53'	66.5'	*(mf=2.67)*
Footcandles	688	172	76	43	28	*(pc=17,200)*

BERKEY COLORTRAN
Model No. 104-202*

SPOT BEAM SPREAD: 50° beam, 68° field

FLOOD BEAM SPREAD: 80° beam, 90° field

FOCUSABLE: Yes

LAMP BASE TYPE: Mogul Screw

STANDARD LAMP: 176-166** - 1.5 KW.

OTHER LAMPS: BWF - 2 KW. *(cf=1.7)*

ACCESSORIES AVAILABLE:
Color Frame

WEIGHT: 10 lbs.

SIZE: 21⅝" x 21¼" x 27⅜" (L x W x H)

PHOTOMETRICS CHART

*(Performance data for this unit are measured using 176-166** - 1.5 KW. lamp.)*

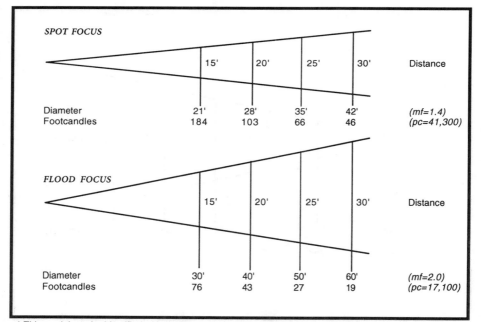

SPOT FOCUS

	15'	20'	25'	30'	Distance
Diameter	21'	28'	35'	42'	*(mf=1.4)*
Footcandles	184	103	66	46	*(pc=41,300)*

FLOOD FOCUS

	15'	20'	25'	30'	Distance
Diameter	30'	40'	50'	60'	*(mf=2.0)*
Footcandles	76	43	27	19	*(pc=17,100)*

* This model can be identified with model numbers -202, -205, -206, or -207 depending on the wire leads and connectors supplied.

** 176-166 is the Colortran catalog number for this lamp. It is described as a tungsten halogen 1.5 KW. 3050°K, 2000 hr. lamp. It is no longer manufactured, and even an ANSI code for it is unavailable.

BERKEY COLORTRAN
Model No. 216-062 *

BEAM ANGLE: 74°

FIELD ANGLE: 105°

FOCUSABLE: No

LAMP BASE TYPE: Mogul Screw

STANDARD LAMP: DKX - 1.5 KW.

OTHER LAMPS: DKZ - 1 KW. *(cf=.59)*

WEIGHT: 12 lbs.

SIZE: 17" x 18½" x 27" (L x W x H)

ACCESSORIES AVAILABLE:
Color Frame

PHOTOMETRICS CHART
(Performance data for this unit are measured using DKX - 1.5 KW. lamp.)

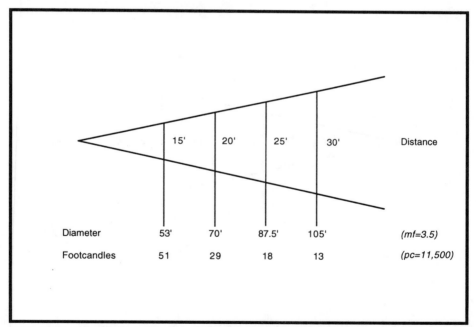

	15'	20'	25'	30'	Distance
Diameter	53'	70'	87.5'	105'	*(mf=3.5)*
Footcandles	51	29	18	13	*(pc=11,500)*

* This model can be identified with model number -062, -065, -066, or -067 depending on the wire leads and connectors supplied.

CENTURY STRAND
Model No. 1318

BEAM ANGLE: 80°

FIELD ANGLE: 110°

FOCUSABLE: No

LAMP BASE TYPE: Mogul Screw

STANDARD LAMP: 2M/PS52/34 - 2 KW.

OTHER LAMPS: 750/IF - 750 W. *(cf=.39)*
             1000/IF - 1 KW. *(cf=.54)*
             1500/IF -1500 W. *(cf=.76)*

ACCESSORIES AVAILABLE:

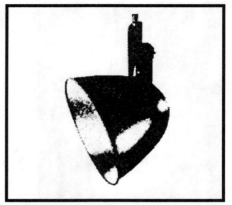

WEIGHT: (not specified in manufacturer's catalog)

SIZE: 18½" x 30¾" (W x H)

PHOTOMETRICS CHART
(Performance data for this unit are measured using 2M/PS52/34 - 2 KW. lamp.)

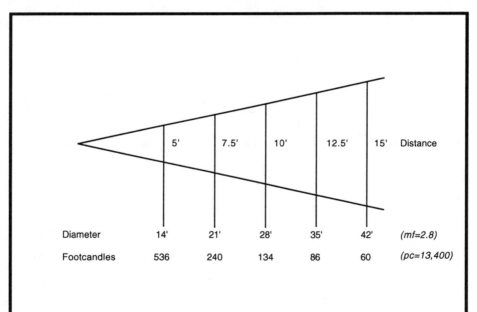

	5'	7.5'	10'	12.5'	15'	Distance
Diameter	14'	21'	28'	35'	42'	*(mf=2.8)*
Footcandles	536	240	134	86	60	*(pc=13,400)*

COLORTRAN
Model No. 104-232*

SPOT BEAM SPREAD: 27.5° beam, 60.5° field

FLOOD BEAM SPREAD: 50.8° beam, 88° field

FOCUSABLE: Yes

LAMP BASE TYPE: Medium Prefocus

STANDARD LAMP: EGJ - 1 KW.

OTHER LAMPS: EGE - 500 W. *(cf=.38)*
EGG - 750 W. *(cf=.57)*

ACCESSORIES AVAILABLE:
Color Frame

WEIGHT: 8.4 lbs.

SIZE: 17¾" x 17¼" x 21⅜" (L x W x H)

PHOTOMETRICS CHART
(Performance data for this unit are measured using EGJ - 1 KW. lamp.)

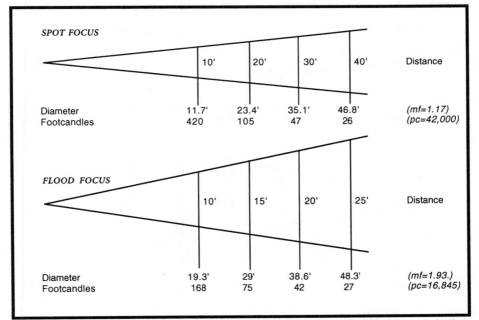

SPOT FOCUS

	10'	20'	30'	40'	Distance
Diameter	11.7'	23.4'	35.1'	46.8'	*(mf=1.17)*
Footcandles	420	105	47	26	*(pc=42,000)*

FLOOD FOCUS

	10'	15'	20'	25'	Distance
Diameter	19.3'	29'	38.6'	48.3'	*(mf=1.93.)*
Footcandles	168	75	42	27	*(pc=16,845)*

* This model can be identified with model numbers -232, -235, -236, or -237 depending on the wire leads and connectors supplied.

COLORTRAN
Model No. 104-242*

SPOT BEAM SPREAD: 28° beam, 68° field

FLOOD BEAM SPREAD: 75° beam, 103° field

FOCUSABLE: Yes

LAMP BASE TYPE: Mogul Screw

STANDARD LAMP: BWF - 2 KW.

OTHER LAMPS: (no other 120 V. lamps listed.)

ACCESSORIES AVAILABLE:
Color Frame

WEIGHT: 8.7 lbs.

SIZE: 17¾" x 17¼" x 21⅜" (L x W x H)

PHOTOMETRICS CHART
(Performance data for this unit are measured using BWF - 2 KW. lamp.)

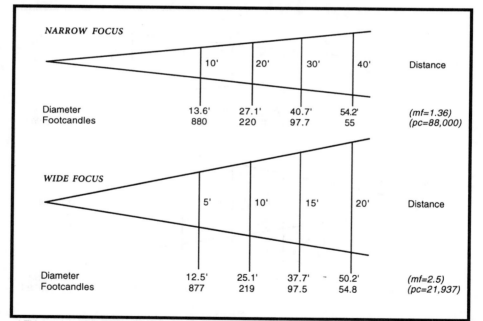

NARROW FOCUS

	10'	20'	30'	40'	Distance
Diameter	13.6'	27.1'	40.7'	54.2'	(mf=1.36)
Footcandles	880	220	97.7	55	(pc=88,000)

WIDE FOCUS

	5'	10'	15'	20'	Distance
Diameter	12.5'	25.1'	37.7'	50.2'	(mf=2.5)
Footcandles	877	219	97.5	54.8	(pc=21,937)

* This model can be identified with model numbers -242, -245, -246, or -247 depending on the wire leads and connectors supplied.

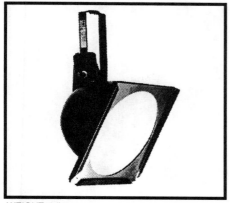

LIGHTING & ELECTRONICS
Model No. 63-18

BEAM ANGLE: 78°

FIELD ANGLE: 110°

FOCUSABLE: No

LAMP BASE TYPE: Mogul Screw

STANDARD LAMP: DKX - 1.5 KW.

OTHER LAMPS: DSE -1 KW. *(cf=.68)*
DSF - 1.5 KW. *(cf=1.0)*

WEIGHT: 9 lbs.

SIZE: 18" x 18" x 20" (L x W x H)

ACCESSORIES AVAILABLE:
Color Frame

PHOTOMETRICS CHART
(Performance data for this unit are measured using DKX - 1.5 KW. lamp.)

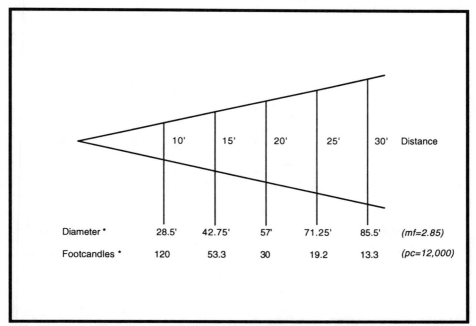

	10'	15'	20'	25'	30'	Distance
Diameter *	28.5'	42.75'	57'	71.25'	85.5'	*(mf=2.85)*
Footcandles *	120	53.3	30	19.2	13.3	*(pc=12,000)*

* The diameter and footcandle information in this chart was calculated using multiplying factor *(mf)* and peak candela *(pc)* data given in catalog. See Introduction for more information about calculating photometric data.

STRAND CENTURY
Model No. 4273A

BEAM ANGLE: 90°

FIELD ANGLE: 110°

FOCUSABLE: No

LAMP BASE TYPE: Mogul Screw

STANDARD LAMP: Q2000/4/95 - 2 KW.

OTHER LAMPS: DSE - 1 KW. *(cf=.41)*
DSF - 1.5 KW. *(cf=.69)*

ACCESSORIES AVAILABLE:
Color Frame

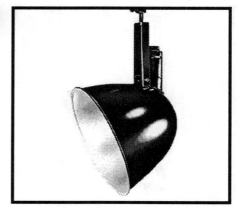

WEIGHT: 9 lbs.

SIZE: 15⁷⁄₁₆" x 18½" x 24¼" (L x W x H)

PHOTOMETRICS CHART

(Performance data for this unit are measured using Q2000/4/95 - 2 KW. lamp.)

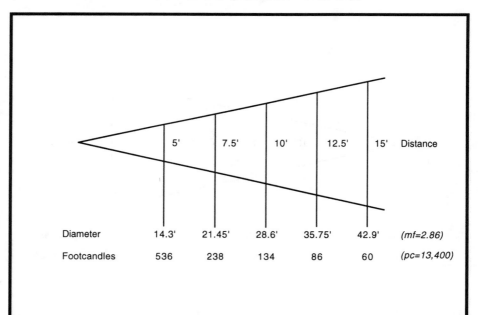

	5'	7.5'	10'	12.5'	15'	Distance
Diameter	14.3'	21.45'	28.6'	35.75'	42.9'	*(mf=2.86)*
Footcandles	536	238	134	86	60	*(pc=13,400)*

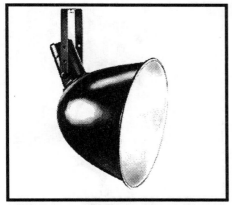

TIMES SQUARE LIGHTING ·
Model No. 185

BEAM ANGLE: (not specified in mfg. catalog)

FIELD ANGLE: 106° *

FOCUSABLE: No

LAMP BASE TYPE: Mogul Screw

STANDARD LAMP: DKX - 1.5 KW.

OTHER LAMPS: 500IF - 500 W. *(cf=.26)*
750IF - 750 W. *(cf=.42)*
1000IF - 1 KW. *(cf=.58)*
DSE - 1 KW. *(cf=.66)*

WEIGHT: (not specified in manufacturer's catalog)

SIZE: (not specified in manufacturer's catalog)

ACCESSORIES AVAILABLE:
Color Frame

PHOTOMETRICS CHART
(Performance data for this unit are measured using DKX - 1.5 KW. lamp.)

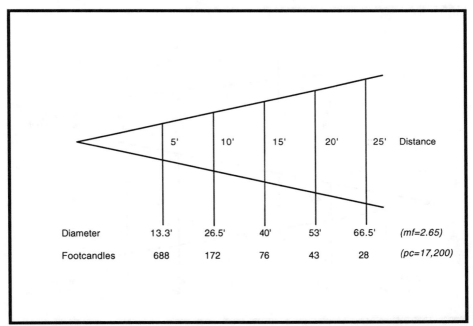

	5'	10'	15'	20'	25'	Distance
Diameter	13.3'	26.5'	40'	53'	66.5'	*(mf=2.65)*
Footcandles	688	172	76	43	28	*(pc=17,200)*

* Beam and field angles not specified in manufacturer's catalog. This field angle is calculated from given distance and field diameter data. See Introduction for more information about calculating photometric data.

ALTMAN STAGE LIGHTING
Model No. 1000Q

MIN. SPOT w/ IRIS: 1.5° (beam), 1.5° (field)

SPOT BEAM SPREAD: 8.5° (beam), 10° (field)

FLOOD BEAM SPREAD: 13° (beam), 14° (field)

COLOR FRAMES: 6

FADER: Yes (blackout frame)

SHUTTERS: Yes (guillotine)

FAN COOLED: Yes

STANDARD LAMP: FEL - 1 KW.

NOTES: Iris, shutters and focus are all rear operated. Relamping is accomplished through hinged door on top.

WEIGHT: without stand – 57.5 lbs.

SIZE: 42" x 17½" x 59½–77½" (L x W x H)

PHOTOMETRICS CHART

(Performance data for this unit are measured using FEL - 1 KW. lamp.)

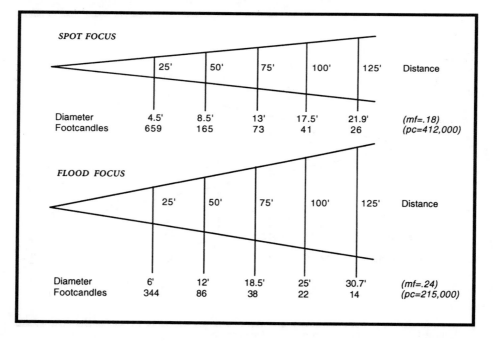

SPOT FOCUS

	25'	50'	75'	100'	125'	Distance
Diameter	4.5'	8.5'	13'	17.5'	21.9'	*(mf=.18)*
Footcandles	659	165	73	41	26	*(pc=412,000)*

FLOOD FOCUS

	25'	50'	75'	100'	125'	Distance
Diameter	6'	12'	18.5'	25'	30.7'	*(mf=.24)*
Footcandles	344	86	38	22	14	*(pc=215,000)*

ALTMAN STAGE LIGHTING
Model No. 902

MIN. SPOT w/ IRIS: 1.9° (beam), 1.9° (field)

SPOT BEAM SPREAD: 23° (beam), 33° (field)

FLOOD BEAM SPREAD: 25° (beam), 45° (field)

COLOR FRAMES: 6

FADER: Yes (blackout frame)

SHUTTERS: Yes (guillotine)

FAN COOLED: No

STANDARD LAMP: DTJ - 1.5 KW.

NOTES: Iris is located on rear of unit. Focus is on the side. Relamping is accomplished through hinged door on top.

WEIGHT: without stand – 40 lbs.

SIZE: 35" x 16" x 55–73" (L x W x H)

PHOTOMETRICS CHART
(Performance data for this unit are measured using DTJ - 1.5 KW. lamp.)

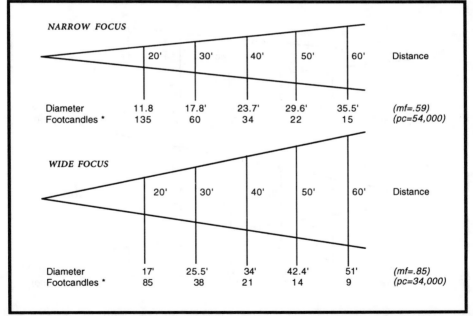

NARROW FOCUS

	20'	30'	40'	50'	60'	Distance
Diameter	11.8	17.8'	23.7'	29.6'	35.5'	(mf=.59)
Footcandles *	135	60	34	22	15	(pc=54,000)

WIDE FOCUS

	20'	30'	40'	50'	60'	Distance
Diameter	17'	25.5'	34'	42.4'	51'	(mf=.85)
Footcandles *	85	38	21	14	9	(pc=34,000)

* The intensity data for this model was published incorrectly in an Altman catalog. Altman confirms that the footcandle information shown here is correct.

FOLLOW SPOT

ALTMAN STAGE LIGHTING
Model No. 902 Jr.

MIN. SPOT w/ IRIS: 1.9° (beam), 1.9° (field)

SPOT BEAM SPREAD: 23° (beam), 33° (field)

FLOOD BEAM SPREAD: 25° (beam), 45° (field)

COLOR FRAMES: 1

FADER: No

SHUTTERS: Yes (guillotine)

FAN COOLED: No

STANDARD LAMP: DTJ - 1.5 KW.

NOTES: Iris and shutters are rear operated. Focus handle is on the side. Relamping is accomplished through hinged door on top.

WEIGHT: without stand – 30.5 lbs.

SIZE: 26½" x 16" x 53¾–71¾" (L x W x H)

PHOTOMETRICS CHART
(Performance data for this unit are measured using DTJ - 1.5 KW. lamp.)

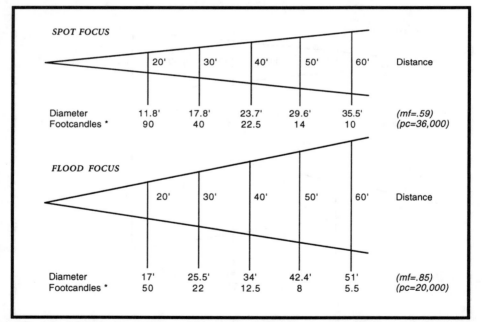

SPOT FOCUS

	20'	30'	40'	50'	60'	Distance
Diameter	11.8'	17.8'	23.7'	29.6'	35.5'	(mf=.59)
Footcandles *	90	40	22.5	14	10	(pc=36,000)

FLOOD FOCUS

	20'	30'	40'	50'	60'	Distance
Diameter	17'	25.5'	34'	42.4'	51'	(mf=.85)
Footcandles *	50	22	12.5	8	5.5	(pc=20,000)

* The intensity data for this model was published incorrectly in an Altman catalog. Altman confirms that the footcandle information shown here is correct.

ALTMAN STAGE LIGHTING
Model No. Comet

MIN. SPOT w/ IRIS: 2.3° (field)

SPOT BEAM SPREAD: 7.2° (field)

FLOOD BEAM SPREAD: 12.2° (field)

COLOR FRAMES: 6

FADER: Yes (blackout frame and 3-leaf dowser)

SHUTTERS: Yes (guillotine)

FAN COOLED: Yes

STANDARD LAMP: FLE (MR-16) - 360 W.

NOTES: Iris and shutters are located on top of unit. Focus is located on the side. Relamping is accomplished through access door at rear of unit.

WEIGHT: without stand – 55 lbs.

SIZE: 34½" x 17½" x 58½" (L x W x H)

PHOTOMETRICS CHART
(Performance data for this unit are measured using FLE (MR-16) - 360 W. lamp.)

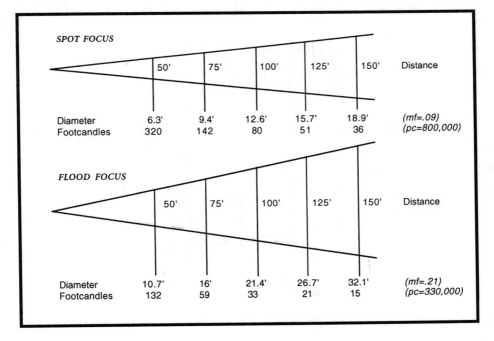

SPOT FOCUS

	50'	75'	100'	125'	150'	Distance
Diameter	6.3'	9.4'	12.6'	15.7'	18.9'	(mf=.09)
Footcandles	320	142	80	51	36	(pc=800,000)

FLOOD FOCUS

	50'	75'	100'	125'	150'	Distance
Diameter	10.7'	16'	21.4'	26.7'	32.1'	(mf=.21)
Footcandles	132	59	33	21	15	(pc=330,000)

ALTMAN STAGE LIGHTING
Model No. Dyna Spot

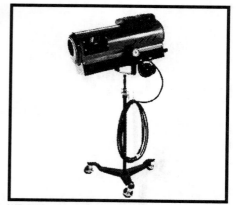

MIN. SPOT w/ IRIS: 1.5° (beam), 1.5° (field)

SPOT BEAM SPREAD: 14° (beam), 18.5° (field)

FLOOD BEAM SPREAD: 14° (beam), 23° (field)

COLOR FRAMES: 6

FADER: Yes (blackout frame)

SHUTTERS: Yes (guillotine)

FAN COOLED: Yes

STANDARD LAMP: DTJ - 1500 W.

NOTES: Iris and shutters are rear operated. Focus handle is on the side. Relamping is accomplished through hinged door on top.

WEIGHT: without stand – 42 lbs.

SIZE: 35" x 16" x 55–73" (L x W x H)

PHOTOMETRICS CHART
(Performance data for this unit are measured using DTJ - 1500 W. lamp.)

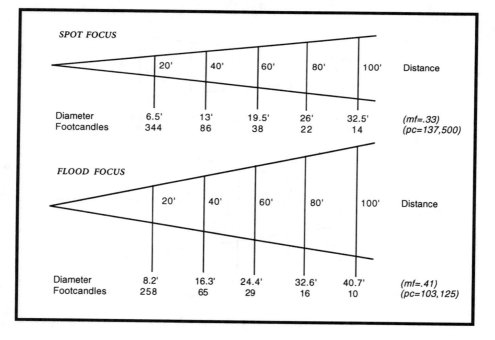

SPOT FOCUS

	20'	40'	60'	80'	100'	Distance
Diameter	6.5'	13'	19.5'	26'	32.5'	(mf=.33)
Footcandles	344	86	38	22	14	(pc=137,500)

FLOOD FOCUS

	20'	40'	60'	80'	100'	Distance
Diameter	8.2'	16.3'	24.4'	32.6'	40.7'	(mf=.41)
Footcandles	258	65	29	16	10	(pc=103,125)

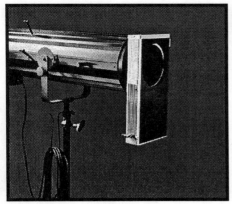

FOLLOW SPOT (incandescent/quartz)

ARIEL DAVIS
Model No. 3100

MIN. SPREAD, IRIS CLOSED: .9° *

MAX. SPREAD, IRIS OPEN: 12° *

COLOR FRAMES: 7

FADER: Yes (blackout frame)

SHUTTERS: Yes (guillotine)

FAN COOLED: No

STANDARD LAMP: 500PAR64NSP - 500 W.

NOTES: Iris and shutters are operated from the right side as is the focus.

WEIGHT: without stand – 55 lbs.

SIZE: 39½" x 48–72" (L x H)

PHOTOMETRICS CHART
(Performance data for this unit are measured using 500PAR64NSP - 500 W. lamp.)

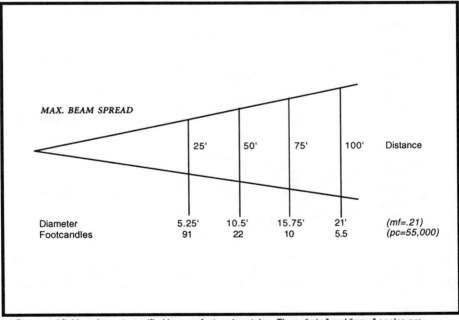

		25'	50'	75'	100'	Distance
MAX. BEAM SPREAD						
Diameter		5.25'	10.5'	15.75'	21'	*(mf=.21)*
Footcandles		91	22	10	5.5	*(pc=55,000)*

* Beam and field angles not specified in manufacturer's catalog. These "min." and "max." angles are calculated from given distance and diameter data. See Introduction for more information about calculating photometric data.

BERKEY COLORTRAN
Model No. 210-011*

MIN. SPOT w/ IRIS: .72° (field)

SPOT BEAM SPREAD: 9° (field)

FLOOD BEAM SPREAD: 9°x20° (field)

COLOR FRAMES: 5

FADER: No

SHUTTERS: Yes (guillotine)

FAN COOLED: No

STANDARD LAMP: FEL - 1 KW.

NOTES: Top rear housing swings back for relamping. Unit includes 20° spread lens for 'stripping' the beam, otherwise unit is lensless, incorporating an optical projection mirror to produce the beam.

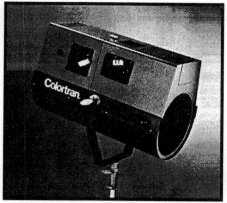

WEIGHT: 114 lbs.

SIZE: 35" x 32" x 57–71" (L x W x H)

PHOTOMETRICS CHART
(Performance data for this unit are measured using FEL - 1 KW. lamp.)

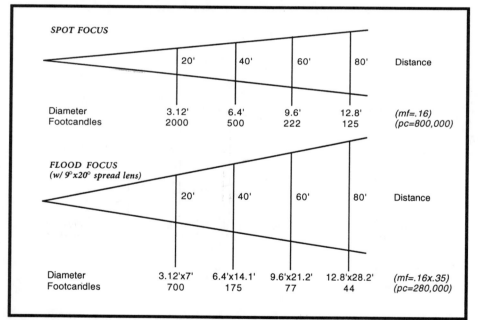

SPOT FOCUS

	20'	40'	60'	80'	Distance
Diameter	3.12'	6.4'	9.6'	12.8'	(mf=.16)
Footcandles	2000	500	222	125	(pc=800,000)

FLOOD FOCUS
(w/ 9°x20° spread lens)

	20'	40'	60'	80'	Distance
Diameter	3.12'x7'	6.4'x14.1'	9.6'x21.2'	12.8'x28.2'	(mf=.16x.35)
Footcandles	700	175	77	44	(pc=280,000)

* This model can be identified with model numbers -011, -012, -016, or -017 depending the wire leads and connectors supplied.

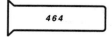

BERKEY COLORTRAN
Model No. LQFS - 1

MIN. SPREAD, IRIS CLOSED: .9°
MAX. SPREAD, IRIS OPEN: 13.4°
COLOR FRAMES: 1 (4 w/ optioinal boomerang)
FADER: No (iris can fully close for dowser effect)
SHUTTERS: No
FAN COOLED: No
STANDARD LAMP: BPS10-32* - 1KW.
NOTES: Iris handle is located on top of unit as is slot for gobo holder. Focus knob is on the side.

WEIGHT: 34 lbs.
SIZE: 41" x 12" (L x W)

PHOTOMETRICS CHART
(Performance data for this unit are measured using BPS10-32 - 1KW. lamp.)*

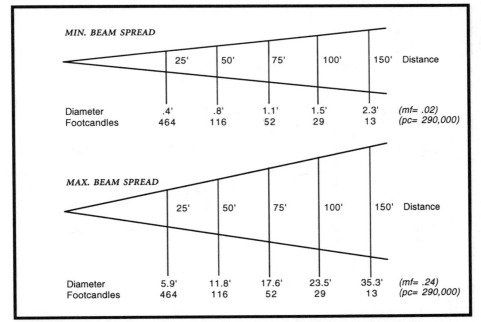

MIN. BEAM SPREAD

	25'	50'	75'	100'	150'	Distance
Diameter	.4'	.8'	1.1'	1.5'	2.3'	*(mf= .02)*
Footcandles	464	116	52	29	13	*(pc= 290,000)*

MAX. BEAM SPREAD

	25'	50'	75'	100'	150'	Distance
Diameter	5.9'	11.8'	17.6'	23.5'	35.3'	*(mf= .24)*
Footcandles	464	116	52	29	13	*(pc= 290,000)*

* BPS10-32 is the Colortran catalog code for this PAR64 NSP 1 KW. lamp. This lamp is obsolete but it is listed as the standard lamp because it is the one used to measure photometric data.

FOLLOW SPOT(incandescent/quartz)

CAPITOL STAGE LIGHTING
Model No. Mark III

MIN. SPOT w/ IRIS: (not specified in mfg. catalog)

SPOT BEAM SPREAD: 8.5° *

FLOOD BEAM SPREAD: (not specified)

COLOR FRAMES: 6

FADER: Yes (blackout frame)

SHUTTERS: Yes (guillotine)

FAN COOLED: Yes

STANDARD LAMP: 1500 W- Q T8 C13D

NOTES: Iris and shutters are rear operated. Focus handle is on the side. Relamping is accomplished through hinged door on top.

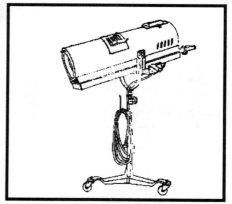

WEIGHT: (not specified in manufacturer's catalog)

SIZE: 43" x 50" (L x H)

PHOTOMETRICS CHART

(Performance data for this unit are measured using QT8C13D - 1500 W. lamp.)

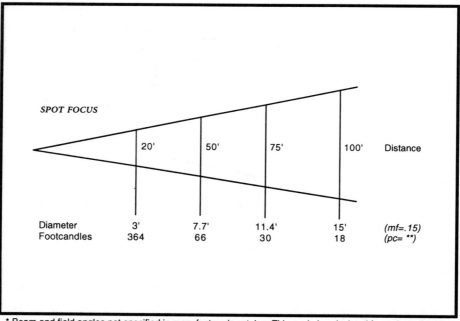

SPOT FOCUS

	20'	50'	75'	100'	Distance
Diameter	3'	7.7'	11.4'	15'	*(mf=.15)*
Footcandles	364	66	30	18	*(pc= **)*

* Beam and field angles not specified in manufacturer's catalog. This angle is calculated from given distance and diameter data. See Introduction for more information about calculating photometric data.
** Footcandle data given in catalog is not consistent enough to calculate a reliable peak candela *(pc)* figure for all throw distances. (At 100' throw and 18 fc, pc=180,000, while at 20' throw and 364 fc, pc=145,600.)

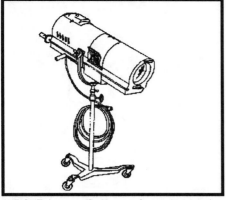

FOLLOW SPOT (incandescent/quartz)

CAPITOL STAGE LIGHTING
Model No. Quartzfollow

MIN. SPOT w/ IRIS: (not specified in mfg. catalog)

SPOT BEAM SPREAD: (not specified)*

FLOOD BEAM SPREAD: (not specified)

COLOR FRAMES: 6

FADER: Yes (blackout frame)

SHUTTERS: Yes (guillotine)

FAN COOLED: Yes

STANDARD LAMP: FEL - 1 KW.

WEIGHT: (not specified in manufacturer's catalog)

SIZE: 43" x 56" (L x H)

NOTES: Iris and shutters are rear operated. Focus handle is on the side. Relamping is accomplished through hinged door on top.

PHOTOMETRICS CHART
(Performance data for this unit are measured using FEL -1 KW. lamp.)

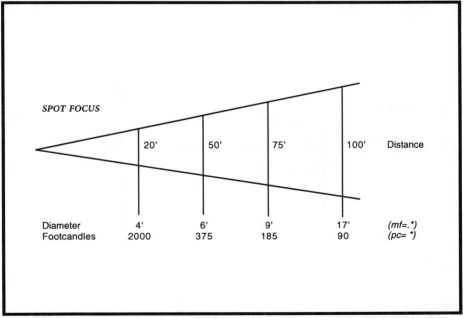

		SPOT FOCUS				
		20'	50'	75'	100'	Distance
Diameter		4'	6'	9'	17'	*(mf=.*)*
Footcandles		2000	375	185	90	*(pc= *)*

* Diameter and footcandle data as given in manufacturer's catalog is not consistent enough to determine a multiplying factor *(mf)* or a peak candela *(pc)* figure for all throw distances. (At 100' throw, 17' diameter and 90 fc, mf=.17 and pc=900,000, while at 20' throw, 4' diameter and 2000 fc, mf=.2 and pc=800,000.) See Introduction for more information about calculating photometric data.

CAPITOL STAGE LIGHTING
Model No. Super Q

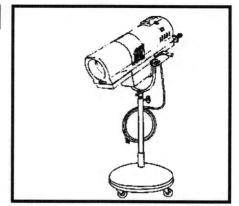

MIN. SPOT w/ IRIS: (not specified in mfg. catalog)

SPOT BEAM SPREAD: (not specified)*

FLOOD BEAM SPREAD: (not specified)

COLOR FRAMES: 6

FADER: Yes (blackout frame)

SHUTTERS: Yes (guillotine)

FAN COOLED: Yes

STANDARD LAMP: Unspecified 1350 W.

NOTES: Iris and shutters are rear operated. Focus handle is on the side. Relamping is accomplished through hinged door on top.

WEIGHT: (not specified in manufacturer's catalog)

SIZE: 42" x 50" (L x H)

PHOTOMETRICS CHART

(Performance data for this unit are measured using unspecified 1350 W. lamp.)

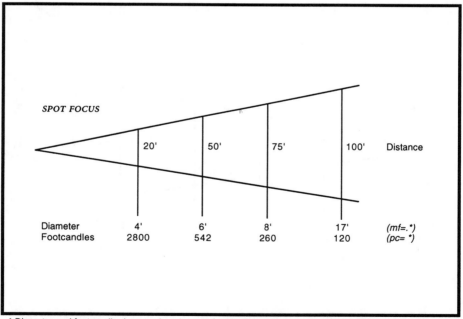

SPOT FOCUS				
	20'	50'	75'	100'	Distance
Diameter	4'	6'	8'	17'	(mf=.*)
Footcandles	2800	542	260	120	(pc= *)

* Diameter and footcandle data as given in manufacturer's catalog is not consistent enough to determine a multiplying factor *(mf)* or a peak candela *(pc)* figure for all throw distances. (At 100' throw, 17' diameter and 120 fc, mf=.17 and pc=1,200,000, while at 20' throw, 4' diameter and 2800 fc, mf=.2 and pc=1,120,000.) See Introduction for more information about calculating photometric data.

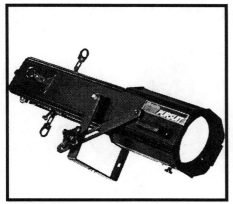

WEIGHT: 16 lbs.

SIZE: 31½" x 9¹³⁄₁₆" x 10¹³⁄₁₆" (L x W x H)

CCT LIGHTING
Model No. Z0607FS

MIN. SPOT w. IRIS: (not specified in mfg. catalog)

SPOT BEAM SPREAD: 5° (beam), 8° (field)

FLOOD BEAM SPREAD: 8° (beam), 10 (field)

COLOR FRAMES: 1

FADER: No

SHUTTERS: No

STANDARD LAMP: FMR - 600 W.

ACCESSORIES AVAILABLE:
4-color Boomerang

PHOTOMETRICS CHART

(Performance data for this unit are measured using FMR - 600 W. lamp.)

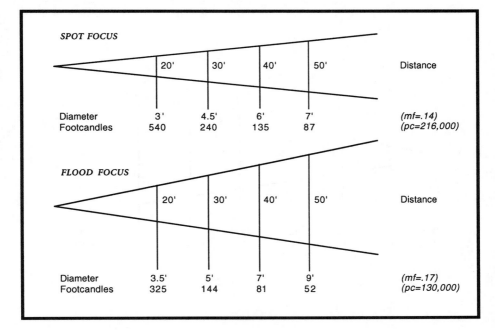

SPOT FOCUS

	20'	30'	40'	50'	Distance
Diameter	3'	4.5'	6'	7'	*(mf=.14)*
Footcandles	540	240	135	87	*(pc=216,000)*

FLOOD FOCUS

	20'	30'	40'	50'	Distance
Diameter	3.5'	5'	7'	9'	*(mf=.17)*
Footcandles	325	144	81	52	*(pc=130,000)*

FOLLOW SPOT (incandescent/quartz)

CENTURY LIGHTING
Model No. 1064

MIN. BEAM w/ IRIS: (not specified in mfg. catalog)

BEAM SPREAD: 11° (beam), 15° (field)

COLOR FRAMES: 1 (6 w/ optional boomerang)

FADER: Yes (fully closing iris)

SHUTTERS: No

FAN COOLED: No

STANDARD LAMP: PAR64 1KW. (28V.) cat. #4556

NOTES: Iris handle is located on top of unit. Focus adjustment is limited to sharp or soft edge by moving lens tube. Relamping is through hinged back of unit.

WEIGHT: (not specified in manufacturer's catalog)

SIZE: 31" x 55–79" (L x H)

PHOTOMETRICS CHART

(Performance data for this unit are measured using PAR 64 - 1 KW. (28V.))

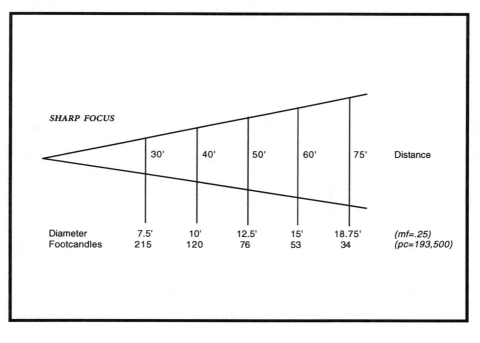

SHARP FOCUS

	30'	40'	50'	60'	75'	Distance
Diameter	7.5'	10'	12.5'	15'	18.75'	*(mf=.25)*
Footcandles	215	120	76	53	34	*(pc=193,500)*

CENTURY LIGHTING
Model No. 1542

MIN. SPOT w. IRIS: (not specified in mfg. catalog)

BEAM SPREAD: 8.5° (beam), 12° (field)

COLOR FRAMES: 1

FADER: No

SHUTTERS: Yes (guillotine)

STANDARD LAMP: 3M/T32/C13D - 3 KW.

NOTES: Iris is located on left side, as is guillotine, and control of the 30° spread roundel. Focus is accomplished by moving the lens barrel in or out.

WEIGHT: (not specified in manufacturer's catalog)

SIZE: 54" x 24" x 64" (L x W x H)

PHOTOMETRICS CHART

(Performance data for this unit are measured using 3M/T32/C13D - 3 KW. lamp.)

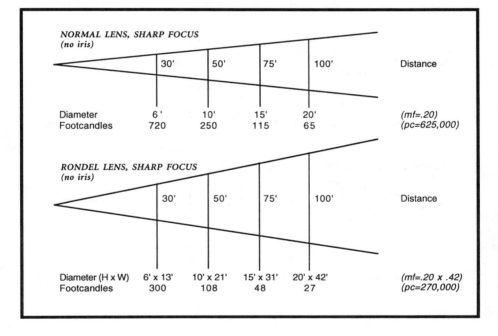

NORMAL LENS, SHARP FOCUS
(no iris)

	30'	50'	75'	100'	Distance
Diameter	6 '	10'	15'	20'	*(mf=.20)*
Footcandles	720	250	115	65	*(pc=625,000)*

RONDEL LENS, SHARP FOCUS
(no iris)

	30'	50'	75'	100'	Distance
Diameter (H x W)	6' x 13'	10' x 21'	15' x 31'	20' x 42'	*(mf=.20 x .42)*
Footcandles	300	108	48	27	*(pc=270,000)*

FOLLOW SPOT (incandescant/quartz)

CENTURY LIGHTING
Model No. 1545

MIN. SPOT w. IRIS: (not specified in mfg. catalog)

BEAM SPREAD: 8.5° (beam), 12° (field)

COLOR FRAMES: 1

FADER: No

SHUTTERS: Yes (guillotine)

STANDARD LAMP: 3M/T32/C13D - 5 KW.

NOTES: Iris is located on left side, as is guillotine, and control of the 30° spread roundel. Focus is accomplished by moving the lens barrel in or out.

WEIGHT: (not specified in manufacturer's catalog)

SIZE: 54" x 24" x 64" (L x W x H)

PHOTOMETRICS CHART
(Performance data for this unit are measured using 3M/T32/C13D - 5 KW. lamp.)

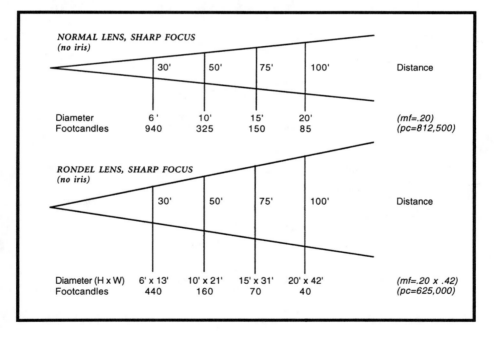

NORMAL LENS, SHARP FOCUS *(no iris)*

	30'	50'	75'	100'	Distance
Diameter	6'	10'	15'	20'	*(mf=.20)*
Footcandles	940	325	150	85	*(pc=812,500)*

RONDEL LENS, SHARP FOCUS *(no iris)*

	30'	50'	75'	100'	Distance
Diameter (H x W)	6' x 13'	10' x 21'	15' x 31'	20' x 42'	*(mf=.20 x .42)*
Footcandles	440	160	70	40	*(pc=625,000)*

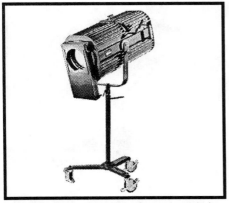

WEIGHT: (not specified in manufacturer's catalog)

SIZE: 37" x 17½" (L x W)

CENTURY LIGHTING
Model No. 4481

MIN. SPOT w/ IRIS: (not specified in mfg. catalog)

SPOT BEAM SPREAD: 7.5° (beam), 13° (field)

FLOOD BEAM SPREAD: 11.5° (beam), 24° (field)

COLOR FRAMES: 6

FADER: No

SHUTTERS: Yes (guillotine)

FAN COOLED: No

STANDARD LAMP: FEL - 1 KW.

NOTES: Iris and shutters are located on top of unit. Focus is located on the left side near the back. Relamping is through socket cap on back of unit.

PHOTOMETRICS CHART

(Performance data for this unit are measured using FEL - 1 KW. lamp.)

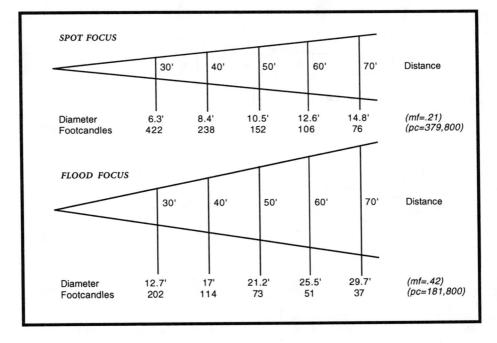

SPOT FOCUS

	30'	40'	50'	60'	70'	Distance
Diameter	6.3'	8.4'	10.5'	12.6'	14.8'	(mf=.21)
Footcandles	422	238	152	106	76	(pc=379,800)

FLOOD FOCUS

	30'	40'	50'	60'	70'	Distance
Diameter	12.7'	17'	21.2'	25.5'	29.7'	(mf=.42)
Footcandles	202	114	73	51	37	(pc=181,800)

FOLLOW SPOT (incandescent/quartz)

CENTURY STRAND
Model No. 4452

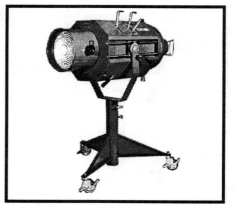

MIN. SPOT w/ IRIS: (not specified in mfg. catalog)

SPOT BEAM SPREAD: 6° (beam), 8° (field)

FLOOD BEAM SPREAD: 8°x16.4° * (field)

COLOR FRAMES: 1 (boomerang option available)

FADER: No

SHUTTERS: Yes (guillotine)

FAN COOLED: Yes

STANDARD LAMP: BWA - 2 KW.

NOTES: Iris and shutter handles are located on top of unit, focus is located at front. Relamping accomplished through hinged socket cap at rear of unit. An internal, horizontal spread rondel is used to "strip" the beam.

WEIGHT: (not specified in manufacturer's catalog)

SIZE: 54" x 24" x 53–64" (L x W x H)

PHOTOMETRICS CHART

(Performance data for this unit are measured using BWA - 2 KW. lamp.)

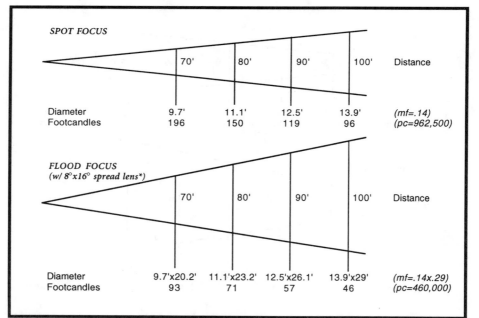

SPOT FOCUS

	70'	80'	90'	100'	Distance
Diameter	9.7'	11.1'	12.5'	13.9'	*(mf=.14)*
Footcandles	196	150	119	96	*(pc=962,500)*

FLOOD FOCUS
(w/ 8°x16° spread lens)*

	70'	80'	90'	100'	Distance
Diameter	9.7'x20.2'	11.1'x23.2'	12.5'x26.1'	13.9'x29'	*(mf=.14x.29)*
Footcandles	93	71	57	46	*(pc=460,000)*

* Manufacturer describes the spread rondel as a "9-1/2", 30° spread lens," but using the given diameter figures, the angle calculates as 16.4°. See Introduction for more information about calculating photometric data.

474

WEIGHT: 94 lbs.
SIZE: 33" x 19" x 60" (L x W x H)

KLIEGL BROS.
Model Nos.1174
1175 (w/ boomerang)

MIN. SPOT w/ IRIS: (not specified in mfg. catalog)

SPOT BEAM SPREAD: 4° *

FLOOD BEAM SPREAD: 13° *

COLOR FRAMES: 1

FADER: No

SHUTTERS: Yes (guillotine)

FAN COOLED: Yes

STANDARD LAMP: T30/C13D - 2 KW.

Accessories: Color Frame (14" sq.)

NOTES: Iris, shutters and focus are rear operated. Relamping is accomplished through hinged casing on the back.

PHOTOMETRICS CHART

(Performance data for this unit are measured using T30/C13D - 2 KW. lamp.)

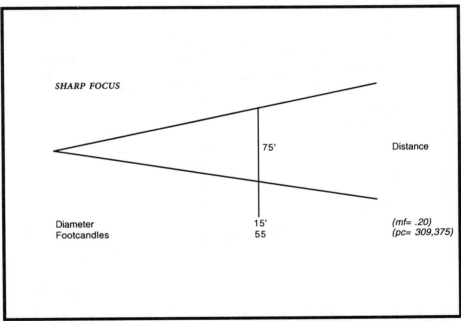

SHARP FOCUS

75'

Distance

Diameter	15'	(mf= .20)
Footcandles	55	(pc= 309,375)

* Beam and field angles not specified in manufacturer's catalog. This angle is calculated from given distance and diameter data. See Introduction for more information about calculating photometric data.

KLIEGL BROS.
Model Nos.1178
1179 (w/ boomerang)

MIN. SPOT w/ IRIS: (not specified in mfg. catalog)

SPOT BEAM SPREAD: 3.5° *

FLOOD BEAM SPREAD: 11.5° *

COLOR FRAMES: 1 (1178), 5 (1179)

FADER: No

SHUTTERS: Yes (guillotine)

FAN COOLED: Yes

STANDARD LAMP: 3M T32/C13D - 3 KW.

NOTES: Iris, shutters, focus and spread lens are rear operated. Relamping is accomplished through hinged casing on the back.

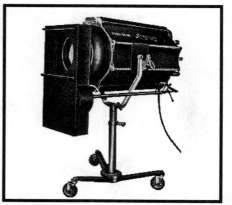

WEIGHT: 196 lbs.

SIZE: 48" x 23" x 54" (L x W x H)

PHOTOMETRICS CHART

(Performance data for this unit are measured using 3M T32/C13D - 3 KW. lamp.)

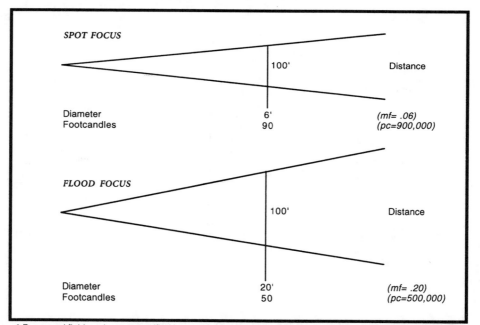

SPOT FOCUS

100'   Distance

| Diameter | 6' | (mf= .06) |
| Footcandles | 90 | (pc=900,000) |

FLOOD FOCUS

100'   Distance

| Diameter | 20' | (mf= .20) |
| Footcandles | 50 | (pc=500,000) |

* Beam and field angles not specified in manufacturer's catalog. This angle is calculated from given distance and diameter data. See Introduction for more information about calculating photometric data.

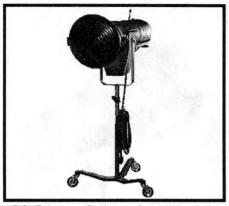

KLIEGL BROS.
Model No. 1393

MIN. SPOT w/ IRIS: (not specified in mfg. catalog)

BEAM SPREAD: 6° (field)*

COLOR FRAMES: 5

FADER: No

SHUTTERS: Yes (guillotine and framing shutters)

FAN COOLED: Yes

STANDARD LAMP: DWT - 1 KW.

NOTES: Iris is located on top of unit. Focus adjustment knob is on lens barrel. Relamping is accomplished through hinged rear body.

WEIGHT: (not specified in manufacturer's catalog)

SIZE: 44" x 15½" (L x W)

PHOTOMETRICS CHART
(Performance data for this unit are measured using DWT - 1 KW. lamp.)

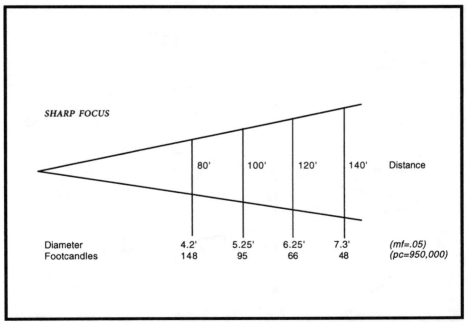

SHARP FOCUS

	80'	100'	120'	140'	Distance
Diameter	4.2'	5.25'	6.25'	7.3'	(mf=.05)
Footcandles	148	95	66	48	(pc=950,000)

* The manufacturer's catalog specifies a 6° field angle, however the given distance and diameter data yields an angle of 3°. See Introduction for more information on calculating photometric data.

LIGHTING & ELECTRONICS
Model No. 64-10, "Mini"

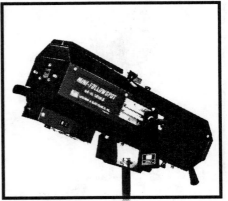

MIN. SPOT w/ IRIS: 4°

SPOT BEAM SPREAD: 17° (beam)

FLOOD BEAM SPREAD: 22.6° (beam)

COLOR FRAMES: 1 (6 frame boomerang optional)

FADER: No (dimmer available as an option)

SHUTTERS: Yes (guillotine)

FAN COOLED: Yes

STANDARD LAMP: DYS - 600 W.

NOTES: Iris is located on top side of unit. Focus knobs are located on top and bottom. Relamping is accomplished through access panel on back.

WEIGHT: without stand – 17 lbs.

SIZE: 27¾" x 13¾" x 48¾–65¾" (L x W x H)

PHOTOMETRICS CHART

(Performance data for this unit are measured using DYS - 600 W. lamp.)

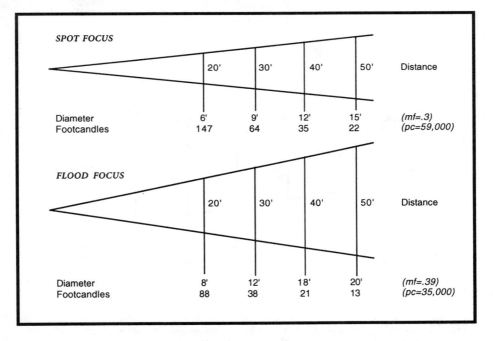

SPOT FOCUS

	20'	30'	40'	50'	Distance
Diameter	6'	9'	12'	15'	*(mf=.3)*
Footcandles	147	64	35	22	*(pc=59,000)*

FLOOD FOCUS

	20'	30'	40'	50'	Distance
Diameter	8'	12'	18'	20'	*(mf=.39)*
Footcandles	88	38	21	13	*(pc=35,000)*

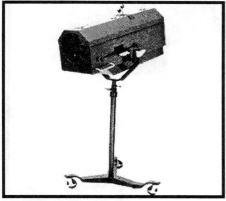

LIGHTING & ELECTRONICS
Model No. 64-14, "Commander"*

MIN. SPOT w/ IRIS: (not specified in mfg. catalog)

SPOT BEAM SPREAD: 11° (beam)

FLOOD BEAM SPREAD: 17° (beam)

COLOR FRAMES: 7

FADER: No

SHUTTERS: Yes (guillotine)

FAN COOLED: Yes

STANDARD LAMP: FEL - 1 KW.

NOTES: Iris and shutter handles are located on top. Focus hande is on the right side.

WEIGHT: without stand – 46 lbs.

SIZE: 33¼" x 12" (L x W)

PHOTOMETRICS CHART
(Performance data for this unit are measured using FEL - 1 KW. lamp.)

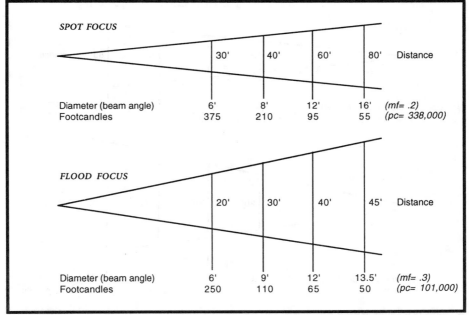

SPOT FOCUS

	30'	40'	60'	80'	Distance
Diameter (beam angle)	6'	8'	12'	16'	(mf= .2)
Footcandles	375	210	95	55	(pc= 338,000)

FLOOD FOCUS

	20'	30'	40'	45'	Distance
Diameter (beam angle)	6'	9'	12'	13.5'	(mf= .3)
Footcandles	250	110	65	50	(pc= 101,000)

* There are two versions of model 64-14, each having different photometrics. This page describes the earlier version.

FOLLOW SPOT (incandescent/quartz)

LIGHTING & ELECTRONICS
Model No. 64-14, "Commander"*

MIN. SPOT w/ IRIS: (not specified in mfg. catalog)

SPOT BEAM SPREAD: 12° (beam)

FLOOD BEAM SPREAD: 18° (beam)

COLOR FRAMES: 7

FADER: Yes (blackout frame)

SHUTTERS: Yes (guillotine)

FAN COOLED: Yes

STANDARD LAMP: FEL - 1 KW.

NOTES: Iris and shutter handles are located on top. Focus hande is on the right side.

WEIGHT: (not specified in manufacturer's catalog)

SIZE: 35" x 12" (L x W)

PHOTOMETRICS CHART
(Performance data for this unit are calculated from manufacturer's beam angle and peak candela data using FEL - 1 KW. lamp.)

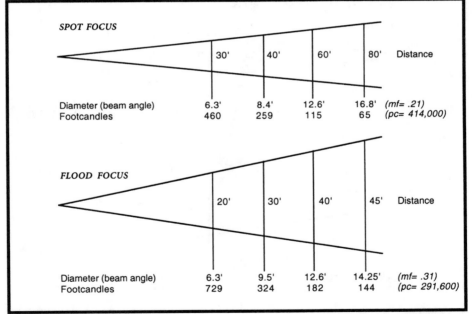

SPOT FOCUS

	30'	40'	60'	80'	Distance
Diameter (beam angle)	6.3'	8.4'	12.6'	16.8'	(mf= .21)
Footcandles	460	259	115	65	(pc= 414,000)

FLOOD FOCUS

	20'	30'	40'	45'	Distance
Diameter (beam angle)	6.3'	9.5'	12.6'	14.25'	(mf= .31)
Footcandles	729	324	182	144	(pc= 291,600)

* There are two versions of model 64-14, each having different photometrics. This page describes the later version.

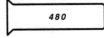

LUDWIG PANI
Model No. HV-1001

MIN. SPOT w/ IRIS: (not specified in mfg. catalog)

NARROW SPREAD LENS: 8° *

WIDE SPREAD LENS: 13.3° *

COLOR FRAMES: 1

FADER: No (but iris can be closed completely)

SHUTTERS: Yes (4-way)

FAN COOLED: No

STANDARD LAMP: Radium RH 1003 P-1 KW. (220V)

NOTES: Iris is located on top of unit and focus handle is on the side. 4-color boomerang available. Beam width is changed by means of four interchangeable lenses with focal lengths of 10cm, 20cm, 35cm, and 40cm.

WEIGHT: 28 lbs. (12.75 kg.)

SIZE: 23½" (600mm) x 16¼" (415mm) (L x H)

PHOTOMETRICS CHART

(Performance data for this unit are measured using RADIUM RH 1003 P - 1 KW. lamp.)

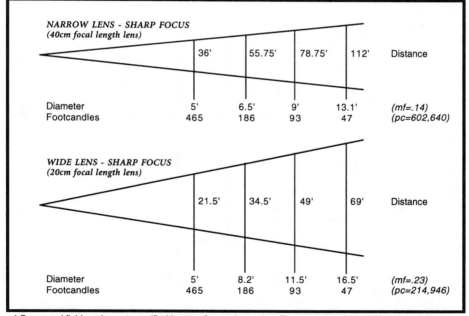

NARROW LENS - SHARP FOCUS
(40cm focal length lens)

	36'	55.75'	78.75'	112'	Distance
Diameter	5'	6.5'	9'	13.1'	(mf=.14)
Footcandles	465	186	93	47	(pc=602,640)

WIDE LENS - SHARP FOCUS
(20cm focal length lens)

	21.5'	34.5'	49'	69'	Distance
Diameter	5'	8.2'	11.5'	16.5'	(mf=.23)
Footcandles	465	186	93	47	(pc=214,946)

* Beam and field angles not specified in manufacturer's catalog. This angle is calculated from given distance and field diameter data. See Introduction for more information about calculating photometric data.
** Photometric data in manufacturer's catalog is given in metric units, and was converted to English units for this book. Also, these diameter and footcandle figures were read from a graph, and as such can only be considered approximate values.

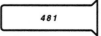

FOLLOW SPOT (incandescent/quartz)

LUDWIG PANI
Model No. HV-2002

MIN. SPOT w/ IRIS: (not specified in mfg. catalog)

NARROW SPREAD LENS: 7.2° *

WIDE SPREAD LENS: 18.5° *

COLOR FRAMES: 1

FADER: No (but iris can be closed completely)

SHUTTERS: Yes (4-way)

FAN COOLED: No

STANDARD LAMP: Radium RH 2003 P-2 KW. (220V)

NOTES: Iris is located on top of unit and focus handle is on the side. 4-color boomerang available. Beam width is changed by means of four interchangeable lenses with focal lengths of 10cm, 20cm, 35cm, and 40cm.

WEIGHT: 32.2 lbs. (14.6 kg.)

SIZE: 24½" (620mm) x 15¾" (400mm) (L x H)

PHOTOMETRICS CHART

(Performance data for this unit are measured using RADIUM RH 2003 P - 2 KW. lamp.)

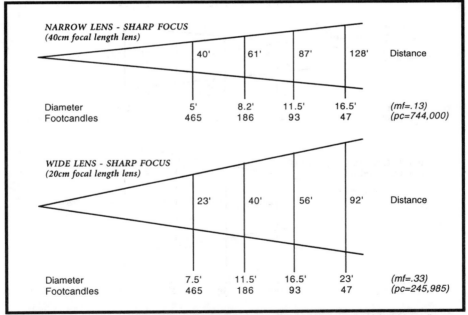

NARROW LENS - SHARP FOCUS
(40cm focal length lens)

	40'	61'	87'	128'	Distance
Diameter	5'	8.2'	11.5'	16.5'	*(mf=.13)*
Footcandles	465	186	93	47	*(pc=744,000)*

WIDE LENS - SHARP FOCUS
(20cm focal length lens)

	23'	40'	56'	92'	Distance
Diameter	7.5'	11.5'	16.5'	23'	*(mf=.33)*
Footcandles	465	186	93	47	*(pc=245,985)*

* Beam and field angles not specified in manufacturer's catalog. This angle is calculated from given distance and field diameter data. See Introduction for more information about calculating photometric data.
** Photometric data in manufacturer's catalog is given in metric units, and was converted to English units for this book. Also, these diameter and footcandle figures were read from a graph, and as such can only be considered approximate values.

LYCIAN STAGE LIGHTING
Model No. 1206

MIN. SPOT w/ IRIS:1° *

SPOT BEAM SPREAD: 9° * (iris open)

FLOOD BEAM SPREAD: 9°x18.5° * (w/ spread lens)

COLOR FRAMES: 6

FADER: Yes (blackout frame)

SHUTTERS: Yes (guillotine)

FAN COOLED: Yes

STANDARD LAMP: FEL - 1 KW.

NOTES: Iris is located on top of unit, focus on side. Relamping is accomplished through top door. Unit has gobo slot, and separately controlled horizontal spread lens.

WEIGHT: without stand – 40 lbs.

SIZE: 32" x 13½" x 54–67" (L x W x H)

PHOTOMETRICS CHART

(Performance data for this unit are measured using FEL - 1 KW. lamp.)

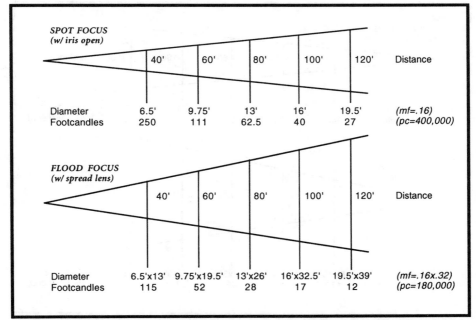

SPOT FOCUS *(w/ iris open)*

Distance	40'	60'	80'	100'	120'	
Diameter	6.5'	9.75'	13'	16'	19.5'	(mf=.16)
Footcandles	250	111	62.5	40	27	(pc=400,000)

FLOOD FOCUS *(w/ spread lens)*

Distance	40'	60'	80'	100'	120'	
Diameter	6.5'x13'	9.75'x19.5'	13'x26'	16'x32.5'	19.5'x39'	(mf=.16x.32)
Footcandles	115	52	28	17	12	(pc=180,000)

* Beam and field angles not specified in manufacturer's catalog. This angle is calculated from given distance and field diameter data. See Introduction for more information about calculating photometric data.

LYCIAN STAGE LIGHTING
Model No. 1207

MIN. SPOT w/ IRIS: 1° *

SPOT BEAM SPREAD: 11° * (iris open)

FLOOD BEAM SPREAD: 11°x18.5° * (w/ spread lens)

COLOR FRAMES: 6

FADER: Yes (blackout frame)

SHUTTERS: Yes (guillotine)

FAN COOLED: Yes

STANDARD LAMP: BWA - 2 KW.

NOTES: Iris is located on top of unit, focus on side. Relamping is accomplished through top door. Unit has gobo slot, and separately controlled horizontal spread lens.

WEIGHT: without stand – 46 lbs.

SIZE: 32" x 13½" x 54–67" (L x W x H)

PHOTOMETRICS CHART

(Performance data for this unit are measured using BWA - 2 KW. lamp.)

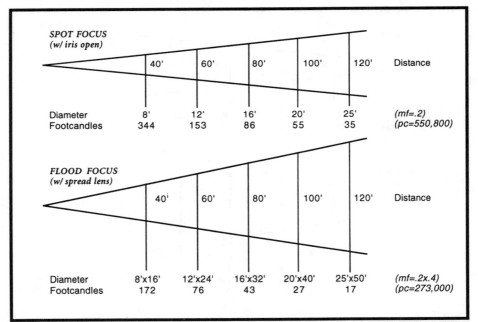

SPOT FOCUS *(w/ iris open)*

Distance	40'	60'	80'	100'	120'	
Diameter	8'	12'	16'	20'	25'	(mf=.2)
Footcandles	344	153	86	55	35	(pc=550,800)

FLOOD FOCUS *(w/ spread lens)*

Distance	40'	60'	80'	100'	120'	
Diameter	8'x16'	12'x24'	16'x32'	20'x40'	25'x50'	(mf=.2x.4)
Footcandles	172	76	43	27	17	(pc=273,000)

* Beam and field angles not specified in manufacturer's catalog. This angle is calculated from given distance and field diameter data. See Introduction for more information about calculating photometric data.

LYCIAN STAGE LIGHTING
Model No. 1236, "Clubspot"

MIN. SPOT w/ IRIS: .95° *

SPOT BEAM SPREAD: 7° *

FLOOD BEAM SPREAD: 14° *

COLOR FRAMES: 6

FADER: Yes (blackout frame)

SHUTTERS: Yes (guillotine)

FAN COOLED: Yes

STANDARD LAMP: FLE (MR-16) - 360 W.

WEIGHT: without stand – 29 lbs.

SIZE: 26¾" x 11" x 57–70" (L x W x H)

NOTES: Iris and shutters are located on the top of the unit. Relamping is accomplished through an access door at the rear of the unit.

PHOTOMETRICS CHART
(Performance data for this unit are measured using FLE (MR-16) - 360 W. lamp.)

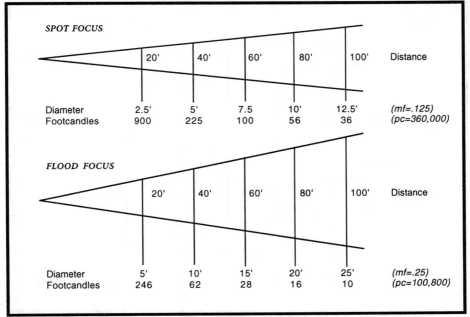

SPOT FOCUS						
	20'	40'	60'	80'	100'	Distance
Diameter	2.5'	5'	7.5	10'	12.5'	(mf=.125)
Footcandles	900	225	100	56	36	(pc=360,000)

FLOOD FOCUS						
	20'	40'	60'	80'	100'	Distance
Diameter	5'	10'	15'	20'	25'	(mf=.25)
Footcandles	246	62	28	16	10	(pc=100,800)

* Beam and field angles not specified in manufacturer's catalog. This angle is calculated from given distance and diameter data. See Introduction for more information about calculating photometric data.

FOLLOW SPOT (incandescent/quartz)

PHOEBUS
Model No. Ultra Quartz

MIN. SPOT w/ IRIS: .64° *

SPOT BEAM SPREAD: 7.6° *

FLOOD BEAM SPREAD: 18.6° *

COLOR FRAMES: 6

FADER: No

SHUTTERS: Yes (guillotine)

FAN COOLED: Yes

STANDARD LAMP: FLE (MR-16) - 360 W.

NOTES: Iris is located on top of unit. Focus is located on side of unit.

WEIGHT: with base – 75.5 lbs.

SIZE: 31.5" x 51" (L x H)

PHOTOMETRICS CHART

(Performance data for this unit are measured using FLE (MR-16) - 360 W. lamp.)

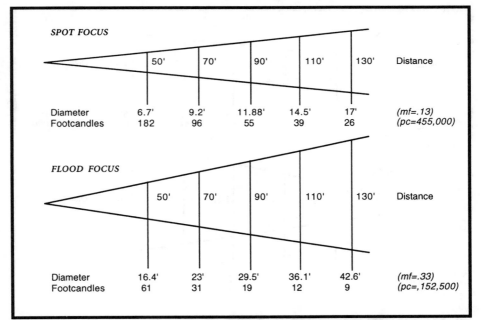

SPOT FOCUS						
	50'	70'	90'	110'	130'	Distance
Diameter	6.7'	9.2'	11.88'	14.5'	17'	(mf=.13)
Footcandles	182	96	55	39	26	(pc=455,000)
FLOOD FOCUS						
	50'	70'	90'	110'	130'	Distance
Diameter	16.4'	23'	29.5'	36.1'	42.6'	(mf=.33)
Footcandles	61	31	19	12	9	(pc=,152,500)

* Beam and field angles not specified in manufacturer's catalog. This angle is calculated from given distance and diameter data. See Introduction for more information about calculating photometric data.

STRAND CENTURY
Model No. 4473

MIN. SPOT w. IRIS: (not specified in mfg. catalog)

BEAM SPREAD: 8° (beam), 13° (field)

COLOR FRAMES: 1

FADER: Yes (blackout disc)

SHUTTERS: No

STANDARD LAMP: CYX - 2KW.

OTHER LAMPS: CP56 - 2KW. *(cf=.92)*

NOTES: Iris is located on right side of the unit, as is the blackout disc.

WEIGHT: 60 lbs.

SIZE: 39⁹⁄₁₆" x 16¹⁵⁄₁₆" x 18⁵⁄₁₆" (L x W x H)

PHOTOMETRICS CHART

(Performance data for this unit are measured using CYX - 2KW. lamp.)

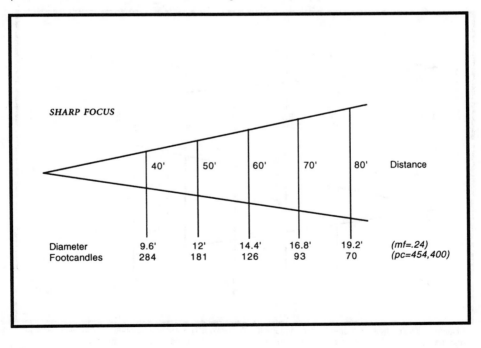

SHARP FOCUS

	40'	50'	60'	70'	80'	Distance
Diameter	9.6'	12'	14.4'	16.8'	19.2'	*(mf=.24)*
Footcandles	284	181	126	93	70	*(pc=454,400)*

STRONG INTERNATIONAL
Model No. Trouperette III

MIN. SPOT w/ IRIS: 1.4° *

SPOT BEAM SPREAD: 7° *

FLOOD BEAM SPREAD: 22° *

COLOR FRAMES: 6

FADER: No

SHUTTERS: Yes (guillotine)

FAN COOLED: No

STANDARD LAMP: FEL - 1 KW.

NOTES: Iris is located at top of unit. Focus handle is on right front.

WEIGHT: 74 lbs.

SIZE: 33" x 52–77" (L x H)

PHOTOMETRICS CHART

(Performance data for this unit are measured using FEL - 1 KW. lamp.)

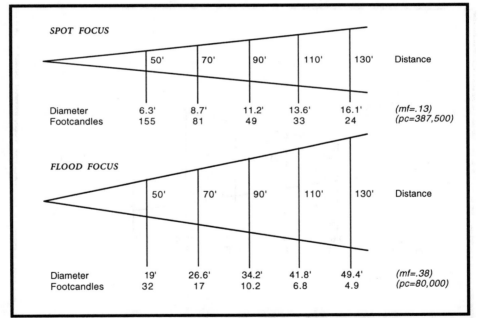

SPOT FOCUS

		50'	70'	90'	110'	130'	Distance
Diameter		6.3'	8.7'	11.2'	13.6'	16.1'	*(mf=.13)*
Footcandles		155	81	49	33	24	*(pc=387,500)*

FLOOD FOCUS

		50'	70'	90'	110'	130'	Distance
Diameter		19'	26.6'	34.2'	41.8'	49.4'	*(mf=.38)*
Footcandles		32	17	10.2	6.8	4.9	*(pc=80,000)*

* Beam and field angles not specified in manufacturer's catalog. This angle is calculated from given distance and diameter data. See Introduction for more information about calculating photometric data.

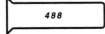

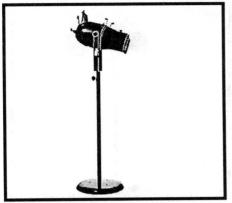

TIMES SQUARE LIGHTING
Model No. 6E4IF

MIN. SPOT w/ IRIS: (not specified in mfg. catalog)

BEAM SPREAD: 49° (field)*

COLOR FRAMES: 1

FADER: No

SHUTTERS: Yes (framing shutters)

FAN COOLED: No

STANDARD LAMP: EGJ - 1 KW.

NOTES: Iris is located on top of unit.

WEIGHT: with stand – 37 lbs.

SIZE: 18" x 10" (L x W)

PHOTOMETRICS CHART

(Information about lamp used to measure performance data is not provided in manufacturer's catalog.)

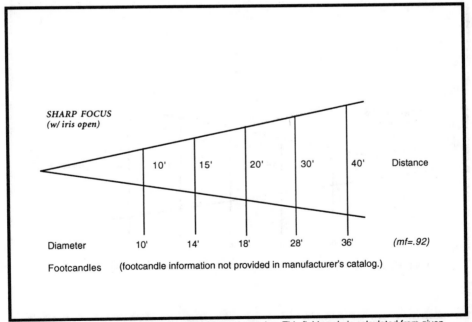

SHARP FOCUS
(w/ iris open)

| | 10' | 15' | 20' | 30' | 40' | Distance |

| Diameter | 10' | 14' | 18' | 28' | 36' | (mf=.92) |

Footcandles   (footcandle information not provided in manufacturer's catalog.)

* Beam and field angles not specified in manufacturer's catalog. This field angle is calculated from given distance and field diameter data. See Introduction for more information about calculating photometric data.

FOLLOW SPOT (incandescent/quartz)

TIMES SQUARE LIGHTING
Model No. 6EIF

MIN. SPOT w/ IRIS: (not specified in mfg. catalog)

BEAM SPREAD: 22° (field)*

COLOR FRAMES: 1

FADER: No

SHUTTERS: Yes (framing shutters)

FAN COOLED: No

STANDARD LAMP: EGJ - 1 KW.

NOTES: Iris is located on top of unit.

WEIGHT: with stand – 37 lbs.

SIZE: 18" x 10" (L x W)

PHOTOMETRICS CHART

(Information about lamp used to measure performance data is not provided in manufacturer's catalog.)

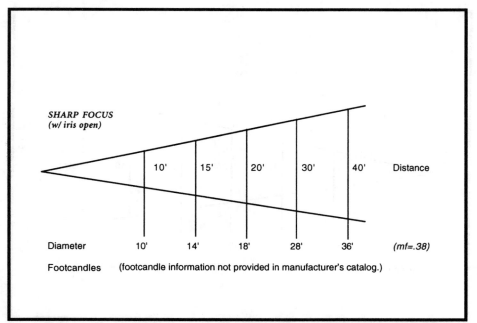

SHARP FOCUS
(w/ iris open)

| Distance | 10' | 15' | 20' | 30' | 40' |
| Diameter | 10' | 14' | 18' | 28' | 36' | (mf=.38) |

Footcandles (footcandle information not provided in manufacturer's catalog.)

* Beam and field angles not specified in manufacturer's catalog. This field angle is calculated from given
distance and field diameter data. See Introduction for more information about calculating photometric data.

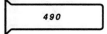

TIMES SQUARE LIGHTING
Model No. IQ3MF

MIN. SPOT w/ IRIS: (not specified in mfg. catalog)

BEAM SPREAD: 28° (field)*

COLOR FRAMES: 1

FADER: No

SHUTTERS: Yes (framing shutters)

FAN COOLED: No

STANDARD LAMP: FAD - 650 W.

NOTES: Iris is located on top of unit.

WEIGHT: with stand – 9 lbs.

SIZE: (not specified in manufacturer's catalog)

PHOTOMETRICS CHART

(Information about lamp used to measure performance data is not provided in manufacturer's catalog.)

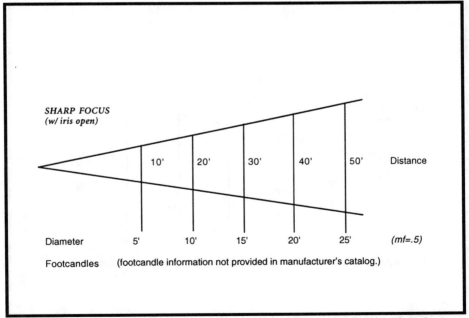

SHARP FOCUS
(w/ iris open)

| | 10' | 20' | 30' | 40' | 50' | Distance |

| Diameter | 5' | 10' | 15' | 20' | 25' | (mf=.5) |

Footcandles   (footcandle information not provided in manufacturer's catalog.)

* Beam and field angles not specified in manufacturer's catalog. This field angle is calculated from given distance and field diameter data. See Introduction for more information about calculating photometric data.

TIMES SQUARE LIGHTING
Model No. IQ3NF

MIN. SPOT w/ IRIS: (not specified in mfg. catalog)

BEAM SPREAD: 15° (field)*

COLOR FRAMES: 1

FADER: No

SHUTTERS: Yes (framing shutters)

FAN COOLED: No

STANDARD LAMP: FAD - 650 W.

NOTES: Iris is located on top of unit.

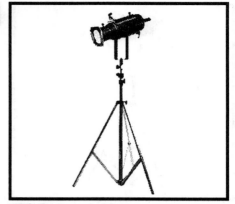

WEIGHT: with stand – 9 lbs.

SIZE: (not specified in manufacturer's catalog)

PHOTOMETRICS CHART

(Information about lamp used to measure performance data is not provided in manufacturer's catalog.)

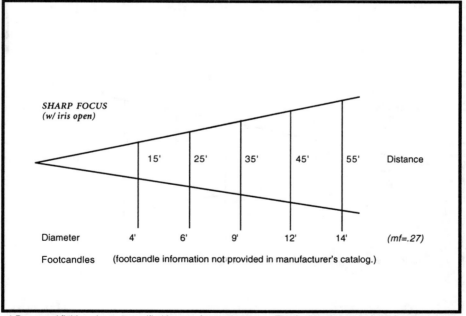

SHARP FOCUS
(w/ iris open)

| | 15' | 25' | 35' | 45' | 55' | Distance |

| Diameter | 4' | 6' | 9' | 12' | 14' | (mf=.27) |

Footcandles   (footcandle information not provided in manufacturer's catalog.)

* Beam and field angles not specified in manufacturer's catalog. This field angle is calculated from given distance and field diameter data. See Introduction for more information about calculating photometric data.

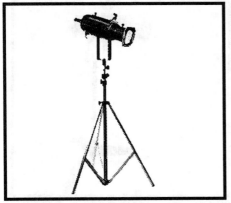

TIMES SQUARE LIGHTING
Model No. IQ3WF

MIN. SPOT w/ IRIS: (not specified in mfg. catalog)

BEAM SPREAD: 44° (field)*

COLOR FRAMES: 1

FADER: No

SHUTTERS: Yes (framing shutters)

FAN COOLED: No

STANDARD LAMP: FAD - 650 W.

NOTES: Iris is located on top of unit.

WEIGHT: with stand – 9 lbs.

SIZE: (not specified in manufacturer's catalog)

PHOTOMETRICS CHART

(Information about lamp used to measure performance data is not provided in manufacturer's catalog.)

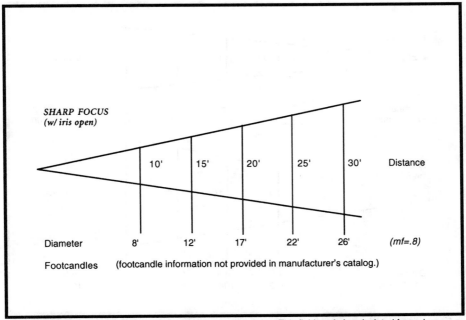

SHARP FOCUS
(w/ iris open)

| | 10' | 15' | 20' | 25' | 30' | Distance |

Diameter 8' 12' 17' 22' 26' (mf=.8)

Footcandles (footcandle information not provided in manufacturer's catalog.)

* Beam and field angles not specified in manufacturer's catalog. This field angle is calculated from given distance and field diameter data. See Introduction for more information about calculating photometric data.

FOLLOW SPOT (incandescent/quartz)

TIMES SQUARE LIGHTING
Model No. IQ4ZMF

MIN. SPOT w/ IRIS: 3° *

SPOT BEAM SPREAD: 33° *

FLOOD BEAM SPREAD: (not specified)

COLOR FRAMES: 1

FADER: No

SHUTTERS: Yes (framing shutters)

FAN COOLED: No

STANDARD LAMP: EVR - 500 W.

NOTES: Iris is located on top of unit. Focus and flood/spot knobs are underneath at the front.

WEIGHT: with stand – 44 lbs.

SIZE: 19" x 6¼" x 51" (L x W x H)

PHOTOMETRICS CHART
(Performance data for this unit are measured using EVR - 500 W. lamp.)

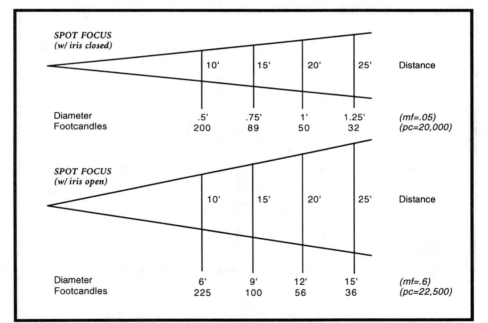

SPOT FOCUS (w/ iris closed)

	10'	15'	20'	25'	Distance
Diameter	.5'	.75'	1'	1.25'	(mf=.05)
Footcandles	200	89	50	32	(pc=20,000)

SPOT FOCUS (w/ iris open)

	10'	15'	20'	25'	Distance
Diameter	6'	9'	12'	15'	(mf=.6)
Footcandles	225	100	56	36	(pc=22,500)

* Beam and field angles not specified in manufacturer's catalog. This angle is calculated from given distance and diameter data. See Introduction for more information about calculating photometric data.

TIMES SQUARE LIGHTING
Model No. QF1000

MIN. SPOT w/ IRIS: 1.1° *

SPOT BEAM SPREAD: 10° *

FLOOD BEAM SPREAD: (not specified)

COLOR FRAMES: 6

FADER: Yes (blackout frame)

SHUTTERS: Yes (guillotine)

FAN COOLED: Yes

STANDARD LAMP: FEL - 1 KW.

WEIGHT: 87 lbs.

SIZE: 31½" x 15⅛" x 54¾" (L x W x H)

NOTES: Iris and shutters are rear operated, as is focus adjustment. Relamping is accomplished through hinged door at back. Unit is equipped with an internal transformer to boost the output of the lamp.

PHOTOMETRICS CHART
(Performance data for this unit are measured using FEL - 1 KW. lamp.)

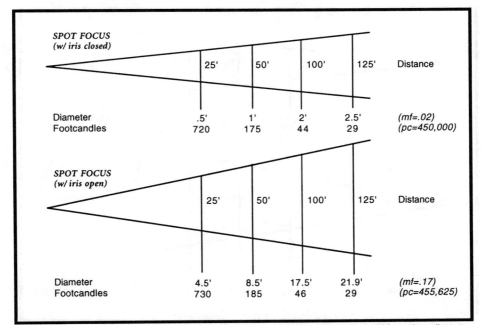

SPOT FOCUS (w/ iris closed)	25'	50'	100'	125'	Distance
Diameter	.5'	1'	2'	2.5'	(mf=.02)
Footcandles	720	175	44	29	(pc=450,000)
SPOT FOCUS (w/ iris open)	25'	50'	100'	125'	Distance
Diameter	4.5'	8.5'	17.5'	21.9'	(mf=.17)
Footcandles	730	185	46	29	(pc=455,625)

* Beam and field angles not specified in manufacturer's catalog. This angle is calculated from given distance and diameter data. See Introduction for more information about calculating photometric data.

TIMES SQUARE LIGHTING
Model No. QF1002

MIN. SPOT w/ IRIS: 1.1° *

SPOT BEAM SPREAD: 10° *

FLOOD BEAM SPREAD: (not specified)

COLOR FRAMES: 6

FADER: Yes (blackout frame)

SHUTTERS: Yes (guillotine)

FAN COOLED: Yes

STANDARD LAMP: FEL - 1 KW.

NOTES: Iris and shutters are rear operated, as is focus adjustment. Relamping is accomplished through hinged door at back.

WEIGHT: 81 lbs.

SIZE: 31½" x 15⅛" x 54¾" (L x W x H)

PHOTOMETRICS CHART
(Performance data for this unit are measured using FEL - 1 KW. lamp.)

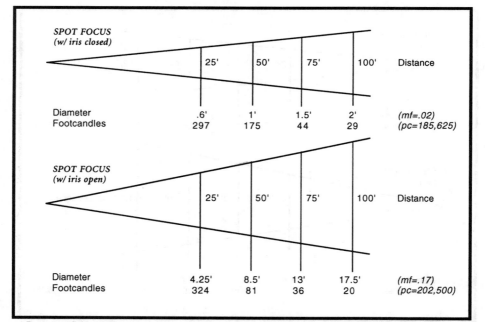

SPOT FOCUS
(w/ iris closed)

	25'	50'	75'	100'	Distance
Diameter	.6'	1'	1.5'	2'	(mf=.02)
Footcandles	297	175	44	29	(pc=185,625)

SPOT FOCUS
(w/ iris open)

	25'	50'	75'	100'	Distance
Diameter	4.25'	8.5'	13'	17.5'	(mf=.17)
Footcandles	324	81	36	20	(pc=202,500)

* Beam and field angles not specified in manufacturer's catalog. This angle is calculated from given distance and diameter data. See Introduction for more information about calculating photometric data.

ALTMAN STAGE LIGHTING
Model No. Explorer (Long Throw)

MIN. SPOT w/ IRIS: (not specified in mfg. catalog)

SPOT BEAM SPREAD: 5.7° (field)

FLOOD BEAM SPREAD: 7° (field)

COLOR FRAMES: 6

FADER: Yes (dowser)

SHUTTERS: Yes

FAN COOLED: Yes

STANDARD LAMP: Osram HMI - 1200 W.

NOTES: Iris, shutters and dowser are top operated. Focus is located on right side of unit.

WEIGHT: (not specified in manufacturer's catalog)

SIZE: (not specified in manufacturer's catalog)

PHOTOMETRICS CHART
(Performance data for this unit are measured using Osram HMI - 1200W. lamp.)

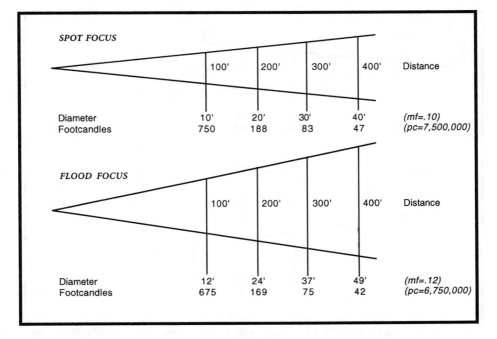

SPOT FOCUS	100'	200'	300'	400'	Distance
Diameter	10'	20'	30'	40'	*(mf=.10)*
Footcandles	750	188	83	47	*(pc=7,500,000)*

FLOOD FOCUS	100'	200'	300'	400'	Distance
Diameter	12'	24'	37'	49'	*(mf=.12)*
Footcandles	675	169	75	42	*(pc=6,750,000)*

FOLLOW SPOT (xenon/HMI/arc/etc.)

ALTMAN STAGE LIGHTING
Model No. Explorer (Medium Throw)

MIN. SPOT w/ IRIS: (not specified in mfg. catalog)

SPOT BEAM SPREAD: 5.25° (field)*

FLOOD BEAM SPREAD: 10.9° (field)*

COLOR FRAMES: 6

FADER: Yes (dowser)

SHUTTERS: Yes

FAN COOLED: Yes

STANDARD LAMP: Osram HMI - 1200 W.

NOTES: Iris, shutters and dowser controls are top operated. Focus is located on right side of unit.

WEIGHT: (not specified in manufacturer's catalog)

SIZE: (not specified in manufacturer's catalog)

PHOTOMETRICS CHART
(Performance data for this unit are measured using Osram HMI - 1200 W. lamp.)

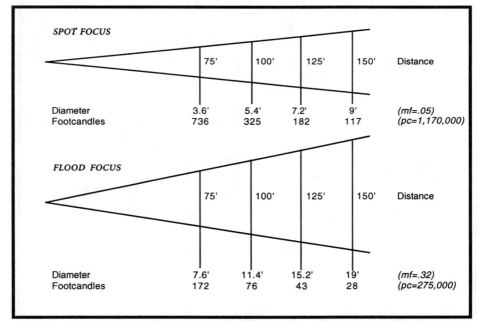

SPOT FOCUS

	75'	100'	125'	150'	Distance
Diameter	3.6'	5.4'	7.2'	9'	(mf=.05)
Footcandles	736	325	182	117	(pc=1,170,000)

FLOOD FOCUS

	75'	100'	125'	150'	Distance
Diameter	7.6'	11.4'	15.2'	19'	(mf=.32)
Footcandles	172	76	43	28	(pc=275,000)

* These field angles are calculated from given distance/diameter data in manufacturer's catalog. The catalog does list field angles of 2.5° (spot) and 5.5° (flood), but Altman confirms that these numbers were published incorrectly.

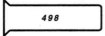

ALTMAN STAGE LIGHTING
Model No. Explorer (Short Throw)

MIN. SPOT w/ IRIS: 2.86° (field)*

SPOT BEAM SPREAD: (not specified in mfg. catalog)

FLOOD BEAM SPREAD: 18.18° (field)*

COLOR FRAMES: 6

FADER: Yes (dowser)

SHUTTERS: Yes

FAN COOLED: Yes

STANDARD LAMP: Osram HMI - 1200 W.

WEIGHT: (not specified in manufacturer's catalog)

SIZE: (not specified in manufacturer's catalog)

NOTES: Iris, shutters and dowser are rear operated. Focus is located on right rear of unit.

PHOTOMETRICS CHART
(Performance data for this unit are measured using Osram HMI - 1200 W. lamp.)

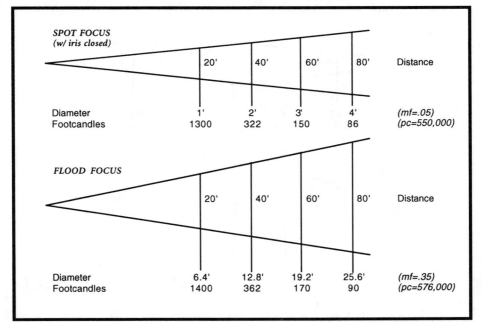

SPOT FOCUS (w/ iris closed)	20'	40'	60'	80'	Distance
Diameter	1'	2'	3'	4'	(mf=.05)
Footcandles	1300	322	150	86	(pc=550,000)

FLOOD FOCUS	20'	40'	60'	80'	Distance
Diameter	6.4'	12.8'	19.2'	25.6'	(mf=.35)
Footcandles	1400	362	170	90	(pc=576,000)

* These field angles are calculated from given distance/diameter data in manufacturer's catalog. The catalog does list field angles of 1.5° (spot) and 17.5° (flood), but Altman confirms that these numbers were published incorrectly.

ALTMAN STAGE LIGHTING
Model No. MARC 350

MIN. SPOT w/ IRIS: (not specified in mfg. catalog)

SPOT BEAM SPREAD: 1.3° (field)

FLOOD BEAM SPREAD: 18° (field)

COLOR FRAMES: 6

FADER: Yes (clamshell type dowser)

SHUTTERS: No

FAN COOLED: Yes

STANDARD LAMP: EZT (MARC 350) - 350 W.

NOTES: Both iris and focus are located on the side of the unit. Relamping is accomplished through a rear access door.

WEIGHT: (not specified in manufacturer's catalog)

SIZE: 29" x 20" x 55½" (L x W x H)

PHOTOMETRICS CHART

(Performance data for this unit are measured using EZT (MARC 350) - 350 W. lamp.)

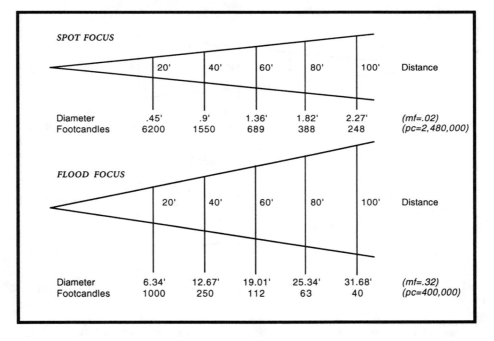

SPOT FOCUS

	20'	40'	60'	80'	100'	Distance
Diameter	.45'	.9'	1.36'	1.82'	2.27'	(mf=.02)
Footcandles	6200	1550	689	388	248	(pc=2,480,000)

FLOOD FOCUS

	20'	40'	60'	80'	100'	Distance
Diameter	6.34'	12.67'	19.01'	25.34'	31.68'	(mf=.32)
Footcandles	1000	250	112	63	40	(pc=400,000)

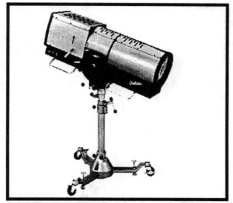

ALTMAN STAGE LIGHTING
Model No. Orbiter (Long Throw)

MIN. SPOT w/ IRIS: (not specified in mfg. catalog)

SPOT BEAM SPREAD: 1° (field)

FLOOD BEAM SPREAD: 8° (field)

COLOR FRAMES: 6

FADER: Yes (clamshell type dowser)

SHUTTERS: No

FAN COOLED: Yes

STANDARD LAMP: EZT (MARC 350) - 350 W.

WEIGHT: without stand – 83 lbs.

SIZE: 44" x 14½" x 53–71" (L x W x H)

NOTES: Both iris and focus are located on the side of the unit. Relamping is accomplished through a rear access door.

PHOTOMETRICS CHART

(Performance data for this unit are measured using EZT (MARC 350) - 350 W. lamp.)

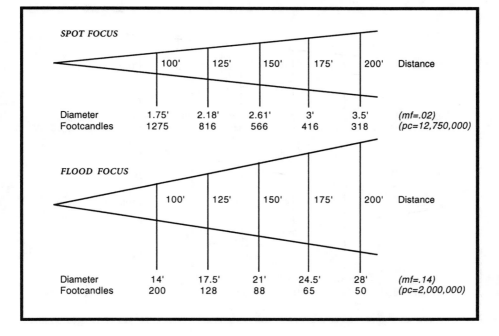

SPOT FOCUS

	100'	125'	150'	175'	200'	Distance
Diameter	1.75'	2.18'	2.61'	3'	3.5'	(mf=.02)
Footcandles	1275	816	566	416	318	(pc=12,750,000)

FLOOD FOCUS

	100'	125'	150'	175'	200'	Distance
Diameter	14'	17.5'	21'	24.5'	28'	(mf=.14)
Footcandles	200	128	88	65	50	(pc=2,000,000)

FOLLOW SPOT (xenon/HMI/arc/etc.)

ALTMAN STAGE LIGHTING
Model No. Orbiter (Short Throw)

MIN. SPOT w/ IRIS: (not specified in mfg. catalog)

SPOT BEAM SPREAD: 1.5° (field)

FLOOD BEAM SPREAD: 18° (field)

COLOR FRAMES: 6

FADER: Yes (clamshell type dowser)

SHUTTERS: No

FAN COOLED: Yes

STANDARD LAMP: EZT (MARC 350) - 350 W.

NOTES: Both iris and focus are located on the side of the unit. Relamping is accomplished through a rear access door.

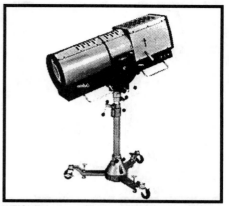

WEIGHT: (not specified in manufacturer's catalog)

SIZE: (not specified in manufacturer's catalog)

PHOTOMETRICS CHART
(Performance data for this unit are measured using EZT (MARC 350) - 350 W. lamp.)

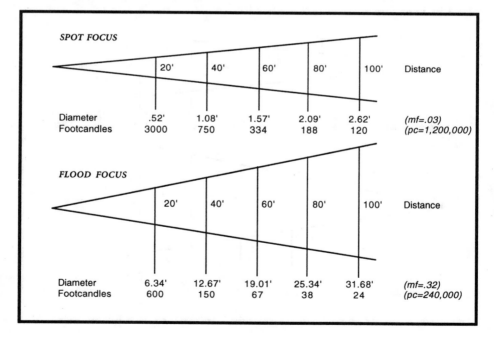

SPOT FOCUS

	20'	40'	60'	80'	100'	Distance
Diameter	.52'	1.08'	1.57'	2.09'	2.62'	(mf=.03)
Footcandles	3000	750	334	188	120	(pc=1,200,000)

FLOOD FOCUS

	20'	40'	60'	80'	100'	Distance
Diameter	6.34'	12.67'	19.01'	25.34'	31.68'	(mf=.32)
Footcandles	600	150	67	38	24	(pc=240,000)

ALTMAN STAGE LIGHTING
Model No. Satellite I

MIN. SPOT w/ IRIS: 1.8°

SPOT BEAM SPREAD: 7.3° (field)

FLOOD BEAM SPREAD: 20° (field)

COLOR FRAMES: 6

FADER: Yes (multi-leaf dowser)

SHUTTERS: Yes (guillotine)

FAN COOLED: Yes

STANDARD LAMP: Osram HMI - 575 W.

WEIGHT: (not specified in manufacturer's catalog)

SIZE: 46" x 12½" x 54" (L x W x H)

NOTES: Iris, shutters and dimmer are rear operated. Relamping is accomplished through an access door in the rear.

PHOTOMETRICS CHART
(Performance data for this unit are measured using Osram HMI - 575 W. lamp.)

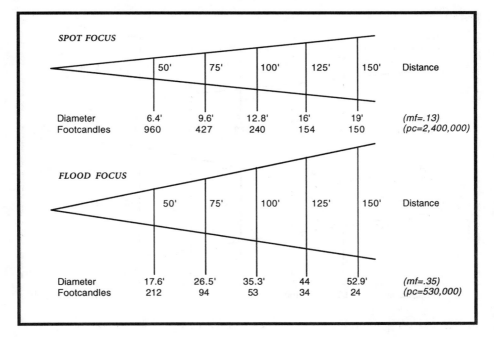

SPOT FOCUS

	50'	75'	100'	125'	150'	Distance
Diameter	6.4'	9.6'	12.8'	16'	19'	(mf=.13)
Footcandles	960	427	240	154	150	(pc=2,400,000)

FLOOD FOCUS

	50'	75'	100'	125'	150'	Distance
Diameter	17.6'	26.5'	35.3'	44	52.9'	(mf=.35)
Footcandles	212	94	53	34	24	(pc=530,000)

FOLLOW SPOT (xenon/HMI/arc/etc.)

ALTMAN STAGE LIGHTING
Model No. Voyager (Long Throw)

MIN. SPOT w/ IRIS: .5°

SPOT BEAM SPREAD: 4° (field)

FLOOD BEAM SPREAD: 9.5° (field)

COLOR FRAMES: 6

FADER: Yes (clamshell type dowser)

SHUTTERS: No

FAN COOLED: Yes

STANDARD LAMP: Osram HTI - 400 W.

NOTES: Iris and focus are located on the side of the unit.

WEIGHT: (not specified in manufacturer's catalog)

SIZE: 49½" x 12¼" x 55–68½" (L x W x H)

PHOTOMETRICS CHART

(Performance data for this unit are measured using Osram HTI - 400 W. lamp.)

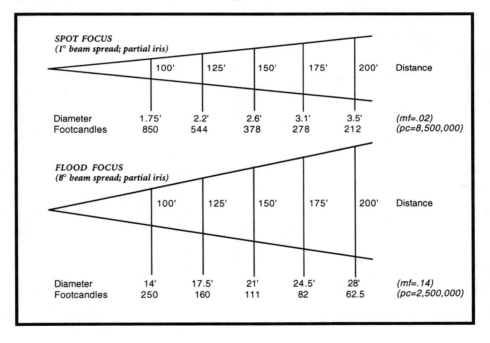

SPOT FOCUS
(1° beam spread; partial iris)

	100'	125'	150'	175'	200'	Distance
Diameter	1.75'	2.2'	2.6'	3.1'	3.5'	(mf=.02)
Footcandles	850	544	378	278	212	(pc=8,500,000)

FLOOD FOCUS
(8° beam spread; partial iris)

	100'	125'	150'	175'	200'	Distance
Diameter	14'	17.5'	21'	24.5'	28'	(mf=.14)
Footcandles	250	160	111	82	62.5	(pc=2,500,000)

WEIGHT: (not specified in manufacturer's catalog)

SIZE: 32³⁄₁₆" x 16½" x 52½–66" (L x W x H)

ALTMAN STAGE LIGHTING
Model No. VOYAGER (Short Throw)

MIN. SPOT w/ IRIS: (not specified in mfg. catalog)

SPOT BEAM SPREAD: 4° (field)

FLOOD BEAM SPREAD: 9.5° (field)

COLOR FRAMES: 6

FADER: Yes (clamshell type dowser)

SHUTTERS: No

FAN COOLED: Yes

STANDARD LAMP: Osram HTI - 400 W.

NOTES: Iris and focus are located on the side of the unit.

PHOTOMETRICS CHART

(Performance data for this unit are measured using Osram HTI - 400 W. lamp.)

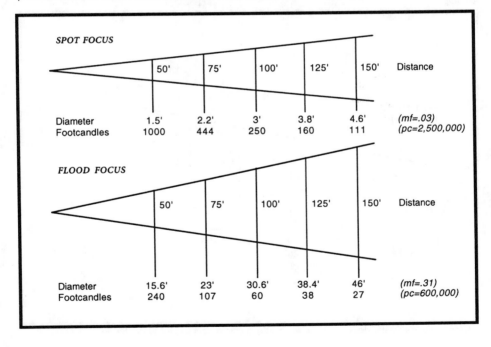

SPOT FOCUS

	50'	75'	100'	125'	150'	Distance
Diameter	1.5'	2.2'	3'	3.8'	4.6'	*(mf=.03)*
Footcandles	1000	444	250	160	111	*(pc=2,500,000)*

FLOOD FOCUS

	50'	75'	100'	125'	150'	Distance
Diameter	15.6'	23'	30.6'	38.4'	46'	*(mf=.31)*
Footcandles	240	107	60	38	27	*(pc=600,000)*

CENTURY LIGHTING
Model No. 202

MIN. SPOT w. IRIS: (not specified in mfg. catalog)

SPOT BEAM SPREAD: 5° *

FLOOD BEAM SPREAD: 45°

COLOR FRAMES: 1

FADER: No

SHUTTERS: No

CARBONS: (not specified in mfg. catalog)

NOTES: Manual carbon feed.

WEIGHT: (not specified in manufacturer's catalog)

SIZE: (not specified in manufacturer's catalog)

PHOTOMETRICS CHART

(Information about carbons used to measure performance data is not provided in manufacturer's catalog.)

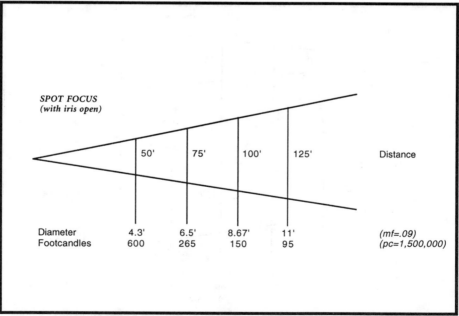

SPOT FOCUS
(with iris open)

	50'	75'	100'	125'	Distance
Diameter	4.3'	6.5'	8.67'	11'	(mf=.09)
Footcandles	600	265	150	95	(pc=1,500,000)

*Manufacturer's catalog doesn't specify whether this measurement was made with the iris open or closed.

CENTURY LIGHTING
Model No. 225

MIN. SPOT w. IRIS: (not specified in mfg. catalog)

SPOT BEAM SPREAD: 1° *

FLOOD BEAM SPREAD: 17°

COLOR FRAMES: 6

FADER: (not specified in mfg. catalog)

SHUTTERS: (not specified in mfg. catalog)

CARBONS: (not specified in mfg. catalog)

WEIGHT: (not specified in manufacturer's catalog)

SIZE: (not specified in manufacturer's catalog)

PHOTOMETRICS CHART

(Information about carbons used to measure performance data is not provided in manufacturer's catalog.)

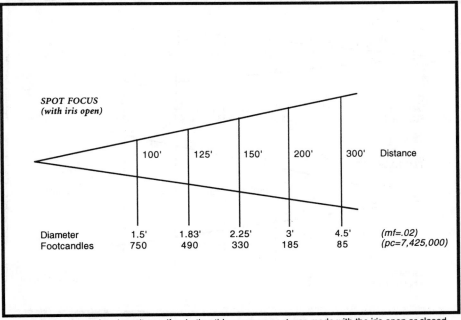

SPOT FOCUS
(with iris open)

	100'	125'	150'	200'	300'	Distance
Diameter	1.5'	1.83'	2.25'	3'	4.5'	*(mf=.02)*
Footcandles	750	490	330	185	85	*(pc=7,425,000)*

* Manufacturer's catalog doesn't specify whether this measurement was made with the iris open or closed.

FOLLOW SPOT (xenon/HMI/arc/etc.)

CENTURY STRAND
Model No. Patt 265 Mark II

MIN. SPOT w/ IRIS: (not specified in mfg. catalog)

SPOT BEAM SPREAD: 14° (beam), 23° (field)

FLOOD BEAM SPREAD: 1.5° (beam), 1.5° (field)

COLOR FRAMES: 2

FADER: No

SHUTTERS: Yes (guillotine)

FAN COOLED: No

STANDARD LAMP: 400/600 W. Mercury-Iodide

NOTES: Iris is located on right side at gate. Focus is located along right side of lens barrel. 4-color boomerang available as an accessory.

WEIGHT: 28½ lbs.

SIZE: 29" x 11½" (L x W)

PHOTOMETRICS CHART
(Performance data for this unit are measured using 400/600 W. Mercury-Iodide lamp at 600 W.)

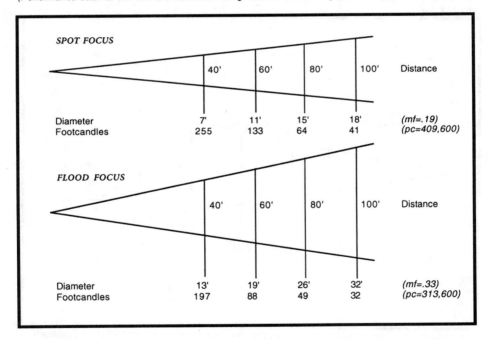

SPOT FOCUS

	40'	60'	80'	100'	Distance
Diameter	7'	11'	15'	18'	*(mf=.19)*
Footcandles	255	133	64	41	*(pc=409,600)*

FLOOD FOCUS

	40'	60'	80'	100'	Distance
Diameter	13'	19'	26'	32'	*(mf=.33)*
Footcandles	197	88	49	32	*(pc=313,600)*

CLAY PAKY
Model No. Shadow Basic*

BEAM ANGLE, SPOT: (not specified in mfg. catalog)

BEAM ANGLE, FLOOD: (not specified in mfg. catalog)

FIELD ANGLE, SPOT: 11°

FIELD ANGLE, FLOOD: 24° **

COLOR FRAMES: 7 dichroic filters, plus 2 color correction filters

FADER: Yes (dowser)

SHUTTERS: Yes

STANDARD LAMP: HMI1200 W/GS***

NOTES: Most functions (save movement) can be remotely controlled either through RS 232, DMX, or 0-10V analog. All controls on unit (electrical sliders) are located on a pivoting panel at the lower rear. Relamping is through the top housing.

WEIGHT: 83.8 lbs

SIZE: 43⅛" x 19½" x 9⅛" (L x W x H)

PHOTOMETRICS CHART

(Performance data for this unit are measured using HMI 1200 W/GS lamp.)

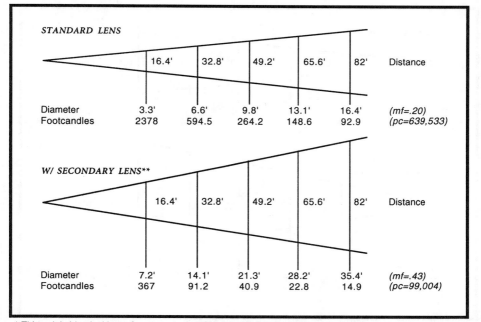

STANDARD LENS

	16.4'	32.8'	49.2'	65.6'	82'	Distance
Diameter	3.3'	6.6'	9.8'	13.1'	16.4'	(mf=.20)
Footcandles	2378	594.5	264.2	148.6	92.9	(pc=639,533)

W/ SECONDARY LENS

	16.4'	32.8'	49.2'	65.6'	82'	Distance
Diameter	7.2'	14.1'	21.3'	28.2'	35.4'	(mf=.43)
Footcandles	367	91.2	40.9	22.8	14.9	(pc=99,004)

* This unit is identical in performance to the "Shadow QS-ST," but differs in weight, length and color controls.
** "Flood" refers to a secondary objective lens which can be put in place by opening the housing, and manually pivoting a pair of lenses in position.
*** Also available in a 575 W. HMI configuration (no photometrics data in mfg. catalog). Weight: 68.3 lbs.

CLAY PAKY
Model No. Shadow QS-LT

BEAM ANGLE, SPOT: (not specified in mfg. catalog)

BEAM ANGLE, FLOOD: (not specified in mfg. catalog)

FIELD ANGLE, SPOT: 3°

FIELD ANGLE, FLOOD: 7° *

COLOR FRAMES: 7 dichroic filters, plus 2 color correction filters

FADER: Yes (dowser)

SHUTTERS: Yes

STANDARD LAMP: HMI1200 W/GS

NOTES: Most functions (save movement) can be remotely controlled either through RS 232, DMX, or 0-10V analog. All controls on unit (electrical sliders) are located on a pivoting panel at the lower rear. Relamping is through the top housing.

WEIGHT: 101.4 lbs

SIZE: 67½" x 19½" x 9⅛" (L x W x H)

PHOTOMETRICS CHART
(Performance data for this unit are measured using HMI 1200 W/GS lamp.)

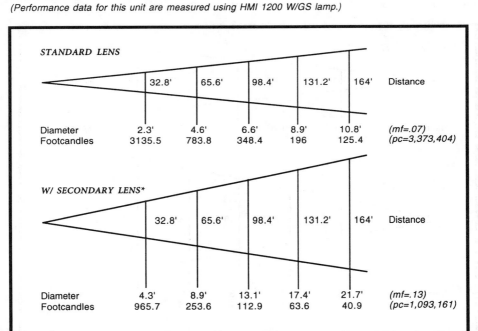

STANDARD LENS

	32.8'	65.6'	98.4'	131.2'	164'	Distance
Diameter	2.3'	4.6'	6.6'	8.9'	10.8'	*(mf=.07)*
Footcandles	3135.5	783.8	348.4	196	125.4	*(pc=3,373,404)*

W/ SECONDARY LENS*

	32.8'	65.6'	98.4'	131.2'	164'	Distance
Diameter	4.3'	8.9'	13.1'	17.4'	21.7'	*(mf=.13)*
Footcandles	965.7	253.6	112.9	63.6	40.9	*(pc=1,093,161)*

* "Flood" refers to a secondary objective lens which can be put in place by opening the housing, and manually pivoting a pair of lenses in position.

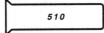

CLAY PAKY
Model No. Shadow QS-ST*

BEAM ANGLE, SPOT: (not specified in mfg. catalog)

BEAM ANGLE, FLOOD: (not specified in mfg. catalog)

FIELD ANGLE, SPOT: 11°

FIELD ANGLE, FLOOD: 24° **

COLOR FRAMES: 7 dichroic filters, plus 2 color correction filters

FADER: Yes (dowser)

SHUTTERS: Yes

STANDARD LAMP: HMI1200 W/GS

NOTES: Most functions (save movement) can be remotely controlled either through RS 232, DMX, or 0-10V analog. All controls on unit (electrical sliders) are located on a pivoting panel at the lower rear. Relamping is through the top housing.

WEIGHT: 83.8 lbs

SIZE: 46⁵⁄₁₆" x 19½" x 9⅛" (L x W x H)

PHOTOMETRICS CHART

(Performance data for this unit are measured using HMI 1200 W/GS lamp.)

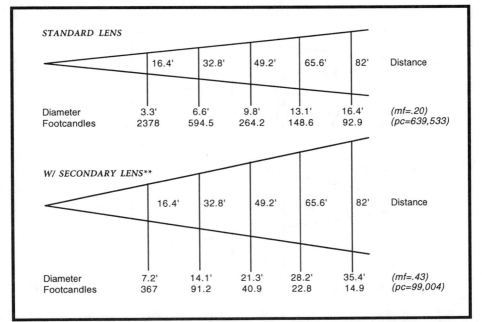

STANDARD LENS

	16.4'	32.8'	49.2'	65.6'	82'	Distance
Diameter	3.3'	6.6'	9.8'	13.1'	16.4'	(mf=.20)
Footcandles	2378	594.5	264.2	148.6	92.9	(pc=639,533)

W/ SECONDARY LENS**

	16.4'	32.8'	49.2'	65.6'	82'	Distance
Diameter	7.2'	14.1'	21.3'	28.2'	35.4'	(mf=.43)
Footcandles	367	91.2	40.9	22.8	14.9	(pc=99,004)

* This unit is identical in performance to the "Shadow Basic," but differs in weight, length and color controls.
** "Flood" refers to a secondary objective lens which can be put in place by opening the housing, and manually pivoting a pair of lenses in position.

FOLLOW SPOT (xenon/HMI/arc/etc.)

COLORTRAN
Model No. 210-100

MIN. SPOT w/ IRIS: .53° (field)

SPOT BEAM SPREAD: 3.5° (field)

FLOOD BEAM SPREAD: 8° (field)

COLOR FRAMES: 6

FADER: Yes (dowser)

SHUTTERS: Yes (guillotine)

FAN COOLED: Yes

STANDARD LAMP: 2 KW. Xenon

NOTES: Iris is located on top of the unit. Focus is located on the side. Relamping is accomplished through access door at rear of unit.

WEIGHT: 240 lbs. (includes stand)

SIZE: 65½" x 43–57" (L x H)

PHOTOMETRICS CHART
(Performance data for this unit are measured using 2 KW. Xenon lamp.)

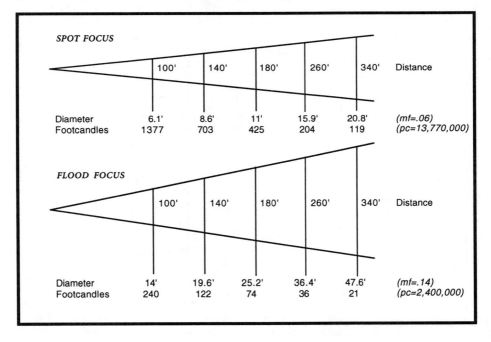

SPOT FOCUS

	100'	140'	180'	260'	340'	Distance
Diameter	6.1'	8.6'	11'	15.9'	20.8'	*(mf=.06)*
Footcandles	1377	703	425	204	119	*(pc=13,770,000)*

FLOOD FOCUS

	100'	140'	180'	260'	340'	Distance
Diameter	14'	19.6'	25.2'	36.4'	47.6'	*(mf=.14)*
Footcandles	240	122	74	36	21	*(pc=2,400,000)*

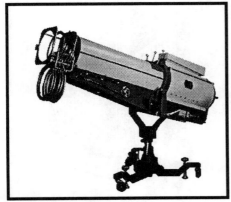

COLORTRAN
Model No. 210-200

MIN. SPOT w/ IRIS: 1.03° (field)

SPOT BEAM SPREAD: 6.9° (field)

FLOOD BEAM SPREAD: 22.6° (field)

COLOR FRAMES: 6

FADER: Yes (dowser)

SHUTTERS: Yes (guillotine)

FAN COOLED: Yes

STANDARD LAMP: 2 KW. Xenon

NOTES: Iris is located on top of the unit. Focus is located on the side of unit. Relamping is accomplished through access door at rear of unit.

WEIGHT: 240 lbs. (includes stand)

SIZE: 65½" x 43–57" (L x H)

PHOTOMETRICS CHART
(Performance data for this unit are measured using 2 KW. Xenon lamp.)

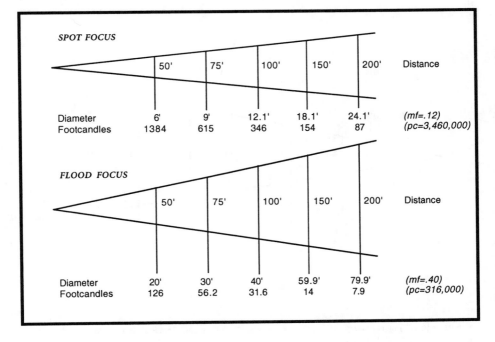

SPOT FOCUS

	50'	75'	100'	150'	200'	Distance
Diameter	6'	9'	12.1'	18.1'	24.1'	*(mf=.12)*
Footcandles	1384	615	346	154	87	*(pc=3,460,000)*

FLOOD FOCUS

	50'	75'	100'	150'	200'	Distance
Diameter	20'	30'	40'	59.9'	79.9'	*(mf=.40)*
Footcandles	126	56.2	31.6	14	7.9	*(pc=316,000)*

LUDWIG PANI
Model No. HMV 1200/20*

MIN. SPOT w/ IRIS: (not specified in mfg. catalog)

MAX. BEAM SPREAD: 10.44° **

COLOR FRAMES: 1

FADER: Yes (shutter with blackout disk)

SHUTTERS: Yes (4-way)

FAN COOLED: No

STANDARD LAMP: Osram HMI - 1200 W.

NOTES: Iris is located on top of unit, focus on the right side. A mechanical dimming shutter is available as an option, as is a 4-color boomerang.

WEIGHT: 56 lbs. (25.5 kg.)

SIZE: 36¾" (935mm) x 60¼" (1530mm) (L x H)

PHOTOMETRICS CHART

(Performance data for this unit are measured using Osram HMI - 1200 W. lamp.)

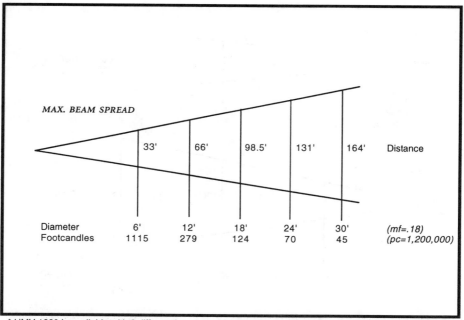

MAX. BEAM SPREAD

Distance	33'	66'	98.5'	131'	164'
Diameter	6'	12'	18'	24'	30'
Footcandles	1115	279	124	70	45

(mf=.18)
(pc=1,200,000)

* HMV 1200 is available with 2 different lenses: model 1200/20 has a 20cm focal length objective lens (this page) and model 1200/35 has a 35cm lens (next page).
** Photometric data in manufacturer's catalog is given in metric units, and was converted to English units for this book. Also, beam and field angles not specified in manufacturer's catalog. This angle is calculated from given distance and diameter data. See Introduction for more information about calculating photometric data.

LUDWIG PANI
Model No. HMV 1200/35*

MIN. SPOT w/ IRIS: (not specified in mfg. catalog)

MAX. BEAM SPREAD: 6.28° **

COLOR FRAMES: 1

FADER: Yes (shutter with blackout disk)

SHUTTERS: Yes (4-way)

FAN COOLED: No

STANDARD LAMP: Osram HMI - 1200 W.

NOTES: Iris is located on top of unit, focus on the right side. A mechanical dimming shutter is available as an option, as is a 4-color boomerang.

WEIGHT: 58.4 lbs. (26.5 kg.)

SIZE: 36¾" (935mm) x 60¼" (1530mm) (L x H)

PHOTOMETRICS CHART
(Performance data for this unit are measured using Osram HMI - 1200 W. lamp.)

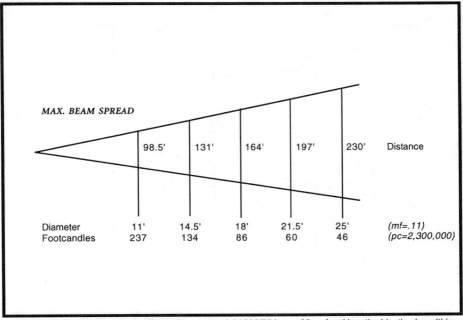

MAX. BEAM SPREAD

	98.5'	131'	164'	197'	230'	Distance
Diameter	11'	14.5'	18'	21.5'	25'	(mf=.11)
Footcandles	237	134	86	60	46	(pc=2,300,000)

* HMV 1200 is available with 2 different lenses: model 1200/35 has a 35cm focal length objective lens (this page) and model 1200/20 has a 20cm lens (prevoius page).

** Photometric data in manufacturer's catalog is given in metric units, and was converted to English units for this book. Also, beam and field angles not specified in manufacturer's catalog. This angle is calculated from given distance and diameter data. See Introduction for more information about calculating photometric data.

FOLLOW SPOT (xenon/HMI/arc/etc.)

LUDWIG PANI
Model No. HMV 2500

MIN. SPOT w/ IRIS: (not specified in mfg. catalog)

SPOT BEAM SPREAD: 6°

FLOOD BEAM SPREAD: 16°

COLOR FRAMES: 1

FADER: No

SHUTTERS: Yes (horizontal & vertical guillotines)

FAN COOLED: Yes

STANDARD LAMP: 2.5 KW. CID

NOTES: Iris and shutters are located on side of unit. A mechanical dimming shutter is available as an option, as is a 6-color boomerang.

WEIGHT: 145 lbs. (65.8 kg.)

SIZE: 47¼" (1200mm) x 20½" (520mm) (L x W)

PHOTOMETRICS CHART

(Performance data for this unit are measured using 2.5 KW. CID lamp.)

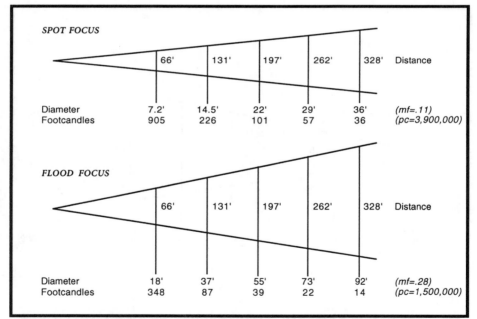

SPOT FOCUS

	66'	131'	197'	262'	328'	Distance
Diameter	7.2'	14.5'	22'	29'	36'	(mf=.11)
Footcandles	905	226	101	57	36	(pc=3,900,000)

FLOOD FOCUS

	66'	131'	197'	262'	328'	Distance
Diameter	18'	37'	55'	73'	92'	(mf=.28)
Footcandles	348	87	39	22	14	(pc=1,500,000)

* Photometric data in manufacturer's catalog is given in metric units, and was converted to English units for this book.

LYCIAN STAGE LIGHTING
Model No. 1209

MIN. SPOT w/ IRIS: .85° *

SPOT BEAM SPREAD: 10.3° *

FLOOD BEAM (w/ spread lens): 10.3°x20.4° *

COLOR FRAMES: 5

FADER: Yes (blackout frame)

SHUTTERS: Yes (guillotine)

FAN COOLED: Yes

STANDARD LAMP: Osram HMI - 575 W.

WEIGHT: without stand – 78 lbs.

SIZE: 37" x 14" x 55½–68½" (L x W x H)

NOTES: Iris and shutters are located on the top of unit. Relamping is accomplished through a lamp drawer at rear.

PHOTOMETRICS CHART
(Performance data for this unit are measured using Osram HMI - 575 W. lamp.)

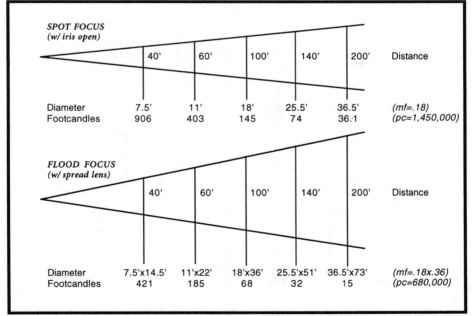

SPOT FOCUS *(w/ iris open)*

	40'	60'	100'	140'	200'	Distance
Diameter	7.5'	11'	18'	25.5'	36.5'	(mf=.18)
Footcandles	906	403	145	74	36.1	(pc=1,450,000)

FLOOD FOCUS *(w/ spread lens)*

	40'	60'	100'	140'	200'	Distance
Diameter	7.5'x14.5'	11'x22'	18'x36'	25.5'x51'	36.5'x73'	(mf=.18x.36)
Footcandles	421	185	68	32	15	(pc=680,000)

* Beam and field angles not specified in manufacturer's catalog. This angle is calculated from given distance and diameter data. See Introduction for more information about calculating photometric data.

LYCIAN STAGE LIGHTING
Model No. 1266

MIN. SPOT w/ IRIS: 1.2° *

SPOT BEAM SPREAD: (not specified in mfg. catalog)

FLOOD BEAM SPREAD: 17° *

COLOR FRAMES: 7

FADER: Yes (dowser)

SHUTTERS: Yes (guillotine)

FAN COOLED: Yes

STANDARD LAMP: Osram HTI - 400 W.

NOTES: Iris and lens focus control are combined into one controller which is located on side of unit. Relamping is accomplished through an access door at the rear.

WEIGHT: without stand – 86 lbs.

SIZE: 32¼" x 15" x 57½–70½" (L x W x H)

PHOTOMETRICS CHART

(Performance data for this unit are measured using Osram HTI - 400 W. lamp.)

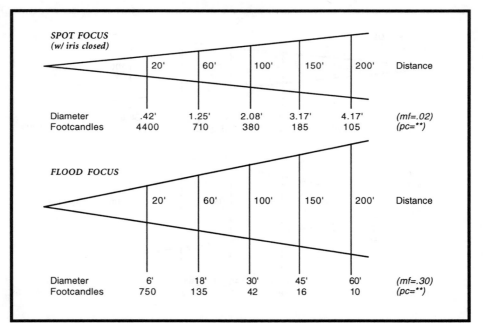

SPOT FOCUS
(w/ iris closed)

	20'	60'	100'	150'	200'	Distance
Diameter	.42'	1.25'	2.08'	3.17'	4.17'	(mf=.02)
Footcandles	4400	710	380	185	105	(pc=**)

FLOOD FOCUS

	20'	60'	100'	150'	200'	Distance
Diameter	6'	18'	30'	45'	60'	(mf=.30)
Footcandles	750	135	42	16	10	(pc=**)

* Beam and field angles not specified in manufacturer's catalog. This field angle is calculated from given distance and field diameter data. See Introduction for more information about calculating photometric data.
** Footcandle data given in manufacturer's catalog is not consistent enough to calculate a reliable peak candela figure for all throw distances. (At spot focus, 100' throw and 380 fc, pc=3,800,000 while at 20' throw and 4400 fc, pc=1,760,000; at flood focus, 100' throw and 42 fc, pc=420,000, while at 20' throw and 750 fc, pc=300,000.)

LYCIAN STAGE LIGHTING
Model No. 1267

MIN. SPOT w/ IRIS: .45° *

SPOT BEAM SPREAD: (not specified in mfg. catalog)

FLOOD BEAM SPREAD: 8.6° *

COLOR FRAMES: 7

FADER: Yes (dowser)
SHUTTERS: Yes (guillotine)

FAN COOLED: Yes

STANDARD LAMP: Osram HTI - 400 W.

WEIGHT: without stand – 101 lbs.

SIZE: 50" x 15" x 57½–70½" (L x W x H)

NOTES: Iris and lens focus control are combined into one controller which is located on side of unit. Relamping is accomplished through an access door at the rear.

PHOTOMETRICS CHART
(Performance data for this unit are measured using Osram HTI - 400 W. lamp.)

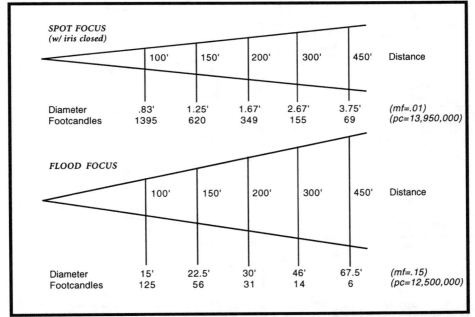

SPOT FOCUS
(w/ iris closed)

Distance	100'	150'	200'	300'	450'	
Diameter	.83'	1.25'	1.67'	2.67'	3.75'	(mf=.01)
Footcandles	1395	620	349	155	69	(pc=13,950,000)

FLOOD FOCUS

Distance	100'	150'	200'	300'	450'	
Diameter	15'	22.5'	30'	46'	67.5'	(mf=.15)
Footcandles	125	56	31	14	6	(pc=12,500,000)

* Beam and field angles not specified in manufacturer's catalog. This angle is calculated from given distance and field diameter data. See Introduction for more information about calculating photometric data.

FOLLOW SPOT (xenon/HMI/arc/etc.)

LYCIAN STAGE LIGHTING
Model No. 1271

MAX. BEAM SPREAD: 23.3° * (w/ wide lens)

MAX. BEAM SPREAD: 10° * (w/ medium lens)

MAX. BEAM SPREAD: 6.5° * (w/ narrow lens)

COLOR FRAMES: 6

FADER: Yes

SHUTTERS: No

FAN COOLED: Yes

STANDARD LAMP: Osram HMI - 1200 W.

NOTES: Iris and fader are rear operated. Focus is located on side of unit. Relamping is accomplished through top of lamphouse. Unit is available with three different lens assemblies.

WEIGHT: without stand – 65 lbs. (approximately)

SIZE: 29½–51½" x 14½" x 56–69" (L x W x H)

PHOTOMETRICS CHART

(Performance data for this unit are measured using Osram HMI - 1200 W. lamp.)

WIDE LENS *(max. beam size)*

	20'	30'	40'	50'	80'	Distance
Diameter	8.33'	12.5'	16.5'	21'	33'	*(mf=.41)*
Footcandles	1094	486	273	175	68	*(pc=436,800)*

MEDIUM LENS *(max. beam size)*

	20'	30'	40'	50'	80'	Distance
Diameter	3.5'	5.4'	7'	9'	14.5'	*(mf=.18)*
Footcandles	2320	1031	580	371	145	*(pc=928,000)*

NARROW LENS *(max. beam size)*

	20'	30'	40'	50'	80'	Distance
Diameter	2.5'	3.5'	4.5'	5.5'	9.5'	*(mf=.11)*
Footcandles	3938	1750	984	630	246	*(pc=1,574,400)*

* Beam and field angles not specified in manufacturer's catalog. This angle is calculated from given distance and diameter data. See Introduction for more information about calculating photometric data.

LYCIAN STAGE LIGHTING
Model No. 1272

MIN. SPOT w/ IRIS: (not specified in mfg. catalog)

MAX. BEAM SPREAD: 11.94° (beam)*

COLOR FRAMES: 6

FADER: Yes

SHUTTERS: Yes (guillotine)

FAN COOLED: Yes

STANDARD LAMP: MSR/HMI - 1200 W.

NOTES: Iris is located on top of unit, as is guillotine, mechanical fader and gobo slot. Relamping is accomplished through lamp drawer located on rear of unit.

WEIGHT: (not specified in manufacturer's catalog)

SIZE: (not specified in manufacturer's catalog)

PHOTOMETRICS CHART
(Performance data for this unit are measured using MSR/HMI - 1200 W. lamp.)

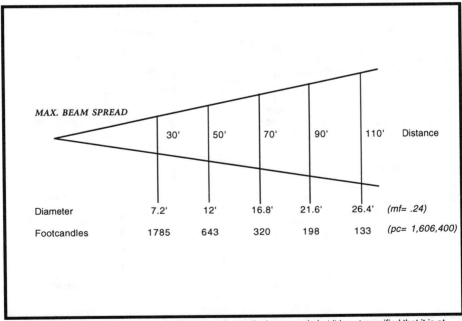

	30'	50'	70'	90'	110'	Distance
Diameter	7.2'	12'	16.8'	21.6'	26.4'	*(mf= .24)*
Footcandles	1785	643	320	198	133	*(pc= 1,606,400)*

* This angle is described in the manufacturer's catalog as the beam angle but it is not specified that it is at 50% of peak intensity. The angle described by the given diameter and distance figures calculates to be 13.685° which may be the field angle, but again it is not specified.

LYCIAN STAGE LIGHTING
Model No. 1275

MIN. SPOT w/ IRIS: .47° *

SPOT BEAM SPREAD: 3.78° *

FLOOD BEAM SPREAD: 9.15° *

COLOR FRAMES: 6

FADER: Yes (dowser)

SHUTTERS: Yes (guillotine)

FAN COOLED: Yes

STANDARD LAMP: 1200-HB - 1200 W.

NOTES: Both iris and focus are located on top of unit. Relamping is accomplished through top of lamphouse.

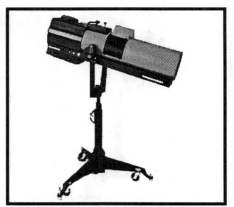

WEIGHT: without stand – 90 lbs.

SIZE: 48" x 17" x 48–61" (L x W x H)

PHOTOMETRICS CHART
(Performance data for this unit are measured using 1200-HB - 1200 W. Metal Halide lamp.)

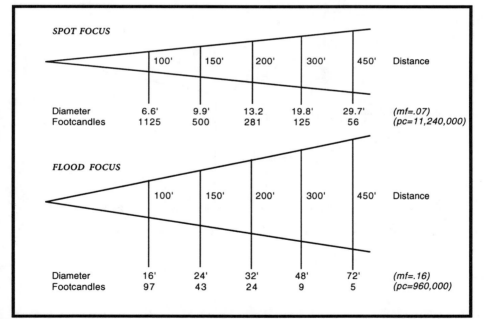

SPOT FOCUS

	100'	150'	200'	300'	450'	Distance
Diameter	6.6'	9.9'	13.2	19.8'	29.7'	(mf=.07)
Footcandles	1125	500	281	125	56	(pc=11,240,000)

FLOOD FOCUS

	100'	150'	200'	300'	450'	Distance
Diameter	16'	24'	32'	48'	72'	(mf=.16)
Footcandles	97	43	24	9	5	(pc=960,000)

* Beam and field angles not specified in manufacturer's catalog. This angle is calculated from given distance and diameter data. See Introduction for more information about calculating photometric data.

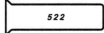

LYCIAN STAGE LIGHTING
Model No. 1278

MIN. SPOT w/ IRIS: .62° (beam)*

SPOT BEAM SPREAD: (not specified in mfg. catalog)

FLOOD BEAM SPREAD: 12.44° (beam)*

COLOR FRAMES: 6

FADER: Yes (dowser)

SHUTTERS: Yes (guillotine)

FAN COOLED: Yes

STANDARD LAMP: MSR-2500 - 2.5 KW.

NOTES: Iris, shutters and focus are located on side of unit. Relamping is accomplished through side access door.

WEIGHT: without stand – 232 lbs.

SIZE: 58" x 20" x 69½"–75½" (L x W x H)

PHOTOMETRICS CHART

(Performance data for this unit are measured using MSR-2500 - 2.5 KW. lamp.)

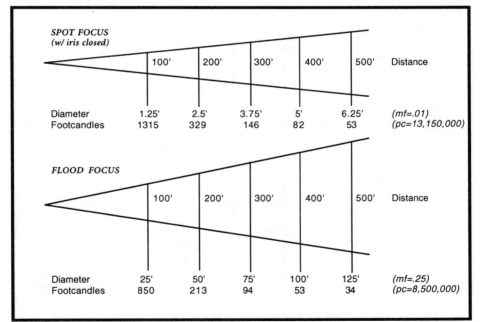

SPOT FOCUS (w/ iris closed)

	100'	200'	300'	400'	500'	Distance
Diameter	1.25'	2.5'	3.75'	5'	6.25'	(mf=.01)
Footcandles	1315	329	146	82	53	(pc=13,150,000)

FLOOD FOCUS

	100'	200'	300'	400'	500'	Distance
Diameter	25'	50'	75'	100'	125'	(mf=.25)
Footcandles	850	213	94	53	34	(pc=8,500,000)

* This angle is described in the manufacturer's catalog as the beam angle but it is not specified that it is at 50% of peak intensity. The angles described by the given diameter and distance figures calculate to be .715° at minimum spot and 14.25° at maximum flood. These angles may be field angles, but again it is not specified.

FOLLOW SPOT (xenon/HMI/arc/etc.)

LYCIAN STAGE LIGHTING
Model No. 1290XLT

MIN. SPOT w. IRIS: (not specified in mfg. catalog)

SPOT BEAM SPREAD: .57° (iris closed), 3.44° (open)

FLOOD BEAM SPREAD: 8.57° (iris open)

COLOR FRAMES: 6

FADER: Yes

SHUTTERS: Yes (guillotine)

NOTES: All controls are either top mounted, mounted on both sides or reversible. Unit has a gobo slot. Unit is fan cooled.

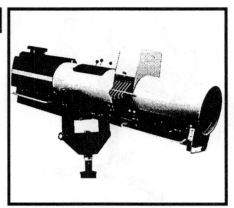

WEIGHT: without stand – 200 lbs.

SIZE: 72" x 19½" x 25" (L x W x H)

PHOTOMETRICS CHART

(Performance data for this unit are measured using 2 KW. Xenon lamp.)

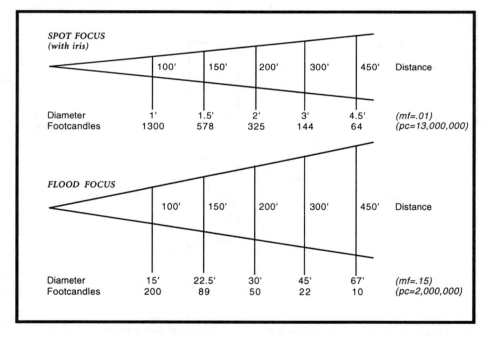

SPOT FOCUS (with iris)

	100'	150'	200'	300'	450'	Distance
Diameter	1'	1.5'	2'	3'	4.5'	*(mf=.01)*
Footcandles	1300	578	325	144	64	*(pc=13,000,000)*

FLOOD FOCUS

	100'	150'	200'	300'	450'	Distance
Diameter	15'	22.5'	30'	45'	67'	*(mf=.15)*
Footcandles	200	89	50	22	10	*(pc=2,000,000)*

WEIGHT: without stand – 65 lbs.

SIZE: 27¾" (L)

LYCIAN STAGE LIGHTING
Model No. L1262, "SuperArc 350"

MIN. SPOT w/ IRIS: (not specified in mfg. catalog)

SPOT BEAM SPREAD: 2.41° *

FLOOD BEAM SPREAD: 21.24° *

COLOR FRAMES: 6

FADER: Yes (clamshell type dowser)

SHUTTERS: Yes (guillotine)

FAN COOLED: (not specified in mfg. catalog)

STANDARD LAMP: EZT (MARC 350) - 350 W.

NOTES: This is the Short Throw model.
Spot/flood (beam size) adjustment wheel is
located on side at rear of unit.

PHOTOMETRICS CHART

(Performance data for this unit are measured using EZT (MARC 350) - 350 W. lamp.)

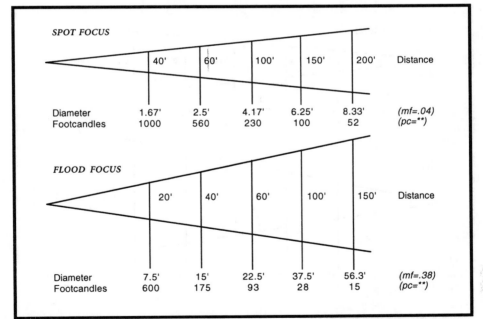

SPOT FOCUS

	40'	60'	100'	150'	200'	Distance
Diameter	1.67'	2.5'	4.17'	6.25'	8.33'	(mf=.04)
Footcandles	1000	560	230	100	52	(pc=**)

FLOOD FOCUS

	20'	40'	60'	100'	150'	Distance
Diameter	7.5'	15'	22.5'	37.5'	56.3'	(mf=.38)
Footcandles	600	175	93	28	15	(pc=**)

* Beam and field angles not specified in manufacturer's catalog. This angle is calculated from given distance
and diameter data. See Introduction for more information about calculating photometric data.
** Footcandle data given in manufacturer's catalog is not consistent enough to calculate a reliable peak
candela figure for all throw distances. (At spot focus, 100' throw and 230 fc, pc=2,300,000 while at 40' throw
and 1000 fc, pc=1,600,000; at flood focus, 100' throw and 28 fc, pc=280,000 while
at 20' throw and 600 fc, pc=240,000.)

FOLLOW SPOT (xenon/HMI/arc/etc.)

LYCIAN STAGE LIGHTING
Model No. L1264, "SuperArc 350"

MIN. SPOT w/ IRIS: (not specified in mfg. catalog)

SPOT BEAM SPREAD: .86° *

FLOOD BEAM SPREAD: 6.7° *

COLOR FRAMES: 6

FADER: Yes (clamshell type dowser)

SHUTTERS: Yes (guillotine)

FAN COOLED: (not specified in mfg. catalog)

STANDARD LAMP: EZT (MARC 350) - 350 W.

NOTES: Spot/flood (beam size) adjustment wheel is located on side at rear of unit.

WEIGHT: without stand – 71 lbs.

SIZE: 46¾" (L)

PHOTOMETRICS CHART

(Performance data for this unit are measured using EZT (MARC 350) - 350 W. lamp.)

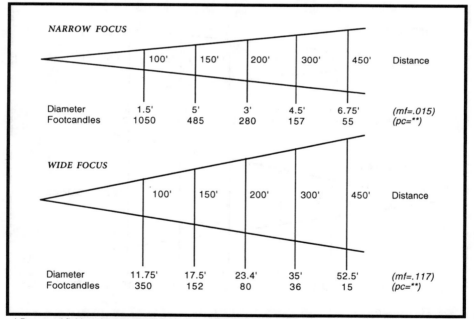

NARROW FOCUS

	100'	150'	200'	300'	450'	Distance
Diameter	1.5'	5'	3'	4.5'	6.75'	*(mf=.015)*
Footcandles	1050	485	280	157	55	*(pc=**)*

WIDE FOCUS

	100'	150'	200'	300'	450'	Distance
Diameter	11.75'	17.5'	23.4'	35'	52.5'	*(mf=.117)*
Footcandles	350	152	80	36	15	*(pc=**)*

* Beam and field angles not specified in manufacturer's catalog. This angle is calculated from given distance and diameter data. See Introduction for more information about calculating photometric data.
** Footcandle data given in manufacturer's catalog is not consistent enough to calculate a reliable peak candela figure for all throw distances. At spot focus, 100' throw and 1050 fc, pc=10,500,000; at flood focus, 100' throw and 350 fc, pc=3,500,000.

PHOEBUS
Model Nos. Mighty Arc II & II/S*

MIN. SPOT w/ IRIS: .57° **

SPOT BEAM SPREAD: 3.55° **

FLOOD BEAM SPREAD: 7.25° **

COLOR FRAMES: 6

FADER: Yes (dowser)

SHUTTERS: Yes (guillotine)

FAN COOLED: Yes

STANDARD LAMP: Osram HTI - 400 W.

NOTES: Iris is located on top of the unit. Focus/zoom handle is located on side.

WEIGHT: 97 lbs.

SIZE: 33" x 19½" x 50" (L x W x H)

PHOTOMETRICS CHART
(Performance data for this unit are measured using Osram HTI - 400 W. lamp.)

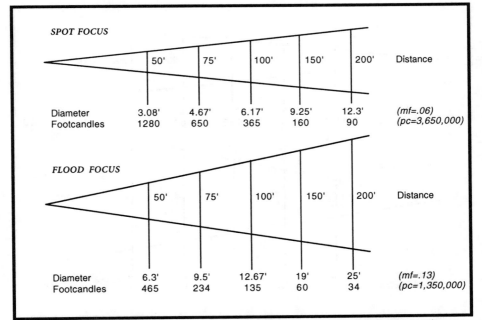

SPOT FOCUS

	50'	75'	100'	150'	200'	Distance
Diameter	3.08'	4.67'	6.17'	9.25'	12.3'	*(mf=.06)*
Footcandles	1280	650	365	160	90	*(pc=3,650,000)*

FLOOD FOCUS

	50'	75'	100'	150'	200'	Distance
Diameter	6.3'	9.5'	12.67'	19'	25'	*(mf=.13)*
Footcandles	465	234	135	60	34	*(pc=1,350,000)*

* Photometrics for these two models are the same. The only difference is in the focus/zoom controls.
** Beam and field angles not specified in manufacturer's catalog. This angle is calculated from given distance and diameter data. See Introduction for more information about calculating photometric data.

FOLLOW SPOT (xenon/HMI/arc/etc.)

PHOEBUS
Model No. Ultra Arc II (Long Throw)

MIN. SPOT w/ IRIS: .48° *

SPOT BEAM SPREAD: 2.86° *

FLOOD BEAM SPREAD: 9.15° *

COLOR FRAMES: 6

FADER: Yes (dowser)

SHUTTERS: Yes (guillotine)

FAN COOLED: Yes

STANDARD LAMP: Osram HTI - 400 W.

NOTES: Iris is located on top of the unit.
Focus/zoom handle is located on side.

WEIGHT: 143 lbs. (includes stand)

SIZE: 56" x 22" x 56½" (L x W x H)

PHOTOMETRICS CHART

(Performance data for this unit are measured using Osram HTI - 400 W. lamp.)

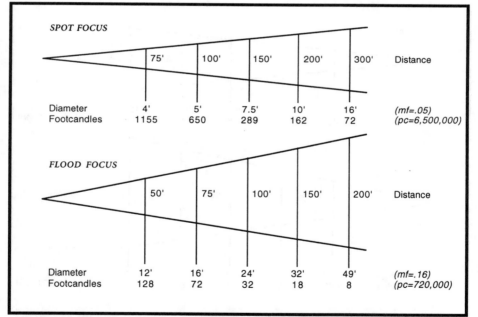

SPOT FOCUS

	75'	100'	150'	200'	300'	Distance
Diameter	4'	5'	7.5'	10'	16'	*(mf=.05)*
Footcandles	1155	650	289	162	72	*(pc=6,500,000)*

FLOOD FOCUS

	50'	75'	100'	150'	200'	Distance
Diameter	12'	16'	24'	32'	49'	*(mf=.16)*
Footcandles	128	72	32	18	8	*(pc=720,000)*

* Beam and field angles not specified in manufacturer's catalog. This angle is calculated from given distance and diameter data. See Introduction for more information about calculating photometric data.

WEIGHT: 128 lbs. (includes stand)

SIZE: 40" x 22" x 56½" (L x W x H)

PHOEBUS
Model No. Ultra Arc II (Short Throw)

MIN. SPOT w/ IRIS: 2.13° *

SPOT BEAM SPREAD: 7.38° *

FLOOD BEAM SPREAD: 21.45° *

COLOR FRAMES: 6

FADER: Yes (dowser)

SHUTTERS: Yes (guillotine)

FAN COOLED: Yes

STANDARD LAMP: Osram HTI - 400 W.

NOTES: Iris is located on top of the unit. Focus/zoom handle is located on side.

PHOTOMETRICS CHART
(Performance data for this unit are measured using Osram HTI - 400 W. lamp.)

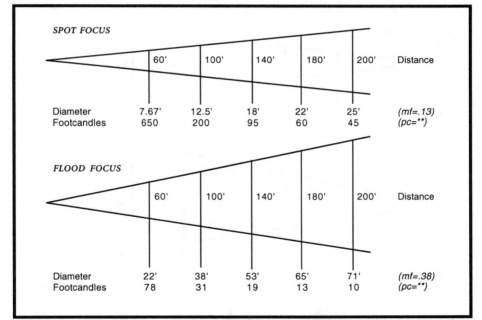

SPOT FOCUS						
	60'	100'	140'	180'	200'	Distance
Diameter	7.67'	12.5'	18'	22'	25'	(mf=.13)
Footcandles	650	200	95	60	45	(pc=**)

FLOOD FOCUS						
	60'	100'	140'	180'	200'	Distance
Diameter	22'	38'	53'	65'	71'	(mf=.38)
Footcandles	78	31	19	13	10	(pc=**)

* Beam and field angles not specified in manufacturer's catalog. This angle is calculated from given distance and diameter data. See Introduction for more information about calculating photometric data.

** Footcandle data given in manufacturer's catalog is not consistent enough to calculate a reliable peak candela figure for all throw distances. (At spot focus, 100' throw and 200 fc, pc=2,000,000, while at 60' throw and 650 fc, pc=2,340,000; at flood focus, 100' throw and 31 fc, pc=310,000, while at 60' throw and 78 fc, pc=280,000.)

FOLLOW SPOT (xenon/HMI/arc/etc.)

PHOEBUS
Model No. Ultra Arc Titan

MIN. SPOT w. IRIS: (not specified in mfg. catalog)

SPOT BEAM SPREAD: 1.9° * (iris closed), 7.6° * (open)

FLOOD BEAM SPREAD: 3.7° * (iris closed), 24° * (open)

COLOR FRAMES: 6

FADER: Yes (dowser)

SHUTTERS: Yes (guillotine)

STANDARD LAMP: 1200 W. HMI

NOTES: Iris, dowser and guillotine controls located on top of unit. Focus is located on right side in front. Unit is fan-cooled.

WEIGHT: 125 lbs (without ballast)

SIZE: 40" x 22" x 22⅛" (L x W x H) (head only)
56½" (overall height with base)

PHOTOMETRICS CHART

(Performance data for this unit are measured at flat field using Osram1200W/ SE lamp.)

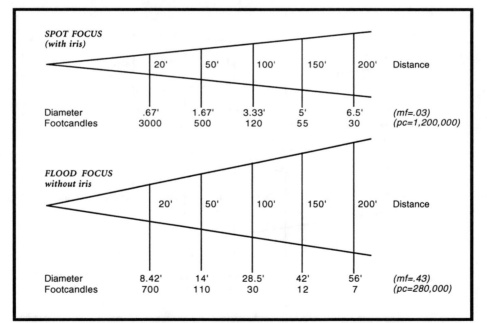

SPOT FOCUS (with iris)

	20'	50'	100'	150'	200'	Distance
Diameter	.67'	1.67'	3.33'	5'	6.5'	(mf=.03)
Footcandles	3000	500	120	55	30	(pc=1,200,000)

FLOOD FOCUS without iris

	20'	50'	100'	150'	200'	Distance
Diameter	8.42'	14'	28.5'	42'	56'	(mf=.43)
Footcandles	700	110	30	12	7	(pc=280,000)

* Beam and field angles not specified in manufacturer's catalog. This angle is calculated from given distance and diameter data. See Introduction for more information about calculating photometric data.

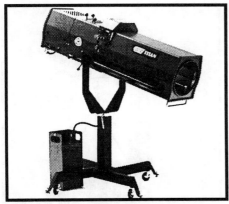

WEIGHT: 133 lbs (without ballast)

SIZE: 56" x 22" x 22⅛" (L x W x H) (head only)
56½" (overall height with base)

PHOEBUS
Model No. Ultra Arc Titan (Long Throw)

MIN. SPOT w. IRIS: (not specified in mfg. catalog)

SPOT BEAM SPREAD: .57° * (iris closed), 2.9° * (open)

FLOOD BEAM SPREAD: 1.15° * (closed), 9.15° * (open)

COLOR FRAMES: 6

FADER: Yes (dowser)

SHUTTERS: Yes (guillotine)

STANDARD LAMP: 1200 W. HMI

NOTES: Iris, dowser and guillotine controls located on top of unit. Focus is located on right side in front. Unit is fan-cooled.

PHOTOMETRICS CHART
(Performance data for this unit are measured at flat field using Osram1200W/ SE lamp.)

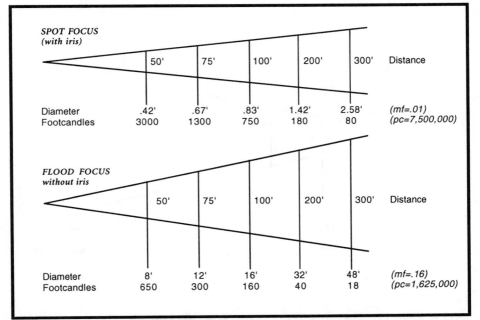

SPOT FOCUS
(with iris)

	50'	75'	100'	200'	300'	Distance
Diameter	.42'	.67'	.83'	1.42'	2.58'	(mf=.01)
Footcandles	3000	1300	750	180	80	(pc=7,500,000)

FLOOD FOCUS
without iris

	50'	75'	100'	200'	300'	Distance
Diameter	8'	12'	16'	32'	48'	(mf=.16)
Footcandles	650	300	160	40	18	(pc=1,625,000)

* Beam and field angles not specified in manufacturer's catalog. This angle is calculated from given distance and diameter data. See Introduction for more information about calculating photometric data.

FOLLOW SPOT (xenon/HMI/arc/etc.)

STRAND CENTURY
Model No. 4411*

MIN. SPOT w/ IRIS: (not specified in mfg. catalog)

BEAM SPREAD: 11° (beam), 14° (field)

COLOR FRAMES: 4 as accessory

FADER: Yes (blackout disc on iris)

SHUTTERS: Yes (horizontal & vertical guillotines)

FAN COOLED: No

STANDARD LAMP: CSI - 1 KW.

NOTES: Iris and shutters are located on the side of the unit. Focus handle at the front on the lens tube.

WEIGHT: 70 lbs.

SIZE: 41⅜" x 16¹⁵⁄₁₆" (L x W)

PHOTOMETRICS CHART
(Performance data for this unit are measured using CSI - 1 KW. lamp.)

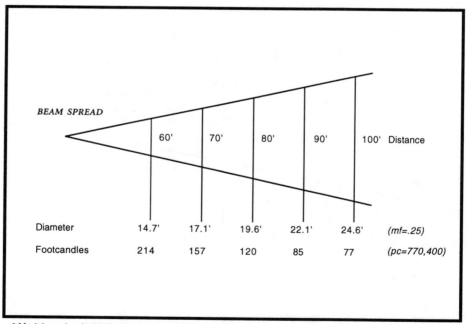

BEAM SPREAD

	60'	70'	80'	90'	100' Distance
Diameter	14.7'	17.1'	19.6'	22.1'	24.6' (mf=.25)
Footcandles	214	157	120	85	77 (pc=770,400)

* Model number is 4412 when unit is wired for 220 V.

STRONG INTERNATIONAL
Model No. 18000, "Roadie"

MIN. SPOT w/ IRIS: .46°

SPOT BEAM SPREAD: 3.5°

FLOOD BEAM SPREAD: 9.32°

COLOR FRAMES: 6

FADER: Yes (dowser)

SHUTTERS: Yes (guillotine)

FAN COOLED: Yes

STANDARD LAMP: Osram HTI - 400 W. SE

NOTES: Iris is located on top of unit. Focus is located on both sides of unit. Unit is equipped with a gobo slot.

WEIGHT: without stand – 65 lbs.

SIZE: 37½" x 52" (L x H)

PHOTOMETRICS CHART

(Performance data for this unit are measured using Osram HTI - 400 W. SE lamp.)

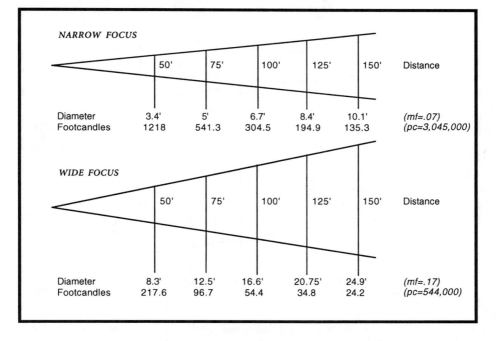

NARROW FOCUS

	50'	75'	100'	125'	150'	Distance
Diameter	3.4'	5'	6.7'	8.4'	10.1'	*(mf=.07)*
Footcandles	1218	541.3	304.5	194.9	135.3	*(pc=3,045,000)*

WIDE FOCUS

	50'	75'	100'	125'	150'	Distance
Diameter	8.3'	12.5'	16.6'	20.75'	24.9'	*(mf=.17)*
Footcandles	217.6	96.7	54.4	34.8	24.2	*(pc=544,000)*

FOLLOW SPOT (xenon/HMI/arc/etc.)

STRONG INTERNATIONAL
Model No. 45050 "Super Trouperette"

MIN. SPOT w/ IRIS: .77° *

SPOT BEAM SPREAD: 4.27° *

FLOOD BEAM SPREAD: 10.15° *

COLOR FRAMES: 6

FADER: No

SHUTTERS: Yes (guillotine)

FAN COOLED: Yes

STANDARD LAMP: Osram HTI - 400 W.

NOTES: Iris is located on top of unit. Focus is located on front right side.

WEIGHT: 94 lbs.

SIZE: 36" x 52" (L x H)

PHOTOMETRICS CHART
(Performance data for this unit are measured using Osram HTI - 400 W. lamp.)

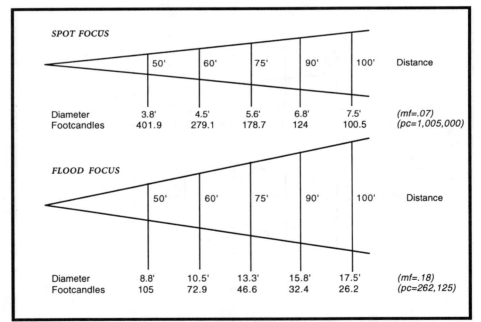

SPOT FOCUS

	50'	60'	75'	90'	100'	Distance
Diameter	3.8'	4.5'	5.6'	6.8'	7.5'	*(mf=.07)*
Footcandles	401.9	279.1	178.7	124	100.5	*(pc=1,005,000)*

FLOOD FOCUS

	50'	60'	75'	90'	100'	Distance
Diameter	8.8'	10.5'	13.3'	15.8'	17.5'	*(mf=.18)*
Footcandles	105	72.9	46.6	32.4	26.2	*(pc=262,125)*

* Beam and field angles not specified in manufacturer's catalog. This angle is calculated from given distance and diameter data. See Introduction for more information about calculating photometric data.

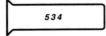

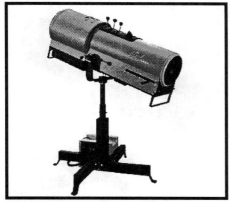

WEIGHT: with power supply – 198 lbs.

SIZE: 46½" x 56–68" (L x H)

STRONG INTERNATIONAL
Model No. 575

MIN. SPOT w/ IRIS: 1.1° *

SPOT BEAM SPREAD: 5.73° *

FLOOD BEAM SPREAD: 17.1° *

COLOR FRAMES: 6

FADER: Yes

SHUTTERS: Yes (guillotine)

FAN COOLED: Yes

STANDARD LAMP: 575 W. Metal Halide

NOTES: Iris is located on top of unit. Focus is located on front right side.

PHOTOMETRICS CHART
(Performance data for this unit are measured using 575 W. Metal Halide lamp.)

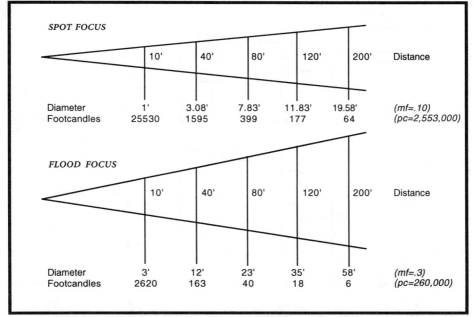

SPOT FOCUS

	10'	40'	80'	120'	200'	Distance
Diameter	1'	3.08'	7.83'	11.83'	19.58'	(mf=.10)
Footcandles	25530	1595	399	177	64	(pc=2,553,000)

FLOOD FOCUS

	10'	40'	80'	120'	200'	Distance
Diameter	3'	12'	23'	35'	58'	(mf=.3)
Footcandles	2620	163	40	18	6	(pc=260,000)

* Beam and field angles not specified in manufacturer's catalog. This angle is calculated from given distance and diameter data. See Introduction for more information about calculating photometric data.

STRONG INTERNATIONAL
Model No. Gladiator II

MIN. SPOT w/ IRIS: .6° *

SPOT BEAM SPREAD: 3.27° *

FLOOD BEAM SPREAD: 9.53° *

COLOR FRAMES: 6

FADER: Yes

SHUTTERS: Yes (guillotine)

FAN COOLED: Yes

STANDARD LAMP: Xenon - 2.5 KW.

NOTES: Iris is located on top of unit. Focus is located on front right side.

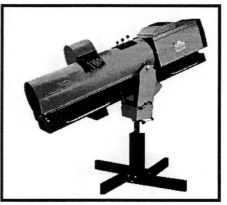

WEIGHT: 671 lbs.

SIZE: 77½" x 69–77" (L x H)

PHOTOMETRICS CHART

(Performance data for this unit are measured using 2.5 KW. Xenon lamp.)

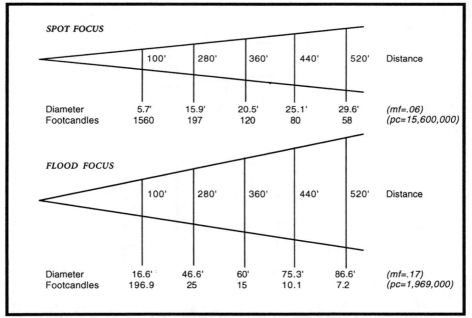

SPOT FOCUS

Distance	100'	280'	360'	440'	520'	
Diameter	5.7'	15.9'	20.5'	25.1'	29.6'	(mf=.06)
Footcandles	1560	197	120	80	58	(pc=15,600,000)

FLOOD FOCUS

Distance	100'	280'	360'	440'	520'	
Diameter	16.6'	46.6'	60'	75.3'	86.6'	(mf=.17)
Footcandles	196.9	25	15	10.1	7.2	(pc=1,969,000)

* Beam and field angles not specified in manufacturer's catalog. This angle is calculated from given distance and diameter data. See Introduction for more information about calculating photometric data.

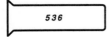

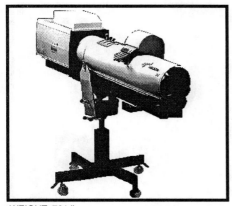

STRONG INTERNATIONAL
Model No. Gladiator III

MIN. SPOT w/ IRIS: .62° *

SPOT BEAM SPREAD: 3.21° *

FLOOD BEAM SPREAD: 10.29° *

COLOR FRAMES: 6

FADER: Yes

SHUTTERS: Yes (guillotine)

FAN COOLED: Yes

STANDARD LAMP: Xenon - 3 KW.

NOTES: Iris is located on top of unit. Focus is located on front right side.

WEIGHT: 721 lbs.

SIZE: 85½" x 69–77" (L x H)

PHOTOMETRICS CHART
(Performance data for this unit are measured using 3 KW. Xenon lamp.)

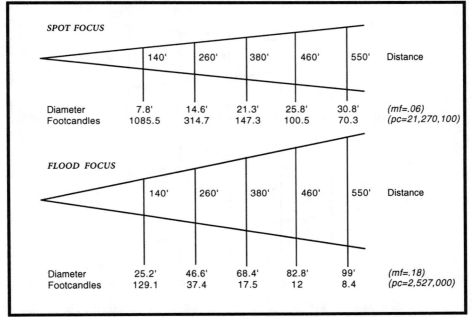

SPOT FOCUS

	140'	260'	380'	460'	550'	Distance
Diameter	7.8'	14.6'	21.3'	25.8'	30.8'	*(mf=.06)*
Footcandles	1085.5	314.7	147.3	100.5	70.3	*(pc=21,270,100)*

FLOOD FOCUS

	140'	260'	380'	460'	550'	Distance
Diameter	25.2'	46.6'	68.4'	82.8'	99'	*(mf=.18)*
Footcandles	129.1	37.4	17.5	12	8.4	*(pc=2,527,000)*

* Beam and field angles not specified in manufacturer's catalog. This angle is calculated from given distance and diameter data. See Introduction for more information about calculating photometric data.

FOLLOW SPOT (xenon/HMI/arc/etc.)

STRONG INTERNATIONAL
Model No. Super Trouper (Arc)

SPOT BEAM SPREAD: 3° (field)

FLOOD BEAM SPREAD: 14° (field)

COLOR FRAMES: 6

FADER: Yes

SHUTTERS: Yes (guillotine)

FAN COOLED: No

CARBONS: 7mm x 12" Positive, 6mm x 9" Negative

NOTES: Iris is located on top of unit. Focus is located on front right side. Carbon trimming and replacement is through access door on the side of the lamphouse.

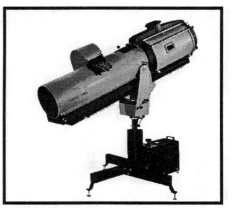

WEIGHT: 395 lbs.

SIZE: 80½" x 67–79" (L x H)

PHOTOMETRICS CHART

(Performance data for this unit are measured using 7mm x 12" Positive, 6mm x 9" Negative carbons.)

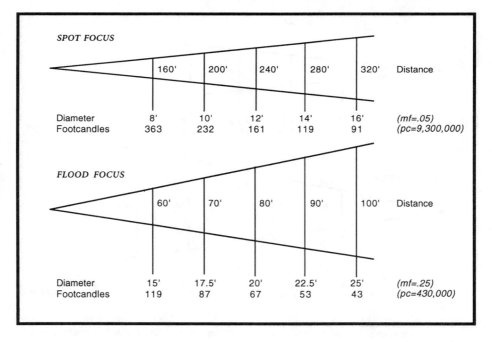

SPOT FOCUS

	160'	200'	240'	280'	320'	Distance
Diameter	8'	10'	12'	14'	16'	(mf=.05)
Footcandles	363	232	161	119	91	(pc=9,300,000)

FLOOD FOCUS

	60'	70'	80'	90'	100'	Distance
Diameter	15'	17.5'	20'	22.5'	25'	(mf=.25)
Footcandles	119	87	67	53	43	(pc=430,000)

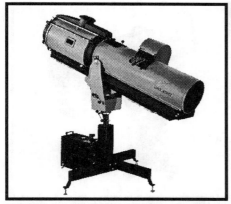

WEIGHT: (not specified in manufacturer's catalog)

SIZE: (not specified in manufacturer's catalog)

STRONG INTERNATIONAL
Model No. Super Trouper-Xenon

MIN. SPOT w/ IRIS: .8° *

SPOT BEAM SPREAD: 4.45° *

FLOOD BEAM SPREAD: 11.42° *

COLOR FRAMES: 6

FADER: No

SHUTTERS: Yes (guillotine)

FAN COOLED: Yes

STANDARD LAMP: 1600 W. XENON

NOTES: A Super Trouper carbon arc follow spot can be retrofitted to use a 1600 W. Xenon lamp. The conversion kit consists of a new reflector, bulb mounts and power supply. The photometric performance with the Xenon lamp is shown below.

PHOTOMETRICS CHART
(Performance data for this unit are measured using 1600 W. Xenon lamp.)

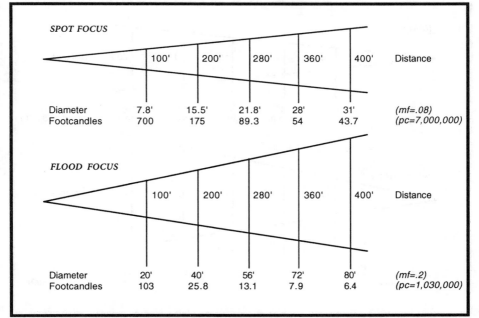

SPOT FOCUS

	100'	200'	280'	360'	400'	Distance
Diameter	7.8'	15.5'	21.8'	28'	31'	(mf=.08)
Footcandles	700	175	89.3	54	43.7	(pc=7,000,000)

FLOOD FOCUS

	100'	200'	280'	360'	400'	Distance
Diameter	20'	40'	56'	72'	80'	(mf=.2)
Footcandles	103	25.8	13.1	7.9	6.4	(pc=1,030,000)

* Beam and field angles not specified in manufacturer's catalog. This angle is calculated from given distance and diameter data. See Introduction for more information about calculating photometric data.

STRONG INTERNATIONAL
Model No. Trouper 1200

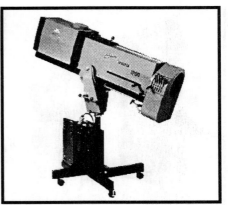

MIN. SPOT w/ IRIS: 1.25° *

SPOT BEAM SPREAD: 4.58° *

FLOOD BEAM SPREAD: 13.68° *

COLOR FRAMES: 6

FADER: Yes

SHUTTERS: Yes (guillotine)

FAN COOLED: Yes

STANDARD LAMP: 1200 W. Metal Halide

NOTES: Iris is located on top of unit. Focus is located on front right side.

WEIGHT: with power supply – 230 lbs.

SIZE: 54" x 56–68" (L x H)

PHOTOMETRICS CHART
(Performance data for this unit are measured using 1200 W. Metal Halide lamp.)

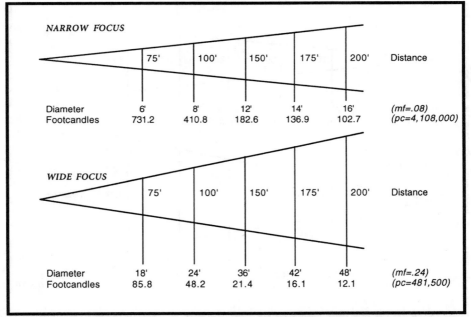

NARROW FOCUS

	75'	100'	150'	175'	200'	Distance
Diameter	6'	8'	12'	14'	16'	*(mf=.08)*
Footcandles	731.2	410.8	182.6	136.9	102.7	*(pc=4,108,000)*

WIDE FOCUS

	75'	100'	150'	175'	200'	Distance
Diameter	18'	24'	36'	42'	48'	*(mf=.24)*
Footcandles	85.8	48.2	21.4	16.1	12.1	*(pc=481,500)*

* Beam and field angles not specified in manufacturer's catalog. This angle is calculated from given distance and diameter data. See Introduction for more information about calculating photometric data.

WEIGHT: 225 lbs.

SIZE: 63" x 53–65" (L x H)

STRONG INTERNATIONAL
Model No. Trouper (Arc)

SPOT BEAM SPREAD: 10° (field)

FLOOD BEAM SPREAD: 28° (field)

COLOR FRAMES: 6

FADER: No

SHUTTERS: Yes (guillotine)

FAN COOLED: No

CARBONS: 6mm x 7" A.C.

NOTES: Iris is located on top of unit. Focus is located on front right side. Carbon trimming and replacement is through access door on the side of the lamphouse.

PHOTOMETRICS CHART

(Performance data for this unit are measured using 6mm x 7" A.C. carbons.)

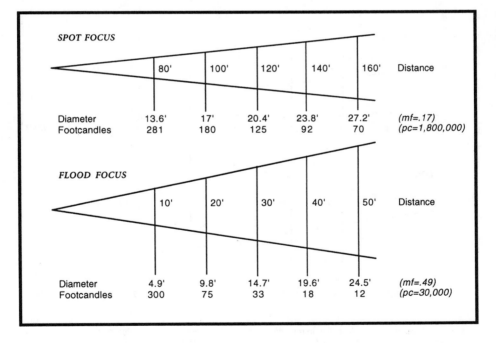

SPOT FOCUS

	80'	100'	120'	140'	160'	Distance
Diameter	13.6'	17'	20.4'	23.8'	27.2'	*(mf=.17)*
Footcandles	281	180	125	92	70	*(pc=1,800,000)*

FLOOD FOCUS

	10'	20'	30'	40'	50'	Distance
Diameter	4.9'	9.8'	14.7'	19.6'	24.5'	*(mf=.49)*
Footcandles	300	75	33	18	12	*(pc=30,000)*

FOLLOW SPOT (xenon/HMI/arc/etc.)

STRONG INTERNATIONAL
Xenon Super Trouper (Long Throw)

MIN. SPOT w/ IRIS: .84° *

SPOT BEAM SPREAD: 3.58° *

FLOOD BEAM SPREAD: 7.96° *

COLOR FRAMES: 6

FADER: Yes

SHUTTERS: Yes (guillotine)

FAN COOLED: Yes

STANDARD LAMP: Xenon - 1.6 KW.

NOTES: Iris is located on top of unit. Focus is located on front right side.

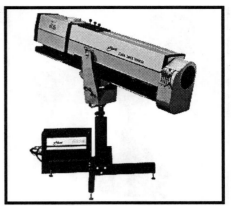

WEIGHT: with power supply – 496 lbs.

SIZE: 77½" x 62–70" (L x H)

PHOTOMETRICS CHART

(Performance data for this unit are measured using 1.6 KW. Xenon lamp.)

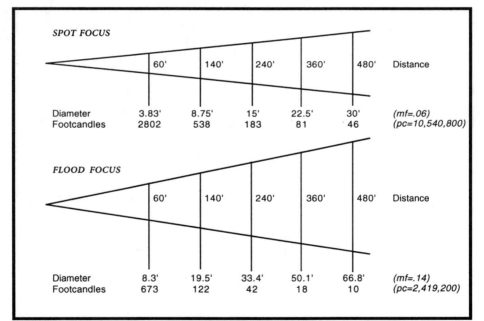

SPOT FOCUS

	60'	140'	240'	360'	480'	Distance
Diameter	3.83'	8.75'	15'	22.5'	30'	(mf=.06)
Footcandles	2802	538	183	81	46	(pc=10,540,800)

FLOOD FOCUS

	60'	140'	240'	360'	480'	Distance
Diameter	8.3'	19.5'	33.4'	50.1'	66.8'	(mf=.14)
Footcandles	673	122	42	18	10	(pc=2,419,200)

* Beam and field angles not specified in manufacturer's catalog. This angle is calculated from given distance and diameter data. See Introduction for more information about calculating photometric data.

542

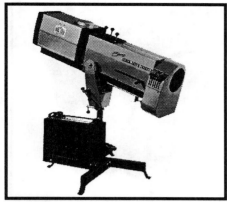

STRONG INTERNATIONAL
Xenon Super Trouper (Short Throw)

MIN. SPOT w/ IRIS: 1.32° *

SPOT BEAM SPREAD: 6.67° *

FLOOD BEAM SPREAD: 23.17° *

COLOR FRAMES: 6

FADER: Yes

SHUTTERS: Yes (guillotine)

FAN COOLED: Yes

STANDARD LAMP: Xenon - 1.6 KW.

NOTES: Iris is located on top of unit. Focus is located on front right side.

WEIGHT: with power supply – 350 lbs.

SIZE: 77½" x 62–70" (L x H)

PHOTOMETRICS CHART
(Performance data for this unit are measured using 1.6 KW. Xenon lamp.)

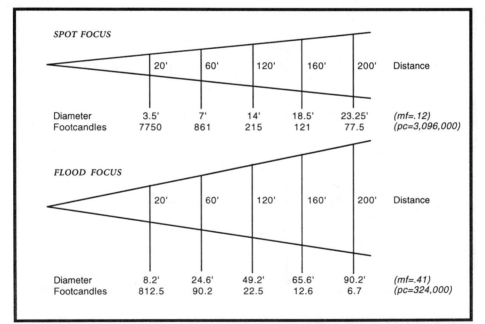

SPOT FOCUS

Distance	20'	60'	120'	160'	200'	
Diameter	3.5'	7'	14'	18.5'	23.25'	(mf=.12)
Footcandles	7750	861	215	121	77.5	(pc=3,096,000)

FLOOD FOCUS

Distance	20'	60'	120'	160'	200'	
Diameter	8.2'	24.6'	49.2'	65.6'	90.2'	(mf=.41)
Footcandles	812.5	90.2	22.5	12.6	6.7	(pc=324,000)

* Beam and field angles not specified in manufacturer's catalog. This angle is calculated from given distance and diameter data. See Introduction for more information about calculating photometric data.

FOLLOW SPOT (xenon/HMI/arc/etc.)

STRONG INTERNATIONAL
Model No. Xenon Trouper

MIN. SPOT w/ IRIS: 1.07° *

SPOT BEAM SPREAD: 6.8° *

FLOOD BEAM SPREAD: 22.69° *

COLOR FRAMES: 6

FADER: No

SHUTTERS: Yes (guillotine)

FAN COOLED: Yes

STANDARD LAMP: Xenon - (wattage not specified)

NOTES: Iris is located on top of unit. Focus is located on front right side.

WEIGHT: with power supply – 322 lbs.

SIZE: 54" x 56–68" (L x H)

PHOTOMETRICS CHART
(Performance data for this unit are measured using unspecified wattage Xenon lamp.)

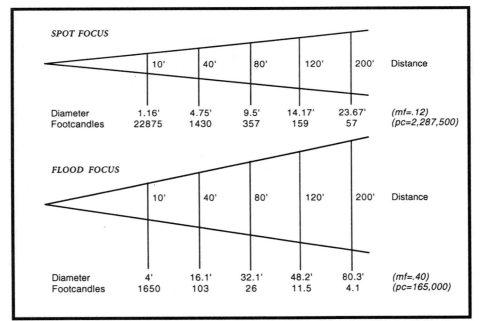

SPOT FOCUS

	10'	40'	80'	120'	200'	Distance
Diameter	1.16'	4.75'	9.5'	14.17'	23.67'	*(mf=.12)*
Footcandles	22875	1430	357	159	57	*(pc=2,287,500)*

FLOOD FOCUS

	10'	40'	80'	120'	200'	Distance
Diameter	4'	16.1'	32.1'	48.2'	80.3'	*(mf=.40)*
Footcandles	1650	103	26	11.5	4.1	*(pc=165,000)*

* Beam and field angles not specified in manufacturer's catalog. This angle is calculated from given distance and diameter data. See Introduction for more information about calculating photometric data.

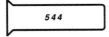

WEIGHT: 135 lbs.

SIZE: 54" x 12" (L x W)

TIMES SQUARE LIGHTING
Model No. QA-300

MIN. SPOT w/ IRIS: (not specified in mfg. catalog)

BEAM SPREAD: 13.2° *

COLOR FRAMES: 6

FADER: Yes (blackout frame)

SHUTTERS: Yes (guillotine)

FAN COOLED: No

STANDARD LAMP: EZM (MARC 300/16) - 300 W.

NOTES: Iris and shutters are rear operated. Unit is equipped with auxilary power outlet.

PHOTOMETRICS CHART

(Information about lamp used to measure performance data is not provided in manufacturer's catalog.)

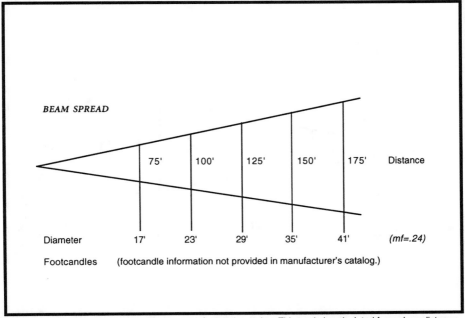

BEAM SPREAD

| | 75' | 100' | 125' | 150' | 175' | Distance |

Diameter   17'   23'   29'   35'   41'   (mf=.24)

Footcandles   (footcandle information not provided in manufacturer's catalog.)

* Beam and field angles not specified in manufacturer's catalog. This angle is calculated from given distance and diameter data. The instrument has adjustable lenses, but only one set of distance/diameter figures is provided in mfg. catalog. It would be logical to assume that the data is for the spot focus, without iris. See Introduction for more information about calculating photometric data.

ALTMAN STAGE LIGHTING
Model No. 520

BEAM ANGLE: 50°

FIELD ANGLE: 75°

STANDARD CONFIGS: 6' - 4 circuit - 16 lamps
6'-9" - 3 circuit - 18 lamps
7'-6" - 4 circuit - 20 lamps
8' - 3 circuit - 21 lamps

LAMP CENTERS: 4.5"

LAMP BASE TYPE: Medium Screw

STANDARD LAMP: 75A/19 - 75 W.

OTHER LAMPS: 60A/19 - 60 W. *(cf=.73)*
100A/19 - 100 W. *(cf=1.47)*

NOTES: Unit takes 4" roundels. Available in custom lengths.

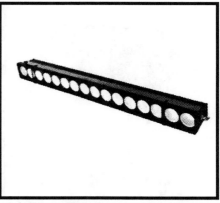

WEIGHT: (not specified in manufacturer's catalog)

SIZE: 8¼" x 9½" (W x H)

PHOTOMETRICS CHART
(Performance data for this unit are measured using a single 75A/19 - 75 W. lamp.)

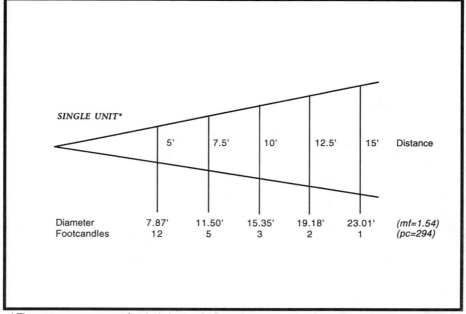

SINGLE UNIT*

	5'	7.5'	10'	12.5'	15'	Distance
Diameter	7.87'	11.50'	15.35'	19.18'	23.01'	*(mf=1.54)*
Footcandles	12	5	3	2	1	*(pc=294)*

* These measurements are for single lamp only. Beam patterns overlap depending on lamp centers and number of circuits per instrument.

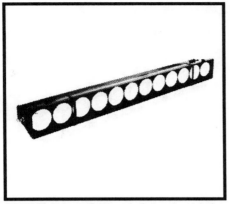

ALTMAN STAGE LIGHTING
Model No. 528

BEAM ANGLE: 52°

FIELD ANGLE: 80°

STANDARD CONFIGS: 6'-0" - 3 or 4 circuit - 12 lamps
7'-6" - 3 circuit - 15 lamps
8'-0" - 4 circuit - 16 lamps

LAMP CENTERS: 6"

LAMP BASE TYPE: Medium Screw

STANDARD LAMP: 150A/23 - 150 W.

OTHER LAMPS: 200A/23 - 200 W. *(cf=1.44)*
300M/IF - 300 W. *(cf=2.28)*

NOTES: Unit takes 5⅝" roundels. Available in custom lengths.

WEIGHT: (not specified in manufacturer's catalog)

SIZE: 9⅛" x 13¼" (W x H)

PHOTOMETRICS CHART

(Performance data for this unit are measured using a single 150A/23 - 150 W. lamp.)

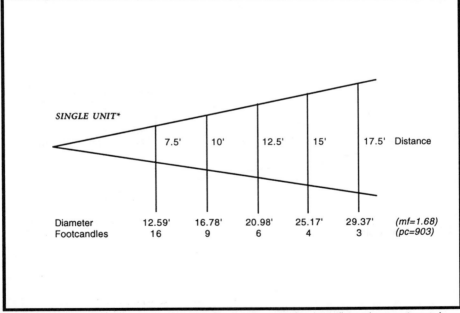

SINGLE UNIT*

	7.5'	10'	12.5'	15'	17.5'	Distance
Diameter	12.59'	16.78'	20.98'	25.17'	29.37'	*(mf=1.68)*
Footcandles	16	9	6	4	3	*(pc=903)*

* These measurements are for single lamp only. Beam patterns overlap depending on lamp centers and number of circuits per instrument.

ALTMAN STAGE LIGHTING
Model No. 537

BEAM ANGLE: 25°

FIELD ANGLE: 65°

STANDARD CONFIGS: 5'-4" - 4 circuit - 8 lamps
6'-0" - 3 circuit - 9 lamps
8'-0" - 3 or 4 circuit - 12 lamps

LAMP CENTERS: 8"

LAMP BASE TYPE: Medium Screw

STANDARD LAMP: 300M/IF - 300 W.

OTHER LAMPS: 200PS/25 - 200 W. *(cf=.6)*

NOTES: Unit takes 7⁹⁄₁₆" roundels. Available in custom lengths.

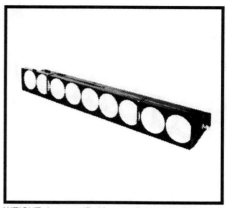

WEIGHT: (not specified in manufacturer's catalog)

SIZE: 11½" x 14" (W x H)

PHOTOMETRICS CHART
(Performance data for this unit are measured using a single 300M/IF - 300 W. lamp.)

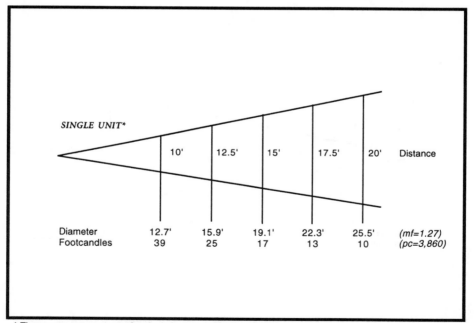

SINGLE UNIT*

	10'	12.5'	15'	17.5'	20'	Distance
Diameter	12.7'	15.9'	19.1'	22.3'	25.5'	*(mf=1.27)*
Footcandles	39	25	17	13	10	*(pc=3,860)*

* These measurements are for single lamp only. Beam patterns overlap depending on lamp centers and number of circuits per instrument.

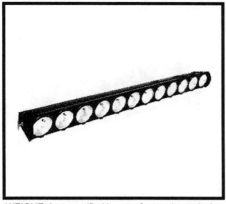

ALTMAN STAGE LIGHTING
Model No. 600

BEAM ANGLE: 20°

FIELD ANGLE: 48°

STANDARD CONFIGS: 5'-4" - 4 circuit - 8 lamps
6'-0" - 3 circuit - 9 lamps
8'-0" - 3 or 4 circuit - 12 lamps

LAMP CENTERS: 8"

LAMP BASE TYPE: Recessed Single-Contact

STANDARD LAMP: EHM - 300 W.

OTHER LAMPS: FCL - 500 W. *(cf=1.87)*

NOTES: Unit takes 7⁹⁄₁₆" roundels. Available in custom lengths.

WEIGHT: (not specified in manufacturer's catalog)

SIZE: (not specified in manufacturer's catalog)

PHOTOMETRICS CHART
(Performance data for this unit are measured using a single EHM - 300 W. lamp.)

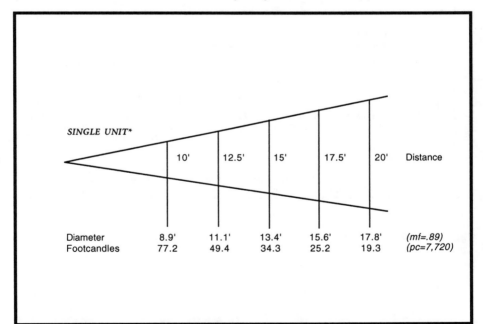

SINGLE UNIT*		10'	12.5'	15'	17.5'	20'	Distance
Diameter		8.9'	11.1'	13.4'	15.6'	17.8'	*(mf=.89)*
Footcandles		77.2	49.4	34.3	25.2	19.3	*(pc=7,720)*

* These measurements are for single lamp only. Beam patterns overlap depending on lamp centers and number of circuits per instrument.

ALTMAN STAGE LIGHTING
Model No. Ground Cyc

BEAM ANGLE (W x H): 50° x 65°
FIELD ANGLE (W x H): 70° x 83°
STANDARD CONFIGS: 0'-11" - 1 circuit - 1 lamp
                   1'-8" - 1 or 2 circuit - 2 lamps
                   2'-5" - 3 circuit - 3 lamps
                   3'-2" - 2 circuit - 4 lamps
                   4'-8" - 2 or 3 circuit - 6 lamps
                   6'-2" - 2 or 4 circuit - 8 lamps
                   6'-11" - 3 circuit - 9 lamps
                   7'-8" - 2 circuit - 10 lamps
                   9'-2" - 3 circuit - 12 lamps

LAMP CENTERS: 9"
LAMP BASE TYPE: Recessed Single-Contact
STANDARD LAMP: FCM - 1000 W.
OTHER LAMPS: (see below)

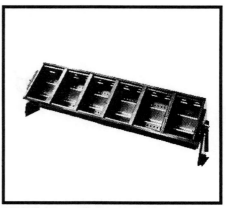

WEIGHT: (not specified in manufacturer's catalog)
SIZE: (not specified in manufacturer's catalog)

PHOTOMETRICS CHART

(Performance data for this unit is measured using a single compartment with FCM - 1000 W. lamp.)

SINGLE UNIT *
70° x 83° field angle

	2.5'	5'	7.5'	10'	Distance
Dimensions (W x H)	3.5'x4.4'	7'x9'	10.5'x13'	14'x18'	*(mf= 1.4x1.8)*
Footcandles	1296	324	144	81	*(pc=81,000)*

Watts	Lamp ANSI Code	Color Temp. °K	Life (hrs)	Lumens	Correction Factor
1000	FCM	3200	300	27,500	1.0
750	EJG	3200	400	20,000	.73
500	FDF	3200	400	13,250	.48
500	FCL	3200	2000	10,950	.40
300	EHM	3000	2000	5,950	.22

* These measurements are for single lamp only. Beam patterns overlap depending on lamp centers and number of circuits per instrument.

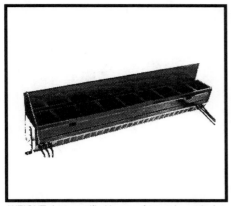

ALTMAN STAGE LIGHTING
Model No. PAR 56B

BEAM ANGLE: (determined by lamp - see below)

FIELD ANGLE: (determined by lamp - see below)

STANDARD CONFIGS: 4'-0" - 3 circuit - 6 lamps
 5'- 4" - 4 circuit - 8 lamps
 6'-0" - 3 circuit - 9 lamps
 8'-0" - 3 or 4 circuit - 12 lamps

LAMP CENTERS: 8"

LAMP BASE TYPE: Extended Mogul End Prong

STANDARD LAMP: PAR 56

OTHER LAMPS: (see below)

NOTES: Unit takes 7⁹⁄₁₆" roundels. Available in custom lengths.

WEIGHT: (not specified in manufacturer's catalog)

SIZE: 8¼" x 10½" (W x H)

PHOTOMETRICS CHART

(Performance data for this unit determined by lamp used. Sample: Q500 PAR56/MFL - 500 W. lamp.)

MEDIUM FLOOD (MFL) *
42° x 20° field angle

	10'	20'	40'	50'	Distance
Dimensions (W x H)	7.7'x3.5'	15.4'x7.1'	30.7'x14.1'	38.4'x17.6'	(mf= .77x.35)
Footcandles	430	108	27	17	(pc=43,000)

Watts	Lamp	Beam Shape	Color Temp. °K	Beam Angle W x H	Field Angle W x H	(mf) diameter	(pc) candela
500	Q500 PAR56/NSP	NSP	3000	13°x8°	32°x15°	.57x.26	96,000
500	Q500 PAR56/MFL	MFL	3000	26°x10°	42°x20°	.77x.35	43,000
500	Q500 PAR56/WFL	WFL	3000	44°x20°	66°x34°	1.30x.61	19,000
300	300 PAR56/NSP	NSP	2800	10°x8°	20°x15°	.35x.26	70,000
300	300 PAR56/MFL	MFL	2800	23°x11°	35°x20°	.63x.35	24,000
300	300 PAR56/WFL	WFL	2800	37°x18°	60°x30°	1.15x.54	10,000

* These measurements are for single lamp only. Beam patterns overlap depending on lamp centers and number of circuits per instrument.

STRIPLIGHT / CYC LIGHT

ALTMAN STAGE LIGHTING
Model No. PAR 64B

BEAM ANGLE: (determined by lamp - see below)

FIELD ANGLE: (determined by lamp - see below)

STANDARD CONFIGS: 4'-6" - 3 circuit - 6 lamps
6'-0" - 4 circuit - 8 lamps
6'-9" - 6 circuit - 9 lamps
9'-0" - 6 or 8 circuit - 12 lamps

LAMP CENTERS: 9"

LAMP BASE TYPE: Mogul End Prong

STANDARD LAMP: PAR 64

OTHER LAMPS: (see below)

NOTES: Unit takes 7⁹⁄₁₆" roundels. Available in custom lengths.

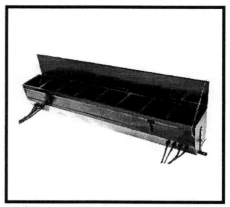

WEIGHT: (not specified in manufacturer's catalog)

SIZE: 8¼" x 10½" (W x H)

PHOTOMETRICS CHART

(Performance data for this unit determined by lamp used. Sample: Q1000 PAR64/MFL - 1 KW. lamp.)

MEDIUM FLOOD (MFL) *
45° x 22° field angle

	10'	20'	40'	50'	Distance
Dimensions (W x H)	8.3'x3.9'	16.6'x7.8'	33.1'x15.6'	41.4'x19.4'	(mf= .83x.39)
Footcandles	800	200	50	32	(pc=80,000)

Watts	Lamp	Beam Shape	Color Temp. °K	Beam Angle W x H	Field Angle W x H	(mf) diameter	(pc) candela
1000	Q1000 PAR64/1	VNSP	3200	12°x6°	24°x10°	.43x.18	400,000
1000	Q1000 PAR64/2	NSP	3200	14°x7°	26°x14°	.46x.25	330,000
1000	Q1000 PAR64/5	MFL	3200	28°x12°	44°x21°	.8x.37	125,000
1000	Q1000 PAR64/6	WFL	3200	48°x24°	71°x45°	1.43x.83	40,000
1000	Q1000 PAR64/NSP	NSP	3000	15°x8°	31°x14°	.55x.25	180,000
1000	Q1000 PAR64/MFL	MFL	3000	28°x12°	45°x22°	.83x.39	80,000
1000	Q1000 PAR64/WFL	WFL	3000	48°x24°	72°x45°	1.45x.83	33,000
500	500 PAR64/NSP	NSP	2800	12°x7°	20°x13°	.35x.23	110,000
500	500 PAR64/MFL	MFL	2800	23°x11°	35°x20°	.63x.35	35,000
500	500 PAR64/WFL	WFL	2800	42°x20°	65°x35°	1.27x.63	12,000

* These measurements are for single lamp only. Beam patterns overlap depending on lamp centers and number of circuits per instrument.

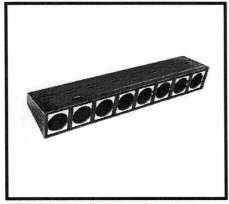

ALTMAN STAGE LIGHTING
Model No. R40

BEAM ANGLE: (not specified in mfg. catalog)

FIELD ANGLE: (determined by lamp - see below)

STANDARD CONFIGS: 6'-0" - 3 or 4 circuit - 12 lamps
7'-6" - 3 circuit - 15 lamps
8'-0" - 4 circuit - 16 lamps

LAMP CENTERS: 6"

LAMP BASE TYPE: Medium Screw

STANDARD LAMP: R40

OTHER LAMPS: (see below)

WEIGHT: (not specified in manufacturer's catalog)

SIZE: 7" x 11" (W x H)

NOTES: Unit takes 5⅝" roundels. Available in custom lengths.

PHOTOMETRICS CHART

(Performance data for this unit determined by lamp used. Sample: 150R/FL - 150 W. lamp.)

150R/FL - 150 W.*
60° field angle

	10'	15'	17.5'	20'	Distance
Diameter	11.5'	17.3'	20.2'	23.1'	(mf= 1.15)
Footcandles	10.4	4.6	3.4	2.6	(pc=1,040)

Watts	Lamp	Rated Life (Hours)	Field Angle	(mf) diameter	(pc) candela
100	100R/FL	2000	60°	1.15	900
100	100R/SP	2000	36°	.65	5,000
150	150R/FL	2000	60°	1.15	1,040
150	150R/SP	2000	49°	.91	5,400
200	200R/FL	2000	60°	1.15	2,000
200	200R/SP	2000	34°	.61	12,000
300	300R/FL	2000	60°	1.15	1,950
300	300R/FL	2000	24°	.43	8,900

* These measurements are for single lamp only. Beam patterns overlap depending on lamp centers and number of circuits per instrument.

ALTMAN STAGE LIGHTING
Model No. SE

BEAM ANGLE: (not specified in mfg. catalog)

FIELD ANGLE: 85°

STANDARD CONFIGS: 6'-0" - 3 or 4 circuit - 18 lamps

LAMP CENTERS: 4"

LAMP BASE TYPE: Medium Screw

STANDARD LAMP: 100A - 100 W.

OTHER LAMPS: 40A - 40 W. *(cf=.3)*
                 60A - 60 W. *(cf=.5)*

NOTES: Unit depends on colored lamps - there are no provisions for rondels or gel. Available in custom lengths.

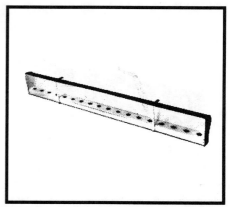

WEIGHT: (not specified in manufacturer's catalog)

SIZE: 4⅞" x 9½" (W x H)

PHOTOMETRICS CHART
(Performance data for this unit are measured using a single 100A - 100 W. lamp.)

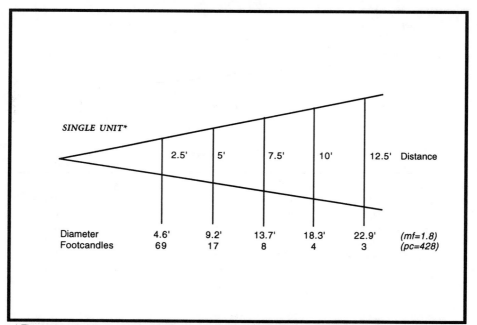

SINGLE UNIT*

	2.5'	5'	7.5'	10'	12.5'	Distance
Diameter	4.6'	9.2'	13.7'	18.3'	22.9'	*(mf=1.8)*
Footcandles	69	17	8	4	3	*(pc=428)*

* These measurements are for single lamp only. Beam patterns overlap depending on lamp centers and number of circuits per instrument.

ALTMAN STAGE LIGHTING
Model No. Sky Cyc

BEAM ANGLE (W x H): 100° x 105°

FIELD ANGLE (W x H): 108° x 107°

STANDARD CONFIGS: 1 circuit - 1 lamp
2 circuit - 2 or 4 lamps
3 circuit - 3 lamps
4 circuit - 4 lamps

LAMP CENTERS: 8"

LAMP BASE TYPE: Recessed Single-Contact

STANDARD LAMP: FDB - 1.5 KW.

OTHER LAMPS: FFT - 1 KW. *(cf=.64)*

NOTES: Manufacturer gives a peak candela figure of 10,924 cd with FDB lamp. Because of the asymetrical reflector, this maximum output is not along the horizontal axis of the unit.

WEIGHT: (not specified in manufacturer's catalog)

SIZE: (not specified in manufacturer's catalog)

PHOTOMETRICS CHART

(Performance graphs below showing single lamp and four lamp configurations using FDB - 1.5 KW. lamp are copied from manufacturer's catalog.)

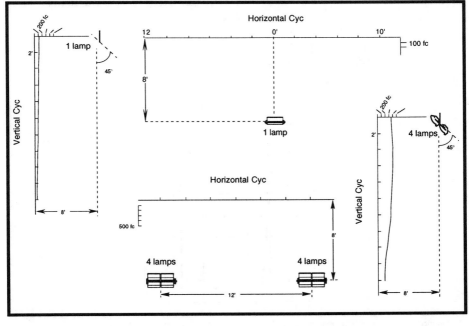

CCT LIGHTING
Model No. Z0C46*

BEAM ANGLE: (not specified in mfg. catalog)

FIELD ANGLE: (not specified in mfg. catalog)

STANDARD CONFIGS: 1 circuit - 1 lamp

LAMP CENTERS: (not specified in mfg. catalog)

LAMP BASE TYPE: Recessed Single-Contact

STANDARD LAMP: FDN - 500 W.

OTHER LAMPS: EHZ - 300 W. *(cf=.45)*
FCZ - 500 W. *(cf=.84)*

ACCESSORIES AVAILABLE: Color Frame

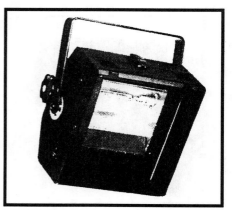

WEIGHT: 7.5 lbs.*

SIZE: 5⅞" x 9" x 11⅞" (L x W x H)

PHOTOMETRICS CHART

(Performance graph below showing single lamp configuration using FDN - 500 W. lamp is copied from manufacturer's catalog.)

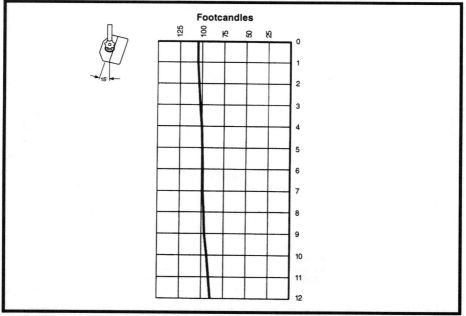

* This model is available in double, triple and quad configurations: Z0C47 (double) weighs 15 lbs., Z0C48 (triple) weighs 22.5 lbs. and Z0C49 (quad) weighs 30 lbs.

WEIGHT: 13.25 lbs.*

SIZE: 8½" x 11" x 20½" (L x W x H)

CCT LIGHTING
Model No. Z0C96*

BEAM ANGLE: (not specified in mfg. catalog)

FIELD ANGLE: (not specified in mfg. catalog)

STANDARD CONFIGS: 1 circuit - 1 lamp

LAMP CENTERS: (not specified in mfg. catalog)

LAMP BASE TYPE: Recessed Single-Contact

STANDARD LAMP: FFT - 1 KW.

OTHER LAMPS: FGV - 1 KW. *(cf=.97)*
FDB - 1.5 KW. *(cf=1.5)*

ACCESSORIES AVAILABLE: Color Frame

PHOTOMETRICS CHART

(Performance graph below showing single lamp configuration using FFT - 1 KW. lamp is copied from manufacturer's catalog.)

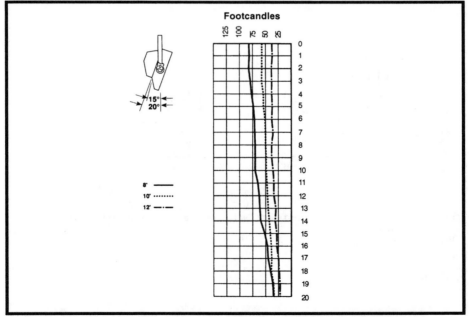

* This model is available in double, triple and quad configurations: Z0C97 (double) weighs 26.5 lbs., Z0C98 (triple) weighs 41 lbs. and Z0C99 (quad) weighs 54 lbs.

STRIPLIGHT / CYC LIGHT

LEE COLORTRAN
Model No. 108-392, "Far Cyc"*

BEAM ANGLE: (not specified in mfg. catalog)

FIELD ANGLE: (not specified in mfg. catalog)

STANDARD CONFIGS: 1 circuit - 1 lamp

LAMP CENTERS: (not specified in mfg. catalog)

LAMP BASE TYPE: Recessed Single-Contact

STANDARD LAMP: FGT - 1.5 KW.

OTHER LAMPS: FFT - 1 KW. *(cf=.6)*

WEIGHT: (not specified in manufacturer's catalog)

SIZE: (not specified in manufacturer's catalog)

PHOTOMETRICS CHART

(Performance graphs below showing multiple lamp configurations using FGT - 1.5 KW. lamps are copied from manufacturer's catalog.)

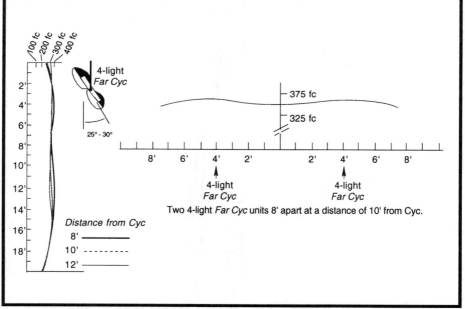

Two 4-light *Far Cyc* units 8' apart at a distance of 10' from Cyc.

* Instruments in this series are equipped as follows: 108-362 is 1 circuit, 1 lamp, 108-382 is 2 circuits, 2 lamps (horizontal), 108-412 is 3 circuits, 3 lamps, 108-392 is 4 circuits, 4 lamps. If the last digit of the model number is a 6 instead of a 2 the unit is equipped with a twist-lock connector rather than a pin connector.

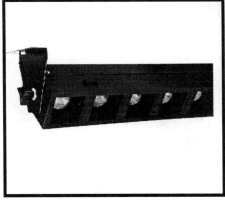

LIGHTING & ELECTRONICS
Model No. 65-05

BEAM ANGLE: (not specified in mfg. catalog)

FIELD ANGLE: (determined by lamp - see below)

STANDARD CONFIGS: 3'-0" - 3 circuits - 9 lamps
4'-0" - 3 or 4 circuits - 12 lamps
5'-0" - 3 circuits - 15 lamps

LAMP CENTERS: 3.91"

LAMP BASE TYPE: Medium Screw

STANDARD LAMP: (see below)

ACCESSORIES AVAILABLE:
Color Frame (3¾" sq.)

WEIGHT: 1.5 lbs. per lamp compartment

SIZE: 4" x 4" times the number of lamp compartments
x 5⅞" (L x W x H)

PHOTOMETRICS CHART

(Performance data for this unit determined by lamp used. Sample: 75PAR16/CAP/NFL lamp.)

75PAR16/CAP/NFL
30° field angle

	10'	15'	17.5'	20'	Distance
Diameter	5.4'	8.1'	9.45'	10.8'	*(mf= .54)*
Footcandles	20	8.9	6.5	5.0	*(pc=2,000)*

Watts	Lamp	Rated Life (Hours)	Field Angle	(mf) diameter	(pc) candela
55	55PAR16/CAP/NFL	2000	30°	.54	1,300
55	55PAR16/CAP/NSP	2000	12°	.21	5,000
75	75PAR16/CAP/NFL	2000	30°	.54	2,000
75	75PAR16/CAP/NSp	2000	12°	.21	7,500
50	50PAR20/CAP/NFL	2000	32°	.57	1,250
50	50PAR20/CAP/NSP	2000	12°	.21	4,600

LIGHTING & ELECTRONICS
Model No. 65-40, "Nano-Strip"

BEAM ANGLE: (determined by lamp - see below)

FIELD ANGLE: (not specified in mfg. catalog)

STANDARD CONFIG: 1'-9" - 1 circuit - 10 lamps

LAMP CENTERS: (not specified in mfg. catalog)

LAMP BASE TYPE: GU4 (2-Pin)

STANDARD LAMP: MR-11

ACCESSORIES AVAILABLE:
Color Frame (3¹¹⁄₁₆" x 2¼")

WEIGHT: 10.5 lbs.

SIZE: 21¼" x 4½" x 3½" (L x W x H)

PHOTOMETRICS CHART

(Performance data for this unit determined by lamp used. Sample: FTE - 35 W. MR-11 lamp.)

FTE - 35 W. MR-11
10° beam angle

	5'	10'	15'	20'	Distance
Diameter	.88'	1.75'	2.6'	3.5'	*(mf= .17)*
Footcandles	600	150	66.7	37.5	*(pc=15,000)*

Watts	Lamp	Rated Life (Hours)	Beam Angle	(mf) diameter	(pc) candela
20	FTB (NSP)	2000	10°	.17	9,700
20	FTC (SPT)	2000	20°	.35	3,100
20	FTD (FLD)	2000	30°	.54	1,200
35	FTE (NSP)	3000	10°	.17	15,000
35	FTF (NFL)	3000	20°	.35	5,200
35	FTH (MFL)	3000	30°	.54	2,600

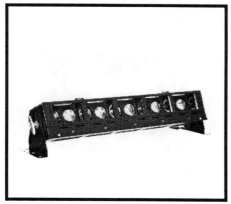

LIGHTING & ELECTRONICS
Model No. 65-60*, "Mini-Strip"

BEAM ANGLE: (determined by lamp - see below)

FIELD ANGLE: (determined by lamp - see below)

STANDARD CONFIGS: 6'-0" - 3 circuit - 30 lamps*

LAMP CENTERS: 2⅜" (2 lamps per compartment)

LAMP BASE TYPE: GX 5.3 (2-Pin)

STANDARD LAMP: MR-16

OTHER LAMPS: (see below)

ACCESSORIES: Color Frame (4¹¹⁄₁₆" x 3⁵⁄₁₆")

WEIGHT: 45 lbs.*

SIZE: 75⅛"* x 5½" x 5" (L x W x H)

PHOTOMETRICS CHART

(Performance data for this unit determined by lamp used. Sample: EYF/C - 75 W. MR-16 lamp.)

EYF/C - 75 W. MR-16
14° beam angle

	5'	10'	15'	20'	Distance
Diameter	1.3'	2.5'	3.8'	5.0'	(mf= .25)
Footcandles	984	246	109	61.5	(pc=24,600)

Watts	Lamp	Rated Life (Hours)	Field Angle	(mf) diameter	Beam Angle	(mf) diameter	(pc) candela
50	EXT/C (NSP)	5000	24°	.42	14°	.25	19,000
50	EXZ (NFL)	5000	49°	.91	27°	.48	5,400
65	FPA (NSP)	3500	30°	.54	14°	.25	22,000
65	FPC (NFL)	3500	40°	.73	27	.48	9,000
65	FPB (FLD)	3500	51°	.95	38°	.69	4,000
75	EYF/C (NSP)	4000	24°	.42	14°	.25	24,600
75	EYJ/C (NFL)	4000	35°	.63	25°	.44	9,200
75	EYC/C (FLD)	4000	64°	1.25	42°	.77	4,200

* This model is also available in 2-foot (65-60/4), 4-foot (65-80/4), and 8-foot (65-80) lengths, as well as a two-up, four foot version (65-80/4) and a three-up, two foot version (65-60/3).

STRIPLIGHT / CYC LIGHT

LIGHTING & ELECTRONICS
Model No. 65-71/1*, "Broad Cyc"

BEAM ANGLE: 72° (horizontal)

FIELD ANGLE: 95° (horizontal)

STANDARD CONFIGS: 1 circuit - 1 lamp*

LAMP CENTERS: (not specified in mfg. catalog)

LAMP BASE TYPE: Recessed Single-Contact

STANDARD LAMP: FHM - 1 KW.

OTHER LAMPS: EHZ - 300 W. *(cf=.22)*
FDN - 500 W. *(cf=.48)*
FCZ - 500 W. *(cf=.36)*
EMD - 750 W. *(cf=.75)*
FDB - 1.5 KW. *(cf=1.58)*
FFW - 2 KW. *(cf=2.19)*
(all lamps are frosted except FDB and FFW)

WEIGHT: 11 lbs.*

SIZE: 11⅞" x 8" x 10¼" (L x W x H)

ACCESSORIES AVAILABLE: Color Frame (10⅛" x 11¾"

PHOTOMETRICS CHART

(Performance graph below showing single lamp configuration using FHM - 1 KW. lamp is copied from manufacturer's catalog.)

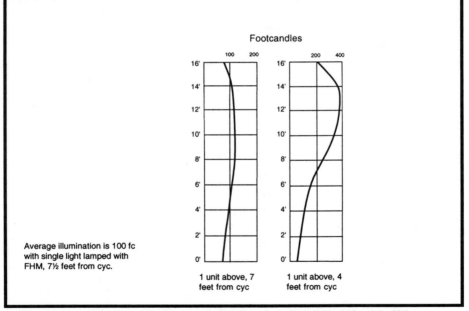

Footcandles

Average illumination is 100 fc with single light lamped with FHM, 7½ feet from cyc.

1 unit above, 7 feet from cyc

1 unit above, 4 feet from cyc

* This model is available in double, triple and quad configurations: 65-71/2 (double) weighs 23 lbs., 65-71/3 (triple) weighs 34 lbs., and 65-71/4 (quad) weighs 48 lbs.

LIGHTING & ELECTRONICS
Model No. 65-90, "Baby Broad
Groundrow"

BEAM ANGLE: 72° (horizontal)

FIELD ANGLE: 95° (horizontal)

STANDARD CONFIGS: 1 circuit - 1 lamp

LAMP CENTERS: (not specified in mfg. catalog)

LAMP BASE TYPE: Recessed Single-Contact

STANDARD LAMP: FHM - 1 KW.

OTHER LAMPS: EHZ - 300 W. *(cf=.22)*
FCZ - 500 W. *(cf=.36)*
EMD - 750 W. *(cf=.75)*

WEIGHT: 8.4 lbs.

SIZE: 11⅞" x 6¼" x 6¾" (L x W x H)

ACCESSORIES AVAILABLE: Color Frame (10½" x 7")

PHOTOMETRICS CHART

(Performance graph below showing single lamp configuration using FHM - 1 KW. lamp is copied from manufacturer's catalog.)

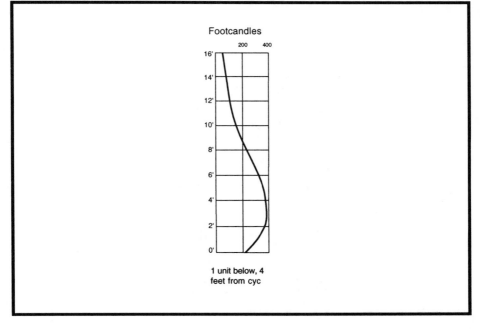

Footcandles

1 unit below, 4 feet from cyc

LIGHTING & ELECTRONICS
Model No. 65-95, "The Runt"

BEAM ANGLE: 72° (horizontal)

FIELD ANGLE: 95° (horizontal)

STANDARD CONFIGS: 1 circuit - 1 lamp

LAMP CENTERS: (not specified in mfg. catalog)

LAMP BASE TYPE: Recessed Single-Contact

STANDARD LAMP: EJG - 750 W.

OTHER LAMPS: EHM - 300 W. *(cf=.29)*
FCL - 500 W. *(cf=.50)*
FCZ - 500 W. *(cf=.47)*
EMD - 750 W. *(cf=.98)*

(FCZ and EMD are frosted lamps)

(Photo not available at press time.)

WEIGHT: 6 lbs.

SIZE: 9" x 6⅛" x 7⅝" (L x W x H)

ACCESSORIES AVAILABLE: Color Frame (8³⁄₁₆" x 7⁹⁄₁₆"

PHOTOMETRICS CHART

(Performance graph below showing single lamp configuration using EJG - 750 W. lamp is copied from manufacturer's catalog.)

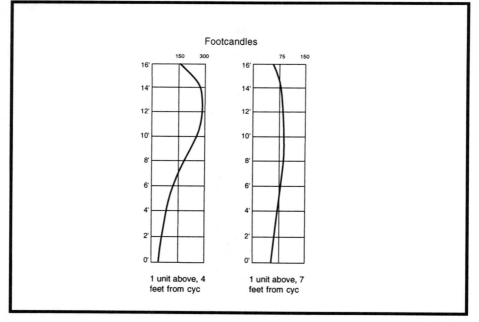

Footcandles

1 unit above, 4 feet from cyc

1 unit above, 7 feet from cyc

WEIGHT: (not specified in manufacturer's catalog)
SIZE: (not specified in manufacturer's catalog)

STRAND CENTURY
Model No. 5911*

BEAM ANGLE: (not specified in mfg. catalog)

FIELD ANGLE: (not specified in mfg. catalog)

STANDARD CONFIGS: 1 circuit - 1 lamp

LAMP CENTERS: (not specified in mfg. catalog)

LAMP BASE TYPE: Recessed Single-Contact

STANDARD LAMP: FFT - 1 KW.

OTHER LAMPS: FDB - 1.5 KW. *(cf=1.49)*

PHOTOMETRICS CHART

(Performance graphs below showing multiple lamp configurations using FFT - 1 KW. lamps are copied from manufacturer's catalog.)

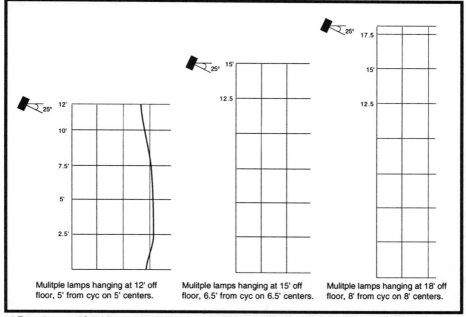

Mulitple lamps hanging at 12' off floor, 5' from cyc on 5' centers.

Mulitple lamps hanging at 15' off floor, 6.5' from cyc on 6.5' centers.

Mulitple lamps hanging at 18' off floor, 8' from cyc on 8' centers.

* Two, three and four circuit versions of this model are numbered 5912, 5913, and 5914 respectively.

STRAND CENTURY
Model No. 5915, "Mini-Cyc"

BEAM ANGLE: (not specified in mfg. catalog)

FIELD ANGLE: (not specified in mfg. catalog)

STANDARD CONFIGS: 1 circuit - 1 lamp

LAMP CENTERS: (not specified in mfg. catalog)

LAMP BASE TYPE: Recessed Single-Contact

STANDARD LAMP: FFT - 1 KW.

OTHER LAMPS: FDB - 1.5 KW. *(cf=1.49)*

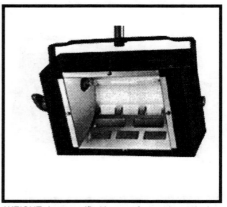

WEIGHT: (not specified in manufacturer's catalog)

SIZE: (not specified in manufacturer's catalog)

PHOTOMETRICS CHART
(Performance data are measured using multiple instruments with FFT - 1 KW. lamps.)

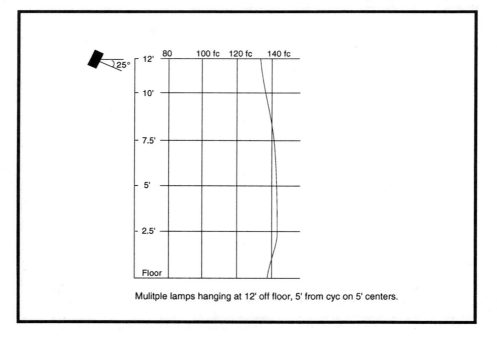

Mulitple lamps hanging at 12' off floor, 5' from cyc on 5' centers.

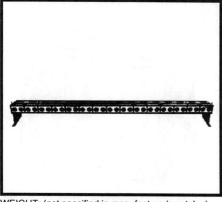

STRIPLIGHT / CYC LIGHT

STRAND LIGHTING
Model No. 5300

BEAM ANGLE: (determined by lamp - see below)

FIELD ANGLE: (determined by lamp - see below)

STANDARD CONFIGS: 6'-0" - 3 circuit - 30 lamps

LAMP CENTERS: 2⅜" (2 lamps per compartment)

LAMP BASE TYPE: GX 5.3 (2-Pin)

STANDARD LAMP: MR-16

OTHER LAMPS: (see below)

WEIGHT: (not specified in manufacturer's catalog)

SIZE: (not specified in manufacturer's catalog)

NOTES: Lamps are 12V., wired in series in circuits of 10 lamps each. Neon indicator lamps make it easy to identify lamps which need replacing.

PHOTOMETRICS CHART

(Performance data for this unit determined by lamp used. Sample performance charts using various combinations of units are copied from manufacturer's catalog.)

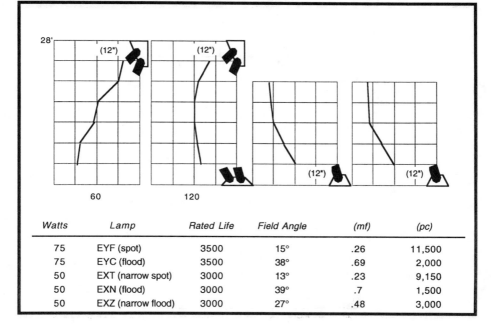

Watts	Lamp	Rated Life	Field Angle	(mf)	(pc)
75	EYF (spot)	3500	15°	.26	11,500
75	EYC (flood)	3500	38°	.69	2,000
50	EXT (narrow spot)	3000	13°	.23	9,150
50	EXN (flood)	3000	39°	.7	1,500
50	EXZ (narrow flood)	3000	27°	.48	3,000

STRAND LIGHTING
Model No. 5940

BEAM ANGLE: (not specified in mfg. catalog)

FIELD ANGLE: (not specified in mfg. catalog)

STANDARD CONFIGS: 1 circuit - 1 lamp

LAMP CENTERS: 12"

LAMP BASE TYPE: Recessed Single-Contact

STANDARD LAMP: FHM - 1 KW. (frosted)

OTHER LAMPS: FDN - 500 W. frosted *(cf=.48)*
FDF - 500 W. *(cf=.51)*
EJG - 750 W. *(cf=.78)*
FCM - 1 KW. *(cf=1.06)*

WEIGHT: (not specified in manufacturer's catalog)

SIZE: (not specified in manufacturer's catalog)

PHOTOMETRICS CHART

(Performance graph below showing multiple lamp configuration using FHM - 1 KW. lamps is copied from manufacturer's catalog.

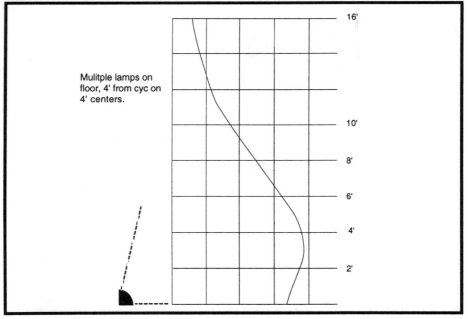

Mulitple lamps on floor, 4' from cyc on 4' centers.

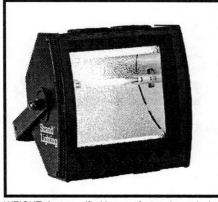

STRAND LIGHTING
Model No. CODA 500/1 MkII*

BEAM ANGLE: (not specified in mfg. catalog)

FIELD ANGLE: (not specified in mfg. catalog)

STANDARD CONFIGS: 1 circuit - 1 lamp
3 circuit - 3 lamps
4 circuit - 4 lamps

LAMP CENTERS: (not specified in mfg. catalog)

LAMP BASE TYPE: Recessed Single-Contact

STANDARD LAMP: FDN - 500 W.

OTHER LAMPS: (none specified in mfg. catalog)

WEIGHT: (not specified in manufacturer's catalog)

SIZE: (not specified in manufacturer's catalog)

PHOTOMETRICS CHART

(Performance graphs below showing multiple configurations using FDN - 500 W. lamp are copied from manufacturer's catalog. Readings taken with phototcell facing Coda units.)

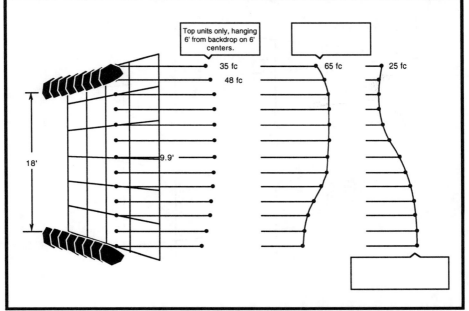

* Three circuit and four circuit versions of this model are numbered Coda 500/3 Mk II and Coda 500/4 MkII respectively.

STRIPLIGHT / CYC LIGHT

TIMES SQUARE LIGHTING
Model No. 701, "Mini X-Ray"*

BEAM ANGLE: (not specified in mfg. catalog)*

FIELD ANGLE: (determined by lamp)**

STANDARD CONFIGS: 2'-4" - 1 circuit - 10 lamps
2'-10" - 3 circuit - 12 lamps
3'-9" - 4 circuit - 16 lamps
5'-7" - 3 or 4 circuit - 24 lamps
7'-0" - 3 circuit - 30 lamps

LAMP CENTERS: 2⅞" (2 lamps per compartment)

LAMP BASE TYPE: Medium Screw

STANDARD LAMP: 75PAR16CAP/NSP and ...NFL

OTHER LAMPS: 55PAR16 (spot & flood)
50PAR20 (spot & flood)

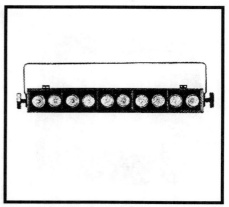

WEIGHT: (not specified in manufacturer's catalog)

SIZE: (not specified in manufacturer's catalog)

NOTES: Model number has suffix which denotes the different sizes: 701-5 is the 5 compartment model, 701-8 is the 8 compartment, etc. Also available in custom lengths

PHOTOMETRICS CHART

(Performance data for this unit are measured using Model 701-5, 1 circuit, 10 lamps, with 75PAR16/CAP/NSP or NFL lamps. Note that the photometrics are for the whole unit, not one lamp.)

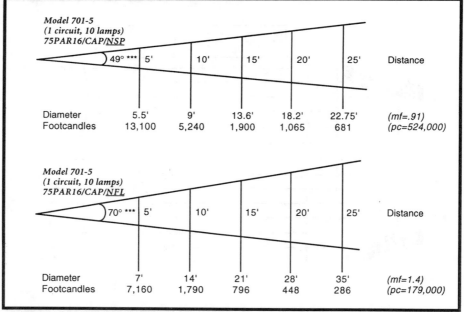

Model 701-5
(1 circuit, 10 lamps)
75PAR16/CAP/NSP

49° *** | 5' | 10' | 15' | 20' | 25' | Distance

Diameter	5.5'	9'	13.6'	18.2'	22.75'	(mf=.91)
Footcandles	13,100	5,240	1,900	1,065	681	(pc=524,000)

Model 701-5
(1 circuit, 10 lamps)
75PAR16/CAP/NFL

70° *** | 5' | 10' | 15' | 20' | 25' | Distance

Diameter	7'	14'	21'	28'	35'	(mf=1.4)
Footcandles	7,160	1,790	796	448	286	(pc=179,000)

* Model numbers in this series indicate the number of lamp compartments in the unit: 701-5 has 5 compartments (with two lamps in each), 701-6 has 6, 701-8 has 8, 701-12 has 12, and 701-15 has 15.
** Photometrics for specific lamps are given in lamp manufacturer's catalogs.
*** Manufacturer's catalog lists field angle for the PAR16/NSP as 24° and the field angle for the PAR16/NFL lamp as 49° however, using the given diameter and distance data, the angles calculate as 49° and 70° respectively. See Introduction for more information on calculating photometrics.

TIMES SQUARE LIGHTING
Model No. 702*

BEAM ANGLE: (not specified in mfg. catalog)

FIELD ANGLE: (determined by lamp - see below)

STANDARD CONFIGS: 3'-0" - 3 circuit - 6 lamps
4'-6" - 3 circuit - 9 lamps
6'-0" - 3 circuit - 12 lamps
7'-6" - 3 circuit - 15 lamps
8'-0" - 3 circuit - 16 lamps

LAMP CENTERS: 6"

LAMP BASE TYPE: Medium Screw

STANDARD LAMP: R40

OTHER LAMPS: (see below)

NOTES: Unit takes 5⅝" roundels. Available in custom lengths.

WEIGHT: (not specified in manufacturer's catalog)

SIZE: 5⅞" (W)

PHOTOMETRICS CHART
(Performance data for this unit determined by lamp used. Sample: 150R/FL. lamp.)

FLOOD (FL)
60° field angle

	10'	15'	17.5'	20'	Distance
Diameter	11.5'	17.3'	20.2'	23.1'	(mf= 1.15)
Footcandles	10.4	4.6	3.4	2.6	(pc=1,040)

Watts	Lamp	Rated Life (Hours)	Field Angle	(mf) diameter	(pc) candela
100	100R/FL	2000	60°	1.15	900
100	100R/SP	2000	36°	.65	5,000
150	150R/FL	2000	60°	1.15	1,040
150	150R/SP	2000	49°	.91	5,400
200	200R/FL	2000	60°	1.15	2,000
200	200R/SP	2000	34°	.61	12,000
300	300R/FL	2000	60°	1.15	1,950
300	300R/FL	2000	24°	.43	8,900

* Model numbers in this series indicate the length of the unit: 702-3 is 3' long, 702-4 is 4.5' long, 702-6 is 6' long, 702-7 is 7.5' long, and 702-8 is 8' long.

Striplights – General Description

Striplights are different in many ways from other types of theatrical lights. They do not have lenses for focusing the light and, except for some contemporary "cyc lights," striplights have multiple lamps and reflectors. Because of these differences, many manufacturers do not give traditional photometric data for their striplights. Those instruments for which the manufacturers provided some sort of illumination and beam width data are given photometric pages in this book (pages 546–571).

The vast majority of catalogs do not provide any photometric data for their striplights, but rather than ignore this large group of instruments, the following list provides general descriptions of most striplights in use. This list *includes* the instruments on the previous pages for which photometric data was available and these units are indicated with an asterisk (*).

Altman Stage Lighting

520*	Circular Reflector	Continuous custom length borderlight, also 6', 6'-9", 7'-6", and 8' lengths; 3 or 4 circuits; uses 75–100 W. A lamps. See p. 546.
528*	Circular Reflector	Same as 520; 6', 7'-6" and 8' lengths; uses 150–300 W. A lamps. See p. 547.
537*	Circular Reflector	Same as 520; 5'-4", 6' and 8' lengths; uses 200–300 W. A lamps. See p. 548.
600*	Circular Reflector	Continuous custom length borderlight, also 5'-4", 6', and 8' lengths; 3 or 4 circuits; uses 300–500 W. recessed single-contact lamp. See p. 549.
990	Circular Reflector	Continuous custom length borderlight; 3 or 4 circuits; uses 300–500 W. lamps.
AB	Circular Reflector	Continuous custom length borderlight; 3 circuits; uses 150–300 W. general service lamps.
D5-40004-L	Trough	5' long footlight; 15 lamps; 3 circuits; uses 40–100 W. lamps; unit is not equipped with individual reflectors, color is achieved through dipped lamps. Inside surface is either "Lume Crome" or painted white.
D5-41623	Circular Reflector	5' long disappearing footlight; 9 lamps; 3 circuits; uses 100 W. lamps.
D5-41625	Circular Reflector	Same; uses 150 W. lamps.
D5-43523	Circular Reflector	Same; uses 12 60–100 W. lamps.
Ground Cyc*	Asymmetrical	Various lengths from 11" to 9'-2"; 1, 2, 3 or 4 circuits; uses 1 KW recessed single contact lamps. See p. 550.
GB	Asymmetrical	Continuous custom length borderlight; 40–100 W. lamps.
PAR-56*	Reflector Lamp	Continuous custom length borderlight, also 4', 5'-4", 6', and 8'; 3 or 4 circuits; uses 300 W. PAR-56 lamps. See p. 551.
PAR-64*	Reflector Lamp	Same; uses 500 W. PAR-64 lamps. See p. 552.
Pinrail	Trough	Continuous custom length borderlight; uses 40–100 W. general service lamps.Unit is not equipped with individual reflectors, inside of hood is white paint, color is achieved through dipped lamps; this unit seems to be intended for worklight use.
PSR	Asymmetrical	Continuous custom length portable footlight; uses 25–100 W. lamps.
R40*	Reflector Lamp	Continuous custom length borderlight; also 6', 7'-6", and 8' lengths; 3 or 4 circuits; uses 75–300 W. R-40 lamps. See p. 553.
SE*	Trough	Continuous custom length borderlight; 3 circuits; uses 40–100 W. general service lamps; unit is not equipped with individual reflectors, inside of hood is white paint, color is achieved through dipped lamps. See p. 554.
Sky Cyc	Asymmetrical	1–4 circuits; uses 1.5 KW recessed single-contact lamps. See p. 555.
ST	Trough	Continuous custom length strip light; uses 25–60 W. lamps; inside of hood is aluminum paint, color is achieved through dipped lamps.

Ariel Davis

211	Reflector Lamp	4.5' long; 9 lamps; 3 circuits; uses 150 – 300 W. R-40 lamps.
212	Reflector Lamp	Same; 6' long; 12 lamps.
213	Reflector Lamp	Same; 4 circuit.
214	Reflector Lamp	Same; 7'-6" long; 15 lamps; 3 circuit.
215	Reflector Lamp	Same; 8' long; 16 lamps; 4 circuit.
221	Circular Reflector	4'-6" long ;9 lamps; 3 circuits; uses 100 W. A-21 lamps.
222	Circular Reflector	Same; 6' long; 12 lamps.
223	Circular Reflector	Same; 4 circuit.
224	Circular Reflector	Same; 7'-6" long; 15 lamps; 3 circuit.
225	Circular Reflector	Same; 8' long; 16 lamps; 4 circuit.
231	Circular Reflector	4'-6" long; 9 lamps; 3 circuits; uses 150 W. A-23 lamps.
232	Circular Reflector	Same; 6' long; 12 lamps.

233	Circular Reflector	Same ; 4 circuit.
234	Circular Reflector	Same; 7'-6" long; 15 lamps; 3 circuit.
235	Circular Reflector	Same; 8' long; 16 lamps; 4 circuit.

Capitol Stage Lighting

520	Circular Reflector	Continuous custom length borderlight; 3 circuits; uses 75 W. lamps.
524	Circular Reflector	Same; uses 100 W. lamps.
528	Circular Reflector	Same; uses 150 W. lamps.
532	Circular Reflector	Same; uses 200 W. lamps.
A.B.	Circular Reflector	Same; uses 150–200 W. lamps.
A.L.	Circular Reflector	Same; uses 300–500 W. lamps.
F.S. 75	Circular Reflector	Continuous custom length footlight.
F.S.R. 100	Circular Reflector	Continuous custom length footlight; 3 circuits; uses 100 W. lamps.
F.S.R. 150	Circular Reflector	Same; uses 150 W. lamps.
G.B.	Asymmetrical	Continuous custom length footlight; 3 or more circuits.
G.F.	Asymmetrical	Continuous custom length footlight; uses 60–100 W. lamps.
P.G.F.	Asymmetrical	6' and 7'-6" lengths; 3 circuits; unit is a footlight.
P.R.40	Reflector Lamp	Continuous custom length borderlight; 3 circuits; uses R-40 lamps; unit also manufactured in portable sections.
P.S.R.	Trough	5', 6', and 7'-6" lengths; 3 circuits; uses 25–100 W. lamps; unit is not equipped with individual reflectors, inside of hood is aluminum paint, color is achieved through dipped lamps; unit is a footlight.
S.E.	Trough	Continuous custom length footlight; 3 or more circuits; unit is not equipped with individual reflectors, inside of hood is aluminum paint, color is achieved through dipped lamps.
S.T.12	Trough	Same; 90" long; 12 lamps.
S.T.6	Asymmetrical	45" long; 6 lamps; 3 circuits; uses small wattage general service lamps; unit is a footlight.
S.T.9	Asymmetrical	Same; 67½" long; 9 lamps.

CCT Lighting

Z0C46*	Asymmetrical	1 lamp; 1 circuit; uses 300–500 W. recessed single-contact lamps. See p. 556.
Z0C47	Asymmetrical	2 lamp; 2 circuits; uses 300–500 W. recessed single-contact lamps.
Z0C48	Asymmetrical	3 lamp; 3 circuits; uses 300–500 W. recessed single-contact lamps.
Z0C49	Asymmetrical	4 lamp; 4 circuits; uses 300–500 W. recessed single-contact lamps.
Z0C96*	Asymmetrical	1 lamp; 1 circuit; uses 1–1.5 KW. recessed single-contact lamps. See p. 557.
Z0C97	Asymmetrical	2 lamp; 2 circuits; uses 1–1.5 KW. recessed single-contact lamps.
Z0C93	Asymmetrical	3 lamp; 3 circuits; uses 1–1.5 KW. recessed single-contact lamps.
Z0C98	Asymmetrical	4 lamp; 4 circuits; uses 1–1.5 KW. recessed single-contact lamps.

Century Lighting

1804	Trough	33" long; 4 lamps; uses 25–100 W. general service lamps; unit is not equipped with individual reflectors, inside of hood is white or aluminum paint, color is achieved through dipped lamps.
1806	Trough	Same; 49" long; 6 lamps.
391	Reflector Lamp	4'-6", 6' or 7'-6" lengths; uses 150–300 W. R-40 or 150 W. PAR-38 lamps; unit is a rolling cyc footlight (also available as an overhead custom length borderlight).
392	Reflector Lamp	6' long; 12 lamps; 3 circuits; uses 75–300 W. R-40 or PAR-38; unit is equipped with casters for floor use.
400	Trough	Continuous custom length borderlights; uses 200 W. lamps; unit is not equipped with individual reflectors, inside of hood is white or aluminum paint, color is achieved through dipped lamps.
401	Trough	Same; inside of hood is matte aluminum stripping.
402	Trough	Same; inside of hood is chromium plated stripping.
403	Trough	Same; inside of hood is Alzak stripping.
404	Circular Reflector	6' long; 16 lamps; 4 circuits; uses 300–500 W. R-40 lamps.
411	Circular Reflector	Continuous custom length borderlights; uses 75–150 W. lamps; Alzak reflector, combination rondel/gel frame holder.
412	Circular Reflector	6' long; 12 lamps; uses 100 W. A-21 lamps.
413	Circular Reflector	Continuous custom length borderlights; uses 200 W. lamps; Alzak reflector, combination rondel/gel frame holder.
414	Circular Reflector	Same; uses 300–500 W. lamps.
416	Circular Reflector	6' long; 12 lamps; 3 circuits; uses 60–100 W. A-19 lamps.

417	Circular Reflector	Continuous custom length borderlight; uses 150 W. lamps.
418	Circular Reflector	6' long; 12 lamps; 4 circuits; uses 60–100 W. A-19 lamps.
423	Circular Reflector	Continuous custom length borderlights; uses 75–150 W. lamps; chromium reflector, combination rondel/gel frame holder.
424	Circular Reflector	Same; uses 200 W. lamps.
425	Circular Reflector	Same; uses 300–500 W. lamps.
426	Circular Reflector	6' long; 12 lamps; 3 circuits; uses 150–200 W. A-23 lamps.
428	Circular Reflector	Same; 4 circuits.
431	Reflector Lamp	6' long; 12 lamps; 3 circuits; uses 150 W. PAR-38 or 150–300 W. R-40 lamps.
434	Reflector Lamp	Same; 4'-6" long; 9 lamps.
436	Reflector Lamp	8' long; 16 lamps; 4 circuits; uses 300–500 W. R-40 lamps.
437	Reflector Lamp	7'-6" long; 15 lamps; uses 150 W. PAR-38 or 150–300 W. R-40 lamps.
438	Reflector Lamp	Same; 8' long; 16 lamps; 4 circuits.
441	Circular Reflector	Continuous custom length borderlights; uses 75–150 W. lamps; aluminum reflector, combination rondel/gel frame holder.
442	Circular Reflector	Same; uses 200 W. lamps.
443	Circular Reflector	Same; uses 300–500 W. lamps.
446	Reflector Lamp	6' long; 12 lamps; 3 circuits; uses 75–300 W. R-40 or PAR-38 lamps.
448	Reflector Lamp	Same; 4 circuits.
450	Circular Reflector	Continuous custom length borderlight; uses 75–150 W. lamps; Alzak reflector, spring ring rondel holder.
451	Circular Reflector	Same; chromium reflector.
452	Circular Reflector	Same; aluminum reflector.
453	Circular Reflector	Continuous custom length borderlight; uses 150 W. lamps; has spring ring rondel holder.
455	Circular Reflector	Continuous custom length borderlights; uses 200 W. lamps; Alzak reflector, spring ring rondel holder.
456	Circular Reflector	Same; chromium reflector.
457	Circular Reflector	Same; aluminum reflector.
458	Reflector Lamp	Continuous custom length borderlight; 3 or 4 circuits; uses 150–300 W. R-40 or PAR-38 lamps; has combination rondel/gel frame.
460	Circular Reflector	Continuous custom length borderlight; 3 or 4 circuits; uses 300–500 W. lamps; has spring ring rondel holder.
461	Circular Reflector	Same; chromium reflector.
462	Circular Reflector	Same; aluminum reflector.
470	Reflector Lamp	8' long; 12 lamps; 3 circuits; uses 300 W. PAR-56 lamp; has arm and clamp hanger.
472	Reflector Lamp	Same; has trunion and strap hanger.
474	Reflector Lamp	2'-8" long; 8 lamps; 4 circuits; uses 300–500 W. R-40 lamps; unit is two strips with a common hanger.
475	Reflector Lamp	9' long; 12 lamps; 3 circuits; uses 500 W. PAR-64 lamp.
478	Reflector Lamp	Same as 474; 5'-4" long; 16 lamps.
800	Trough	Continuous custom length footlights; uses 25–100 W. lamps; unit is not equipped with individual reflectors, inside of hood is white or aluminum paint, color is achieved through dipped lamps.
801	Trough	Same; inside of hood is aluminum stripping.
802	Trough	Same; inside of hood is chromium stripping.
803	Trough	Same; inside of hood is Alzak stripping.
808	Circular Reflector	Continuous custom length footlights; uses 75–100 W. lamps; aluminum reflector, combination rondel/gel frame.
809	Circular Reflector	Same; chromium reflector.
810	Circular Reflector	Same; Alzak reflector.
811	Trough	5'-2" long; 15 lamps; 3 circuits; uses 25–100 W. general service lamps; unit is a disappearing footlight; color is achieved through dipped lamps.
816	Circular Reflector	5' long; 12 lamps; 3 circuits; uses 100 W. lamps; unit is a disappearing footlight.
817	Circular Reflector	Same; 9 lamps; uses 150 W. lamps.
823	Circular Reflector	Continuous custom length footlight; uses 75–150 W. lamps in a double (over and under) row; chromium reflector, spring ring rondel holder.
824	Circular Reflector	Same; Alzak reflector.
843	Circular Reflector	5'-2" long; 9 lamps; 3 circuits; uses 150 W. general service lamps; unit is a disappearing footlight.
846	Circular Reflector	Same; 12 lamps; uses 75–100 W. general service lamps.
856	Circular Reflector	5' long; 12 lamps; 3 circuits; uses 100 W. lamps; unit is a motorized disappearing footlight.
857	Circular Reflector	Same; 9 lamps; uses 150 W. lamps.

860	Circular Reflector	Continuous custom length footlights; 3 circuits; uses 75–150 W. lamps; Alzak reflector, spring ring rondel holder.
861	Circular Reflector	Same; chromium reflector.
862	Circular Reflector	Same; aluminum reflector.

Century Strand

409	Circular Reflector	6' long; 12 lamps; 3 circuits; uses 60–100 W. A-19 lamps.
410	Circular Reflector	Same; 4 color.
419	Circular Reflector	6' long; 12 lamps; 3 circuits; uses 150–200 W. A-23 lamps.
420	Circular Reflector	Same; 4 circuits.
439	Reflector Lamp	Continuous (6' multiples) borderlight; 4 circuits; uses 75–300 W. R-40 or PAR-38 lamps.
440	Reflector Lamp	Same; 3 circuits.
446	Reflector Lamp	6' long; 12 lamps; 3 circuits; uses 75–300 W. R-40 or PAR-38 lamps.
448	Reflector Lamp	Same; 4 circuits.
816	Circular Reflector	5'-2" long; 12 lamp; 3 circuits; uses 100 W. A-19 lamps; unit is a disappearing footlight.
817	Circular Reflector	Same; uses 150 W. PS-25 lamps.
Patt QGR 4	Asymmetrical	4' long; 4 lamps; 4 circuits; uses 750 W. K/3 or 625 W. P2/10 lamps.

Electro Controls

36631	Circular Reflector	4'-6" long; 9 lamps; 3 circuits; uses 150 W. A-23 lamps.
36632	Circular Reflector	Same; 6' long; 12 lamps.
36633	Circular Reflector	Same; 4 circuits.
36634	Circular Reflector	Same; 7'-6" long; 15 lamps; 3 circuits.
36635	Circular Reflector	Same; 8' long; 16 lamps; 4 circuits.
7661A	Reflector Lamp	4'-6" long; 9 lamps; 3 circuits; uses 150–300 W. R-40 lamps.
7662A	Reflector Lamp	Same; 6' long; 12 lamps.
7663A	Reflector Lamp	Same; 4 circuits.
7664A	Reflector Lamp	Same; 7'-6" long; 15 lamps; 3 circuits.
7665A	Reflector Lamp	Same; 8' long; 16 lamps; 4 circuits.
7671A	Circular Reflector	4'-6" long; 9 lamps; 3 circuits; uses 150 W. A-23 lamps.
7672A	Circular Reflector	Same; 6' long; 12 lamps.
7673A	Circular Reflector	Same; 4 color.
7674A	Circular Reflector	Same; 7'-6" long; 15 lamps; 3 circuits.
7675A	Circular Reflector	Same; 8' long; 16 lamps; 4 circuits.

Kliegl Bros.

3466P, 3466PP	Asymmetrical	6' long; 9 lamps; 3 circuits; uses 200–300 W. recessed single-contact lamps; #3466P has pigtails on one end, #3466PP has pigtails on both ends.
3468	Asymmetrical	Custom lengths; 3 circuits; uses 200–300 W. recessed single-contact lamps.
3469	Asymmetrical	Same; 4 circuits
3468P, 3468PP	Asymmetrical	Same; 8' long; 12 lamps; #3468P has pigtails on one end, #3468PP has pigtails on both ends.
3469P, 3469PP	Asymmetrical	Same; 8' long; 4 circuits; #3469P has pigtails on one end, #3469PP has pigtails on both ends.
3480P, 3480PP	Asymmetrical	6' long; 8 lamps; 4 circuits; uses 300–500 W. recessed single-contact lamps; unit is a rolling cyclorama footlight, and is also available in custom lengths; #3480P has pigtails on one end, #3480PP has pigtails on both ends.
3481P, 3481PP	Asymmetrical	6'-9" long; 9 lamps; 3 circuits; uses 300–500 W. recessed single-contact lamps; #3481P has pigtails on one end, #3481PP has pigtails on both ends.
3485P, 3485PP	Asymmetrical	Same; 6' long; 8 lamps; 4 circuits; #3485P has pigtails on one end, #3485PP has pigtails on both ends.
3486P, 3486PP	Asymmetrical	Same; 6'-9" long; 9 lamps;3 circuits; #3486P has pigtails on one end, #3486PP has pigtails on both ends.
3490P, 3490PP	Asymmetrical	6' long; 16 lamps; 4 circuits; uses 300–500 W. recessed single-contact lamps; unit is a double row of cyc lights with a common mounting hanger; #3490P has pigtails on one end, #3490PP has pigtails on both ends.
3491P, 3491PP	Asymmetrical	Same; 6'-9" long; 18 lamps; 3 circuits; #3491P has pigtails on one end, #3491PP has pigtails on both ends.
3494P, 3494PP	Asymmetrical	6' long; 8 lamps; 4 circuits; uses 200–500 W. recessed single-contact lamps; #3494P has pigtails on one end, #3494PP has pigtails on both ends.
3495	Asymmetrical	Same; 9' long; 12 lamps; 3 circuits; can be hung or used on floor.

3496P, 3496PP	Asymmetrical	Same; 6'-9" long; 9 lamps; #3496P has pigtails on one end, #3496PP has pigtails on both ends.
3497	Asymmetrical	Same; 9' long; 12 lamps.
3498P, 3498PP	Asymmetrical	Same; 6' long; 8 lamps; 4 circuits; #3498P has pigtails on one end, #3498PP has pigtails on both ends.
3499P, 3499PP	Asymmetrical	Same; 6'-9" long; 9 lamps; 3 circuits; #3499P has pigtails on one end, #3499PP has pigtails on both ends.
3500ABC	Asymmetrical	3'-6" long; 2 lamps; 1 circuit; uses 300–500 W. recessed single-contact lamps; unit is a cyc striplight.
3500AFC	Asymmetrical	3'-6 long; 2 lamps; 1 circuit; uses 300–500 W. recessed single-contact lamps; unit is a cyc footlight.
3500BBC	Asymmetrical	Same; 1' long; 1 lamp; 1 circuit.
3500BC	Asymmetrical	Same; 7' long; 4 lamps; 1 circuit.
3500BFC	Asymmetrical	Same; 1' long; 1 lamp; 1 circuit.
3500FC	Asymmetrical	Same; 7' long; 4 lamps; 1 circuit.
3501AF	Asymmetrical	Same; 3'-6" long; 2 circuits.
3501F	Asymmetrical	Same; 7' long; 8 lamps; 2 circuits.
3504AB	Asymmetrical	3'-6" long; 4 lamps; 2 circuits; uses 300–500 W. recessed single-contact lamps; unit is a cyc striplight.
3504B	Asymmetrical	Same; 7' long; 8 lamps.
590	Reflector Lamp	8' long; 15 lamps; 3 circuits; uses 150 W. R lamps.
590C	Reflector Lamp	Continuous custom length borderlight; 3 or more circuits; uses 150 W. R lamps.
590CP	Reflector Lamp	8' long; 3 circuits; uses 150 W. R-40 lamps.
597	Circular Reflector	Continuous custom height proscenium striplight; mounted vertically on either side of proscenium arch; uses 200 W. A21 lamps.
598	Circular Reflector	Same; uses 150 W. lamps.
599	Circular Reflector	Same; uses 60–100 W. lamps.
600	Trough	Continuous custom length borderlights; 3 or more circuits; uses 60–100 W. lamps; unit is not equipped with individual reflectors, inside of hood is white paint, color is achieved through dipped lamps.
602	Circular Reflector	Continuous custom length borderlight; uses 200–300 W. lamps, has spring ring rondel holder.
602C	Circular Reflector	Same; uses 200 W. lamps; has combination rondel/gel holder.
602CP	Circular Reflector	8' long; 3 circuits; uses 200 W. A lamps.
603	Circular Reflector	Continuous custom length borderlight; uses 300–500 W. lamps; has spring ring rondel holder.
610	Circular Reflector	Continuous custom length borderlight; uses 100 W. A lamps, has spring ring rondel holders.
610A	Circular Reflector	Same; uses 150 W. lamps.
610AC	Circular Reflector	Same; has combination rondel/gel holder.
610ACP	Circular Reflector	8' long; 3 color; uses 150 W. A lamps.
610C	Circular Reflector	Continuous custom length borderlight; uses 100 W. A lamps, has combination rondel/gel holder.
610CP	Circular Reflector	8' long; 3 circuits; uses 100 W. A lamps.
620	Circular Reflector	Continuous custom length footlight; uses 75–100 W. lamps.
620A	Circular Reflector	Same; uses 100 W. lamps.
620B	Circular Reflector	Same; uses 150 W. lamps.
620BCP	Circular Reflector	8' long; 20 lamps; 3 circuits; uses 150 W. lamps; unit is a portable footlight.
620CP	Circular Reflector	Same; uses 100 W. lamps.
621	Circular Reflector	Continuous custom length footlight with two rows of lamps; uses 75–100 W. lamps.
621A	Circular Reflector	Same; uses 100 W. lamps.
621B	Circular Reflector	Same; uses 150 W. lamps.
629	Trough	Continuous custom length footlights; 3 circuits; uses 60–100 W. lamps; unit is not equipped with individual reflectors, inside of hood is white paint, color is achieved through dipped lamps.
649	Trough	18" long; 2 lamps; 1 circuit; uses 40–150 W. lamps; unit is not equipped with individual reflectors, inside of hood is white or aluminum paint, color is achieved through dipped lamps.
650	Asymmetrical	Same; 3' long; 4 lamps.
651	Asymmetrical	Same; 5' long; 6 lamps.
652	Asymmetrical	Same; 8' long; 10 lamps.
829	Trough	5' long; 15 lamps; 3 circuits; uses 60–100 W. lamps; unit is not equipped with individual reflectors, inside of hood is white or aluminum paint, color is achieved through dipped lamps; unit is a disappearing footlight.

830	Circular Reflector	5' long; 12 lamps; 3 circuits; uses 75–100 W. A-21 lamps; unit is a disappearing footlight.
831	Circular Reflector	5' long; 24 lamps; 3 circuits; uses 75–100 W. A-21 lamps; unit is a disappearing footlight with a double row of lamps.
832	Circular Reflector	5' long; 12 lamps; 3 circuits; uses 100 W. A lamps; unit is a disappearing footlight.
833-A	Circular Reflector	Same; uses 75–100 W. lamps.
833	Circular Reflector	5' long; 9 lamps; 3 circuits; uses 150 W. lamps; unit is a disappearing footlight.

Lee Colortran

108-362, -366*	Asymmetrical	1 lamp; 1 circuit; uses 1.5 KW. recessed single-contact lamps; #108-366 is the same unit with different connectors. See p. 558.
108-382, -386*	Asymmetrical	Same; 2 lamps; 2 circuits.See p. 558.
108-392, -396*	Asymmetrical	Same; 3 lamps; 3 circuits. See p. 558.
108-412, -416*	Asymmetrical	Same; 4 lamps; 4 circuits. See p. 558.
216-002, -005, -006	Reflector Lamp	6'-3" long; 12 lamps; 3 circuits; uses 150–250 PAR-38 or 150–300 W. R-40; #216-005, #216-006 are the same unit with different connectors.
216-012, -015, -016	Reflector Lamp	Same; 4 circuits. #216-015, #216-016 are the same unit with different connectors.
6901 - TV	Asymmetrical	1 compartment; 1 circuit; uses 1–2 KW.recessed single-contact lamps.
6902 - TV	Asymmetrical	Same; 2 compartments; 2 circuits.
6903 - TV	Asymmetrical	Same; 3 compartments; 3 circuits.
6904 - TV	Asymmetrical	Same; 4 compartments; 4 circuits.

Lighting, Electronics (L&E)

A-Lamp Border	Circular Reflector	Continuous custom length borderlights; uses 100–300 W. A lamps.
PAR-56 Border	Reflector Lamp	6', 8' or continuous custom length borderlight; uses PAR-56 lamps.
PAR-56 Border	Reflector Lamp	9', or continuous custom length borderlight; uses PAR-64 lamps.
R-40/PAR-38 Border	Reflector Lamp	6', 7'-6", 8', or continuous custom length borderlights; uses 150–300 W. R-40 or PAR-38 lamps.
65-05*	Reflector Lamp	3', 4' or 5' lengths; 9, 12 or 15 lamps; 3 or 4 circuits; uses 50–75 W. PAR-16 or PAR-20 lamps. See p. 559.
65-40*	Reflector Lamp	21¼" length; 10 lamps; 1 circuit; 20–35 W. 12 volt MR-11 lamps; unit is commonly known as a "Nano-Strip." See p. 560 for photometrics information.
65-50	Asymmetrical	3, 6, 9, or 12 lamp units; uses 300–100 W. lamps.
65-60*	Reflector Lamp	6' and 8' lengths; 30 or 40 lamps; 3 or 4 circuits; uses 12V. MR-16 lamps; unit is commonly known as a Mini-Strip. See p. 561.
65-60/2	Reflector Lamp	6' long; 60 lamps; 3 circuits; uses 12 V. MR-16 lamps; unit is a pair of Mini-Strips sharing a common hanger.
65-60/3	Reflector Lamp	3 units; 2' long each; 30 lamps; 3 circuits; uses MR-16 lamps; units are continuously wired but not attached together to permit following a curve.
65-70	Asymmetrical	Length unknown; 1 lamp; 1 circuit; uses 300–1000 W. lamps.
65-70/3	Asymmetrical	Same; 3 units hinged together to permit following a curve, or rigidly attached.
65-70/4	Asymmetrical	Same; 4 units, hinged or rigid.
65-71/1*	Asymmetrical	11⅞" long; 1 lamp; 1 circuit unit; uses 300–750 W. recessed single-contact lamps; unit is known as a "Broad Cyc." See p. 562.
65-71/2	Asymmetrical	23¾" long; 2 lamps; 2 circuits. See p. 562 for photometrics for single unit.
65-71/3	Asymmetrical	35⅝" long; 3 lamps; 3 circuits. See p. 562 for photometrics for single unit.
65-71/4	Asymmetrical	23¾" long (2 over 2); 4 lamps; 4 circuits. See p. 562 for photometrics for single unit.
65-80/2	Reflector Lamp	8' long; 80 lamps; 4 circuits; uses MR-16 lamps; unit is a pair of Mini-Strips sharing a common hanger.
65-80/3	Reflector Lamp	2 units; 4' long each; 40 lamps; 4 circuits; uses MR-16 lamps; units are continuously wired but not attached together to permit following a curve.
65-90*	Asymmetrical	11⅞" long; 1 lamp; 1 circuit unit; uses 300–1000 W. recessed single-contact lamps; unit is known as a "Baby Broad Groundrow." See p. 563.
65-95*	Asymmetrical	9" long; 1 lamp; 1 circuit unit; uses 300–750 W. recessed single-contact lamps; unit is known as a "The Runt." See p. 564.

Major Equipment Co.

B-103	Trough	Continuous custom borderlight; uses up to 100 W. A-23 lamps; unit is not equipped with individual reflectors, inside of hood is Alzak, color is achieved through dipped lamps.
B-106	Circular Reflector	Continuous custom length borderlight; uses 75–100 W. A-23 lamps.
B-156	Circular Reflector	Same; uses 150 W. PS-25 lamps.
B-208	Circular Reflector	Continuous custom length borderlight ; uses 200 W. PS-30 lamps.

B-2R	Circular Reflector	Continuous custom length borderlight; could be made for any lamp from 60–500 W; unit uses two rows of lamps.
B-512	Circular Reflector	Same; uses 300–500 W. PS-35 or PS-40 lamps.
D-156	Circular Reflector	Same; 9 lamps; uses 150 W. lamps.
D-2R104	Circular Reflector	5' long; 24 lamps; 3 circuits; uses 100 W. lamps; unit is a disappearing footlight with two rows of lamps.
D-63-B	Trough	5' long; 15 lamps standard (12 or 18 lamp versions also available); 3 circuits; unit is not equipped with individual reflectors, inside of hood is Alzak, color is achieved through dipped lamps; unit is a disappearing footlight.
D-64	Circular Reflector	5' long; 12 lamps; 3 circuits; uses 60 W. lamps; unit is a disappearing footlight.
F-106	Circular Reflector	Continuous custom length footlight; uses 75–100 W. lamps.
F-156	Circular Reflector	Same; uses 150 W. PS-25 lamps.
F-63	Trough	Continuous custom footlight; unit is not equipped with individual reflectors, inside of hood is either painted or Alzak, color is achieved through dipped lamps.
F-63-2	Asymmetrical	Same; uses double row of lamps.
F2R-156	Circular Reflector	Continuous custom length footlight; uses 150 W. PS-25 lamps in two rows.

Strand Century

5276	Reflector Lamp	6' long; 12 lamps; 3 circuits; uses 75–150 W. PAR-38 or 150–300 W. R-40 lamps.
5278	Reflector Lamp	8' long; 12 lamps; 3 circuits; uses 300 W. PAR-56 lamps.
5286	Reflector Lamp	6' long; 12 lamps; 4 circuits; uses 75–150 W. PAR-38 or 150–300 W. R-40 lamps.
5288	Reflector Lamp	8' long; 12 lamps; 4 circuits; uses 300 W. PAR-56 lamps.
5911*	Asymmetrical	1 lamp; 1 circuit; uses 1.5 KW recessed single contact lamp. See p. 565.
5912*	Asymmetrical	Same; 2 lamps; 2 circuits. See p. 565.
5913*	Asymmetrical	Same; 3 lamps; 3 circuits. See p. 565.
5914*	Asymmetrical	Same; 4 lamps; 4 circuits. See p. 565.
5915*	Asymmetrical	1 lamp; 1 circuit; uses 1 KW recessed single contact lamp. See p. 566.

Strand Lighting

5276	Reflector Lamp	6' long; 12 lamps; 3 circuits; uses 75–150 W. PAR-38 or 150–300 W. R-40 lamps.
5286	Reflector Lamp	Same; 4 circuits.
5300*	Reflector Lamp	6' long; 30 lamps; 3 circuits; uses 12V. MR-16 lamps. See p. 567.
5940*	Asymmetrical	1 lamp; 1 circuit; uses 1 KW recessed single-contact lamps. See p. 568.
Coda 500/1 MkII*	Asymmetrical	1 lamp; 1 circuit; uses 500 W. recessed single-contact lamps. See p. 569.
Coda 500/3 MkII*	Asymmetrical	Same; 3 lamps; 3 circuits. See p. 569.
Coda 500/4 MkII*	Asymmetrical	Same; 4 lamps; 4 circuits. See p. 569.

Times Square

701-5*	Reflector Lamp	2'-4" long; 10 lamps; 1 circuit; uses 75 W. PAR-16 lamps. See p. 570.
701-6*	Reflector Lamp	Same; 2'-10" long; 12 lamps; 3 circuits. See p. 570.
701-8*	Reflector Lamp	Same; 3'-9" long; 16 lamps; 4 circuits. See p. 570.
701-12*	Reflector Lamp	Same; 5'-7" long; 24 lamps; 3 or 4 circuits. See p. 570.
701-15*	Reflector Lamp	Same; 7' long; 30 lamps; 3 circuits. See p. 570.
702-3, 703-3*	Reflector Lamp	3' long; 6 lamps; 3 circuits; uses 75–150 W. R-40 or PAR-38 lamps. #702 has rondels supplied; #703 does not. See p. 571.
702-4, 703-4*	Reflector Lamp	Same; 4'-6" long; 9 lamps; #702 has rondels supplied; #703 does not. See p. 571.
702-6, 703-6*	Reflector Lamp	Same; 6' long; 12 lamps; #702 has rondels supplied; #703 does not. See p. 571.
702-7, 703-7*	Reflector Lamp	Same; 7'-6" long; 15 lamps. #702 has rondels supplied; #703 does not. See p. 571.
702-8, 703-8*	Reflector Lamp	Same; 8' long; 16 lamps; 4 circuits. #702 has rondels supplied; #703 does not. See p. 571.
Diluvio A 1000F	Asymmetrical	9½" long; 1 lamp; 1 circuit; uses 500–1000 W. recessed single-contact lamps; unit is a modular striplight; numerous units may be grouped together to form a larger strip, with a c-clamp arm attaching in a similar way.

ELECTRONIC THEATRE CONTROLS (ETC)
Model No. S4PAR-CL*

BEAM ANGLE: 8°

FIELD ANGLE: 14°

LAMP BASE TYPE: Medium 2-Pin

LAMP MOUNT: Axial

STANDARD LAMP: HPL 575/115 - 575 W.

OTHER LAMPS: HPL5375/115X - 575 W.
HPL 575/120V - 575W.
HPL 550/77V - 575W.**
HPL 550/77X - 550W.**

INTEGRAL PATTERN SLOT: Yes

ACCESSORIES AVAILABLE:
Color Frame (7½" sq.)

WEIGHT: 7.5 lbs.

SIZE: 10¾" x 10¼" x 11" (L x W x H) (without c-clamp)

PHOTOMETRICS CHART

(Performance data for this unit are measured at cosine focus using HPL 575/115 lamp.)

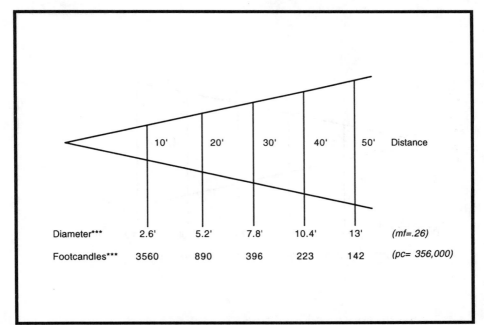

	10'	20'	30'	40'	50'	Distance
Diameter***	2.6'	5.2'	7.8'	10.4'	13'	*(mf=.26)*
Footcandles***	3560	890	396	223	142	*(pc= 356,000)*

* Lens assemblies are interchangeable within the same fixture body. See following pages.
** The HPL 550/77 lamp is used with an ETC multiplexing system, used to allow multiple units to share a common circuit, with separate dimming of each.
*** The diameter and footcandle information in this chart was calculated using multiplying factor (*mf*) and peak candela (*pc*) data given in catalog. See Introduction for more information about calculating photometric data.

ELECTRONIC THEATRE CONTROLS (ETC)
Model No. S4PAR-VNSP*

BEAM ANGLE: 8°

FIELD ANGLE: 15°

LAMP BASE TYPE: Medium 2-Pin

LAMP MOUNT: Axial

STANDARD LAMP: HPL 575/115 - 575 W.

OTHER LAMPS: HPL5375/115X - 575 W.
HPL 575/120V - 575W.
HPL 550/77V - 575W.**
HPL 550/77X - 550W.**

INTEGRAL PATTERN SLOT: Yes

ACCESSORIES AVAILABLE:
Color Frame (7½" sq.)

WEIGHT: 7.5 lbs.

SIZE: 10¾" x 10¼" x 11" (L x W x H) (without c-clamp)

PHOTOMETRICS CHART
(Performance data for this unit are measured at cosine focus using HPL 575/115 lamp.)

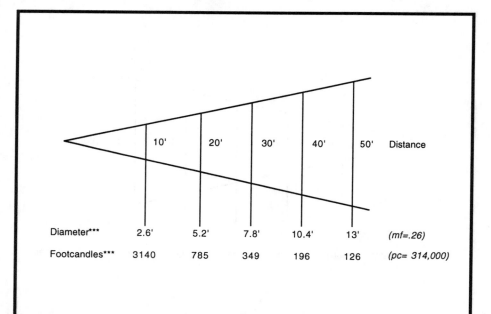

Distance	10'	20'	30'	40'	50'	
Diameter***	2.6'	5.2'	7.8'	10.4'	13'	(mf=.26)
Footcandles***	3140	785	349	196	126	(pc= 314,000)

* Lens assemblies are interchangeable within the same fixture body. See preceding and following pages.
** The HPL 550/77 lamp is used with an ETC multiplexing system, used to allow multiple units to share a common circuit, with separate dimming of each.
*** The diameter and footcandle information in this chart was calculated using multiplying factor (*mf*) and peak candela (*pc*) data given in catalog. See Introduction for more information about calculating photometric data.

ELECTRONIC THEATRE CONTROLS (ETC)
Model No. S4PAR-NSP*

BEAM ANGLE: 10°

FIELD ANGLE: 19°

LAMP BASE TYPE: Medium 2-Pin

LAMP MOUNT: Axial

STANDARD LAMP: HPL 575/115 - 575 W.

OTHER LAMPS: HPL5375/115X - 575 W.
HPL 575/120V - 575W.
HPL 550/77V - 575W.**
HPL 550/77X - 550W.**

INTEGRAL PATTERN SLOT: Yes

ACCESSORIES AVAILABLE:
Color Frame (7½" sq.)

WEIGHT: 7.5 lbs.

SIZE: 10¾" x 10¼" x 11" (L x W x H) (without c-clamp)

PHOTOMETRICS CHART

(Performance data for this unit are measured at cosine focus using HPL 575/115 lamp.)

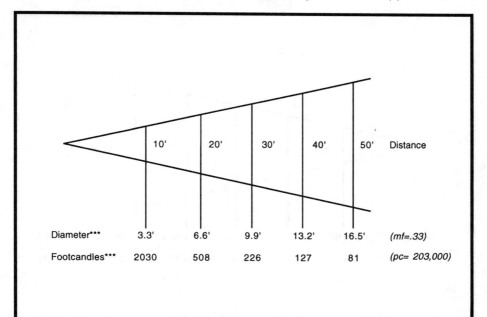

	10'	20'	30'	40'	50'	Distance
Diameter***	3.3'	6.6'	9.9'	13.2'	16.5'	*(mf=.33)*
Footcandles***	2030	508	226	127	81	*(pc= 203,000)*

* Lens assemblies are interchangeable within the same fixture body. See preceding and following pages.
** The HPL 550/77 lamp is used with an ETC multiplexing system, used to allow multiple units to share a common circuit, with separate dimming of each.
*** The diameter and footcandle information in this chart was calculated using multiplying factor (*mf*) and peak candela (*pc*) data given in catalog. See Introduction for more information about calculating photometric data.

ELECTRONIC THEATRE CONTROLS (ETC)
Model No. S4PAR-MFL*

BEAM ANGLE: 12° x 19°

FIELD ANGLE: 21° x 34°

LAMP BASE TYPE: Medium 2-Pin

LAMP MOUNT: Axial

STANDARD LAMP: HPL 575/115 - 575 W.

OTHER LAMPS: HPL5375/115X - 575 W.
HPL 575/120V - 575W.
HPL 550/77V - 575W.**
HPL 550/77X - 550W.**

INTEGRAL PATTERN SLOT: Yes

ACCESSORIES AVAILABLE:
Color Frame (7½" sq.)

WEIGHT: 7.5 lbs.

SIZE: 10¾" x 10¼" x 11" (L x W x H) (without c-clamp)

PHOTOMETRICS CHART
(Performance data for this unit are measured at cosine focus using HPL 575/115 lamp.)

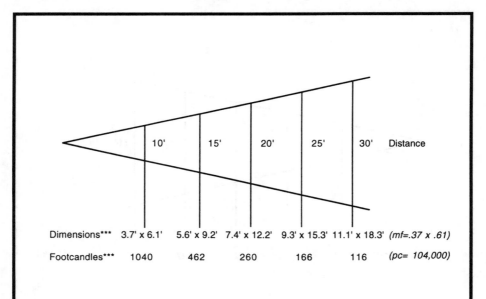

	10'	15'	20'	25'	30'	Distance
Dimensions***	3.7' x 6.1'	5.6' x 9.2'	7.4' x 12.2'	9.3' x 15.3'	11.1' x 18.3'	*(mf=.37 x .61)*
Footcandles***	1040	462	260	166	116	*(pc= 104,000)*

* Lens assemblies are interchangeable within the same fixture body. See preceding and following pages.
** The HPL 550/77 lamp is used with an ETC multiplexing system, used to allow multiple units to share a common circuit, with separate dimming of each.
*** The diameter and footcandle information in this chart was calculated using multiplying factor (*mf*) and peak candela (*pc*) data given in catalog. See Introduction for more information about calculating photometric data.

WEIGHT: 7.5 lbs.

SIZE: 10¾" x 10¼" x 11" (L x W x H) (without c-clamp)

ELECTRONIC THEATRE CONTROLS (ETC) Model No. S4PAR-WFL*

BEAM ANGLE: 16° x 33°

FIELD ANGLE: 30° x 50°

LAMP BASE TYPE: Medium 2-Pin

LAMP MOUNT: Axial

STANDARD LAMP: HPL 575/115 - 575 W.

OTHER LAMPS: HPL5375/115X - 575 W.
HPL 575/120V - 575W.
HPL 550/77V - 575W.**
HPL 550/77X - 550W.**

INTEGRAL PATTERN SLOT: Yes

ACCESSORIES AVAILABLE:
Color Frame (7½" sq.)

PHOTOMETRICS CHART

(Performance data for this unit are measured at cosine focus using HPL 575/115 lamp.)

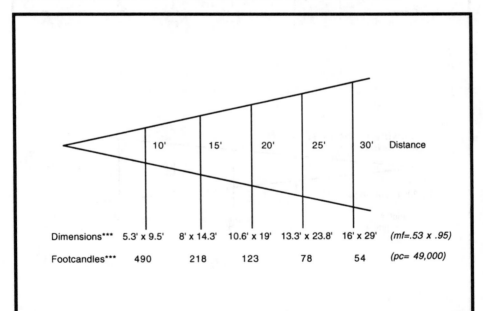

	10'	15'	20'	25'	30'	Distance
Dimensions***	5.3' x 9.5'	8' x 14.3'	10.6' x 19'	13.3' x 23.8'	16' x 29'	*(mf=.53 x .95)*
Footcandles***	490	218	123	78	54	*(pc= 49,000)*

* Lens assemblies are interchangeable within the same fixture body. See preceding pages.
** The HPL 550/77 lamp is used with an ETC multiplexing system, used to allow multiple units to share a common circuit, with separate dimming of each.
*** The diameter and footcandle information in this chart was calculated using multiplying factor (*mf*) and peak candela (*pc*) data given in catalog. See Introduction for more information about calculating photometric data.

MR-11, 20 W. LAMPS (12 V.)

LAMP BASE TYPE: 2-Pin (GU4)
COLOR TEMPERATURE: 2900° K

PHOTOMETRICS CHART

(Performance data from GE Lighting Stage & Studio Lighting Lamp Catalog, SS-123P dated 11/95.)

SPOT (SPT)
ANSI Code: FTB

	3'	6'	9'	12'	Distance
10° beam angle					
Diameter	.5'	1.1'	1.6'	2.1'	*(mf= .18)*
Footcandles	611	153	68	38	*(pc=5,500)*

NARROW FLOOD (NFL)
ANSI Code: FTD

	3'	6'	9'	12'	Distance
30° beam angle					
Diameter	1.6'	3.2'	4.8'	6.4'	*(mf= .54)*
Footcandles	68	17	7	4	*(pc=600)*

LAMP BASE TYPE: 2-Pin (GU4)

COLOR TEMPERATURE: 2900° K

PHOTOMETRICS CHART

(Performance data from GE Lighting Stage & Studio Lighting Lamp Catalog, SS-123P dated 11/95.)

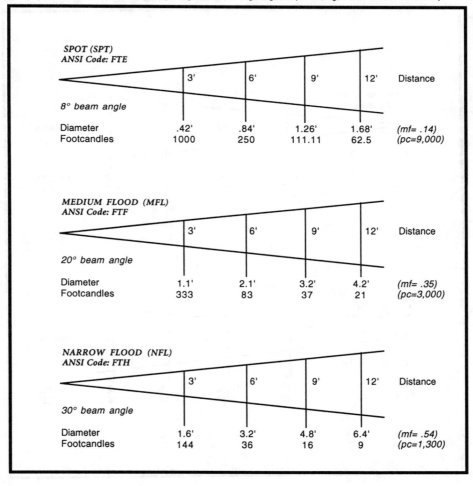

SPOT (SPT)
ANSI Code: FTE

	3'	6'	9'	12'	Distance
8° beam angle					
Diameter	.42'	.84'	1.26'	1.68'	(mf= .14)
Footcandles	1000	250	111.11	62.5	(pc=9,000)

MEDIUM FLOOD (MFL)
ANSI Code: FTF

	3'	6'	9'	12'	Distance
20° beam angle					
Diameter	1.1'	2.1'	3.2'	4.2'	(mf= .35)
Footcandles	333	83	37	21	(pc=3,000)

NARROW FLOOD (NFL)
ANSI Code: FTH

	3'	6'	9'	12'	Distance
30° beam angle					
Diameter	1.6'	3.2'	4.8'	6.4'	(mf= .54)
Footcandles	144	36	16	9	(pc=1,300)

MR-16, 20 W. LAMPS (12 V.)

LAMP BASE TYPE: 2-Pin (GX5.3)
COLOR TEMPERATURE: 2900° K

PHOTOMETRICS CHART

(Performance data from GE Lighting Stage & Studio Lighting Lamp Catalog, SS-123P dated 11/95.)

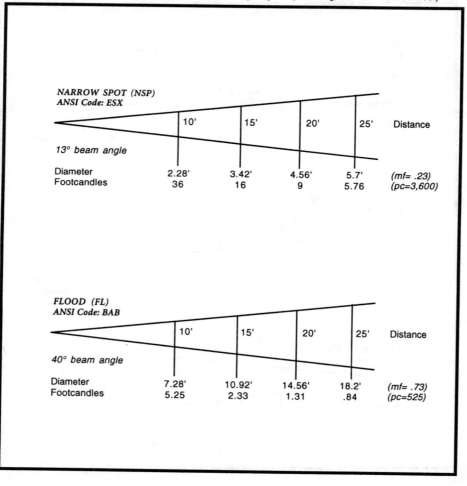

NARROW SPOT (NSP)
ANSI Code: ESX

	10'	15'	20'	25'	Distance
13° beam angle					
Diameter	2.28'	3.42'	4.56'	5.7'	*(mf= .23)*
Footcandles	36	16	9	5.76	*(pc=3,600)*

FLOOD (FL)
ANSI Code: BAB

	10'	15'	20'	25'	Distance
40° beam angle					
Diameter	7.28'	10.92'	14.56'	18.2'	*(mf= .73)*
Footcandles	5.25	2.33	1.31	.84	*(pc=525)*

LAMP BASE TYPE: 2-Pin (GX5.3)
COLOR TEMPERATURE: 3000° K

PHOTOMETRICS CHART

(Performance data from GE Lighting Stage & Studio Lighting Lamp Catalog, SS-123P dated 11/95.)

NARROW SPOT (NSP)
ANSI Code: FRB

	5'	10'	15'	20'	Distance

12° beam angle

Diameter	1.1'	2.1'	3.2'	4.2'	*(mf= .21)*
Footcandles	348	87	38.67	21.75	*(pc=8,700)*

SPOT (SP)
ANSI Code: FRA

	5'	10'	15'	20'	Distance

20° beam angle

Diameter	1.8'	3.5'	5.3'	7'	*(mf= .35)*
Footcandles	156	39	17.33	9.75	*(pc=3,900)*

FLOOD (FL)
ANSI Code: FMW

	5'	10'	15'	20'	Distance

40° beam angle

Diameter	3.6'	7.3'	10.9'	14.5'	*(mf= .73)*
Footcandles	40	10	4.44	2.5	*(pc=1,000)*

MR-16, 42 W. LAMPS (12 V.)

LAMP BASE TYPE: 2-Pin (GX5.3)

COLOR TEMPERATURE: 3000° K *

PHOTOMETRICS CHART

(Performance data from GE Lighting Stage & Studio Lighting Lamp Catalog, SS-123P dated 11/95.)

VERY NARROW SPOT (VNSP)
ANSI Code: EZY

	2'	4'	6'	8'	Distance

9° beam angle

Diameter	.32'	.64'	.96'	1.28'	(mf= .16)
Footcandles	3275	819	364	205	(pc=13,100)

SPOT (SP)
ANSI Code: EYS

	2'	4'	6'	8'	Distance

27° beam angle

Diameter	.96'	1.92'	2.88'	3.84'	(mf= .48)
Footcandles	600	150	67	38	(pc=2,400)

LAMP BASE TYPE: 2-Pin (GX5.3)
COLOR TEMPERATURE: 3050° K

PHOTOMETRICS CHART

(Performance data from GE Lighting Stage & Studio Lighting Lamp Catalog, SS-123P dated 11/95.)

NARROW SPOT (NSP)
ANSI Code: EXT

	5'	10'	15'	20'	Distance
14° beam angle					
Diameter	1.23'	2.45'	3.68'	4.9'	(mf= .25)
Footcandles	408	102	45.33	25.5	(pc=10,200)

NARROW FLOOD (NFL)
ANSI Code: EXZ

	5'	10'	15'	20'	Distance
27° beam angle					
Diameter	2.4'	4.8'	7.2'	9.6'	(mf= .48)
Footcandles	136	34	15.11	8.5	(pc=3,400)

FLOOD (FL)
ANSI Code: EXN

	5'	10'	15'	20'	Distance
40° beam angle					
Diameter	3.64'	7.28'	10.92'	14.56'	(mf= .73)
Footcandles	74	18.5	8.22	4.63	(pc=1,850)

MR-16, 50 W. LAMPS (12 V.)

LAMP BASE TYPE: 2-Pin (GX5.3)
COLOR TEMPERATURE: 3050° K

PHOTOMETRICS CHART

(Performance data from GE Lighting Stage & Studio Lighting Lamp Catalog, SS-123P *dated 11/95.)*

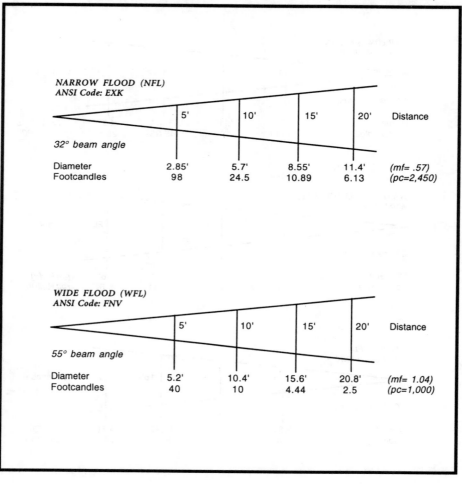

LAMP BASE TYPE: 2-Pin (GX5.3)
COLOR TEMPERATURE: 3050° K

PHOTOMETRICS CHART

(Performance data from Sylvania / GTE Lighting Handbook, 8th edition, dated 1989.)

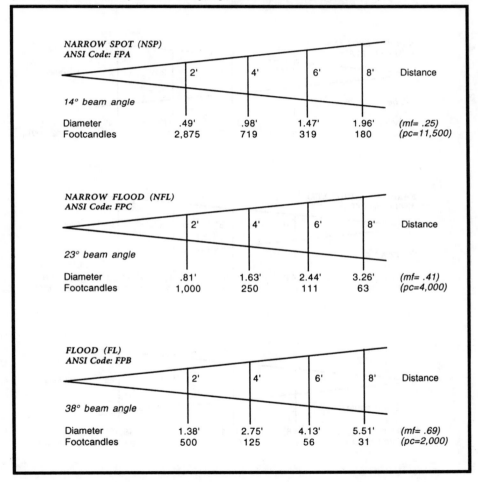

NARROW SPOT (NSP)
ANSI Code: FPA

Distance	2'	4'	6'	8'
14° beam angle				

	2'	4'	6'	8'	
Diameter	.49'	.98'	1.47'	1.96'	*(mf= .25)*
Footcandles	2,875	719	319	180	*(pc=11,500)*

NARROW FLOOD (NFL)
ANSI Code: FPC

23° beam angle

	2'	4'	6'	8'	
Diameter	.81'	1.63'	2.44'	3.26'	*(mf= .41)*
Footcandles	1,000	250	111	63	*(pc=4,000)*

FLOOD (FL)
ANSI Code: FPB

38° beam angle

	2'	4'	6'	8'	
Diameter	1.38'	2.75'	4.13'	5.51'	*(mf= .69)*
Footcandles	500	125	56	31	*(pc=2,000)*

MR-16, 75 W. LAMPS (12 V.)

LAMP BASE TYPE: 2-Pin (GX5.3)
COLOR TEMPERATURE: 3050° K

PHOTOMETRICS CHART

(Performance data from GE Lighting Stage & Studio Lighting Lamp Catalog, SS-123P dated 11/95.)

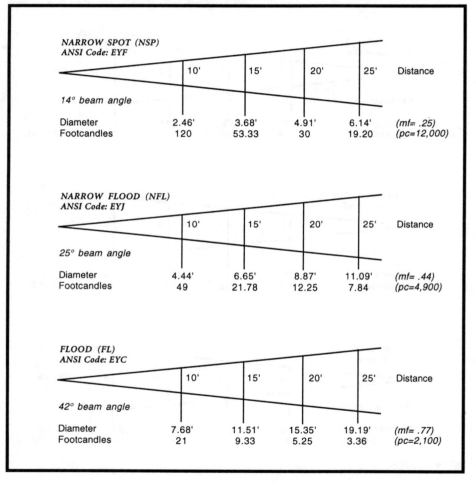

NARROW SPOT (NSP)
ANSI Code: EYF

	10'	15'	20'	25'	Distance

14° beam angle

	10'	15'	20'	25'	
Diameter	2.46'	3.68'	4.91'	6.14'	(mf= .25)
Footcandles	120	53.33	30	19.20	(pc=12,000)

NARROW FLOOD (NFL)
ANSI Code: EYJ

	10'	15'	20'	25'	Distance

25° beam angle

	10'	15'	20'	25'	
Diameter	4.44'	6.65'	8.87'	11.09'	(mf= .44)
Footcandles	49	21.78	12.25	7.84	(pc=4,900)

FLOOD (FL)
ANSI Code: EYC

	10'	15'	20'	25'	Distance

42° beam angle

	10'	15'	20'	25'	
Diameter	7.68'	11.51'	15.35'	19.19'	(mf= .77)
Footcandles	21	9.33	5.25	3.36	(pc=2,100)

LAMP BASE TYPE: Oval 2-Pin (GY5.3)

COLOR TEMPERATURE: see each lamp below

PHOTOMETRICS CHART

(Performance data from GE Lighting Stage & Studio Lighting Lamp Catalog, SS-123P dated 11/95.)

FLOOD (FL)
ANSI Code: EZK
150 Watts

3200° K

60° beam angle

	2'	4'	6'	8'	Distance
Diameter	2.31'	4.62'	6.93'	9.24'	(mf= 1.16)
Footcandles	901	225	100	56	(pc=3,600)

FLOOD (FL)
ANSI Code: ENH
250 Watts

3250° K

32° beam angle

	2'	4'	6'	8'	Distance
Diameter	1.15'	2.29'	3.44'	4.59'	(mf= .57)
Footcandles	2,925	731	325	183	(pc=11,700)

FLOOD (FL)
ANSI Code: EXX
250 Watts

3300° K

57° beam angle

	2'	4'	6'	8'	Distance
Diameter	2.17'	4.34'	6.52'	8.69'	(mf= 1.09)
Footcandles	1,690	422	188	106	(pc=6,750)

PAR-38, 150 W. LAMPS

LAMP BASE TYPE: Medium Skirted
COLOR TEMPERATURE: 2750° K

PHOTOMETRICS CHART

(Performance data from GE Lighting Stage & Studio Lighting Lamp Catalog, SS-123 dated 15/90.)*

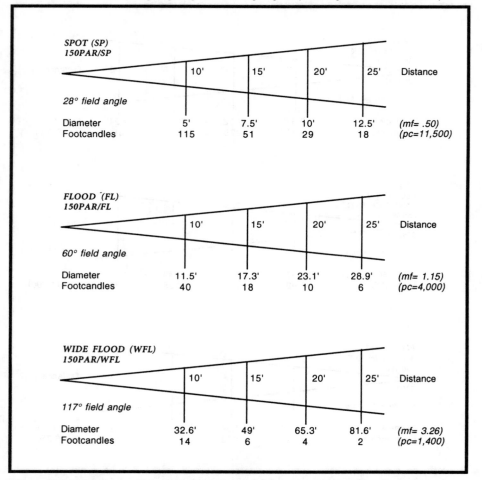

SPOT (SP)
150PAR/SP

	10'	15'	20'	25'	Distance

28° field angle

| Diameter | 5' | 7.5' | 10' | 12.5' | (mf= .50) |
| Footcandles | 115 | 51 | 29 | 18 | (pc=11,500) |

FLOOD (FL)
150PAR/FL

	10'	15'	20'	25'	Distance

60° field angle

| Diameter | 11.5' | 17.3' | 23.1' | 28.9' | (mf= 1.15) |
| Footcandles | 40 | 18 | 10 | 6 | (pc=4,000) |

WIDE FLOOD (WFL)
150PAR/WFL

	10'	15'	20'	25'	Distance

117° field angle

| Diameter | 32.6' | 49' | 65.3' | 81.6' | (mf= 3.26) |
| Footcandles | 14 | 6 | 4 | 2 | (pc=1,400) |

* These PAR-38 lamps are not listed in the 1995 Stage & Studio catalog. Check with your local supplier for availability.

LAMP BASE TYPE: Medium Skirted
COLOR TEMPERATURE: 2900° K

PHOTOMETRICS CHART

(Performance data from GE Lighting Stage & Studio Lighting Lamp Catalog, SS-123P dated 11/95.)

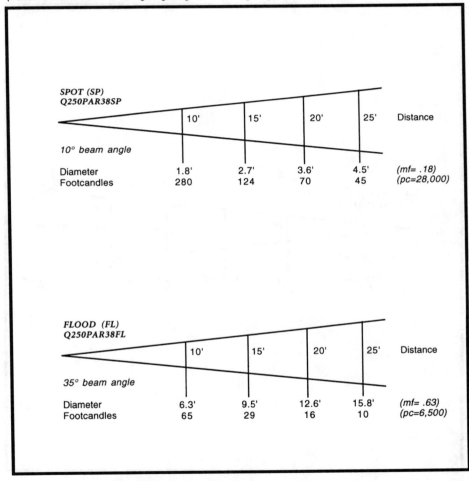

SPOT (SP)
Q250PAR38SP

	10'	15'	20'	25'	Distance

10° beam angle

	10'	15'	20'	25'	
Diameter	1.8'	2.7'	3.6'	4.5'	(mf= .18)
Footcandles	280	124	70	45	(pc=28,000)

FLOOD (FL)
Q250PAR38FL

	10'	15'	20'	25'	Distance

35° beam angle

	10'	15'	20'	25'	
Diameter	6.3'	9.5'	12.6'	15.8'	(mf= .63)
Footcandles	65	29	16	10	(pc=6,500)

* The 1995 Stage & Studio catalog lists these as "halogen projector lamps."

PAR-46, 150 W. LAMPS

LAMP BASE TYPE: Medium Side Prong
COLOR TEMPERATURE: 2750° K

PHOTOMETRICS CHART
(Performance data from GE Lighting Stage & Studio Lighting Lamp Catalog, SS-123P dated 11/95.)

MEDIUM FLOOD (MFL)
150PAR46/3MFL

	10'	20'	30'	40'	Distance

39°x25° field angle

Dimensions (WxH)	7.1'x 4.4'	14.2'x 8.9'	21.2'x 13.3'	26.3'x 17.7'	*(mf= .71 x.44)*
Footcandles	80	20	9	5	*(pc=8,000)*

LAMP BASE TYPE: Medium Side Prong
COLOR TEMPERATURE: 2750° K

PHOTOMETRICS CHART

(Performance data from GE Lighting Stage & Studio Lighting Lamp Catalog, SS-123P dated 11/95.)

NARROW SPOT (NSP)
200PAR46/3NSP

	10'	20'	30'	40'	Distance

23°x19° field angle

Dimensions (WxH)	4.1'x 3.3'	8.1'x 6.7'	12.2'x 10'	16.3'x 13.4'	*(mf= .41 x.33)*
Footcandles	310	77	34	19	*(pc=31,000)*

MEDIUM FLOOD (MFL)
200PAR46/3MFL

	10'	20'	30'	40'	Distance

40°x24° field angle

Dimensions (WxH)	7.3'x 4.3'	14.6'x 8.5'	21.8'x 12.8'	29.1'x 17'	*(mf= .73 x.43)*
Footcandles	115	29	13	7	*(pc=11,500)*

597

PAR-56, 300 W. LAMPS

LAMP BASE TYPE: Mogul End Prong
COLOR TEMPERATURE: 2750° K

PHOTOMETRICS CHART

(Performance data from GE Lighting Stage & Studio Lighting Lamp Catalog, SS-123P dated 11/95.)

NARROW SPOT (NSP)
300PAR56/NSP

	10'	20'	30'	40'	Distance

20°x14° field angle

| Dimensions (WxH) | 3.5'x 2.5' | 7.1'x 4.9' | 10.6'x 7.4' | 14.1'x 9.8' | (mf= .35 x.25) |
| Footcandles | 680 | 170 | 76 | 43 | (pc=68,000) |

MEDIUM FLOOD (MFL)
300PAR56/MFL

	10'	20'	30'	40'	Distance

34°x19° field angle

| Dimensions (WxH) | 6.1'x 3.3' | 12.2'x 6.7' | 18.3'x 10' | 24.5'x13.4' | (mf= .61 x.33) |
| Footcandles | 240 | 60 | 27 | 15 | (pc=24,000) |

WIDE FLOOD (WFL)
300PAR56/WFL

	10'	20'	30'	40'	Distance

57°x27° field angle

| Dimensions (WxH) | 10.9'x 4.8' | 21.7'x9.6' | 32.6'x14.4' | 43.4'x19.2' | (mf= 1.09x.48) |
| Footcandles | 110 | 28 | 12 | 7 | (pc=11,000) |

LAMP BASE TYPE: Mogul End Prong
COLOR TEMPERATURE: 2950° K

PHOTOMETRICS CHART

(Performance data from GE Lighting Stage & Studio Lighting Lamp Catalog, SS-123P *dated 11/95.)*

NARROW SPOT (NSP)
Q500PAR56NSP

	10'	20'	30'	40'	Distance

32°x15° field angle

Dimensions (WxH)	5.7'x2.6'	11.5'x5.3'	17.2'x7.9'	22.9'x10.5'	*(mf= .57x.26)*
Footcandles	960	240	107	60	*(pc=96,000)*

MEDIUM FLOOD (MFL)
Q500PAR56MFL

	10'	20'	30'	40'	Distance

42°x20° field angle

Dimensions (WxH)	7.7'x3.5'	15.4'x7.1'	23'x10.6'	30.7'x14.1'	*(mf= .77x.35)*
Footcandles	430	108	48	27	*(pc=43,000)*

WIDE FLOOD (WFL)
Q500PAR56WFL

	10'	20'	30'	40'	Distance

66°x34° field angle

Dimensions (WxH)	13'x6.1'	26'x12.2'	39'x18.3'	52'x 24.5'	*(mf= 1.3x.61)*
Footcandles	190	48	21	12	*(pc=19,000)*

PAR-64, 1 KW. LAMPS

LAMP BASE TYPE: Extended Mogul End Prong
COLOR TEMPERATURE: 3000° K

PHOTOMETRICS CHART

(Performance data from GE Lighting Stage & Studio Lighting Lamp Catalog, SS-123P dated 11/95.)

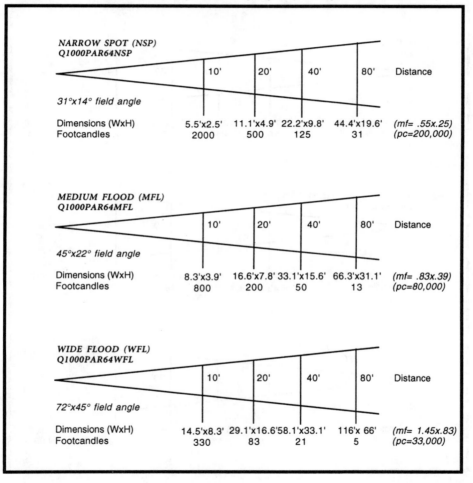

NARROW SPOT (NSP)
Q1000PAR64NSP

	10'	20'	40'	80'	Distance

31°x14° field angle

Dimensions (WxH)	5.5'x2.5'	11.1'x4.9'	22.2'x9.8'	44.4'x19.6'	*(mf= .55x.25)*
Footcandles	2000	500	125	31	*(pc=200,000)*

MEDIUM FLOOD (MFL)
Q1000PAR64MFL

	10'	20'	40'	80'	Distance

45°x22° field angle

Dimensions (WxH)	8.3'x3.9'	16.6'x7.8'	33.1'x15.6'	66.3'x31.1'	*(mf= .83x.39)*
Footcandles	800	200	50	13	*(pc=80,000)*

WIDE FLOOD (WFL)
Q1000PAR64WFL

	10'	20'	40'	80'	Distance

72°x45° field angle

Dimensions (WxH)	14.5'x8.3'	29.1'x16.6'	58.1'x33.1'	116'x 66'	*(mf= 1.45x.83)*
Footcandles	330	83	21	5	*(pc=33,000)*

LAMP BASE TYPE: Extended Mogul End Prong

COLOR TEMPERATURE: 3200° K

PHOTOMETRICS CHART

(Performance data from GE Lighting Stage & Studio Lighting Lamp Catalog, SS-123P dated 11/95.)

VERY NARROW SPOT (VNSP)
ANSI Code: FFN

	10'	20'	40'	80'	Distance
24°x10° field angle					
Dimensions (WxH)	4.3'x1.7'	8.5'x3.5'	17'x7'	34'x14'	*(mf= .43x.17)*
Footcandles	4000	1000	250	63	*(pc=400,000)*

NARROW SPOT (NSP)
ANSI Code: FFP

	10'	20'	40'	80'	Distance
26°x14° field angle					
Dimensions (WxH)	4.6'x2.5'	9.2'x4.9'	18.5'x 9.8'	36.9'x19.6'	*(mf= .46x.25)*
Footcandles	3300	825	206	52	*(pc=330,000)*

MEDIUM FLOOD (MFL)
ANSI Code: FFR

	10'	20'	40'	80'	Distance
44°x21° field angle					
Dimensions (WxH)	8.1'x3.7'	16.2'x7.4'	32.3'x 14.8'	64.6'x27.7'	*(mf=.81x.37)*
Footcandles	1250	313	78	20	*(pc=125,000)*

WIDE FLOOD (WFL)
ANSI Code: FFS

	10'	20'	40'	80'	Distance
71°x45° field angle					
Dimensions (WxH)	14.3'x8.3'	28.5'x16.6'	57'x33'	114'x66'	*(mf=1.43x.83)*
Footcandles	400	100	25	6	*(pc=40,000)*

PAR-64, 1 KW. LAMPS

LAMP BASE TYPE: Extended Mogul End Prong
COLOR TEMPERATURE: 5200° K (daylight)

PHOTOMETRICS CHART
(Performance data from GE Lighting Stage & Studio Lighting Lamp Catalog, SS-123P dated 11/95.)

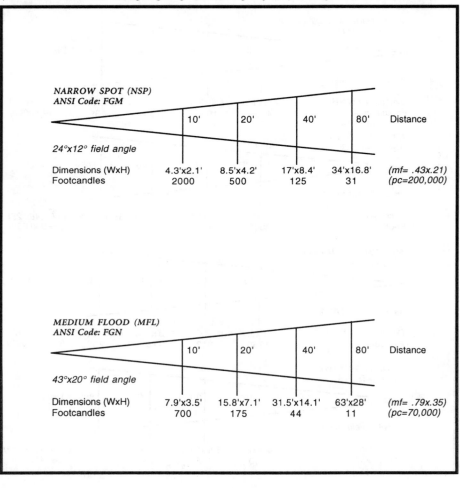

NARROW SPOT (NSP)
ANSI Code: FGM

	10'	20'	40'	80'	Distance
24°x12° field angle					
Dimensions (WxH)	4.3'x2.1'	8.5'x4.2'	17'x8.4'	34'x16.8'	*(mf= .43x.21)*
Footcandles	2000	500	125	31	*(pc=200,000)*

MEDIUM FLOOD (MFL)
ANSI Code: FGN

	10'	20'	40'	80'	Distance
43°x20° field angle					
Dimensions (WxH)	7.9'x3.5'	15.8'x7.1'	31.5'x14.1'	63'x28'	*(mf= .79x.35)*
Footcandles	700	175	44	11	*(pc=70,000)*

LAMP BASE TYPE: Extended Mogul End Prong
COLOR TEMPERATURE: 2800° K

PHOTOMETRICS CHART

(Performance data from GE Lighting Stage & Studio Lighting Lamp Catalog, SS-123P dated 11/95.)

NARROW SPOT (NSP)
500PAR64/NSP

	10'	20'	30'	40'	Distance

19°x14° field angle

Dimensions (WxH)	3.3'x2.5'	6.7'x4.9'	10'x7.4'	13.4'x9.8'	*(mf= .33x.25)*
Footcandles	1100	275	122	69	*(pc=110,000)*

MEDIUM FLOOD (MFL)
500PAR64/MFL

	10'	20'	30'	40'	Distance

35°x19° field angle

Dimensions (WxH)	6.3'x3.3'	12.6'x6.7'	18.9'x 10'	25.2'x 13.4'	*(mf= .63x.33)*
Footcandles	370	93	41	23	*(pc=37,000)*

WIDE FLOOD (WFL)
500PAR64/WFL

	10'	20'	40'	40'	Distance

55°x32° field angle

Dimensions (WxH)	10.4'x5.7'	208'x11.5'	31.2'x17.2'	41.6'x22.9'	*(mf= 1.04x.57)*
Footcandles	130	33	14	8	*(pc=13,000)*

R-40, 150 W. LAMPS

LAMP BASE TYPE: Medium Screw
COLOR TEMPERATURE: 2750° K

PHOTOMETRICS CHART

(Performance data from GE Lighting Stage & Studio Lighting Lamp Catalog, SS-123 dated 5/90*.)

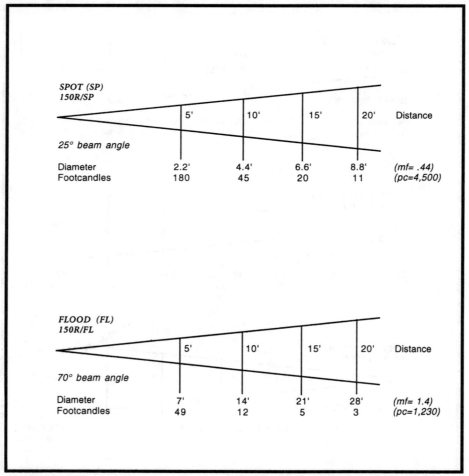

SPOT (SP) 150R/SP	5'	10'	15'	20'	Distance
25° beam angle					
Diameter	2.2'	4.4'	6.6'	8.8'	(mf= .44)
Footcandles	180	45	20	11	(pc=4,500)

FLOOD (FL) 150R/FL	5'	10'	15'	20'	Distance
70° beam angle					
Diameter	7'	14'	21'	28'	(mf= 1.4)
Footcandles	49	12	5	3	(pc=1,230)

*These R-40 lamps are not listed in the 1995 Stage & Studio catalog. Check with your local supplier for availability.

LAMP BASE TYPE: Medium Screw
COLOR TEMPERATURE: 2750° K

PHOTOMETRICS CHART

(Performance data from GE Lighting Stage & Studio Lighting Lamp Catalog, SS-123 dated 5/90.)*

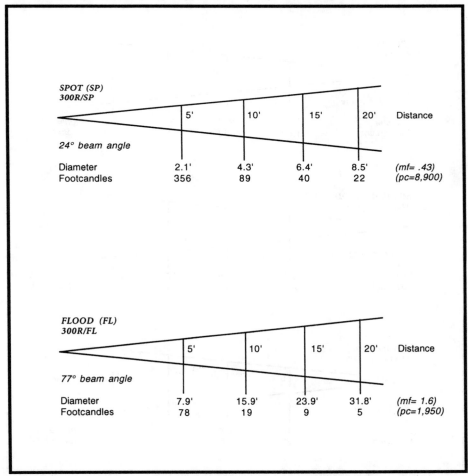

SPOT (SP)
300R/SP

| | 5' | 10' | 15' | 20' | Distance |

24° beam angle

| Diameter | 2.1' | 4.3' | 6.4' | 8.5' | (mf= .43) |
| Footcandles | 356 | 89 | 40 | 22 | (pc=8,900) |

FLOOD (FL)
300R/FL

| | 5' | 10' | 15' | 20' | Distance |

77° beam angle

| Diameter | 7.9' | 15.9' | 23.9' | 31.8' | (mf= 1.6) |
| Footcandles | 78 | 19 | 9 | 5 | (pc=1,950) |

*The 1995 Stage & Studio catalog does not specify the beam or field angle for these lamps, just general beam characteristics (spot, medium flood, etc.).

LAMP BASE TYPE: Medium Screw

COLOR TEMPERATURE: 3200° K

PHOTOMETRICS CHART

(Performance data from GE Lighting Stage & Studio Lighting Lamp Catalog, SS-123 dated 5/90.)*

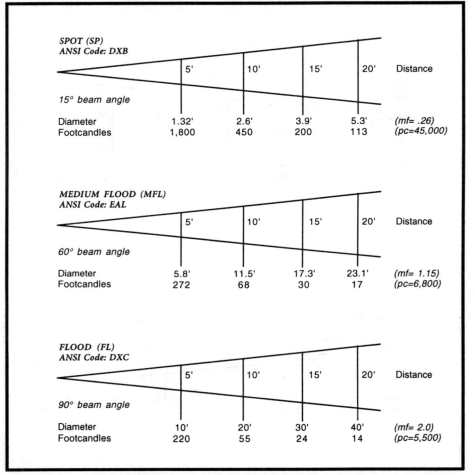

SPOT (SP)
ANSI Code: DXB

	5'	10'	15'	20'	Distance
15° beam angle					
Diameter	1.32'	2.6'	3.9'	5.3'	*(mf= .26)*
Footcandles	1,800	450	200	113	*(pc=45,000)*

MEDIUM FLOOD (MFL)
ANSI Code: EAL

	5'	10'	15'	20'	Distance
60° beam angle					
Diameter	5.8'	11.5'	17.3'	23.1'	*(mf= 1.15)*
Footcandles	272	68	30	17	*(pc=6,800)*

FLOOD (FL)
ANSI Code: DXC

	5'	10'	15'	20'	Distance
90° beam angle					
Diameter	10'	20'	30'	40'	*(mf= 2.0)*
Footcandles	220	55	24	14	*(pc=5,500)*

*The 1995 Stage & Studio catalog does not specify the beam or field angle for these lamps, just general beam characteristics (spot, medium flood, etc.).

R-40, 500 W. LAMPS

LAMP BASE TYPE: Mogul Screw
COLOR TEMPERATURE: 2800° K

PHOTOMETRICS CHART
(Performance data from GE Lighting Stage & Studio Lighting Lamp Catalog, SS-123 dated 5/90*.)

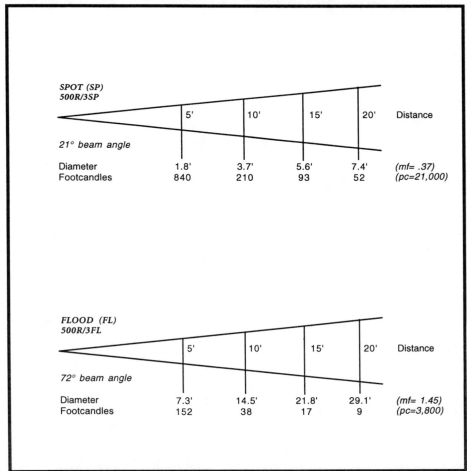

SPOT (SP)
500R/3SP

| | 5' | 10' | 15' | 20' | Distance |

21° beam angle

	5'	10'	15'	20'	
Diameter	1.8'	3.7'	5.6'	7.4'	*(mf= .37)*
Footcandles	840	210	93	52	*(pc=21,000)*

FLOOD (FL)
500R/3FL

72° beam angle

	5'	10'	15'	20'	
Diameter	7.3'	14.5'	21.8'	29.1'	*(mf= 1.45)*
Footcandles	152	38	17	9	*(pc=3,800)*

*The 1995 Stage & Studio catalog does not specify the beam or field angle for these lamps, just general beam characteristics (spot, medium flood, etc.).

(Mc)	Mini-Can Screw (Mc)	GX5.3
(Med Skt)	Medium Skirted (Med Skt)	GY5.3
E26	Medium Screw (Med)	GX7.9
E39	Mogul Screw (Mog)	——
BA15s	Single-Contact Bayonet Candelabra (SC Bay)	R7s
BA15d	Double-Contact Bayonet Candelabra (DC Bay)	GY9.5
		G9.5
P40s	Mogul Prefocus (Mog Pf)	GX9.5
P28s	Medium Prefocus (Med Pf)	GY16
G22	Medium Bipost (Med Bp)	(MSP)
G38	Mogul Bipost (Mog Bp)	(EMEP)
		(MEP)

(Mc) Mini-Can Screw (Mc)
(Med Skt) Medium Skirted (Med Skt)
E26 Medium Screw (Med)
E39 Mogul Screw (Mog)
BA15s Single-Contact Bayonet Candelabra (SC Bay)
BA15d Double-Contact Bayonet Candelabra (DC Bay)
P40s Mogul Prefocus (Mog Pf)
P28s Medium Prefocus (Med Pf)
G22 Medium Bipost (Med Bp)
G38 Mogul Bipost (Mog Bp)

GX5.3 2-Pin (MR-16 lamp)
GY5.3 Oval 2-Pin (MR-16 120 V. lamp)
GX7.9 (MARC 300 lamp)
—— special 2-pin plug (MARC 350 lamp)
R7s Recessed Single-Contact
GY9.5 (no common name)
G9.5 Medium 2-Pin (Med 2P)
GX9.5 (no common name)
GY16 (no common name)
(MSP) Medium Side Prong (MSP)
(EMEP) Extended Mogul End Prong (EMEP)
(MEP) Mogul End Prong (MEP)

Illustration by George Chiang (except those marked with asterisk which are by Robert Mumm)
From BACKSTAGE HANDBOOK by Paul Carter
Reprinted with permission.

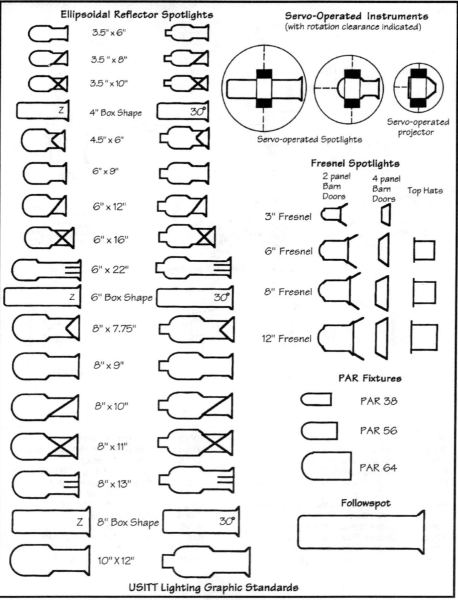

Ellipsoidal Reflector Spotlights

3.5" x 6"
3.5" x 8"
3.5" x 10"
4" Box Shape
4.5" x 6"
6" x 9"
6" x 12"
6" x 16"
6" x 22"
6" Box Shape
8" x 7.75"
8" x 9"
8" x 10"
8" x 11"
8" x 13"
8" Box Shape
10" X 12"

Servo-Operated Instruments
(with rotation clearance indicated)

Servo-operated Spotlights

Servo-operated projector

Fresnel Spotlights

2 panel Barn Doors | 4 panel Barn Doors | Top Hats

3" Fresnel
6" Fresnel
8" Fresnel
12" Fresnel

PAR Fixtures

PAR 38
PAR 56
PAR 64

Followspot

USITT Lighting Graphic Standards

Reprinted with permission. © USITT (United States Institute for Theatre Technology, Inc.)

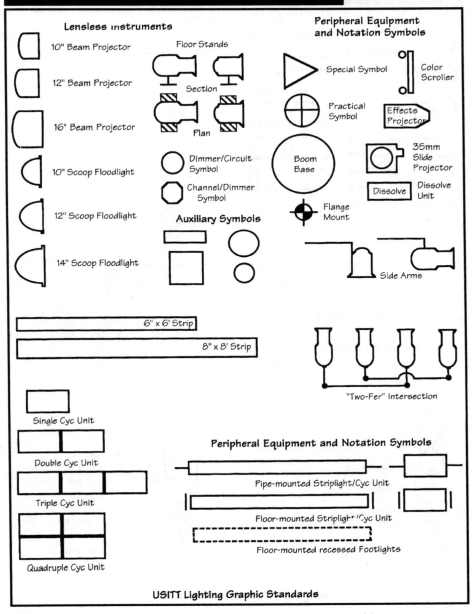

Lensless instruments

10" Beam Projector

12" Beam Projector

16" Beam Projector

10" Scoop Floodlight

12" Scoop Floodlight

14" Scoop Floodlight

Floor Stands

Section

Plan

Dimmer/Circuit Symbol

Channel/Dimmer Symbol

Auxiliary Symbols

Peripheral Equipment and Notation Symbols

Special Symbol

Practical Symbol

Boom Base

Flange Mount

Color Scroller

Effects Projector

35mm Slide Projector

Dissolve   Dissolve Unit

Side Arms

6" x 6' Strip

8" x 8' Strip

Single Cyc Unit

Double Cyc Unit

Triple Cyc Unit

Quadruple Cyc Unit

"Two-Fer" Intersection

Peripheral Equipment and Notation Symbols

Pipe-mounted Striplight/Cyc Unit

Floor-mounted Striplight/Cyc Unit

Floor-mounted recessed Footlights

USITT Lighting Graphic Standards

Instrument Notation

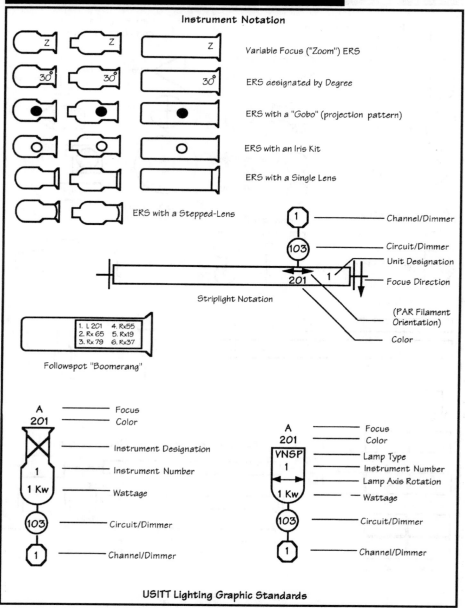

Variable Focus ("Zoom") ERS

ERS aesignated by Degree

ERS with a "Gobo" (projection pattern)

ERS with an Iris Kit

ERS with a Single Lens

ERS with a Stepped-Lens

Channel/Dimmer

Circuit/Dimmer

Unit Designation

Focus Direction

(PAR Filament Orientation)

Color

Striplight Notation

1. L 201	4. Rx55
2. Rx 65	5. Rx19
3. Rx 79	6. Rx37

Followspot "Boomerang"

A
201
Focus
Color

Instrument Designation

Instrument Number

1 Kw — Wattage

103 — Circuit/Dimmer

1 — Channel/Dimmer

A
201
Focus
Color

VNSP
1
Lamp Type
Instrument Number
Lamp Axis Rotation

1 Kw — Wattage

103 — Circuit/Dimmer

1 — Channel/Dimmer

USITT Lighting Graphic Standards

Throughout this book you will find multiplying factors *(mf)* asssociated with the beam diameter data presented as part of the photometric charts. Multiplying factors are simply numbers you can use to calculate the diameter of a given beam of light at *any* distance. If you don't happen to know the multiplying factor for an instrument but you do know its field angle (or beam angle for that matter) the table below lets you look up the multiplying factor associated with that particular angle.

For example: Let's say you have a 10" fresnel and your catalog says it has an 18° field angle in spot focus. To determine the diameter of that 18° beam of light at say 48 feet throw distance, simply look up the multiplying factor for 18° (.3168) and multiply it times 48 feet (.3168 x 48 = 15.2064). If on the other hand you only know the beam angle for a given instrument, remember, the multiplying factor will give you the diameter of that beam angle or "hot spot."

A word of caution: beam angle/field angle information as published by various manufacturers isn't always what you expect. Some manufacturers are very precise about measuring this kind of performance data, some are not. Recently manufacturers have begun using a flat field or cosine distribution lamp focus when taking their photometric measurements. Flat field focus is often more representative of real life situations than setting the lamp at peak focus, however changing the lamp focus also changes the beam angle characteristics of the instrument. So, don't be surprised if calculated beam diameters which work so nicely on your plot, don't work so well when you get around to hanging and focusing your lights.

Angle	mf.	Angle	mf.	Angle	mf.
1°	0.0175	37°	0.6692	73°	1.4799
2°	0.0349	38°	0.6887	74°	1.5071
3°	0.0524	39°	0.7082	75°	1.5347
4°	0.0698	40°	0.7279	76°	1.5626
5°	0.0873	41°	0.7478	77°	1.5909
6°	0.1048	42°	0.7677	78°	1.6196
7°	0.1223	43°	0.7878	79°	1.6487
8°	0.1399	44°	0.8081	80°	1.6782
9°	0.1574	45°	0.8284	81°	1.7082
10°	0.1750	46°	0.8489	82°	1.7386
11°	0.1926	47°	0.8696	83°	1.7695
12°	0.2102	48°	0.8905	84°	1.8008
13°	0.2279	49°	0.915	85°	1.8327
14°	0.2456	50°	0.9326	86°	1.8650
15°	0.2633	51°	0.9540	87°	1.8979
16°	0.2811	52°	0.9755	88°	1.9314
17°	0.2989	53°	0.9972	89°	1.9654
18°	0.3168	54°	1.0191	90°	2.0000
19°	0.3347	55°	1.0411	91°	2.0352
20°	0.3527	56°	1.0634	92°	2.0711
21°	0.3707	57°	1.0859	93°	2.1076
22°	0.3888	58°	1.1086	94°	2.1447
23°	0.4069	59°	1.1315	95°	2.1826
24°	0.4251	60°	1.1547	96°	2.2212
25°	0.4434	61°	1.1781	97°	2.2606
26°	0.4617	62°	1.2017	98°	2.3007
27°	0.4802	63°	1.2256	99°	2.3417
28°	0.4987	64°	1.2497	100°	2.3835
29°	0.5172	65°	1.2741	105°	2.6065
30°	0.5359	66°	1.2988	110	2.8563
31°	0.5546	67°	1.3238	115°	3.1394
32°	0.5735	68°	1.3490	120°	3.4641
33°	0.5924	69°	1.3746	125°	3.8420
34°	0.6115	70°	1.4004	130°	4.2890
35°	0.6306	71°	1.4266		
36°	0.6498	72°	1.4531		

Throw Distance Lookup Table

This table enables you to look up the actual *throw distance at head height* for any lighting positions if you know the vertical and horizontal measurements of the position from the point on stage which you want to light.

Horizontal distance (in feet) from hanging position to point on stage.

Vertical distance (in feet) from stage floor to hanging position.

	1	2	3	4	5	6	7	8	9	10	11	12	13	14	15
10	4.6	4.9	5.4	6.0	6.7	7.5	8.3	9.2	10.1	11.0	11.9	12.8	13.8	14.7	15.7
11	5.6	5.9	6.3	6.8	7.4	8.1	8.9	9.7	10.6	11.4	12.3	13.2	14.1	15.0	16.0
12	6.6	6.8	7.2	7.6	8.2	8.9	9.6	10.3	11.1	11.9	12.8	13.7	14.5	15.4	16.4
13	7.6	7.8	8.1	8.5	9.0	9.6	10.3	11.0	11.7	12.5	13.3	14.2	15.0	15.9	16.8
14	8.6	8.7	9.0	9.4	9.9	10.4	11.0	11.7	12.4	13.1	13.9	14.7	15.5	16.4	17.2
15	9.6	9.7	10.0	10.3	10.7	11.2	11.8	12.4	13.1	13.8	14.5	15.3	16.1	16.9	17.8
16	10.6	10.7	10.9	11.2	11.6	12.1	12.6	13.2	13.8	14.5	15.2	16.0	16.7	17.5	18.3
17	11.5	11.7	11.9	12.2	12.5	13.0	13.5	14.0	14.6	15.2	15.9	16.6	17.4	18.1	18.9
18	12.5	12.7	12.9	13.1	13.5	13.9	14.3	14.8	15.4	16.0	16.7	17.3	18.0	18.8	19.5
19	13.5	13.7	13.8	14.1	14.4	14.8	15.2	15.7	16.2	16.8	17.4	18.1	18.7	19.5	20.2
20	14.5	14.6	14.8	15.0	15.3	15.7	16.1	16.6	17.1	17.6	18.2	18.8	19.5	20.2	20.9
21	15.5	15.6	15.8	16.0	16.3	16.6	17.0	17.4	17.9	18.5	19.0	19.6	20.2	20.9	21.6
22	16.5	16.6	16.8	17.0	17.2	17.6	17.9	18.3	18.8	19.3	19.8	20.4	21.0	21.6	22.3
23	17.5	17.6	17.8	18.0	18.2	18.5	18.9	19.2	19.7	20.2	20.7	21.2	21.8	22.4	23.1
24	18.5	18.6	18.7	18.9	19.2	19.5	19.8	20.2	20.6	21.0	21.5	22.1	22.6	23.2	23.8
25	19.5	19.6	19.7	19.9	20.1	20.4	20.7	21.1	21.5	21.9	22.4	22.9	23.4	24.0	24.6
26	20.5	20.6	20.7	20.9	21.1	21.4	21.7	22.0	22.4	22.8	23.3	23.8	24.3	24.8	25.4
27	21.5	21.6	21.7	21.9	22.1	22.3	22.6	22.9	23.3	23.7	24.2	24.6	25.1	25.7	26.2
28	22.5	22.6	22.7	22.9	23.1	23.3	23.6	23.9	24.2	24.6	25.0	25.5	26.0	26.5	27.0
29	23.5	23.6	23.7	23.8	24.0	24.3	24.5	24.8	25.2	25.5	26.0	26.4	26.9	27.4	27.9
30	24.5	24.6	24.7	24.8	25.0	25.2	25.5	25.8	26.1	26.5	26.9	27.3	27.7	28.2	28.7
31	25.5	25.6	25.7	25.8	26.0	26.2	26.4	26.7	27.0	27.4	27.8	28.2	28.6	29.1	29.6
32	26.5	26.6	26.7	26.8	27.0	27.2	27.4	27.7	28.0	28.3	28.7	29.1	29.5	30.0	30.5
33	27.5	27.6	27.7	27.8	28.0	28.2	28.4	28.6	28.9	29.3	29.6	30.0	30.4	30.9	31.3
34	28.5	28.6	28.7	28.8	28.9	29.1	29.4	29.6	29.9	30.2	30.6	30.9	31.3	31.8	32.2
35	29.5	29.6	29.7	29.8	29.9	30.1	30.3	30.6	30.8	31.2	31.5	31.9	32.2	32.7	33.1
36	30.5	30.6	30.7	30.8	30.9	31.1	31.3	31.5	31.8	32.1	32.4	32.8	33.2	33.6	34.0
37	31.5	31.6	31.6	31.8	31.9	32.1	32.3	32.5	32.8	33.1	33.4	33.7	34.1	34.5	34.9
38	32.5	32.6	32.6	32.8	32.9	33.1	33.3	33.5	33.7	34.0	34.3	34.6	35.0	35.4	35.8
39	33.5	33.6	33.6	33.7	33.9	34.0	34.2	34.4	34.7	35.0	35.3	35.6	35.9	36.3	36.7
40	34.5	34.6	34.6	34.7	34.9	35.0	35.2	35.4	35.7	35.9	36.2	36.5	36.9	37.2	37.6
42	36.5	36.6	36.6	36.7	36.8	37.0	37.2	37.4	37.6	37.9	38.1	38.4	38.8	39.1	39.5
44	38.5	38.6	38.6	38.7	38.8	39.0	39.1	39.3	39.5	39.8	40.0	40.3	40.6	41.0	41.3
46	40.5	40.6	40.6	40.7	40.8	40.9	41.1	41.3	41.5	41.7	42.0	42.2	42.5	42.9	43.2
48	42.5	42.6	42.6	42.7	42.8	42.9	43.1	43.3	43.4	43.7	43.9	44.2	44.4	44.8	45.1
50	44.5	44.5	44.6	44.7	44.8	44.9	45.1	45.2	45.4	45.6	45.8	46.1	46.4	46.7	47.0
52	46.5	46.5	46.6	46.7	46.8	46.9	47.0	47.2	47.4	47.6	47.8	48.0	48.3	48.6	48.9
54	48.5	48.5	48.6	48.7	48.8	48.9	49.0	49.2	49.3	49.5	49.7	50.0	50.2	50.5	50.8
56	50.5	50.5	50.6	50.7	50.8	50.9	51.0	51.1	51.3	51.5	51.7	51.9	52.2	52.4	52.7
58	52.5	52.5	52.6	52.7	52.7	52.8	53.0	53.1	53.3	53.4	53.6	53.9	54.1	54.3	54.6
60	54.5	54.5	54.6	54.7	54.7	54.8	55.0	55.1	55.2	55.4	55.6	55.8	56.0	56.3	56.5
62	56.5	56.5	56.6	56.6	56.7	56.8	56.9	57.1	57.2	57.4	57.6	57.8	58.0	58.2	58.5
64	58.5	58.5	58.6	58.6	58.7	58.8	58.9	59.0	59.2	59.4	59.5	59.7	59.9	60.2	60.4
66	60.5	60.5	60.6	60.6	60.7	60.8	60.9	61.0	61.2	61.3	61.5	61.7	61.9	62.1	62.3
68	62.5	62.5	62.6	62.6	62.7	62.8	62.9	63.0	63.1	63.3	63.5	63.6	63.8	64.1	64.3
70	64.5	64.5	64.6	64.6	64.7	64.8	64.9	65.0	65.1	65.3	65.4	65.6	65.8	66.0	66.2
72	66.5	66.5	66.6	66.6	66.7	66.8	66.9	67.0	67.1	67.3	67.4	67.6	67.8	68.0	68.2
74	68.5	68.5	68.6	68.6	68.7	68.8	68.9	69.0	69.1	69.2	69.4	69.5	69.7	69.9	70.1
76	70.5	70.5	70.6	70.6	70.7	70.8	70.9	71.0	71.1	71.2	71.4	71.5	71.7	71.9	72.1
78	72.5	72.5	72.6	72.6	72.7	72.8	72.8	72.9	73.1	73.2	73.3	73.5	73.7	73.8	74.0
80	74.5	74.5	74.6	74.6	74.7	74.7	74.8	74.9	75.0	75.2	75.3	75.5	75.6	75.8	76.0
82	76.5	76.5	76.6	76.6	76.7	76.7	76.8	76.9	77.0	77.2	77.3	77.4	77.6	77.8	78.0
84	78.5	78.5	78.6	78.6	78.7	78.7	78.8	78.9	79.0	79.1	79.3	79.4	79.6	79.7	79.9
86	80.5	80.5	80.6	80.6	80.7	80.7	80.8	80.9	81.0	81.1	81.3	81.4	81.5	81.7	81.9
88	82.5	82.5	82.6	82.6	82.7	82.7	82.8	82.9	83.0	83.1	83.2	83.4	83.5	83.7	83.9
90	84.5	84.5	84.6	84.6	84.7	84.7	84.8	84.9	85.0	85.1	85.2	85.4	85.5	85.7	85.8

614

Horizontal distance (in feet) from hanging position to point on stage.

	16	17	18	19	20	21	22	23	24	25	26	27	28	29	30
10	16.6	17.6	18.6	19.5	20.5	21.5	22.5	23.4	24.4	25.4	26.4	27.4	28.4	29.4	30.3
11	16.9	17.9	18.8	19.8	20.7	21.7	22.7	23.7	24.6	25.6	26.6	27.6	28.5	29.5	30.5
12	17.3	18.2	19.1	20.1	21.0	22.0	22.9	23.9	24.9	25.8	26.8	27.8	28.7	29.7	30.7
13	17.7	18.6	19.5	20.4	21.4	22.3	23.2	24.2	25.1	26.1	27.1	28.0	29.0	30.0	30.9
14	18.1	19.0	19.9	20.8	21.7	22.7	23.6	24.5	25.5	26.4	27.4	28.3	29.3	30.2	31.2
15	18.6	19.5	20.4	21.2	22.1	23.1	24.0	24.9	25.8	26.7	27.7	28.6	29.6	30.5	31.5
16	19.1	20.0	20.8	21.7	22.6	23.5	24.4	25.3	26.2	27.1	28.0	29.0	29.9	30.8	31.8
17	19.7	20.5	21.4	22.2	23.1	23.9	24.8	25.7	26.6	27.5	28.4	29.4	30.3	31.2	32.1
18	20.3	21.1	21.9	22.7	23.6	24.4	25.3	26.2	27.1	28.0	28.9	29.8	30.7	31.6	32.5
19	20.9	21.7	22.5	23.3	24.1	25.0	25.8	26.7	27.5	28.4	29.3	30.2	31.1	32.0	32.9
20	21.6	22.3	23.1	23.9	24.7	25.5	26.4	27.2	28.0	28.9	29.8	30.7	31.5	32.4	33.3
21	22.3	23.0	23.8	24.5	25.3	26.1	26.9	27.7	28.6	29.4	30.3	31.1	32.0	32.9	33.8
22	23.0	23.7	24.4	25.2	25.9	26.7	27.5	28.3	29.1	30.0	30.8	31.6	32.5	33.4	34.2
23	23.7	24.4	25.1	25.8	26.6	27.3	28.1	28.9	29.7	30.5	31.3	32.2	33.0	33.9	34.7
24	24.5	25.1	25.8	26.5	27.2	28.0	28.7	29.5	30.3	31.1	31.9	32.7	33.6	34.4	35.3
25	25.2	25.9	26.5	27.2	27.9	28.7	29.4	30.2	30.9	31.7	32.5	33.3	34.1	35.0	35.8
26	26.0	26.6	27.3	28.0	28.6	29.4	30.1	30.8	31.6	32.3	33.1	33.9	34.7	35.5	36.3
27	26.8	27.4	28.1	28.7	29.4	30.1	30.8	31.5	32.2	33.0	33.7	34.5	35.3	36.1	36.9
28	27.6	28.2	28.8	29.5	30.1	30.8	31.5	32.2	32.9	33.6	34.4	35.2	35.9	36.7	37.5
29	28.4	29.0	29.6	30.2	30.9	31.5	32.2	32.9	33.6	34.3	35.1	35.8	36.6	37.3	38.1
30	29.3	29.8	30.4	31.0	31.6	32.3	32.9	33.6	34.3	35.0	35.7	36.5	37.2	38.0	38.7
31	30.1	30.7	31.2	31.8	32.4	33.0	33.7	34.3	35.0	35.7	36.4	37.1	37.9	38.6	39.4
32	31.0	31.5	32.0	32.6	33.2	33.8	34.4	35.1	35.8	36.4	37.1	37.8	38.6	39.3	40.0
33	31.8	32.3	32.9	33.4	34.0	34.6	35.2	35.9	36.5	37.2	37.9	38.5	39.3	40.0	40.7
34	32.7	33.2	33.7	34.3	34.8	35.4	36.0	36.6	37.3	37.9	38.6	39.3	40.0	40.7	41.4
35	33.6	34.1	34.6	35.1	35.6	36.2	36.8	37.4	38.0	38.7	39.3	40.0	40.7	41.4	42.1
36	34.4	34.9	35.4	35.9	36.5	37.0	37.6	38.2	38.8	39.4	40.1	40.7	41.4	42.1	42.8
37	35.3	35.8	36.3	36.8	37.3	37.9	38.4	39.0	39.6	40.2	40.8	41.5	42.2	42.8	43.5
38	36.2	36.7	37.2	37.7	38.2	38.7	39.3	39.8	40.4	41.0	41.6	42.3	42.9	43.6	44.2
39	37.1	37.6	38.0	38.5	39.0	39.5	40.1	40.6	41.2	41.8	42.4	43.0	43.7	44.3	45.0
40	38.0	38.5	38.9	39.4	39.9	40.4	40.9	41.5	42.0	42.6	43.2	43.8	44.4	45.1	45.7
42	39.9	40.3	40.7	41.2	41.6	42.1	42.6	43.1	43.7	44.2	44.8	45.4	46.0	46.6	47.3
44	41.7	42.1	42.5	42.9	43.4	43.9	44.3	44.9	45.4	45.9	46.5	47.0	47.6	48.2	48.8
46	43.6	43.9	44.3	44.7	45.2	45.6	46.1	46.6	47.1	47.6	48.1	48.7	49.2	49.8	50.4
48	45.4	45.8	46.2	46.6	47.0	47.4	47.9	48.3	48.8	49.3	49.8	50.4	50.9	51.5	52.0
50	47.3	47.6	48.0	48.4	48.8	49.2	49.6	50.1	50.6	51.0	51.5	52.1	52.6	53.1	53.7
52	49.2	49.5	49.9	50.2	50.6	51.0	51.4	51.9	52.3	52.8	53.3	53.8	54.3	54.8	55.3
54	51.1	51.4	51.7	52.1	52.5	52.9	53.3	53.7	54.1	54.6	55.0	55.4	56.0	56.5	57.0
56	53.0	53.3	53.6	54.0	54.3	54.7	55.1	55.5	55.9	56.3	56.8	57.3	57.7	58.2	58.7
58	54.9	55.2	55.5	55.8	56.2	56.5	56.9	57.3	57.7	58.1	58.6	59.0	59.5	60.0	60.5
60	56.8	57.1	57.4	57.7	58.1	58.4	58.8	59.2	59.6	60.0	60.4	60.8	61.3	61.7	62.2
62	58.7	59.0	59.3	59.6	59.9	60.3	60.6	61.0	61.4	61.8	62.2	62.6	63.1	63.5	64.0
64	60.7	60.9	61.2	61.5	61.8	62.2	62.5	62.9	63.2	63.6	64.0	64.4	64.9	65.3	65.7
66	62.6	62.8	63.1	63.4	63.7	64.0	64.4	64.7	65.1	65.5	65.9	66.3	66.7	67.1	67.5
68	64.5	64.8	65.0	65.3	65.6	65.9	66.3	66.6	67.0	67.3	67.7	68.1	68.5	68.9	69.3
70	66.5	66.7	67.0	67.2	67.5	67.8	68.2	68.5	68.8	69.2	69.5	69.9	70.3	70.7	71.1
72	68.4	68.6	68.9	69.2	69.4	69.7	70.0	70.4	70.7	71.0	71.4	71.8	72.2	72.5	73.0
74	70.3	70.6	70.8	71.1	71.4	71.7	72.0	72.3	72.6	72.9	73.3	73.6	74.0	74.4	74.8
76	72.3	72.5	72.8	73.0	73.3	73.6	73.9	74.2	74.5	74.8	75.1	75.5	75.9	76.2	76.6
78	74.2	74.5	74.7	75.0	75.2	75.5	75.8	76.1	76.4	76.7	77.0	77.4	77.7	78.1	78.5
80	76.2	76.4	76.6	76.9	77.1	77.4	77.7	78.0	78.3	78.6	78.9	79.2	79.6	79.9	80.3
82	78.2	78.4	78.6	78.8	79.1	79.3	79.6	79.9	80.2	80.5	80.8	81.1	81.5	81.8	82.2
84	80.1	80.3	80.5	80.8	81.0	81.3	81.5	81.8	82.1	82.4	82.7	83.0	83.3	83.7	84.0
86	82.1	82.3	82.5	82.7	83.0	83.2	83.5	83.7	84.0	84.3	84.6	84.9	85.2	85.6	85.9
88	84.0	84.2	84.4	84.7	84.9	85.1	85.4	85.6	85.9	86.2	86.5	86.8	87.1	87.5	87.8
90	86.0	86.2	86.4	86.6	86.8	87.1	87.3	87.6	87.8	88.1	88.4	88.7	89.0	89.3	89.7

Vertical distance (in feet) from stage floor to hanging position.

Horizontal distance (in feet) from hanging position to point on stage.

	31	32	33	34	35	36	37	38	39	40	41	42	43	44	45
10	31.3	32.3	33.3	34.3	35.3	36.3	37.3	38.3	39.3	40.3	41.3	42.2	43.2	44.2	45.2
11	31.5	32.5	33.5	34.4	35.4	36.4	37.4	38.4	39.4	40.4	41.4	42.4	43.4	44.3	45.3
12	31.7	32.7	33.6	34.6	35.6	36.6	37.6	38.6	39.5	40.5	41.5	42.5	43.5	44.5	45.5
13	31.9	32.9	33.8	34.8	35.8	36.8	37.8	38.7	39.7	40.7	41.7	42.7	43.7	44.6	45.6
14	32.1	33.1	34.1	35.1	36.0	37.0	38.0	38.9	39.9	40.9	41.9	42.9	43.8	44.8	45.8
15	32.4	33.4	34.3	35.3	36.3	37.2	38.2	39.2	40.1	41.1	42.1	43.1	44.0	45.0	46.0
16	32.7	33.7	34.6	35.6	36.5	37.5	38.5	39.4	40.4	41.4	42.3	43.3	44.3	45.2	46.2
17	33.1	34.0	35.0	35.9	36.8	37.8	38.8	39.7	40.7	41.6	42.6	43.6	44.5	45.5	46.5
18	33.4	34.4	35.3	36.2	37.2	38.1	39.1	40.0	41.0	41.9	42.9	43.8	44.8	45.7	46.7
19	33.8	34.7	35.7	36.6	37.5	38.5	39.4	40.3	41.3	42.2	43.2	44.1	45.1	46.0	47.0
20	34.2	35.1	36.1	37.0	37.9	38.8	39.7	40.7	41.6	42.6	43.5	44.4	45.4	46.3	47.3
21	34.7	35.6	36.5	37.4	38.3	39.2	40.1	41.0	42.0	42.9	43.8	44.8	45.7	46.7	47.6
22	35.1	36.0	36.9	37.8	38.7	39.6	40.5	41.4	42.4	43.3	44.2	45.1	46.1	47.0	47.9
23	35.6	36.5	37.4	38.2	39.1	40.0	40.9	41.8	42.8	43.7	44.6	45.5	46.4	47.4	48.3
24	36.1	37.0	37.8	38.7	39.6	40.5	41.4	42.3	43.2	44.1	45.0	45.9	46.8	47.7	48.7
25	36.6	37.5	38.3	39.2	40.1	40.9	41.8	42.7	43.6	44.5	45.4	46.3	47.2	48.1	49.0
26	37.2	38.0	38.9	39.7	40.6	41.4	42.3	43.2	44.1	45.0	45.8	46.7	47.6	48.5	49.5
27	37.7	38.6	39.4	40.2	41.1	41.9	42.8	43.7	44.5	45.4	46.3	47.2	48.1	49.0	49.9
28	38.3	39.1	39.9	40.8	41.6	42.5	43.3	44.2	45.0	45.9	46.8	47.7	48.5	49.4	50.3
29	38.9	39.7	40.5	41.3	42.2	43.0	43.8	44.7	45.5	46.4	47.3	48.1	49.0	49.9	50.8
30	39.5	40.3	41.1	41.9	42.7	43.6	44.4	45.2	46.1	46.9	47.8	48.6	49.5	50.4	51.2
31	40.1	40.9	41.7	42.5	43.3	44.1	44.9	45.8	46.6	47.4	48.3	49.1	50.0	50.9	51.7
32	40.8	41.6	42.3	43.1	43.9	44.7	45.5	46.3	47.2	48.0	48.8	49.7	50.5	51.4	52.2
33	41.4	42.2	43.0	43.7	44.5	45.3	46.1	46.9	47.7	48.5	49.4	50.2	51.0	51.9	52.7
34	42.1	42.9	43.6	44.4	45.1	45.9	46.7	47.5	48.3	49.1	49.9	50.8	51.6	52.4	53.3
35	42.8	43.5	44.3	45.0	45.8	46.5	47.3	48.1	48.9	49.7	50.5	51.3	52.2	53.0	53.8
36	43.5	44.2	44.9	45.7	46.4	47.2	48.0	48.7	49.5	50.3	51.1	51.9	52.7	53.5	54.4
37	44.2	44.9	45.6	46.4	47.1	47.8	48.6	49.4	50.1	50.9	51.7	52.5	53.3	54.1	54.9
38	44.9	45.6	46.3	47.0	47.8	48.5	49.3	50.0	50.8	51.5	52.3	53.1	53.9	54.7	55.5
39	45.6	46.3	47.0	47.7	48.5	49.2	49.9	50.7	51.4	52.2	53.0	53.7	54.5	55.3	56.1
40	46.4	47.1	47.7	48.4	49.2	49.9	50.6	51.3	52.1	52.8	53.6	54.4	55.1	55.9	56.7
42	47.9	48.5	49.2	49.9	50.6	51.3	52.0	52.7	53.4	54.2	54.9	55.6	56.4	57.2	57.9
44	49.4	50.1	50.7	51.4	52.0	52.7	53.4	54.1	54.8	55.5	56.2	57.0	57.7	58.5	59.2
46	51.0	51.6	52.2	52.9	53.5	54.2	54.9	55.5	56.2	56.9	57.6	58.4	59.1	59.8	60.5
48	52.6	53.2	53.8	54.4	55.1	55.7	56.4	57.0	57.7	58.4	59.1	59.8	60.5	61.2	61.9
50	54.2	54.8	55.4	56.0	56.6	57.2	57.9	58.5	59.2	59.8	60.5	61.2	61.9	62.6	63.2
52	55.9	56.4	57.0	57.6	58.2	58.8	59.4	60.1	60.7	61.3	62.0	62.7	63.3	64.0	64.7
54	57.6	58.1	58.7	59.2	59.8	60.4	61.0	61.6	62.2	62.9	63.5	64.2	64.8	65.5	66.2
56	59.3	59.8	60.3	60.9	61.4	62.0	62.6	63.2	63.8	64.4	65.0	65.7	66.3	67.0	67.6
58	61.0	61.5	62.0	62.5	63.1	63.7	64.2	64.8	65.4	66.0	66.6	67.2	67.9	68.5	69.2
60	62.7	63.2	63.7	64.2	64.8	65.3	65.9	66.4	67.0	67.6	68.2	68.8	69.4	70.0	70.7
62	64.4	64.9	65.4	65.9	66.5	67.0	67.5	68.1	68.7	69.2	69.8	70.4	71.0	71.6	72.2
64	66.2	66.7	67.2	67.7	68.2	68.7	69.2	69.8	70.3	70.9	71.4	72.0	72.6	73.2	73.8
66	68.0	68.4	68.9	69.4	69.9	70.4	70.9	71.4	72.0	72.5	73.1	73.6	74.2	74.8	75.4
68	69.8	70.2	70.7	71.1	71.6	72.1	72.6	73.1	73.7	74.2	74.7	75.3	75.9	76.4	77.0
70	71.6	72.0	72.5	72.9	73.4	73.9	74.4	74.9	75.4	75.9	76.4	77.0	77.5	78.1	78.6
72	73.4	73.8	74.2	74.7	75.1	75.6	76.1	76.6	77.1	77.6	78.1	78.7	79.2	79.7	80.3
74	75.2	75.6	76.0	76.5	76.9	77.4	77.9	78.3	78.8	79.3	79.8	80.4	80.9	81.4	82.0
76	77.0	77.4	77.8	78.3	78.7	79.2	79.6	80.1	80.6	81.1	81.6	82.1	82.6	83.1	83.6
78	78.8	79.2	79.7	80.1	80.5	80.9	81.4	81.9	82.3	82.8	83.3	83.8	84.3	84.8	85.3
80	80.7	81.1	81.5	81.9	82.3	82.7	83.2	83.6	84.1	84.6	85.0	85.5	86.0	86.5	87.0
82	82.5	82.9	83.3	83.7	84.1	84.5	85.0	85.4	85.9	86.3	86.8	87.3	87.8	88.3	88.8
84	84.4	84.8	85.2	85.5	85.9	86.4	86.8	87.2	87.7	88.1	88.6	89.0	89.5	90.0	90.5
86	86.3	86.6	87.0	87.4	87.8	88.2	88.6	89.0	89.4	89.9	90.3	90.8	91.3	91.7	92.2
88	88.1	88.5	88.9	89.2	89.6	90.0	90.4	90.8	91.3	91.7	92.1	92.6	93.0	93.5	94.0
90	90.0	90.4	90.7	91.1	91.5	91.8	92.2	92.7	93.1	93.5	93.9	94.4	94.8	95.3	95.7

Vertical distance (in feet) from stage floor to hanging position.

Horizontal distance (in feet) from hanging position to point on stage.

Vertical distance (in feet) from stage floor to hanging position.

	46	47	48	49	50	51	52	53	57	55	56	57	58	59	60
10	46.2	47.2	48.2	49.2	50.2	51.2	52.2	53.2	54.2	55.2	56.2	57.2	58.2	59.2	60.2
11	46.3	47.3	48.3	49.3	50.3	51.3	52.3	53.3	54.3	55.3	56.3	57.3	58.3	59.3	60.3
12	46.5	47.5	48.4	49.4	50.4	51.4	52.4	53.4	54.4	55.4	56.4	57.4	58.4	59.4	60.4
13	46.6	47.6	48.6	49.6	50.6	51.6	52.5	53.5	54.5	55.5	56.5	57.5	58.5	59.5	60.5
14	46.8	47.8	48.8	49.7	50.7	51.7	52.7	53.7	54.7	55.7	56.6	57.6	58.6	59.6	60.6
15	47.0	48.0	48.9	49.9	50.9	51.9	52.9	53.8	54.8	55.8	56.8	57.8	58.8	59.8	60.8
16	47.2	48.2	49.1	50.1	51.1	52.1	53.1	54.0	55.0	56.0	57.0	58.0	58.9	59.9	60.9
17	47.4	48.4	49.4	50.3	51.3	52.3	53.3	54.2	55.2	56.2	57.2	58.2	59.1	60.1	61.1
18	47.7	48.6	49.6	50.6	51.5	52.5	53.5	54.5	55.4	56.4	57.4	58.4	59.3	60.3	61.3
19	47.9	48.9	49.9	50.8	51.8	52.8	53.7	54.7	55.7	56.6	57.6	58.6	59.6	60.5	61.5
20	48.2	49.2	50.1	51.1	52.1	53.0	54.0	55.0	55.9	56.9	57.9	58.8	59.8	60.8	61.7
21	48.5	49.5	50.4	51.4	52.4	53.3	54.3	55.2	56.2	57.1	58.1	59.1	60.0	61.0	62.0
22	48.9	49.8	50.8	51.7	52.7	53.6	54.6	55.5	56.5	57.4	58.4	59.3	60.3	61.3	62.2
23	49.2	50.2	51.1	52.0	53.0	53.9	54.9	55.8	56.8	57.7	58.7	59.6	60.6	61.5	62.5
24	49.6	50.5	51.4	52.4	53.3	54.3	55.2	56.1	57.1	58.0	59.0	59.9	60.9	61.8	62.8
25	50.0	50.9	51.8	52.7	53.7	54.6	55.5	56.5	57.4	58.4	59.3	60.2	61.2	62.1	63.1
26	50.4	51.3	52.2	53.1	54.0	55.0	55.9	56.8	57.8	58.7	59.6	60.6	61.5	62.5	63.4
27	50.8	51.7	52.6	53.5	54.4	55.4	56.3	57.2	58.1	59.1	60.0	60.9	61.9	62.8	63.7
28	51.2	52.1	53.0	53.9	54.8	55.7	56.7	57.6	58.5	59.4	60.4	61.3	62.2	63.1	64.1
29	51.7	52.6	53.4	54.3	55.3	56.2	57.1	58.0	58.9	59.8	60.7	61.7	62.6	63.5	64.4
30	52.1	53.0	53.9	54.8	55.7	56.6	57.5	58.4	59.3	60.2	61.1	62.0	63.0	63.9	64.8
31	52.6	53.5	54.4	55.2	56.1	57.0	57.9	58.8	59.7	60.6	61.5	62.4	63.4	64.3	65.2
32	53.1	54.0	54.8	55.7	56.6	57.5	58.4	59.3	60.2	61.1	62.0	62.9	63.8	64.7	65.6
33	53.6	54.5	55.3	56.2	57.1	57.9	58.8	59.7	60.6	61.5	62.4	63.3	64.2	65.1	66.0
34	54.1	55.0	55.8	56.7	57.6	58.4	59.3	60.2	61.1	62.0	62.8	63.7	64.6	65.5	66.4
35	54.7	55.5	56.3	57.2	58.1	58.9	59.8	60.7	61.5	62.4	63.3	64.2	65.1	66.0	66.9
36	55.2	56.0	56.9	57.7	58.6	59.4	60.3	61.2	62.0	62.9	63.8	64.7	65.5	66.4	67.3
37	55.8	56.6	57.4	58.3	59.1	59.9	60.8	61.7	62.5	63.4	64.3	65.1	66.0	66.9	67.8
38	56.3	57.1	58.0	58.8	59.6	60.5	61.3	62.2	63.0	63.9	64.8	65.6	66.5	67.4	68.2
39	56.9	57.7	58.5	59.4	60.2	61.0	61.9	62.7	63.6	64.4	65.3	66.1	67.0	67.9	68.7
40	57.5	58.3	59.1	59.9	60.8	61.6	62.4	63.2	64.1	64.9	65.8	66.6	67.5	68.4	69.2
42	58.7	59.5	60.3	61.1	61.9	62.7	63.5	64.4	65.2	66.0	66.8	67.7	68.5	69.4	70.2
44	60.0	60.8	61.5	62.3	63.1	63.9	64.7	65.5	66.3	67.1	68.0	68.8	69.6	70.5	71.3
46	61.3	62.0	62.8	63.6	64.3	65.1	65.9	66.7	67.5	68.3	69.1	69.9	70.7	71.6	72.4
48	62.6	63.4	64.1	64.9	65.6	66.4	67.2	67.9	68.7	69.5	70.3	71.1	71.9	72.7	73.5
50	64.0	64.7	65.5	66.2	66.9	67.7	68.4	69.2	70.0	70.8	71.5	72.3	73.1	73.9	74.7
52	65.4	66.1	66.8	67.6	68.3	69.0	69.8	70.5	71.3	72.0	72.8	73.6	74.3	75.1	75.9
54	66.8	67.5	68.2	68.9	69.7	70.4	71.1	71.8	72.6	73.3	74.1	74.8	75.6	76.4	77.2
56	68.3	69.0	69.7	70.4	71.1	71.8	72.5	73.2	73.9	74.7	75.4	76.2	76.9	77.7	78.4
58	69.8	70.5	71.1	71.8	72.5	73.2	73.9	74.6	75.3	76.0	76.8	77.5	78.2	79.0	79.7
60	71.3	72.0	72.6	73.3	74.0	74.6	75.3	76.0	76.7	77.4	78.1	78.9	79.6	80.3	81.1
62	72.9	73.5	74.1	74.8	75.4	76.1	76.8	77.5	78.2	78.8	79.6	80.3	81.0	81.7	82.4
64	74.4	75.0	75.7	76.3	77.0	77.6	78.3	78.9	79.6	80.3	81.0	81.7	82.4	83.1	83.8
66	76.0	76.6	77.2	77.9	78.5	79.1	79.8	80.4	81.1	81.8	82.4	83.1	83.8	84.5	85.2
68	77.6	78.2	78.8	79.4	80.0	80.7	81.3	81.9	82.6	83.3	83.9	84.6	85.3	85.9	86.6
70	79.2	79.8	80.4	81.0	81.6	82.2	82.9	83.5	84.1	84.8	85.4	86.1	86.7	87.4	88.1
72	80.9	81.4	82.0	82.6	83.2	83.8	84.4	85.0	85.7	86.3	86.9	87.6	88.2	88.9	89.6
74	82.5	83.1	83.6	84.2	84.8	85.4	86.0	86.6	87.2	87.8	88.5	89.1	89.8	90.4	91.1
76	84.2	84.7	85.3	85.9	86.4	87.0	87.6	88.2	88.8	89.4	90.0	90.7	91.3	91.9	92.6
78	85.9	86.4	86.9	87.5	88.1	88.6	89.1	89.8	90.4	91.0	91.6	92.2	92.8	93.5	94.1
80	87.6	88.1	88.6	89.2	89.7	90.3	90.9	91.4	92.0	92.6	93.2	93.8	94.4	95.0	95.7
82	89.3	89.8	90.3	90.8	91.4	91.9	92.5	93.1	93.6	94.2	94.8	95.4	96.0	96.6	97.2
84	91.0	91.5	92.0	92.5	93.1	93.6	94.2	94.7	95.3	95.9	96.4	97.0	97.6	98.2	98.8
86	92.7	93.2	93.7	94.2	94.8	95.3	95.8	96.4	96.9	97.5	98.1	98.6	99.2	99.8	100.4
88	94.5	94.9	95.4	96.0	96.5	97.0	97.5	98.1	98.6	99.2	99.7	100.3	100.8	101.4	102.0
90	96.2	96.7	97.2	97.7	98.2	98.7	99.2	99.7	100.3	100.8	101.4	101.9	102.5	103.1	103.6

Horizontal distance (in feet) from hanging position to point on stage.

Vertical distance (in feet) from stage floor to hanging position.

	61	62	63	64	65	66	67	68	69	70	71	72	73	74	75
10	61.2	62.2	63.2	64.2	65.2	66.2	67.2	68.2	69.2	70.1	71.1	72.1	73.1	74.1	75.1
11	61.3	62.2	63.2	64.2	65.2	66.2	67.2	68.2	69.2	70.2	71.2	72.2	73.2	74.2	75.2
12	61.4	62.3	63.3	64.3	65.3	66.3	67.3	68.3	69.3	70.3	71.3	72.3	73.3	74.3	75.3
13	61.5	62.5	63.4	64.4	65.4	66.4	67.4	68.4	69.4	70.4	71.4	72.4	73.4	74.4	75.4
14	61.6	62.6	63.6	64.6	65.6	66.6	67.5	68.5	69.5	70.5	71.5	72.5	73.5	74.5	75.5
15	61.7	62.7	63.7	64.7	65.7	66.7	67.7	68.7	69.7	70.6	71.6	72.6	73.6	74.6	75.6
16	61.9	62.9	63.9	64.9	65.8	66.8	67.8	68.8	69.8	70.8	71.8	72.8	73.8	74.7	75.7
17	62.1	63.1	64.0	65.0	66.0	67.0	68.0	69.0	70.0	70.9	71.9	72.9	73.9	74.9	75.9
18	62.3	63.3	64.2	65.2	66.2	67.2	68.2	69.1	70.1	71.1	72.1	73.1	74.1	75.1	76.0
19	62.5	63.5	64.4	65.4	66.4	67.4	68.4	69.3	70.3	71.3	72.3	73.3	74.2	75.2	76.2
20	62.7	63.7	64.7	65.6	66.6	67.6	68.6	69.5	70.5	71.5	72.5	73.5	74.4	75.4	76.4
21	62.9	63.9	64.9	65.9	66.8	67.8	68.8	69.7	70.7	71.7	72.7	73.7	74.6	75.6	76.6
22	63.2	64.2	65.1	66.1	67.1	68.0	69.0	70.0	71.0	71.9	72.9	73.9	74.8	75.8	76.8
23	63.5	64.4	65.4	66.4	67.3	68.3	69.3	70.2	71.2	72.2	73.1	74.1	75.1	76.0	77.0
24	63.7	64.7	65.7	66.6	67.6	68.5	69.5	70.5	71.4	72.4	73.4	74.3	75.3	76.3	77.3
25	64.0	65.0	66.0	66.9	67.9	68.8	69.8	70.7	71.7	72.7	73.6	74.6	75.6	76.5	77.5
26	64.4	65.3	66.3	67.2	68.2	69.1	70.1	71.0	72.0	72.9	73.9	74.9	75.8	76.8	77.8
27	64.7	65.6	66.6	67.5	68.5	69.4	70.4	71.3	72.3	73.2	74.2	75.1	76.1	77.1	78.0
28	65.0	66.0	66.9	67.8	68.8	69.7	70.7	71.6	72.6	73.5	74.5	75.4	76.4	77.4	78.3
29	65.4	66.3	67.2	68.2	69.1	70.1	71.0	72.0	72.9	73.8	74.8	75.7	76.7	77.6	78.6
30	65.7	66.7	67.6	68.5	69.5	70.4	71.3	72.3	73.2	74.2	75.1	76.1	77.0	78.0	78.9
31	66.1	67.0	68.0	68.9	69.8	70.8	71.7	72.6	73.6	74.5	75.4	76.4	77.3	78.3	79.2
32	66.5	67.4	68.4	69.3	70.2	71.1	72.1	73.0	73.9	74.9	75.8	76.7	77.7	78.6	79.5
33	66.9	67.8	68.7	69.7	70.6	71.5	72.4	73.4	74.3	75.2	76.1	77.1	78.0	78.9	79.9
34	67.3	68.2	69.2	70.1	71.0	71.9	72.8	73.7	74.7	75.6	76.5	77.4	78.4	79.3	80.2
35	67.8	68.7	69.6	70.5	71.4	72.3	73.2	74.1	75.0	76.0	76.9	77.8	78.7	79.7	80.6
36	68.2	69.1	70.0	70.9	71.8	72.7	73.6	74.5	75.4	76.4	77.3	78.2	79.1	80.0	81.0
37	68.7	69.5	70.4	71.3	72.2	73.1	74.0	74.9	75.9	76.8	77.7	78.6	79.5	80.4	81.4
38	69.1	70.0	70.9	71.8	72.7	73.6	74.5	75.4	76.3	77.2	78.1	79.0	79.9	80.8	81.7
39	69.6	70.5	71.4	72.2	73.1	74.0	74.9	75.8	76.7	77.6	78.5	79.4	80.3	81.2	82.1
40	70.1	71.0	71.8	72.7	73.6	74.5	75.4	76.3	77.1	78.0	78.9	79.8	80.7	81.7	82.6
42	71.1	72.0	72.8	73.7	74.6	75.4	76.3	77.2	78.1	78.9	79.8	80.7	81.6	82.5	83.4
44	72.1	73.0	73.8	74.7	75.6	76.4	77.3	78.1	79.0	79.9	80.8	81.7	82.5	83.4	84.3
46	73.2	74.1	74.9	75.7	76.6	77.4	78.3	79.2	80.0	80.9	81.7	82.6	83.5	84.4	85.2
48	74.4	75.2	76.0	76.8	77.7	78.5	79.3	80.1	81.0	81.9	82.8	83.6	84.5	85.3	86.2
50	75.5	76.3	77.1	78.0	78.8	79.6	80.4	81.3	82.1	83.0	83.8	84.6	85.5	86.4	87.2
52	76.7	77.5	78.3	79.1	79.9	80.7	81.6	82.4	83.2	84.0	84.9	85.7	86.6	87.4	88.2
54	77.9	78.7	79.5	80.3	81.1	81.9	82.7	83.5	84.3	85.2	86.0	86.8	87.6	88.5	89.3
56	79.2	80.0	80.7	81.5	82.3	83.1	83.9	84.7	85.5	86.3	87.1	87.9	88.8	89.6	90.4
58	80.5	81.2	82.0	82.8	83.6	84.3	85.1	85.9	86.7	87.5	88.3	89.1	89.9	90.7	91.5
60	81.8	82.5	83.3	84.1	84.8	85.6	86.4	87.1	87.9	88.7	89.5	90.3	91.1	91.9	92.7
62	83.1	83.9	84.6	85.4	86.1	86.9	87.6	88.4	89.2	90.0	90.7	91.5	92.3	93.1	93.9
64	84.5	85.2	86.0	86.7	87.4	88.2	88.9	89.7	90.5	91.2	92.0	92.8	93.5	94.3	95.1
66	85.9	86.6	87.3	88.1	88.8	89.5	90.3	91.0	91.8	92.5	93.3	94.0	94.8	95.6	96.4
68	87.3	88.0	88.7	89.5	90.2	90.9	91.6	92.4	93.1	93.8	94.6	95.3	96.1	96.9	97.6
70	88.8	89.5	90.2	90.9	91.6	92.3	93.0	93.7	94.5	95.2	95.9	96.7	97.4	98.2	98.9
72	90.2	90.9	91.6	92.3	93.0	93.7	94.4	95.1	95.8	96.6	97.3	98.0	98.7	99.5	100.2
74	91.7	92.4	93.1	93.7	94.4	95.1	95.8	96.5	97.2	97.9	98.7	99.4	100.1	100.8	101.6
76	93.2	93.9	94.5	95.2	95.9	96.6	97.3	98.0	98.6	99.3	100.1	100.8	101.5	102.2	102.9
78	94.7	95.4	96.0	96.7	97.4	98.0	98.7	99.4	100.1	100.8	101.5	102.2	102.9	103.6	104.3
80	96.3	96.9	97.6	98.2	98.9	99.5	100.2	100.9	101.5	102.2	102.9	103.6	104.3	105.0	105.7
82	97.8	98.5	99.1	99.7	100.4	101.0	101.7	102.4	103.0	103.7	104.4	105.1	105.7	106.4	107.1
84	99.4	100.0	100.7	101.3	101.9	102.6	103.2	103.9	104.5	105.2	105.8	106.5	107.2	107.9	108.6
86	101.0	101.6	102.2	102.8	103.5	104.1	104.7	105.4	106.0	106.7	107.3	108.0	108.7	109.3	110.0
88	102.6	103.2	103.8	104.4	105.0	105.7	106.3	106.9	107.6	108.2	108.8	109.5	110.2	110.8	111.5
90	104.2	104.8	105.4	106.0	106.6	107.2	107.8	108.5	109.1	109.7	110.4	111.0	111.7	112.3	113.0

Horizontal distance (in feet) from hanging position to point on stage.

Vertical distance (in feet) from stage floor to hanging position.

	76	77	78	79	80	81	82	83	84	85	86	87	88	8/9	90
10	76.1	77.1	78.1	79.1	80.1	81.1	82.1	83.1	84.1	85.1	86.1	87.1	88.1	89.1	90.1
11	76.2	77.2	78.2	79.2	80.2	81.2	82.2	83.2	84.2	85.2	86.2	87.2	88.2	89.2	90.2
12	76.3	77.3	78.3	79.3	80.3	81.3	82.3	83.3	84.3	85.3	86.3	87.2	88.2	89.2	90.2
13	76.4	77.4	78.4	79.4	80.4	81.4	82.3	83.3	84.3	85.3	86.3	87.3	88.3	89.3	90.3
14	76.5	77.5	78.5	79.5	80.5	81.4	82.4	83.4	84.4	85.4	86.4	87.4	88.4	89.4	90.4
15	76.6	77.6	78.6	79.6	80.6	81.6	82.6	83.5	84.5	85.5	86.5	87.5	88.5	89.5	90.5
16	76.7	77.7	78.7	79.7	80.7	81.7	82.7	83.7	84.7	85.7	86.6	87.6	88.6	89.6	90.6
17	76.9	77.9	78.8	79.8	80.8	81.8	82.8	83.8	84.8	85.8	86.8	87.8	88.8	89.7	90.7
18	77.0	78.0	79.0	80.0	81.0	82.0	83.0	83.9	84.9	85.9	86.9	87.9	88.9	89.9	90.9
19	77.2	78.2	79.2	80.2	81.1	82.1	83.1	84.1	85.1	86.1	87.1	88.0	89.0	90.0	91.0
20	77.4	78.4	79.3	80.3	81.3	82.3	83.3	94.3	85.2	86.2	87.2	88.2	89.2	90.2	91.2
21	77.6	78.5	79.5	80.5	81.5	82.5	83.5	84.4	85.4	86.4	87.4	88.4	89.4	90.3	91.3
22	77.8	78.8	79.7	80.7	81.7	82.7	83.6	84.6	85.6	86.6	87.6	88.6	89.5	90.5	91.5
23	78.0	79.0	79.9	80.9	81.9	82.9	83.9	84.8	85.8	86.8	87.8	88.7	89.7	90.7	91.7
24	78.2	79.2	80.2	81.1	82.1	83.1	84.1	85.0	86.0	87.0	88.0	89.0	89.9	90.9	91.9
25	78.5	79.4	80.4	81.4	82.3	83.3	84.3	85.3	86.2	87.2	88.2	89.2	90.1	91.1	92.1
26	78.7	79.7	80.7	81.6	82.6	83.6	84.5	85.5	86.5	87.4	88.4	89.4	90.4	91.3	92.3
27	79.0	80.0	80.9	81.9	82.8	83.8	84.8	85.7	86.7	87.7	88.7	89.6	90.6	91.6	92.5
28	79.3	80.2	81.2	82.1	83.1	84.1	85.0	86.0	87.0	87.9	88.9	89.9	90.8	91.8	92.8
29	79.6	80.5	81.5	82.4	83.4	84.3	85.3	86.3	87.2	88.2	89.2	90.1	91.1	92.1	93.0
30	79.9	80.8	81.8	82.7	83.7	84.6	85.6	86.5	87.5	88.5	89.4	90.4	91.4	92.3	93.3
31	80.2	81.1	82.1	83.0	84.0	84.9	85.9	86.8	87.8	88.7	89.7	90.7	91.6	92.6	93.5
32	80.5	81.4	82.4	83.3	84.3	85.2	86.2	87.1	88.1	89.0	90.0	91.0	91.9	92.9	93.8
33	80.8	81.8	82.7	83.7	84.6	85.5	86.5	87.4	88.4	89.3	90.3	91.2	92.2	93.2	94.1
34	81.2	82.1	83.0	84.0	84.9	85.9	86.8	87.8	88.7	89.7	90.6	91.6	92.5	93.5	94.4
35	81.5	82.5	83.4	84.3	85.3	86.2	87.1	88.1	89.0	90.0	90.9	91.9	92.8	93.8	94.7
36	81.9	82.8	83.8	84.7	85.6	86.6	87.5	88.4	89.4	90.3	91.3	92.2	93.1	94.1	95.0
37	82.3	83.2	84.1	85.1	86.0	86.9	87.8	88.8	89.7	90.7	91.6	92.5	93.5	94.4	95.4
38	82.7	83.6	84.5	85.4	86.4	87.3	88.2	89.1	90.1	91.0	91.9	92.9	93.8	94.8	95.7
39	83.1	84.0	84.9	85.8	86.7	87.7	88.6	89.5	90.4	91.4	92.3	93.2	94.2	95.1	96.0
40	83.5	84.4	85.3	86.2	87.1	88.0	89.0	89.9	90.8	91.7	92.7	93.6	94.5	95.5	96.4
42	84.3	85.2	86.1	87.0	87.9	88.8	89.8	90.7	91.6	92.5	93.4	94.4	95.3	96.2	97.1
44	85.2	86.1	87.0	87.9	88.8	89.7	90.6	91.5	92.4	93.3	94.2	95.1	96.1	97.0	97.9
46	86.1	87.0	87.9	88.8	89.7	90.6	91.5	92.4	93.3	94.2	95.1	96.0	96.9	97.8	98.7
48	87.1	88.0	88.8	89.7	90.6	91.5	92.4	93.3	94.1	95.0	95.9	96.8	97.7	98.6	99.5
50	88.1	88.9	89.8	90.7	91.5	92.4	93.3	94.2	95.1	95.9	96.8	97.7	98.6	99.5	100.4
52	89.1	90.0	90.8	91.7	92.5	93.4	94.3	95.1	96.0	96.9	97.8	98.6	99.5	100.4	101.3
54	90.2	91.0	91.8	92.7	93.6	94.4	95.3	96.1	97.0	97.9	98.7	99.6	100.5	101.4	102.2
56	91.2	92.1	92.9	93.8	94.6	95.5	96.3	97.2	98.0	98.9	99.7	100.6	101.5	102.3	103.2
58	92.4	93.2	94.0	94.9	95.7	96.5	97.4	98.2	99.1	99.9	100.8	101.6	102.5	103.3	104.2
60	93.5	94.3	95.2	96.0	96.8	97.6	98.5	99.3	100.1	101.0	101.8	102.7	103.5	104.4	105.2
62	94.7	95.5	96.3	97.1	97.9	98.8	99.6	100.4	101.2	102.1	102.9	103.7	104.6	105.4	106.3
64	95.9	96.7	97.5	98.3	99.1	99.9	100.7	101.5	102.4	103.2	104.0	104.8	105.7	106.5	107.3
66	97.1	97.9	98.7	99.5	100.3	101.1	101.9	102.7	103.5	104.3	105.1	106.0	106.8	107.6	108.4
68	98.4	99.2	100.0	100.7	101.5	102.3	103.1	103.9	104.7	105.5	106.3	107.1	107.9	108.8	109.6
70	99.7	100.4	101.2	102.0	102.8	103.5	104.3	105.1	105.9	106.7	107.5	108.3	109.1	109.9	110.7
72	101.0	101.7	102.5	103.3	104.0	104.8	105.6	106.4	107.1	107.9	108.7	109.5	110.3	111.1	111.9
74	102.3	103.1	103.8	104.6	105.3	106.1	106.8	107.6	108.4	109.2	109.9	110.7	111.5	112.3	113.1
76	103.7	104.4	105.1	105.9	106.6	107.4	108.1	108.9	109.7	110.4	111.2	112.0	112.8	113.5	114.3
78	105.0	105.8	106.5	107.2	108.0	108.7	109.5	110.2	111.0	111.7	112.5	113.2	114.0	114.8	115.6
80	106.4	107.1	107.9	108.6	109.3	110.1	110.8	111.5	112.3	113.0	113.8	114.5	115.3	116.1	116.8
82	107.8	108.5	109.3	110.0	110.7	111.4	112.1	112.9	113.6	114.4	115.1	115.9	116.6	117.4	118.2
84	109.3	110.0	110.7	111.4	112.1	112.8	113.5	114.2	115.0	115.7	116.4	117.2	117.9	118.7	119.4
86	110.7	111.4	112.1	112.8	113.5	114.2	114.9	115.6	116.3	117.1	117.8	118.5	119.3	120.0	120.7
88	112.2	112.9	113.5	114.2	114.9	115.6	116.3	117.0	117.7	118.5	119.2	119.9	120.6	121.4	122.1
90	113.7	114.3	115.0	115.7	116.4	117.1	117.7	118.4	119.1	119.9	120.6	121.3	122.0	122.7	123.5

Incidence Angle Table

This table can be used in conjunction with the preceding throw distance table to give you a thorough picture of your lighting position. It is laid out with the same horizontal and vertical axes, and will tell you the approximate angle of incidence, or throw angle of your lighting position.

To use this table, find the height of your position (from your section) on the vertical axis of the table and the horizontal distance from your subject to the position (from your plan) on the horizontal axis of the table. The intersection will fall within one of the radiating bands giving you the approximate angle of incidence. As in the Throw Distance Table, this table assumes the throw angle is measured from the lighting position to the head of your subject (5.5' off the deck).

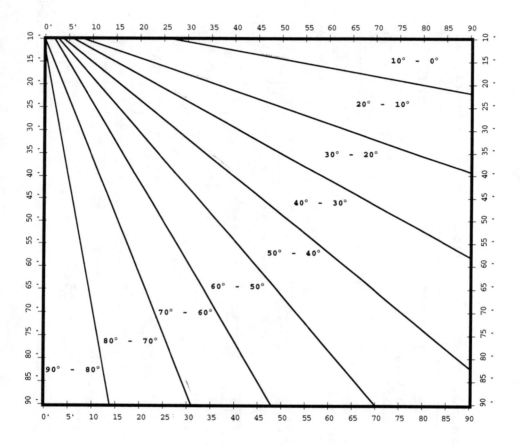

Manufacturer

Model No.

Angle information

Angle information

Lamp base

Standard lamp

Other lamps

Accessories

PHOTOMETRICS CHART

(Performance data for this unit are measured using _____ *lamp.)*

Distance

Diameter
Footcandles
(mf= _____ *)*
(pc= _____ *)*

Distance

Diameter
Footcandles
(mf= _____ *)*
(pc= _____ *)*

(Use this blank form to collect information on instruments not found in this book.) **621**

© Copyright Broadway Press – photocopy permission given for personal use only.

_____

_____ *Manufacturer*

_____ *Model No.*

_____ *Angle information*

_____ *Angle information*

_____ *Lamp base*

_____ *Standard lamp*

_____ *Other lamps*

_____

_____

_____

_____

_____ *Accessories*

PHOTOMETRICS CHART

(Performance data for this unit are measured using _____ *lamp.)*

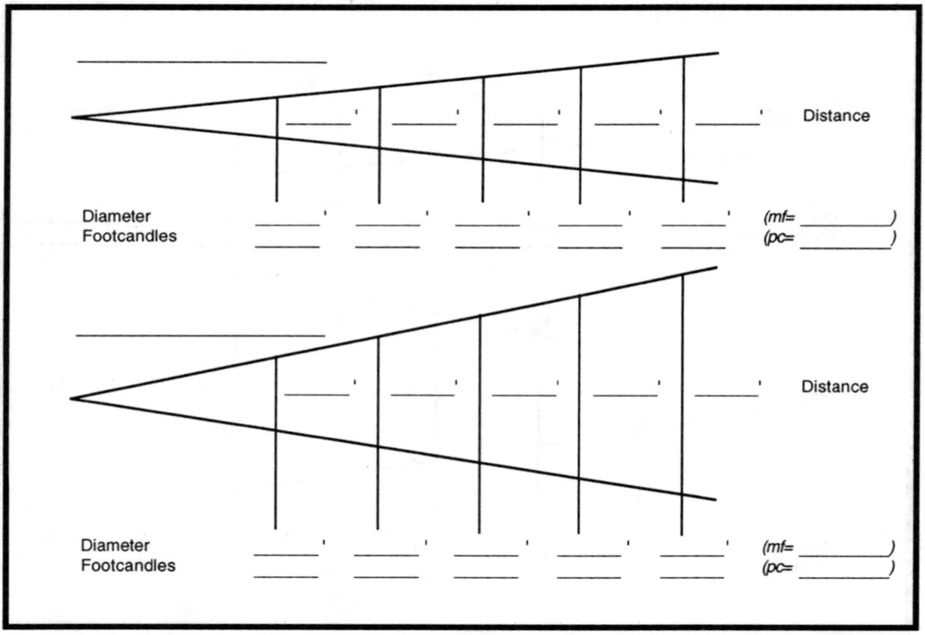

Distance

Diameter
Footcandles   *(mf= _____)*
     (pc= _____)

Distance

Diameter
Footcandles   *(mf= _____)*
     (pc= _____)

_____

_____

_____

622   **(Use this blank form to collect information on instruments not found in this book.)**

© Copyright Broadway Press – photocopy permission given for personal use only.

Manufacturer

Model No.

Angle information

Angle information

Lamp base

Standard lamp

Other lamps

Accessories

PHOTOMETRICS CHART

(Performance data for this unit are measured using _____ *lamp.)*

Distance

Diameter
Footcandles

(mf= _____ *)*
(pc= _____ *)*

Distance

Diameter
Footcandles

(mf= _____ *)*
(pc= _____ *)*

(Use this blank form to collect information on instruments not found in this book.)   **623**

© Copyright Broadway Press – photocopy permission given for personal use only.

_____ Manufacturer

_____ Model No.

_____ Angle information

_____ Angle information

_____

_____ Lamp base

_____ Standard lamp

_____ Other lamps

_____

_____

_____

_____

_____ Accessories

PHOTOMETRICS CHART

(Performance data for this unit are measured using _____ *lamp.)*

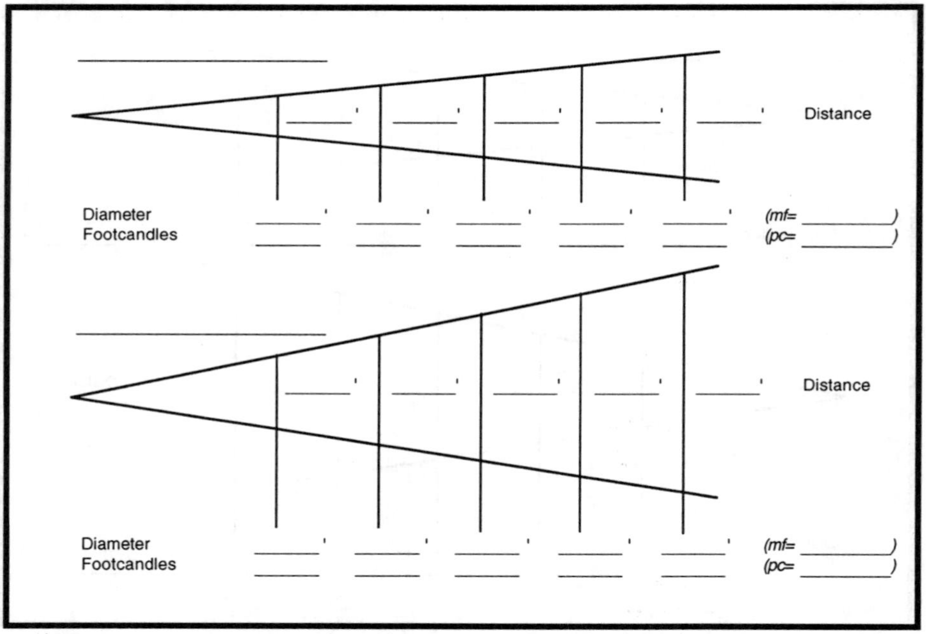

Distance

Diameter
Footcandles

(mf= _____ *)*
(pc= _____ *)*

Distance

Diameter
Footcandles

(mf= _____ *)*
(pc= _____ *)*

(Use this blank form to collect information on instruments not found in this book.)

© Copyright Broadway Press – photocopy permission given for personal use only.

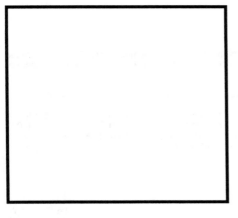

Manufacturer

Model No.

Angle information

Angle information

Lamp base

Standard lamp

Other lamps

Accessories

PHOTOMETRICS CHART

(Performance data for this unit are measured using _____ *lamp.)*

Distance

Diameter
Footcandles

(mf= _____)
(pc= _____)

Distance

Diameter
Footcandles

(mf= _____)
(pc= _____)

(Use this blank form to collect information on instruments not found in this book.)

© Copyright Broadway Press – photocopy permission given for personal use only.

Instruments — No Photometrics

This is a list of instruments with known model numbers, but no known photometric data. Either photometric data was not included in the manufacturer's catalog or incomplete catalog data was available to the author at time of publication.

Striplights are not included in this list. For complete list of striplights and cyc lights see pages 572-578.

If you have photometric data on any of these instruments, and you would like to share it with your fellow lighting designers, please send copies of your catalog pages to the author, care of Broadway Press (3001 Springcrest Dr., Louisville KY 40241 or via e-mail at mumm@broadwaypress.com). Anyone who sends photometric data which can be used in the next edition of this book will be given a complimentary copy of the new edition.

Altman Stage Lighting

1	Olivette
10	6" plano-convex spotlight, 1.5 KW.
105	16" fresnel, 5 KW.
11	8" plano-convex spotlight, 2 KW.
360-4.5PC	4.5x6.5 ERS, 250-500 W.
360-4.5S	4.5x3.5 ERS, 250-500 W.
360-PC	6x9 ERS, 250-750 W.
362	6x12 ERS, 750 W.
362	6x9 ERS, 750 W.
365/8	8x8 ERS, 750 W.
365	6x9 ERS, 750 W.
368	10x14 ERS, 1 KW.
760	16" beam projector, 2 KW.
85	10" fresnel, 2 KW.
9	4.5" plano-convex spotlight, 400 W.
95	12" fresnel, 1 or 2 KW.

Capitol Stage Lighting

1	Olivette, 1.5 KW.
12	6" plano-convex spotlight, 1 KW.
14	6" plano-convex spotlight, 1.5 KW.
15	6" plano-convex spotlight, 2 KW.
16	6" plano-convex spotlight, 2 KW.
19	6" plano-convex spotlight, 1 KW.
2	Olivette, 1.5 KW.
20	scoop, 500 W.
280	6" fresnel
29	hanging floodlight, 1 KW.
32	18" hanging floodlight
490	16" beam projector, 1.5 KW.
5	4.5" plano-convex spotlight, 250 W.
501	8" fresnel
502	10" fresnel, 2 KW.
503	14" fresnel, 5 KW.
6	4.5" plano-convex spotlight, 400 W.
650	ERS, 500-750 W.
655	6x9 ERS, 750 W.
656	6x12 ERS, 750 W.
657	6x16 ERS, 750 W.
661	18" scoop
7	3.5" plano-convex spotlight, 100 W.
8	5" plano-convex spotlight, 400-500 W.
904	3" fresnel
905	8" fresnel, 1 KW.
906	10" fresnel, 2 KW.
907	14" fresnel, 2 KW.

Century Lighting

1532, 1533, 1534	6x11 ERS, 250-750 W.
1536, 1537, 1538	6x8+6x11 ERS, 250-750 W.
1551, 1552	12x12 ERS, 2 KW.
1553, 1554	12x12 ERS, 3 KW.
1560	8x10 ERS, 2 KW.
1560	8x8 ERS, 2 KW.
1562, 1563	8x9 ERS, 3 KW.
1570	8x10 ERS, 500 W.
1570	8x8 ERS, 500 W.
1596, 1597, 1598	6x8 ERS, 250-750 W.
505-S	6" fresnel, 500 W.
506	8" fresnel, 1.5 KW.
507	5" fresnel, 400 W.
510	10" fresnel, 2 KW.
514	14" fresnel, 5 KW.

Display Lighting

100	6" plano-convex spotlight, carbon arc
1101	14" floodlight, 1 KW.
1130	14"x15" floodlight, 500 W.
502	6" plano-convex spotlight, 2 KW.
536	8" plano-convex spotlight, 1-2 KW.
564	6" plano-convex spotlight, 500-1000 W.
600	6" plano-convex spotlight, 500-1000 W.
687	8" plano-convex spotlight, 500-1000 W.
700	5" plano-convex spotlight, 250-400 W.
805	10" beam projector, 400 W.
820	window floodlight, 200 W.
900	4.5" plano-convex spotlight, 250-500 W.

Kliegl Bros.

Major Equipment

Strand Century

Manufacturer Index

This index provides easy access to the information in this book by grouping together instruments from each manufacturer. If you know the manufacturer and the model number, you can quickly find its page of photometric information. Even if you only know the manufacturer and the type of instrument, you can narrow your search down considerably with this index.

The instruments are listed alphabetically by model number. The sort order, which might seem strange, follows these rules: model numbers are alphabetized as though they are words with numerals coming before letters in the sort order. In many cases where variations on the same instrument have been assigned different model numbers, you will find two or more model numbers listed on a line, separated by commas.

If you don't find an instrument in this index, it may be because photometric information on that particular instrument was not available to the author when this edition was published. There is a list of instruments without photometric data preceding this index.

Altman Stage Lighting

Century Strand (See also Century Lighting, Strand Century and Strand Lighting)

Clay Paky

Colortran (See also Berkey Colortran and Lee Colortran)

Display Lighting

Electro Controls

Electronic Theatre Controls (ETC)

High End Systems

Kliegl Bros. (See also Kliegl Bros./ CCT and Kliegl Bros./ RDS)

Kliegl Bros./ CCT (See also Kliegl Bros. and Kliegl Bros./ RDS)

Kliegl Bros./RDS (See also Kliegl Bros. and Kliegl Bros. / CCT)

Lee Colortran (See also Berkey Colortran and Colortran)

Lighting & Electronics (L&E)

Little Stage Lighting

Ludwig Pani

Lycian Stage Lighting

Martin Professional A/S

Phoebus

Strand Century (See also Century Lighting, Century Strand and Strand Lighting)

Vari-Lite, Inc.

KING ALFRED'S COLLEGE
LIBRARY